Troels Andreasen   Henning Christiansen
Juan-Carlos Cubero   Zbigniew W. Raś (Eds.)

Foundations
of Intelligent Systems

21st International Symposium, ISMIS 2014
Roskilde, Denmark, June 25-27, 2014
Proceedings

 Springer

Volume Editors

Troels Andreasen
Roskilde University, Denmark
E-mail: troels@ruc.dk

Henning Christiansen
Roskilde University, Denmark
E-mail: henning@ruc.dk

Juan-Carlos Cubero
University of Granada, Spain
E-mail: jc.cubero@decsai.ugr.es

Zbigniew W. Raś
University of North Carolina, Charlotte, NC, USA
and Warsaw University of Technology, Poland
E-mail: ras@uncc.edu

ISSN 0302-9743   e-ISSN 1611-3349
ISBN 978-3-319-08325-4   e-ISBN 978-3-319-08326-1
DOI 10.1007/978-3-319-08326-1
Springer Cham Heidelberg New York Dordrecht London

Library of Congress Control Number: 2014941587

LNCS Sublibrary: SL 7 – Artificial Intelligence

© Springer International Publishing Switzerland 2014
This work is subject to copyright. All rights are reserved by the Publisher, whether the whole or part of
the material is concerned, specifically the rights of translation, reprinting, reuse of illustrations, recitation,
broadcasting, reproduction on microfilms or in any other physical way, and transmission or information
storage and retrieval, electronic adaptation, computer software, or by similar or dissimilar methodology
now known or hereafter developed. Exempted from this legal reservation are brief excerpts in connection
with reviews or scholarly analysis or material supplied specifically for the purpose of being entered and
executed on a computer system, for exclusive use by the purchaser of the work. Duplication of this publication
or parts thereof is permitted only under the provisions of the Copyright Law of the Publisher's location,
in ist current version, and permission for use must always be obtained from Springer. Permissions for use
may be obtained through RightsLink at the Copyright Clearance Center. Violations are liable to prosecution
under the respective Copyright Law.
The use of general descriptive names, registered names, trademarks, service marks, etc. in this publication
does not imply, even in the absence of a specific statement, that such names are exempt from the relevant
protective laws and regulations and therefore free for general use.
While the advice and information in this book are believed to be true and accurate at the date of publication,
neither the authors nor the editors nor the publisher can accept any legal responsibility for any errors or
omissions that may be made. The publisher makes no warranty, express or implied, with respect to the
material contained herein.

Typesetting: Camera-ready by author, data conversion by Scientific Publishing Services, Chennai, India

Printed on acid-free paper

Springer is part of Springer Science+Business Media (www.springer.com)

Preface

This volume contains the papers presented at ISMIS 2014: 21st International Symposium on Methodologies for Intelligent Systems held during June 25–27, 2014, in Roskilde, Denmark. The symposium was organized by members of the PLIS research group: Programming, Logic and Intelligent Systems, of the Department of Communication, Business and Information Technologies at Roskilde University.

ISMIS is a conference series that started in 1986 and has developed into an established and prestigious conference for exchanging the latest research results in building intelligent systems. The scope of ISMIS represents a wide range of topics on applying artificial intelligence techniques to areas as diverse as decision support, automated deduction, reasoning, knowledge-based systems, machine learning, computer vision, robotics, planning, databases, information retrieval, and so on. ISMIS provides a forum and a means for exchanging information for those interested purely in theory, those interested primarily in implementation, and those interested in specific research and industrial applications.

We want to express our special thanks to the Program Committee members and everyone who contributed at any level to the organization of ISMIS 2014. Also, special thanks to our invited speakers, Matthias Jarke, Xavier Serra and Steffen Staab. We would like to thank every author who submitted a paper to ISMIS 2014 and finally the team of EasyChair, without whose free software the handling of submissions and editing of the proceedings could not have been managed so smoothly by a small group of people. Last but not the least, we thank Alfred Hofmann of Springer for his continuous support.

April 2014

Troels Andreasen
Henning Christiansen
Juan-Carlos Cubero
Zbigniew W. Raś

Organization

The symposium was organized by members of the PLIS research group: Programming, Logic and Intelligent Systems, of the Department of Communication, Business and Information Technologies at Roskilde University.

General Chair

Zbigniew W. Raś   University of North Carolina, Charlotte, USA
and Warsaw University of Technology,
Poland

Symposium Chair

Troels Andreasen   Roskilde University, Denmark

Program Co-chairs

Henning Christiansen   Roskilde University, Denmark
Juan Carlos Cubero   University of Granada, Spain

Steering Committee

Aijun An	York University, Canada
Alexander Felfernig	Graz University of Technology, Austria
Andrzej Skowron	University of Warsaw, Poland
Dominik Ślęzak	Infobright Inc., Canada; and University of Warsaw, Poland
Henryk Rybinski	Warsaw University of Technology, Poland
Jaime Carbonell	CMU, USA
Jan Rauch	University of Economics, Prague, Czech Republic
Jiming Liu	Hong Kong Baptist University, Hong Kong, SAR China
Li Chen	Hong Kong Baptist University, Hong Kong, SAR China
Lorenza Saitta	University of Piemonte Orientale, Italy
Maria Zemankova	NSF, USA
Marzena Kryszkiewicz	Warsaw University of Technology, Poland
Nick Cercone	York University, Canada

Petr Berka	University of Economics, Prague, Czech Republic
Stan Matwin	University of Ottawa, Canada
Tapio Elomaa	Tampere University of Technology, Finland
Zbigniew W. Raś	UNC-Charlotte, USA; and Warsaw University of Technology, Poland

Program Committees

ISMIS Regular Papers and Posters

Luigia Carlucci Aiello	Sapienza Università di Roma, Italy
Aijun An	York University, Canada
Troels Andreasen	Roskilde University, Denmark
Annalisa Appice	University of Bari, Italy
Salima Benbernou	University of Paris V, France
Marenglen Biba	University of New York Tirana, Albania
Maria Bielikova	Slovak University of Technology, Slovakia
Ivan Bratko	University of Ljubljana, Slovenia
Francois Bry	University of Munich, Germany
Henrik Bulskov	Roskilde University, Denmark
Sandra Carberry	University of Delaware, USA
Michelangelo Ceci	University of Bari, Italy
Jianhua Chen	Louisiana State University, USA
Henning Christiansen	Roskilde University (Co-chair), Denmark
William J. Clancey	Florida Institute for Human and Machine Cognition, USA
Luca Console	Università di Torino, Italy
Bruno Cremilleux	Université de Caen Basse-Normandie, France
Juan Carlos Cubero	University of Granada (Co-chair), Spain
Alfredo Cuzzocrea	ICAR-CNR, University of Calabria, Italy
Ramon Lopez de Mantaras	Spanish National Research Council, Spain
Nicola Di Mauro	Università degli Studi di Bari "Aldo Moro", Italy
Jørgen Fischer Nilsson	Technical University of Denmark
Vladimir A. Fomichov	National Research University, France
Laura Giordano	University of Piemonte Orientale, Italy
Jacek Grekow	Bialystok University of Technology, Poland
Jerzy Grzymala-Busse	University of Kansas, USA
Hakim Hacid	Alcatel-Lucent Bell Labs
Allel Hadjali	Université de Rennes 1, France
Shoji Hirano	Shimane Medical University, Japan
Lothar Hotz	University of Hamburg, Germany
Manfred Jaeger	Aalborg University, Denmark
Nathalie Japkowicz	University of Ottawa, Canada
Mieczyslaw Klopotek	Polish Academy of Sciences, Poland

Bozena Kostek	Gdansk University of Technology, Poland
Patrick Lambrix	Linkoping University, Sweden
Rory Lewis	University of Colorado at Colorado Springs, USA
Michael Lowry	NASA Ames, USA
Donato Malerba	University of Bari, Italy
Giuseppe Manco	Universitá della Calabria, Italy
Krzysztof Marasek	Polish-Japanese Institute of Information Technology, Poland
Maria José Martín Bautista	University of Granada, Spain
Nicolas Marín Ruiz	University of Granada, Spain
Elio Masciari	Università della Calabria, Italy
Paola Mello	University of Bologna, Italy
Ernestina Menasalvas Ruiz	Universidad Politécnica de Madrid, Spain
Neil Murray	University at Albany, State University of New York, USA
Agnieszka Mykowiecka	Polish Academy of Sciences, Poland
John Mylopoulos	University of Toronto, Canada
Thomas D. Nielsen	Aalborg University, Denmark
Jean-Marc Petit	Université de Lyon, France
Olivier Pivert	University of Rennes 1, France
Henri Prade	National Center for Scientific Research, France
Vijay Raghavan	University of Louisiana at Lafayette, USA
Hiroshi Sakai	Kyushu Institute of Technology, Japan
Daniel Sanchez Fernandez	University of Granada, Spain
Dominik Slezak	University of Warsaw, Poland
Jerzy Stefanowski	Poznan Technical University, Poland
Jaroslaw Stepaniuk	Bialystok University of Technology, Poland
Marcin Sydow	Polish-Japanese Institute of Information Technology, Poland
Erich Teppan	University of Klagenfurt, Germany
K. Thirunarayan	Wright State University, USA
Christel Vrain	Orleans University, France
Alicja Wieczorkowska	Polish-Japanese Institute of Information Technology, Poland
Franz Wotawa	Graz University of Technology, Poland
Yiyu Yao	University of Regina, Canada
Slawomir Zadrozny	University of North Carolina at Charlotte, USA
Wlodek Zadrozny	University of North Carolina at Charlotte, USA
Ning Zhong	Maebashi Institute of Technology, Japan
Djamel Zighed	University of Lyon 2, France

Special Session: Challenges in Text Mining and Semantic Information Retrieval

Piotr Andruszkiewicz	Warsaw University of Technology, Poland
Robert Bembenik	Warsaw University of Technology, Poland

Philipp Cimiano	Bielefeld University, Germany
Tomasz Gambin	Warsaw University of Technology, Poland
Piotr Gawrysiak	Warsaw University of Technology, Poland
Marzena Kryszkiewicz	Warsaw University of Technology (Co-chair), Poland)
Evangelos Milios	Dalhousie University, Canada
Mikołaj Morzy	Poznań University of Technology, Poland
Robert Nowak	University of Technology
Grzegorz Protaziuk	Warsaw University of Technology, Poland
Henryk Rybinski	University of Technology (Co-chair), Poland
Łukasz Skonieczny	Warsaw University of Technology (Co-chair), Poland
Jerzy Stefanowski	Poznań Technical University, Poland
Julian Szymanski	Gdansk University of Technology, Poland
Krzysztof Walczak	Warsaw University of Technology, Poland
Wlodek Zadrozny	University of North Carolina at Charlotte, USA

Special Session: Warehousing and OLAPing Complex, Spatial and Spatio-Temporal Data

Michelangelo Ceci	University of Bari, Italy
Alfredo Cuzzocrea	ICAR-CNR & University of Calabria (Chair), Italy
Sergio Flesca	University of Calabria, Italy
Filippo Furfaro	University of Calabria, Italy
Carson Leung	University of Manitoba, Canada

Additional Reviewers

Agrawal, Ameeta	Kapanipathi, Pavan
Aiello, Marco	Klec, Mariusz
Boulkrinat, Samia	Lanotte, Pasqua Fabiana
Béchet, Nicolas	Liberatore, Paolo
Cancelliere, Rossella	Lyu, Siwei
Chesani, Federico	Pio, Gianvito
Choiref, Zahira	Protaziuk, Grzegorz
Corby, Olivier	Pusala, Murali
Dragisic, Zlatan	Serafino, Francesco
Fumarola, Fabio	Sharif, Mohammad
Grisetti, Giorgio	Spillane, Sean
Guarascio, Massimo	Thion, Virginie
Hose, Katja	Wang, Wenbo
Ivanova, Valentina	Zihayat, Morteza

Invited Talks

Exploiting Cultural Specificity in Music Information Research

Xavier Serra

Music Technology Group
Universitat Pompeu Fabra, Barcelona
xavier.serra@upf.edu

Music Information Research (MIR) is a discipline that aims to understand and model music from an information processing perspective and one of its major challenges relates to the automatic generation of musically meaningful information with which to better describe and exploit audio music recordings. The goal is to integrate and process a variety of data sources, like the actual audio recordings, plus editorial metadata and contextual information, to obtain structured information that is semantically and musically meaningful and that is of use in search, retrieval and discovery tasks [1].

A piece of music is an information entity that makes sense specially within a particular social and cultural context. Its analysis and description has to take that into account and thus the data-driven approaches have to incorporate domain knowledge from that particular context in order to make sense of the available information on that piece of music.

In this presentation I will introduce the research currently being done in CompMusic (http://compmusic.upf.edu), a project funded by the European Research Council that focuses on a number of MIR problems through the study of five music cultures: Hindustani (North India), Carnatic (South India), Turkish-makam (Turkey), Arab-Andalusian (Maghreb), and Beijing Opera (China). We work on the extraction of musically relevant features from audio music recordings related to melody and rhythm, and on the semantic analysis of the contextual information of those recordings [2].

Given that most of the research in MIR has been based on studying the western commercial music of the last few decades, our claim is that the technologies developed have a strong bias towards that music, thus not being appropriate for other music repertories. We want to identify the current limitations and propose information processing approaches that can go beyond those boundaries. For that we selected a few music cultures that had personalities contrasting with the popular western music, that had alive performance practices and strong social and cultural relevance, for which there were musicological and cultural studies, and for which it was feasible to collect sufficient and coherent machine-readable music data. At the same time we wanted to have a diverse set of music repertoires with which to study a variety of new and diverse MIR problems.

A major effort in CompMusic has been the creation of research corpora. The types of data that we have gathered are mainly audio recordings and editorial metadata, which are then complemented with descriptive information about the items we have, and in some cases with music scores and/or lyrics. In order to evaluate our research results we have defined a user scenario and have developed a complete system-level application with which users can interact and with which we can evaluate most of the research results from a user perspective. The system, Dunya (http://dunya.compmusic.upf.edu), is a web-based application to explore music collections aimed at music connoisseurs of the particular music traditions. It uses the technologies developed for melodic and rhythmic description and semantic analysis to navigate through the audio recordings and the information items available. This navigation promotes the discovery of relationships between the different information items.

References

1. Serra, X., Magas, M., Benetos, E., Chudy, M., Dixon, S., Flexer, A., Gómez, E., Gouyon, F., Herrera, P., Jordà, S., Paytuvi, O., Peeters, G., Schlüter, J., Vinet, H., Widmer, G.: Roadmap for Music Information ReSearch (2013) ISBN: 978-2-9540351-1-6
2. Serra, X.: A Multicultural Approach in Music Information Research. In: Int. Soc. for Music Information Retrieval Conf. (ISMIR), pp. 151–156 (2011)

Big Data Workflows: Issues and Challenges

Matthias Jarke

RWTH Aachen University, Informatik 5 & Fraunhofer FIT
Ahornstr. 55m 52074 Aachen, Germany
jarke@cs.rwth-aachen.de

Abstract. Big Data is often seen as a rather uniform, if not well understood conglomerate of research and practice issues related to massively increased Volume, Velocity, and Variety of data. In reality, there is an enormous diversity of requirements, architectural and algorithmic settings, in which key success factors can range from efficient sensor fusion to rapid query processing and stream mining, to careful semantic-preserving data integration, to aspect such as data protection, provenance maintenance, novel business models, and digital rights management. In the Fraunhofer Big Data [1], twenty-four research institutes from different disciplines in science and engineering have joined forces to explore big data in the six domains of production, logistics, life sciences/healthcare, energy management, security, business and finance.

This presentation will illustrate this variety from the perspective of intelligent, automated workflow assistance in different big data settings. We start with an overview of attempts for automating the integration of heterogeneous structured and semi-structured data, e.g. in corporate as well as cross-organizational multi-database settings [2]. One set of application projects we are currently engaged in comprises different aspects of research data management, focusing on prevention of scientific fraud and traceability, while ensuring correct data ownership and shared understanding. Another important aspect in this context is the evolution analysis of contributions to scientific communities [3], be it open source communities or simply publication and citation networks. As another extreme, we look at highly scalable digital rights policies and management workflows within very large-scale video databases, contrasting automated techniques and "wisdom of the crowd" [4].

References

1. http://www.bigdata.fraunhofer.de
2. Jarke, M., Jeusfeld, M.A., Quix, C.: Data-centric intelligent information integration – from concepts to automation. Journal of Intelligent Information Systems (to appear, 2014)
3. Pham, M.C., Klamma, R., Jarke, M.: Development of computer science disciplines - a social network analysis approach. Social Network Analysis and Mining (SNAM) 1(4), 321–340 (2011)

4. Rashed, K., Renzel, D., Klamma, R., Jarke, M.: Community and trust-aware fake media detection. Multimedia Tools and Applications (2012), doi:10.1007/s11042-012-1103-3; Special issue on Multimedia on the Web 2012

Table of Contents

Complex Networks and Data Stream Mining

Community Detection by an Efficient Ant Colony Approach 1
  *Lúcio Pereira de Andrade, Rogério Pinto Espíndola, and
  Nelson Francisco Favilla Ebecken*

Adaptive XML Stream Classification Using Partial Tree-Edit
Distance 10
  Dariusz Brzezinski and Maciej Piernik

RILL: Algorithm for Learning Rules from Streaming Data with
Concept Drift 20
  Magdalena Deckert and Jerzy Stefanowski

Community Detection for Multiplex Social Networks Based on
Relational Bayesian Networks 30
  Jiuchuan Jiang and Manfred Jaeger

Mining Dense Regions from Vehicular Mobility in Streaming Setting.... 40
  Corrado Loglisci and Donato Malerba

Mining Temporal Evolution of Entities in a Stream of Textual
Documents 50
  *Gianvito Pio, Pasqua Fabiana Lanotte, Michelangelo Ceci, and
  Donato Malerba*

An Efficient Method for Community Detection Based on Formal
Concept Analysis 61
  *Selmane Sid Ali, Fadila Bentayeb, Rokia Missaoui, and
  Omar Boussaid*

Data Mining Methods

On Interpreting Three-Way Decisions through Two-Way Decisions 73
  Xiaofei Deng, Yiyu Yao, and JingTao Yao

FHM: Faster High-Utility Itemset Mining Using Estimated Utility
Co-occurrence Pruning 83
  *Philippe Fournier-Viger, Cheng-Wei Wu, Souleymane Zida, and
  Vincent S. Tseng*

Automatic Subclasses Estimation for a Better Classification
with HNNP 93
  Ruth Janning, Carlotta Schatten, and Lars Schmidt-Thieme

A Large-Scale, Hybrid Approach for Recommending Pages Based on
Previous User Click Pattern and Content   103
    Mohammad Amir Sharif and Vijay V. Raghavan

EverMiner Prototype Using LISp-Miner Control Language   113
    Milan Šimůnek and Jan Rauch

Local Characteristics of Minority Examples in Pre-processing of
Imbalanced Data.   123
    Jerzy Stefanowski, Krystyna Napierała, and Małgorzata Trzcielińska

Visual-Based Detection of Properties of Confirmation Measures   133
    Robert Susmaga and Izabela Szczęch

Intelligent Systems Applications

A Recursive Algorithm for Building Renovation in Smart Cities   144
    *Andrés Felipe Barco, Elise Vareilles, Michel Aldanondo, and
    Paul Gaborit*

Spike Sorting Based upon PCA over DWT Frequency Band Selection . . .   154
    Konrad Ciecierski, Zbigniew W. Raś, and Andrzej W. Przybyszewski

Neural Network Implementation of a Mesoscale Meteorological
Model   164
    Robert Firth and Jianhua Chen

Spectral Machine Learning for Predicting Power Wheelchair Exercise
Compliance   174
    *Robert Fisher, Reid Simmons, Cheng-Shiu Chung, Rory Cooper,
    Garrett Grindle, Annmarie Kelleher, Hsinyi Liu, and Yu Kuang Wu*

Mood Tracking of Radio Station Broadcasts   184
    Jacek Grekow

Evidential Combination Operators for Entrapment Prediction in
Advanced Driver Assistance Systems   194
    Alexander Karlsson, Anders Dahlbom, and Hui Zhong

Influence of Feature Sets on Precision, Recall, and Accuracy of
Identification of Musical Instruments in Audio Recordings.   204
    Elżbieta Kubera, Alicja A. Wieczorkowska, and Magdalena Skrzypiec

Multi-label Ferns for Efficient Recognition of Musical Instruments in
Recordings   214
    Miron B. Kursa and Alicja A. Wieczorkowska

Computer-Supported Polysensory Integration Technology for
Educationally Handicapped Pupils 224
  *Michal Lech, Andrzej Czyzewski, Waldemar Kucharski, and
  Bozena Kostek*

Integrating Cluster Analysis to the ARIMA Model for Forecasting
Geosensor Data....................................... 234
  Sonja Pravilovic, Annalisa Appice, and Donato Malerba

Unsupervised and Hybrid Approaches for On-line RFID Localization
with Mixed Context Knowledge................................. 244
  Christoph Scholz, Martin Atzmueller, and Gerd Stumme

Mining Surgical Meta-actions Effects with Variable Diagnoses'
Number 254
  *Hakim Touati, Zbigniew W. Raś, James Studnicki, and
  Alicja A. Wieczorkowska*

Knowledge Representation in Databases and Systems

A System for Computing Conceptual Pathways in Bio-medical Text
Models 264
  *Troels Andreasen, Henrik Bulskov, Jørgen Fischer Nilsson, and
  Per Anker Jensen*

Putting Instance Matching to the Test: Is Instance Matching Ready for
Reliable Data Linking? 274
  Silviu Homoceanu, Jan-Christoph Kalo, and Wolf-Tilo Balke

Improving Personalization and Contextualization of Queries to
Knowledge Bases Using Spreading Activation and Users' Feedback 285
  Ana Belen Pelegrina, Maria J. Martin-Bautista, and Pamela Faber

Plethoric Answers to Fuzzy Queries: A Reduction Method Based on
Query Mining 295
  Olivier Pivert and Grégory Smits

Generating Description Logic ALC from Text in Natural Language 305
  *Ryan Ribeiro de Azevedo, Fred Freitas, Rodrigo Rocha,
  José Antônio Alves de Menezes, and Luis F. Alves Pereira*

DBaaS-Expert: A Recommender for the Selection of the Right Cloud
Database 315
  Soror Sahri, Rim Moussa, Darrell D.E. Long, and Salima Benbernou

Context-Aware Decision Support in Dynamic Environments:
Methodology and Case Study.................................... 325
  *Alexander Smirnov, Tatiana Levashova, Alexey Kashevnik, and
  Nikolay Shilov*

Textual Data Analysis and Mining

Unsupervised Aggregation of Categories for Document Labelling   335
  Piotr Borkowski, Krzysztof Ciesielski, and Mieczysław A. Kłopotek

Classification of Small Datasets: Why Using Class-Based Weighting
Measures?   345
  Flavien Bouillot, Pascal Poncelet, and Mathieu Roche

Improved Factorization of a Connectionist Language Model for
Single-Pass Real-Time Speech Recognition   355
  Łukasz Brocki, Danijel Koržinek, and Krzysztof Marasek

Automatic Extraction of Logical Web Lists   365
  *Pasqua Fabiana Lanotte, Fabio Fumarola, Michelangelo Ceci,
  Andrea Scarpino, Michele Damiano Torelli, and Donato Malerba*

Combining Formal Logic and Machine Learning for Sentiment
Analysis   375
  Niklas Christoffer Petersen and Jørgen Villadsen

Clustering View-Segmented Documents via Tensor Modeling   385
  Salvatore Romeo, Andrea Tagarelli, and Dino Ienco

Searching XML Element Using Terms Propagation Method   395
  Samia Berchiche-Fellag and Mohamed Mezghiche

Special Session: Challenges in Text Mining and Semantic Information Retrieval

AI Platform for Building University Research Knowledge Base   405
  *Jakub Koperwas, Łukasz Skonieczny, Marek Kozłowski,
  Piotr Andruszkiewicz, Henryk Rybiński, and Wacław Struk*

A Seed Based Method for Dictionary Translation   415
  Robert Krajewski, Henryk Rybiński, and Marek Kozłowski

SAUText — A System for Analysis of Unstructured Textual Data   425
  Grzegorz Protaziuk, Jacek Lewandowski, and Robert Bembenik

Evaluation of Path Based Methods for Conceptual Representation of
the Text   435
  Łukasz Kucharczyk and Julian Szymański

Special Session: Warehousing and OLAPing Complex, Spatial and Spatio-Temporal Data

Restructuring Dynamically Analytical Dashboards Based on Usage
Profiles   445
    Orlando Belo, Paulo Rodrigues, Rui Barros, and Helena Correia

Enhancing Traditional Data Warehousing Architectures with Real-Time
Capabilities   456
    Alfredo Cuzzocrea, Nickerson Ferreira, and Pedro Furtado

Inference on Semantic Trajectory Data Warehouse Using an Ontological
Approach   466
    Thouraya Sakouhi, Jalel Akaichi, Jamal Malki, Alain Bouju, and
    Rouaa Wannous

Combining Stream Processing Engines and Big Data Storages for Data
Analysis   476
    Thomas Steinmaurer, Patrick Traxler, Michael Zwick,
    Reinhard Stumptner, and Christian Lettner

ISMIS Posters

Representation and Evolution of User Profile in Information Retrieval
Based on Bayesian Approach   486
    Farida Achemoukh and Rachid Ahmed-Ouamer

Creating Polygon Models for Spatial Clusters   493
    Fatih Akdag, Christoph F. Eick, and Guoning Chen

Skeleton Clustering by Autonomous Mobile Robots for Subtle Fall Risk
Discovery   500
    Yutaka Deguchi and Einoshin Suzuki

Sonar Method of Distinguishing Objects Based on Reflected Signal
Specifics   506
    Teodora Dimitrova-Grekow and Marcin Jarczewski

Endowing Semantic Query Languages with Advanced Relaxation
Capabilities   512
    Géraud Fokou, Stéphane Jean, and Allel Hadjali

A Business Intelligence Solution for Monitoring Efficiency of
Photovoltaic Power Plants   518
    Fabio Fumarola, Annalisa Appice, and Donato Malerba

WBPL: An Open-Source Library for Predicting Web Surfing
Behaviors   524
  Ted Gueniche, Philippe Fournier-Viger, Roger Nkambou, and
  Vincent S. Tseng

Data-Quality-Aware Skyline Queries   530
  Hélène Jaudoin, Olivier Pivert, Grégory Smits, and Virginie Thion

Neuroscience Rough Set Approach for Credit Analysis of Branchless
Banking   536
  Rory Lewis

Collective Inference for Handling Autocorrelation in Network
Regression   542
  Corrado Loglisci, Annalisa Appice, and Donato Malerba

On Predicting a Call Center's Workload: A Discretization-Based
Approach   548
  Luis Moreira-Matias, Rafael Nunes, Michel Ferreira,
  João Mendes-Moreira, and João Gama

Improved Approximation Guarantee for Max Sum Diversification with
Parameterised Triangle Inequality   554
  Marcin Sydow

Learning Diagnostic Diagrams in Transport-Based Data-Collection
Systems   560
  Vu The Tran, Peter Eklund, and Chris Cook

Author Index   567

Community Detection by an Efficient Ant Colony Approach

Lúcio Pereira de Andrade[1], Rogério Pinto Espíndola[2],
and Nelson Francisco Favilla Ebecken[3]

[1] UFF, Federal Fluminense University, Rio de Janeiro, Brazil
lucio@vm.uff.br
[2] UEZO, State University of West Side, Rio de Janeiro, Brazil
rpespindola@uezo.rj.gov.br
[3] UFRJ, Federal University of Rio de Janeiro (COPPE), Rio de Janeiro, Brazil
nelson@ntt.ufrj.br

Abstract.. Community detection is an efficient tool to analyze large complex networks offering new insights about their structures and functioning. A community is a significant organizational unity formed by nodes with more connections between them. Ant colony algorithms have been used to detect communities on a fast and efficient way. In this work, changes are performed on an ant colony algorithm for community detection by means of modularity optimization. The changes rely on the way an ant moves and on the adopted stopping criteria. To assess the proposed strategy, benchmark networks are studied and preliminary results indicate that the suggested changes make the original algorithm more robust, reaching higher values of modularity of the detected communities.

Keywords: complex networks, community detection, ant colony optimization.

1 Introduction

Many systems can be observed as networks: the Internet, social networks of people with common interests, business networks of companies, logistics networks, metabolic networks and food webs [1].

A large and diverse series of collective natural or artificial structures have network behavior in which two components are noted: the set of instances of one or more entities and the set of relations between the instances. The complexity of network did not rely on its structure but on the challenge of determining the roles of the nodes, individually or collectively; on the inter-relationships between the nodes and their consequences; on the extraction, analysis and interpretation of their main characteristics.

Networks are capable of transforming a wide system in an abstract, reduced and powerful structure, capturing basic information from tangled connections. The usual implementation of complex networks – and its most common graphical visualization approach – is done by means of graphs in which the entities are represented by vertices and the relations by edges [2].

T. Andreasen et al. (Eds.): ISMIS 2014, LNAI 8502, pp. 1–9, 2014.
© Springer International Publishing Switzerland 2014

On complex networks, social ones in special, community effect is frequently noted, which is a tendency the actors have to organize themselves in cohesive groups. According to the context in which it is being studied, groups are called modules, clusters and communities [3]. Although the amount of papers and perspectives about communities is significant, there is no consensus on its concept. Mathematically, given a network represented by graph $G = (V, E)$, a community is defined as a subgraph of the network formed by a subset $V_c \subseteq V$ of entities related by a common interest [4]. Consequently, a community is also formed by a subset of edges $E_c \subseteq E$.

In the simplest way, a community is a group of nodes with many edges between them and few relationships with other groups. As pointed out by Newman [5], the problem is how to quantify a community beyond edge count. So, the basic aim of communities detection is similar of graph partitioning, in which a network is segmented by allocating highly connected nodes in groups with few interconnections. The principal difference between these tasks is that the amount of groups and their lengths are not previously determined on the first one [6].

A good partition is function of two elements: the edges between groups – the fewer the better – and the edges inside the groups – the more the better. To evaluate the quality of the structures found, Newman and Girvan [5] developed a measure based on the grade of distribution of the nodes or modularity, which is used widely in objective functions of partitioning optimization algorithms. Therefore, the higher the modularity the better is the partitioning.

Methods to detect communities are as diverse as the approaches used. They may be organized in four categories in accordance with their focus, as described by Tang and Liu [7]: a) centered on nodes, in which each node in a group has to satisfy some restrictions; b) centered on groups, considering the connections inside the groups; c) centered on networks, observing the entire network's connections and discovering interdependent groups; d) centered on hierarchy, building communities based on the topology of the network in a hierarchical manner.

The ant colony optimization algorithm [8] has been studied on a wide range of variations and applications, including community detection. He et al. [9] developed a two-level ant colony algorithm which employs simulated annealing on the decision of changing the community of a node when the modularity does not improve. This approach achieved good performance on large real networks when compared to other algorithms found in literature. Recent works reveal the suitability of ant colony methods on community detection on diverse networks and approaches like e-mails networks [10], dynamic routing on large road networks with fuzzy logic [11], and to discover unbalanced communities in directed graphs by using an optimization function based on nodes grades [12].

In this work, the algorithm proposed in [9] is modified to find the structure of communities in complex networks by allowing an ant to move even when its entire neighborhood is occupied by other ants. The strategy adopted is similar to the two iterative stages described in [13] and does not request the amount of communities previously. The modularity of a network is calculated whenever a change of the community of a node is performed and the best partitioning found is held. Moreover, a new stopping criterion by stagnation was adopted, which evaluates the stabilization of the optimization algorithm and avoids unproductive iterations.

The remainder of this paper is organized as follows. The next section describes the model proposed. The experiments and results are presented in Section 3. The conclusions and future works are discussed in Section 4.

2   Ant Colonies for Community Detection

2.1   Multi-layer Ant Colony Optimization

Blondel et al. [13] presented a strategy for community detection with two iterative stages. On the first one, each node is its community and the association between two nodes is evaluated in terms of gains in modularity. If the modularity improves, new associations are performed by changing the communities of the nodes on the frontiers between them. When no modularity improvement is obtained at all, the second stage starts with a new network formed by the communities found on the first stage as the nodes, and a new search for communities is performed. At this stage, each node has positive loop equal to the sum of the edges between its nodes (first stage), and the edge between two nodes is the sum of the edges associating nodes of both communities on first stage.

As the strategy above, He et al. [9] presented an approach for community detection on large scale networks based on layers in which the modularity is optimized by an ant colony algorithm [8]. Their work also applies a simulated annealing heuristic [14] on the process of allocation of a node in a community making possible the change of communities even with no improvements on the modularity. So, the approach is divided in two modules called single-layer ant colony optimization and multi-layer ant colony optimization. The first module is responsible to find communities of nodes. The latter receives the communities' structure, transforms communities in nodes and applies it again to find higher levels of communities (agglomerative step). The process continues until no improvements in modularity are observed. When compared to other community detection algorithms on large scale networks, good performances were observed.

2.2   A New Model of Ants for Community Detection

In this work, a small but relevant change on the approach was done inspired on the behavior of an ant species named Harpegnathos saltator. Descripted originally by Thomas Jerdon in 1851, this species has the ability to jump when it is necessary [17]. Based on this behavior, a new ant colony for community detection was created in which an ant is able to jump to another node when other ant assumes its node due to a fully occupied neighborhood. The jump is random and is performed expecting to reallocate the jumper ant in a neighborhood of a node not evaluated yet and with diverse communities on its nodes. These ants are referred as saltator ants in this work.

Beyond the new model of ant colony, other changes on the original algorithm were performed aiming better communities' structures. Firstly, instead of a constant number of iterations for the ant colony search, it was created a parameter of stagnation which stops the search when there is no modularity improvement after few steps.

Other modification deals with the modularity evaluation: here it is done as soon as there is a positive change of the community of a node. This action permits the search to hold the configuration with best modularity.

This work keeps the modularity as the measure of quality of a partitioning and it is calculated by [3]:

$$Q = \frac{1}{2e}\sum_{l=1}^{k}\sum_{i,j\in c_l}\left(E_{ij} - \frac{g_i \cdot g_j}{2e}\right) \qquad (1)$$

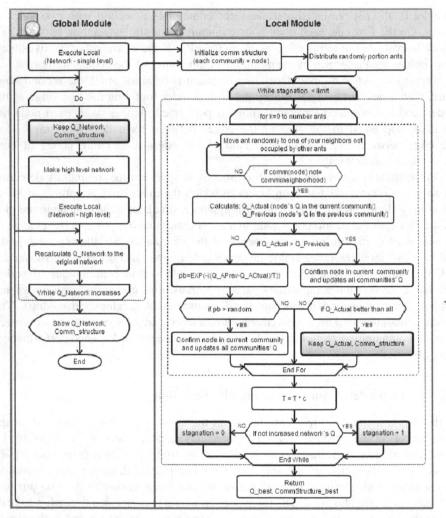

Fig. 1. The proposed algorithm

In which k, e, gi and gj are, respectively, the number of communities, the amount of edges, and the grades of nodes i and j. Eij is the value of the edge relating nodes i and j and the expression inside the summations estimates the community effect, that is, how much the interaction between the nodes is far from a random behavior. The coefficient $1/2e$ is introduced to normalize the modularity inside interval $[-1,1]$.

So, the modularity of a partitioning is the sum of modularities of all communities. A higher value of modularity indicates stronger internal interaction. Negative value means bad partitioning and positive modularity indicates that some degree of community structure is identified.

The algorithm is presented on Fig. 1. Grey elements indicate the changes that were performed on He et al algorithms. The local module is responsible to find communities of nodes until stagnation limit is reached and it keeps the best community structure found. The global module receives the best communities' structure found, transforms them in nodes and applies local module again to perform the agglomerative step. The process continues until no improvements in total modularity are observed.

3   Experiments and Results

3.1   Datasets Studied

Some real networks were studied to evaluate the efficiency of the saltator ant colony model proposed in this work. They are presented on Table 1 and are widely used in literature [18].

Table 1. Studied Networks

Network	karate[1]	dolphin[1]	polbooks[1]	football[1]	jazz[2]	e-mail[2]
|Vertices|	34	62	105	115	198	1,133
|Edges|	78	100	441	613	2,742	5,451

[1] From http://www-personal.umich.edu/~mejn/netdata/.
Access in 20/09/2013.
[2] From http://deim.urv.cat/~aarenas/data/welcome.htm/.
Access in 20/09/2013.

3.2   Algorithm Parameters

The parameterization used is presented on Table 2.

Table 2. Parameter Setting

Technique	Parameter	Value
Ant colony	Stagnation limit	10
	Number of ants	0.6N [*]
Simulated annealing	Initial temperature	500 or 250+0.5N [*]
	Annealing coefficient	0.9

[*] N is the number of nodes

The number of ants was suggested by He et al. [9], although several tests have revealed that values above 50% of number of nodes usually permit good performance and there are no significant impacts on the modularity or on processing time, maybe due to the stagnation stopping criteria. The initial temperature for the simulated annealing may be 500 for networks with more than 500 nodes. Smaller networks are more sensible to this parameter and it was defined as a function of the number of nodes, which has been proved to be a good strategy.

3.3   Results and Discussion

The results of community detection task performed by the saltator ant colony proposed in this work were compared with those obtained by Jin et al. [18] and are presented on Fig. 3. For each problem, the task was performed 50 times and the average performances were obtained. As modularity should be maximized, it can be seen that the saltator ant colony achieved very good performances on all networks studied. When compared with the original algorithm, in only one experiment the saltator model was surpassed.

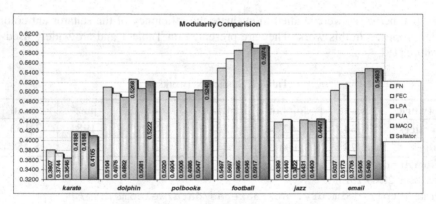

Fig. 2. Modularities of the detected communities

Beyond these results, two popular graph analysis tools were employed to comparatively evaluate the performance of the saltator ants: Gephi[1] e NodeXL[2]. In Fig. 4, the characteristics of the best communities obtained from this work and from Gephi are presented. As it can be seen, the saltator model achieved results as good as the ones obtained from Gephi, which adopts the Blondel et al. [13] approach for community detection. The structures of the communities obtained from both approaches are also very similar. On the e-mail network, although the significant difference of the

---

[1] www.gephi.org/
[2] http://nodexl.codeplex.com/

amounts of communities found, it is possible to note some correspondence between them by considering the bigger communities.

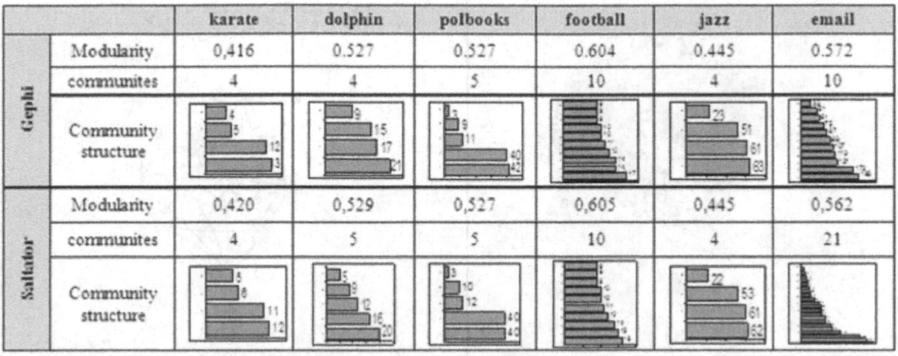

Fig. 3. Communities' structures: saltator x Gephi

Table 3. Results from *Saltator* x Gephi x NodeXL

		celegans[1]	polblogs[1]	geom[2]	powergrid[1]
	|Vertices|	453	1222	3621	4941
	|Edges|	2025	16714	9461	6594
Gephi	modularity	0.437	0.427	0.738	0.933
	# communities	10	277	37	37
NodeXL	modularity	0.427	0.361	0.718	0.932
	# communities	15	304	91	62
Saltator	modularity	0.442	0.427	0.732	0.931
	# communities	14	277	30	40

[1] From http://www-personal.umich.edu/~mejn/netdata/.
Access in 20/09/2013.
[2] From http://vlado.fmf.uni-lj.si/pub/networks/data/collab/
geom.htm/. Acess in 20/09/2013.

To evaluate the saltator model on larger and more complex networks, the results from it is compared to the Gephi and NodeXL ones, as presented in Table 3. Again, the performances of the saltator model are comparatively good. On Fig. 5, as example, the original Caenorhabditis elegans network (celegans) and the communities found by saltator model are illustrated. The visualization of the original network suggests a formation of a large community and some small ones. This suggestion is verified with the image from the communities discovered by the saltator model, in which it can be seen communities with different sizes. Moreover, is this figure loops can be also observed.

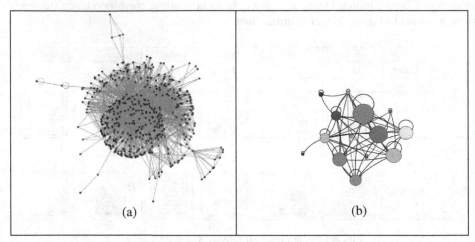

Fig. 4. Celegans network: a) original nodes; b) communities' structure by *saltator* ants

4   Conclusion

Community detection constitutes a significant tool for the analysis of complex networks by enabling the study of mesoscopic structures that are often associated with organizational and functional characteristics.

Social networks analysis has many challenges and one of them is the detection of cohesive communities. Beyond the search space may be extremely large, which imposes high computational costs, there is the need of adaptation of traditional measures of quality of partitioning and the development of new proper ones.

As the results indicate, the use of ant colony in this task has been achieving very good performance on many networks. In particular, the detection of community in stages or layers has been proved to be a good strategy because it has found more compact structures in less time, as in large scale networks as in smaller ones, yielding good quality communities.

The problem of extracting communities from a graph, or of dividing the nodes of a graph into distinct communities, has been approached from several different directions. In fact, algorithms for community extraction have appeared in practically all fields: social network analysis, physics, and computer science among others.

In this work, some changes were proposed on the usual algorithm for community detection and the results obtained from real networks reveal that the changes were positive. As future developments, large scale networks are going to be studied, mainly those representative of the interactions among planktonic organisms.

References

1. Newman, M.E.J.: The structure and function of complex networks. SIAM Review 45, 167–256 (2010)
2. Loscalzo, S., Yu, L.: Social Network Analysis: Tasks and Tools. In: Liu, H., Salerno, J.J., Young, M.J. (eds.) Social Computing, Behavioral Modeling, and Prediction, ch. 17. Springer, New York (2008)

3. Tang, L., Liu, H.: Community Detection and Mining in Social Media, Synthesis Lectures on Data Mining and Knowledge Discovery. Morgan & Claypool (2010)
4. Fan, T.F., Liau, C.J., Lin, T.Y.: Positional analysis in fuzzy social networks. In: Proceedings of the IEEE International Conference on Granular Computing, San Jose, pp. 423–428 (2007)
5. Newman, M.E.J., Girvan, M.: Finding and evaluating community structure in networks. Physical Review E 69(2), 26113–26127 (2004)
6. Newman, M.E.J.: Networks: An Introduction. Oxford University Press, New York (2010)
7. Tang, L., Liu, H.: Graph Mining Applications to Social Network Analysis. In: Aggarwal, C.C., Wang, H. (eds.) Managing and Mining Graph Data, Advances in Database Systems, vol. 40. Springer, New York (2010)
8. Dorigo, M., Stützle, T.: Ant Colony Optimization. MIT Press, Massachusetts (2004)
9. He, D., Liu, J., Liu, D., et al.: Ant colony optimization for community detection in large-scale complex networks. In: 7th International Conference on Natural Computation, Shangai, China, pp. 1151–1155 (2011)
10. Liu, Y., Liu, L., Luo, J.: Adaptive Ant Colony Clustering Method Applied to Finding Closely Communicating Community. Journal of Networks 7(2), 249–258 (2012)
11. Geetha, M., Nawaz, G.M.K.: Hierarchical Community-Fuzzy Ant Based Dynamic Routing on Large Road Networks. Research Journal of Applied Sciences 8(1), 65–71 (2013)
12. Romdhane, L.B., Chaabani, Y., Zardi, H.: A robust ant colony optimization-based algorithm for community mining in large scale oriented social graphs. Expert Systems with Applications 40(14), 5709–5718 (2013)
13. Blondel, V.D., Guillaume, J.-L., Lambiotte, R., Lefebvre, E.: Fast unfolding of communities in large networks. Journal of Statistical Mechanics 2008(10), P10008 (2008)
14. Kirkpatrick, S., Gelatt, C.D., Vecchi, M.P.: Optimization by Simulated Annealing. Science 220(4598), 671–680 (1983)
15. Lumer, E.D., Faieta, B.: Diversity and adaptation in populations of clustering ants. In: Meyer, J.-A., Wilson, S.W. (eds.) Proceedings of the Third International Conference on Simulation of Adaptive Behavior: From Animals to Animats 3, pp. 501–508. MIT Press, Massachusetts (1994)
16. Deneubourg, J.-L., Goss, S., Franks, N., et al.: The dynamics of collective sorting: Robot-like ants and ant-like robots. In: Meyer, J.-A., Wilson, S.W. (eds.) Proceedings of the First International Conference on Simulation of Adaptive Behavior: From Animals to Animats, pp. 356–363. MIT Press, Massachusetts (1991)
17. Urbani, C.B., Boyan, G.S., Blarer, A., et al.: A novel mechanism for jumping in the Indian ant Harpegnathos saltator (Jerdon) (Formicidae, Ponerinae). Experientia 50(1), 63–71 (1994)
18. Jin, D., Liu, D., Yang, B., et al.: Ant Colony Optimization with a New Random Walk Model for Community Detection in Complex Networks. Advances in Complex Systems 14(5), 795–815 (2011)

Adaptive XML Stream Classification
Using Partial Tree-Edit Distance

Dariusz Brzczinski and Maciej Piernik

Institute of Computing Science, Poznan University of Technology,
ul. Piotrowo 2, 60–965 Poznan, Poland
{dariusz.brzezinski,maciej.piernik}@cs.put.poznan.pl

Abstract. XML classification finds many applications, ranging from data integration to e-commerce. However, existing classification algorithms are designed for static XML collections, while modern information systems frequently deal with streaming data that needs to be processed on-line using limited resources. Furthermore, data stream classifiers have to be able to react to concept drifts, i.e., changes of the streams underlying data distribution. In this paper, we propose XStreamClass, an XML classifier capable of processing streams of documents and reacting to concept drifts. The algorithm combines incremental frequent tree mining with partial tree-edit distance and associative classification. XStreamClass was experimentally compared with four state-of-the-art data stream ensembles and provided best average classification accuracy on real and synthetic datasets simulating different drift scenarios.

Keywords: XML, data stream, classification, concept drift.

1 Introduction

In the past few years, several data mining algorithms have been proposed to discover knowledge from XML data [1,2,3,4]. However, these algorithms were almost exclusively analyzed in the context of static datasets, while in many new applications one faces the problem of processing massive data volumes in the form of transient data streams. Example applications involving processing XML data generated at very high rates include monitoring messages exchanged by web-services, management of complex event streams, distributed ETL processes, and publish/subscribe services for RSS feeds [4].

The processing of streaming data implies new requirements concerning limited amount of memory, short processing time, and single scan of incoming examples, none of which are sufficiently handled by traditional XML data mining algorithms. Furthermore, due to the nonstationary nature of data streams, target concepts tend to change over time in an event called *concept drift*. Concept drift occurs when the concept about which data is being collected shifts from time to time after a minimum stability period [5]. Drifts can be reflected by class assignment changes, attribute distribution changes, or an introduction of new classes (*concept evolution*), all of which deteriorate the accuracy of algorithms.

T. Andreasen et al. (Eds.): ISMIS 2014, LNAI 8502, pp. 10–19, 2014.
© Springer International Publishing Switzerland 2014

Although several general data stream classifiers have been proposed [5,6,7,8], they do not take into account the semi-structural nature of XML, like e.g. the XRules algorithm does for static data [1]. To the best of our knowledge, the only available XML stream classification algorithm was proposed by Bifet and Gavaldà [9]. However, this proposal focuses only on incorporating incremental subtree mining to the learning process and, therefore, does not fully utilize the structural similarities between XML documents. Furthermore, the classification method proposed by Bifet and Gavaldà is only capable of dealing with sudden concept drifts, but will not react to gradual drifts or concept evolution.

In this paper, we propose XStreamClass, a stream classification algorithm which employs incremental subtree mining and partial tree-edit distance to classify XML documents online. By dynamically creating separate models for each class, the proposed method is capable of dealing with concept evolution and gradual drift. Moreover, XStreamClass can be easily extended to a cost sensitive model, allowing it to handle skewed class distributions. We will show that the resulting system performs favorably when compared with existing stream classifiers, additionally being able to cope with different types of concept drift.

The remainder of the paper is organized as follows. Section 2 presents related work. In Section 3, we introduce a new incremental XML classification algorithm, which uses maximal frequent induced subtrees and partial tree-edit distance to perform predictions. Furthermore, we analyze possible variations of the proposed algorithm for different stream settings. The algorithm is later experimentally evaluated on real and synthetic datasets in Section 4. Finally, in Section 5 we draw conclusions and discuss lines of future research.

2   Related Work

As an increasingly important data mining technique, data stream classification has been widely studied by different communities; a detailed survey can be found in [5]. In our study, we focus on methods that adaptively learn from blocks of examples. One of the first of such block-based classifiers was the Accuracy Weighted Ensemble algorithm (AWE) [10], which trained a new classifier with each incoming block of examples to form a dynamically weighted and rebuilt classifier ensemble. More recently proposed block-based methods include Learn++NSE [11] which uses a sophisticated accuracy-based weighting mechanism and the Accuracy Updated Ensemble (AUE) [8] which incrementally trains its component classifiers after every processed block of examples.

However, all of the aforementioned algorithms are general classification methods, which are not designed to deal with semi-structural documents. On the other hand, although there exists a number of XML classifiers for static data [1,2], none of them is capable of incrementally processing streams of documents. To the best of our knowledge, the only streaming XML classifier is that proposed by Bifet and Gavaldà [9]. In this approach, the authors propose to adaptively mine closed frequent induced subtrees on batches of XML documents. The discovered subtrees are later used in the learning process, where labeled documents are encoded as tuples with attributes representing the occurrence/absence of frequent trees in a given document. Such tuples are later fed to a bagging or boosting ensemble of decision trees.

The proposed XStreamClass algorithm uses the AdaTreeNat [9] algorithm to incrementally mine maximal frequent induced subtrees and Partial Tree-edit Distance [12] to perform classification. Partial Tree-edit Distance (PTED) is an approximate subtree matching algorithm, which measures how much one tree needs to be modified to become a subtree of another tree. PTED is a combination of subtree matching [13] and tree-edit distance algorithms [14], and was designed specifically for XML classification.

3   The XStreamClass Algorithm

Existing data stream classification algorithms are not designed to process structural data. The algorithm of Bifet and Gavaldà [9] transforms XML documents into vector representations in order to process them using standard classification algorithms and, therefore, neglects the use of similarity measures designed strictly for XML. Furthermore, the cited approach is capable of dealing with sudden drifts, but not gradual changes or concept evolution. The aim of our research is to put forward an XML stream classifier that will use structural similarity measures and be capable of reacting to different types of drift. To achieve this goal, we propose to combine associative classification with partial tree-edit distance, in an algorithm called XStreamClass.

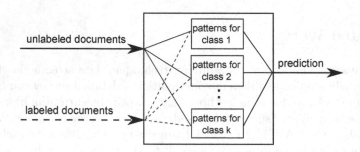

Fig. 1. XStreamClass processing flow

The XStreamClass algorithm maintains a pool of maximal frequent induced subtrees for each class and predicts the label of each incoming document by associating it with the class of the closest of all maximal frequent induced subtrees. The functioning of the algorithm can be divided into two subprocesses: training and learning. It is important to notice that, in accordance with the *anytime prediction* requirement of data stream classification [5], the training and learning processes can occur simultaneously and the algorithm is always capable of giving a prediction (Fig. 1). Algorithm 1 presents the details of the training, while Algorithm 2 summarizes the classification process.

Algorithm 1. XStreamClass: training

Input: $\mathcal{D}$: stream of labeled XML documents, m: buffer size, $minsup$: minimal support
Output: $\mathcal{P}$: set of patterns for each class

1. **for all** documents $d \in \mathcal{D}$ **do**
2. $B \leftarrow B \cup \{d\}$;
3.   **if** $|B| = m$ **then**
4.   split documents into batches B_i ($i = 1, 2, ..., k$) according to class labels;
5. $P_i \leftarrow AdaTreeNat_i(P_i, B_i, minsup)$;
6. $\mathcal{P} \leftarrow P_1 \cup P_2 \cup ... \cup P_k$;
7. $B \leftarrow \emptyset$

In the training process, labeled documents are collected into a buffer B. When the buffer reaches a user-defined size m, documents are separated according to class labels into batches B_i ($i = 1, 2, ..., k$), were k is the number of classes. Each batch B_i is then incrementally mined for maximal frequent subtrees P_i by separate instances of the AdaTreeNat algorithm [9]. Since AdaTreeNat can mine trees incrementally, existing tree miners for each class are reused with each new batch of documents. Furthermore, in case of concept-evolution, a new tree miner can be created without modifying previous models. After the mining phase, frequent subtrees are combined to form a set of *patterns* $\mathcal{P}$, which is used during classification.

It is worth noticing that the training procedure can be slightly altered to achieve a more fluid update procedure. Instead of maintaining a single buffer B and waiting for m documents to update the model, one could create independent buffers for each class label. This would introduce the possibility of defining a different batch size for each class and enable better control of the training process for class-imbalanced streams. The influence of this fluid update strategy will be discussed in Section 4.

Algorithm 2. XStreamClass: classification

Input: $\mathcal{D}$: stream of unlabeled XML documents
Output: $\mathcal{Y}$: stream of class predictions

1. **for all** documents $d \in \mathcal{D}$ **do**
2.   calculate $\Delta(p, d)$ for each pattern $p \in \mathcal{P}$ using (1);
3. $y \leftarrow$ class of $p = \arg\min \Delta(p, d)$ (or p calculated according to (2));
4. $\mathcal{Y} \leftarrow \mathcal{Y} \cup \{y\}$;

Classification is performed incrementally for each document using the set of current patterns $\mathcal{P}$. To assign a class label to a given document d, the algorithm calculates the partial tree-edit distance between d and each pattern $p \in \mathcal{P}$. The partial tree-edit distance is defined as follows [12]. Let s be a sequence of deletion or relabeling operations on leaf nodes or on the root node of a tree. A *partial tree-edit sequence* s between two trees p and d is a sequence which transforms p into any induced subtree of d. The cost $c(s)$ of a partial tree-edit sequence s is the total cost of all operations in s. *Partial tree-edit distance* $\Delta(p, d)$ between a pattern tree p and a document tree d is the minimal cost of all possible partial tree-edit sequences between p and d.

$$\Delta(p, d) = \min\{c(s) : s \text{ is a partial tree-edit sequence between } p \text{ and } d\} \quad (1)$$

After using (1) to calculate distances between d and each pattern $p \in \mathcal{P}$, XStreamClass assigns the class of the pattern closest to d. If there is more than one closest pattern, we propose a weighted voting measure to decide on the most appropriate class. In this scenario, each pattern is granted a weight based on its normalized support and size, and uses this value to vote for its corresponding class. The document is assigned the class c with the highest score, as presented in (2).

$$c = \underset{c_i, i=1,2,\ldots,k}{\arg\max} \sum_{\{p:class(p)=c_i\}} (support(p) \times size(p)) \quad (2)$$

In contrast to general block-based stream classifiers like AWE [10], AUE [8], or Learn++NSE [11], the proposed algorithm is designed to work strictly with structural data. Compared to XML classification algorithms for static data, such as XRules [1] or X-Class [2], we process documents incrementally, use PTED, and do not use default rules or rule priority lists. In contrast to [9], XStream-Class does not encode documents into tuples and calculates pattern-document similarity instead of pattern-document inclusion. Moreover, since XStreamClass mines for maximal frequent subtrees for each class separately, it can have different model refresh rates for each class. In case of class imbalance, independent refresh rates allow the algorithm to mine for patterns of the minority class using more documents than would be found in a single batch containing all classes. Additionally, this feature helps to handle concept-evolution without the need of rebuilding the entire classification model. Finally, because of its modular nature, the proposed algorithm can be easily implemented in distributed stream environments like Storm[1], which would enable high-throughput processing.

4   Experimental Evaluation

The aim of our experiments was to evaluate the XStreamClass algorithm on static and drifting data and compare it against four streaming classifiers employing the methodology presented in [9]: Online Bagging (Bag) [7], Accuracy Weighted Ensemble (AWE) [10], Learn++.NSE [11], and Accuracy Updated Ensemble (AUE) [8]. Bagging was chosen as the algorithm used by Bifet and Gavaldà to test their methodology and the remaining algorithms were chosen as strong representatives of block-based stream classifiers. We tested two versions of XStreamClass: one that synchronously updates all class models using a single batch (XSC) and one that updates each class model independently using separate batches for each class (XSC$_F$).

---

[1] http://storm-project.net/

XStreamClass was implemented in C# [2], tree mining was performed using a C++ implementation of AdaTreeNat [9], while the remaining classifiers were implemented in Java as part of the MOA framework [15]. The experiments were conducted on a machine equipped with a dual-core Intel i7-2640M CPU, 2.8Ghz processor and 16 GB of RAM. To make the comparison more meaningful, we set the same parameter values for all the algorithms. For ensemble methods we set the number of component classifiers to 10: AUE, NSE, and Bag have ten Hoeffding Trees, and since AWE uses static learners it has ten J48 trees. We decided to use 10 component classifiers as this was the number suggested and tested in [9]. The data block size used for block-based classifiers (AWE, AUE, NSE) was the same as the maximal frequent subtree mining batch size (see Section 4.1). For ensemble components we used Hoeffding Trees enhanced with Naive Bayes leaf predictions with a grace period $n_{min} = 100$, split confidence $\delta = 0.01$, and tie-threshold $\tau = 0.05$ [6].

4.1 Data Sets

During our experiments we used 4 real and 8 synthetic datasets. The real datasets were the CSLOG documents, which consist of web logs categorized into two classes, as described in [1,9]. The first four synthetic datasets were the DS XML documents generated and analyzed by Zaki and Aggarwal [1]. The additional four synthetic datasets were generated using the tree generation program of Zaki, as described in [9]. NoDrift contains no drift, Sudden contains 3 sudden drifts every 250k examples, Gradual gradually drifts from the 250k to 750k example, and Evolution contains a sudden introduction of a new class after the 1M example[2]. All of the used datasets are summarized in Table 1.

Table 1. Dataset characteristics

Dataset	#Documents	#Classes	#Drifts	Drift type	Minsup	Batch	Window size
CSLOG12	7628	2	-	unknown	5%	1000	-
CSLOG123	15037	2	-	unknown	4%	1000	-
CSLOG23	15702	2	-	unknown	6%	1000	-
CSLOG31	23111	2	-	unknown	3%	1500	-
DS1	91288	2	-	unknown	1%	5000	-
DS2	67893	2	-	unknown	1%	5000	-
DS3	100000	2	-	unknown	1%	5000	-
DS4	75037	2	-	unknown	0%	5000	-
Evolution	2000000	3	1	mixed	1%	10000	-
Gradual	1000000	2	1	gradual	1%	1000	1000
NoDrift	1000000	2	0	none	1%	10000	-
Sudden	1000000	2	3	sudden	1%	1000	1000

---

[2] Source code, test scripts, and datasets available at:
http://www.cs.put.poznan.pl/dbrzezinski/software.php

As shown in Table 1, minimal support and batch size used for tree mining varied depending on the dataset size and characteristics. For datasets without drift or with concept evolution, we used incremental tree mining without any forgetting mechanism; for datasets with drift, a sliding window equal to the batch size [9]. All of the tested algorithms used the same patterns for classification.

4.2 Results

All of the analyzed algorithms were tested in terms of accuracy, classification time, and memory usage. The results were obtained using the test-then-train procedure [5], with pattern mining (model updates) occurring after each batch of examples. Tables 2–4 present average memory usage, batch classification time, and accuracy, obtained by the tested algorithms on all datasets, respectively.

In terms of memory usage, XStreamClass is the most efficient solution out of all the tested algorithms. This is especially visible for larger datasets, were general stream classifiers grow large classification models, while the proposed algorithm only needs to maintain a list of current maximal frequent subtrees.

As Table 2 shows, low memory consumption is achieved at the cost of relatively high classification time. XStreamClass needs to calculate the PTED between a document and each pattern, which is computationally more expensive than simple subtree matching. However, it is worth noticing that for larger streams XStreamClass offers comparable and sometimes better prediction speed due to its compact classification model.

Concerning accuracy, XStreamClass is the best algorithm on all but one dataset. This is a direct result of using of a tree similarity measure, which, in contrast to simple subtree matching, classifies a document even if it does not contain any of the patterns in the model. Moreover, since the classification model of XStreamClass depends strictly on the current set of frequent patterns, the algorithm has a forgetting model steered directly by adaptive pattern mining. This allows the proposed algorithm to react to changes as soon as they are visible in the patterns, in contrast to the compared algorithms, which would require a separate drift detection model. This was especially visible on the accuracy plot of the Gradual dataset presented in Fig. 2. One can see that around the 250k example accuracy slightly drops as gradual drift starts, but with time XStreamClass recovers from the drift.

It is also worth mentioning that independent batches for each class and, thus, asynchronous pattern mining for each class, offers better accuracy on practically all datasets. This is connected to the fact that in the XSC_F processing scenario a larger number of examples is used for pattern mining for each class, which often allows to find better patterns. Furthermore, the CSLOG datasets are imbalanced with the majority class occurring three times as often as the minority class. Since XSC_F waits for a larger sample of the minority class before pattern mining, it can achieve higher accuracy even on datasets as small as CSLOG.

To verify if the results of the compared classifiers are significantly different, we carried out statistical tests for comparing multiple classifiers over multiple datasets. We used the non-parametric Friedman test along with the Bonferroni-Dunn

Table 2. Average model memory usage [kB]

	AUE	AWE	Bag	NSE	XSC	XSC$_F$
CSLOG12	179.45	2376.80	441.36	38.11	**4.07**	4.78
CSLOG123	564.45	5226.85	889.44	55.04	**4.10**	4.82
CSLOG23	418.03	3642.58	749.76	53.86	**3.94**	5.09
CSLOG31	255.77	3642.58	453.46	48.41	**4.96**	6.50
DS1	1375.00	17114.75	2531.67	232.95	2.48	**2.47**
DS2	1331.89	15411.12	2631.16	613.50	**2.73**	2.80
DS3	1801.53	17574.12	3152.19	174.08	2.45	**2.39**
DS4	1510.63	16273.34	2604.43	123.87	**2.90**	3.53
Evolution	466.25	40979.98	1446.82	1539.28	2.63	**2.51**
Gradual	318.00	4746.09	1230.47	14522.56	**2.53**	2.63
NoDrift	500.81	40287.60	951.82	798.65	**2.54**	2.63
Sudden	263.39	4746.09	1093.75	13048.44	2.69	**2.67**

Table 3. Average block/batch classification time [s]

	AUE	AWE	Bag	NSE	XSC	XSC$_F$
CSLOG12	0.02	0.02	0.03	**0.01**	0.04	0.07
CSLOG123	0.06	**0.05**	0.07	0.05	0.14	0.15
CSLOG23	0.04	**0.04**	0.05	0.04	0.06	0.10
CSLOG31	0.02	0.01	0.03	**0.01**	0.03	0.04
DS1	0.35	0.31	0.43	**0.30**	1.22	0.31
DS2	0.33	**0.30**	0.41	0.30	1.39	0.35
DS3	0.41	0.36	0.47	0.35	1.08	**0.35**
DS4	0.57	0.54	0.63	**0.50**	2.06	0.62
Evolution	0.42	0.22	1.68	0.23	1.12	**0.11**
Gradual	0.03	**0.01**	0.12	0.02	0.07	0.12
NoDrift	0.44	0.25	1.03	0.27	0.83	**0.08**
Sudden	0.02	**0.01**	0.12	0.01	0.06	0.12

Table 4. Average classification accuracy [%]

	AUE	AWE	Bag	NSE	XSC	XSC$_F$
CSLOG12	76.91	76.91	75.60	76.91	75.77	**76.93**
CSLOG123	73.50	76.00	73.50	74.68	76.44	**78.80**
CSLOG23	77.14	77.01	74.71	76.06	78.19	**78.27**
CSLOG31	**76.52**	**76.52**	75.78	75.50	76.03	76.50
DS1	58.02	58.14	57.64	58.30	63.22	**64.95**
DS2	75.29	75.65	74.67	75.05	75.08	**79.78**
DS3	53.75	54.21	53.96	52.00	55.54	**59.51**
DS4	58.90	59.11	58.84	53.28	59.68	**61.31**
Evolution	53.79	54.07	53.81	52.05	**99.49**	**99.49**
Gradual	52.91	53.05	52.46	52.68	94.44	**97.65**
NoDrift	53.78	54.02	53.68	50.94	99.50	**99.99**
Sudden	52.58	52.49	52.45	51.20	96.33	**99.30**

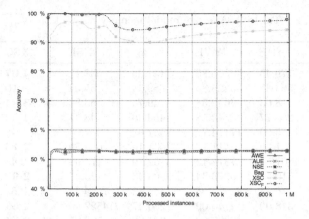

Fig. 2. Accuracy on the `Gradual` dataset

post-hoc test [16] to verify whether the performance of XSC/XSC$_F$ is statistically different from the remaining algorithms. The average ranks of the analyzed algorithms are presented in Table 5 (the lower the rank the better).

Table 5. Average algorithm ranks used in Friedman tests

	AUE	AWE	Bag	NSE	XSC	XSC$_F$
Accuracy	3.83	3.04	5.29	5.08	2.54	**1.21**
Memory	3.67	5.83	4.75	3.75	**1.33**	1.67
Testing time	3.33	**1.75**	4.92	1.83	5.17	4.00

By using the Friedman test [16] to verify the differences between accuracies, we obtain $FF_{Acc} = 25.38$. As the critical value for comparing 6 algorithms over 12 datasets for $p = 0.05$ is 2.38, the null hypothesis can be rejected and we can state that algorithm accuracies significantly differ from each other. Additionally, since the critical difference chosen by the Bonferroni-Dunn test is $CD = 1.97$, we can state that XSC$_F$ is significantly more accurate than AUE, Bag, and NSE. An additional one-tailed Wilcoxon test [16] shows that with $p = 0.001$ XSC$_F$ is also on average more accurate than AWE. A similar analysis ($FF_{Mem} = 71.01$, $FF_{Time} = 18.18$) shows that XSC is the most memory efficient algorithm and AWE and NSE are faster than XSC and XSC$_F$.

5   Conclusions

In this paper, we presented XStreamClass, the first algorithm to use tree similarity in classifying streams of XML documents. The algorithm combines incremental tree mining with partial tree-edit distance and associative classification.

Furthermore, we investigated different processing strategies to address the problem of class imbalance and concept evolution. Finally, we experimentally compared the proposed algorithm with the only competitive XML stream classification methodology. XStreamClass provided best classification accuracy and memory usage in environments with different types of drift as well as in static environments. Future lines of research will include different classification schemes using partial tree-edit distance and implementations for distributed environments.

Acknowledgments. D. Brzezinski's and M. Piernik's research is funded by the Polish National Science Center under Grants No. DEC-2011/03/N/ST6/00360 and DEC-2011/01/B/ST6/05169, respectively.

References

1. Zaki, M.J., Aggarwal, C.C.: Xrules: An effective algorithm for structural classification of xml data. Machine Learning 62(1-2), 137–170 (2006)
2. Costa, G., et al.: X-class: Associative classification of xml documents by structure. ACM Trans. Inf. Syst. 31(1), 1–3 (2013)
3. Brzezinski, D., et al.: XCleaner: A new method for clustering XML documents by structure. Control and Cybernetics 40(3), 877–891 (2011)
4. Mayorga, V., Polyzotis, N.: Sketch-based summarization of ordered XML streams. In: Ioannidis, Y.E., Lee, D.L., Ng, R.T. (eds.) ICDE, pp. 541–552. IEEE (2009)
5. Gama, J.: Knowledge Discovery from Data Streams. Chapman and Hall (2010)
6. Domingos, P., Hulten, G.: Mining high-speed data streams. In: Proc. 6th ACM SIGKDD Int. Conf. Knowl. Disc. Data Min., pp. 71–80 (2000)
7. Oza, N.C., Russell, S.J.: Experimental comparisons of online and batch versions of bagging and boosting. In: Proc. 7th ACM SIGKDD Int. Conf. Knowl. Disc. Data Min., pp. 359–364 (2001)
8. Brzezinski, D., Stefanowski, J.: Reacting to different types of concept drift: The accuracy updated ensemble algorithm. IEEE Trans. on Neural Netw. Learn. Syst. 25(1), 81–94 (2014)
9. Bifet, A., Gavaldà, R.: Adaptive xml tree classification on evolving data streams. In: Buntine, W., Grobelnik, M., Mladenić, D., Shawe-Taylor, J. (eds.) ECML PKDD 2009, Part I. LNCS, vol. 5781, pp. 147–162. Springer, Heidelberg (2009)
10. Wang, H., et al.: Mining concept-drifting data streams using ensemble classifiers. In: Proc. 9th ACM SIGKDD Int. Conf. Knowl. Disc. Data Min., pp. 226–235 (2003)
11. Elwell, R., Polikar, R.: Incremental learning of concept drift in nonstationary environments. IEEE Trans. Neural Netw. 22(10), 1517–1531 (2011)
12. Piernik, M., Morzy, T.: Partial tree-edit distance. Technical Report RA-10/2013, Poznan University of Technology (2013),
http://www.cs.put.poznan.pl/mpiernik/publications/PTED.pdf
13. Valiente, G.: Constrained tree inclusion. J. Discrete Alg. 3(2-4), 431–447 (2005)
14. Pawlik, M., Augsten, N.: RTED: A robust algorithm for the tree edit distance. PVLDB 5(4), 334–345 (2011)
15. Bifet, A., et al.: MOA: Massive Online Analysis. J. Mach. Learn. Res. 11, 1601–1604 (2010)
16. Demsar, J.: Statistical comparisons of classifiers over multiple data sets. J. Machine Learning Research 7, 1–30 (2006)

RILL: Algorithm for Learning Rules
from Streaming Data with Concept Drift

Magdalena Deckert and Jerzy Stefanowski

Institute of Computing Science, Poznań University of Technology,
60-965 Poznań, Poland
{magdalena.deckert,jerzy.stefanowski}@cs.put.poznan.pl

Abstract. Incremental learning of classification rules from data streams
with concept drift is considered. We introduce a new algorithm RILL,
which induces rules and single instances, uses bottom-up rule generalization based on nearest rules, and performs intensive pruning of the obtained rule set. Its experimental evaluation shows that it achieves better
classification accuracy and memory usage than the related rule algorithm
VFDR and it is also competitive to decision trees VFDT-NB.

1 Introduction

Mining data streams has received a growing research interest. Massive volumes
of data, their rapid arrival rate, and changing characteristics impose new computational requirements for algorithms, which are not fulfilled by standard solutions developed for static data repositories. Moreover, the greatest challenge
for classifiers learning from data streams is to properly react to *concept drifts*,
i.e. changes in definitions of target concepts over time. Depending on the rate
of these changes, concept drifts are divided into sudden, gradual or recurrent
ones [6]. There exists a need for a new type of classifier, that, besides stream
requirements on constrained memory usage, limited learning time and efficient
incremental scanning of incoming data, should be able to track drifts and effectively adapt to them.

Most of data stream classifiers are based on either implementing window-
ing forgetting mechanisms, applying drift detectors or include adaptive ensembles [6]. The most popular single classifiers are Hoeffding Trees [3] and Very Fast
Decision Trees which can handle numerical data and concept drifts [6]. However,
decision rules have not received enough attention in the data stream research
community so far. For static data, they are often equivalent to considering trees
as they can provide better interpretability and flexibility for applying them in
various systems [5]. Moreover, individual rules can be considered independently
and in case of concept drift the single outdated rules can be adapted more easily than rebuilding the complete classifier or even changing the structure of the
tree [7]. Besides the pioneering work on FLORA [11], only three other rule approaches have been introduced: AQ11-PM-WAH [9], FACIL [4], and VFDR [7].

We claim that there is still need for a new rule algorithm that fulfills stream
computational requirements and has better reaction to different types of drifts.

T. Andreasen et al. (Eds.): ISMIS 2014, LNAI 8502, pp. 20–29, 2014.
© Springer International Publishing Switzerland 2014

In our proposal we want to consider two-fold knowledge representation—*rules and single instances*. It can be more appropriate for dealing with difficult decision boundaries and decomposition of the concepts into many sub-part with outlying examples which may occur in changing data. Here we are inspired by positive experiences of BRACID algorithm for handling static imbalanced data [10]. Unlike divide and conquer general to specific strategy from VFDR we propose to consider stepwise *bottom-up generalization* based on the *nearest rule* idea. We will also promote other solutions for coping changes.

The main aim of our paper is to introduce this algorithm, called RILL, and to evaluate it in the comparative study on several data containing different types of concept drifts.

2 Related Works

Fundamental works on incremental rule learning with concept drift has been started with FLORA family of algorithms [11]. Their main idea is based on successive modifications of nominal attribute conditions in positive rules (having the same label as the new incoming example), *boundary*, and negative rules with each incoming example. FLORA was also equipped in forgetting mechanism using global window with learning examples. The next proposal is the AQ11-PM-WAH system [9], which stores positive examples from the boundaries of the current rule set inside the partial memory. When new examples arrive, the AQ11 learner combines them with those from the partial memory to modify the rules. Moreover, it dynamically tune the period for keeping examples in the partial memory using the forgetting heuristic from FLORA2.

However, according to [4] these algorithms were not effective for larger data streams, especially with numerical attributes, and the authors introduced the FACIL algorithm. Conditions in FACIL's rules are expressed as intervals over numerical attributes defining a hyper-rectangle in the normalized real numbered space. The algorithm starts from very specific rules. When a new incoming example is not covered by any rule, FACIL tries first to generalize positive rules by looking for the smallest extensions of intervals inside rules. However, it is accepted if each extension is smaller the user-defined threshold and the candidate for generalization does not intersect with any negative rules. The core of FACIL is to store with each rule both negative and selected positive examples (two positive per one negative example covered). This local set with examples is updated when the rule covers the new example. FACIL's rules may be inconsistent (covering both positive and negative examples), however the possible purity is controlled by the user defined threshold. When this threshold is reached, the rule and its generalizations are blocked, and the examples associated with the rule are used to induce new positive and negative rules.

The completely different induction mechanism is applied in Very Fast Decision Rules (VFDR) learner, which generates either ordered or unordered set of rules [7]. Following divide and conquer strategy [5] it starts from the most general rules and successively specialize them by adding new conditions. The specialization is based on the Hoeffding bound adapted from VFDT [3]. VFDR has been extended in [8] to cope with concept drifts by incorporation of the drift detection method (DDM [6]) with each individual rule. This improves the adaptation to changes and enables pruning of the rule set. Worth of notice is fact that only VFDR was more extensively evaluated on massive data streams with concept drift showing that it is competitive to VFDT.

As our proposal follows bottom-up rule induction and integrates rules with single instances it is more similar to FACIL than to VFDR. However, we identify several critical issues and differences in comparison to FACIL which motivate our proposal. Firstly, FACIL does not directly track drifts (only in a limited range by updating stored examples, although in a too computationally complicated way). It also insufficiently prunes rules (which can be critical for sudden drifts). Furthermore, its rule generalization is very limited and it can favor quite pure rules with small supports. Finally, its classification strategy seems to be too complex and not intuitive.

We hypothesize that it is beneficial to: (1) use single instances besides rules to better model complex concepts and their changes; (2) allow much stronger and simpler generalization of rule conditions based on the well known distance measure; (3) keep a simple forgetting mechanism with the global sliding window; (4) implement more aggressive rule pruning depending on monitoring their predictive abilities and (5) use the nearest rule / instance classification strategy.

3   Rule-Based IncrementaL Learner

RILL is an acronym of words **R**ule-based **I**ncrementa**L** **L**earner. It incrementally processes learning instances described by nominal and numerical conditional attributes and the decision class label. The RILL algorithm induces unordered set of decision rules and single instances. Each rule is represented in a form:

$$\textbf{if } (attr_{num} \text{ in } [b_l; b_u]) \text{ and } (attr_{nom} = value) \textbf{ then } class,$$

where $attr_{num}$ is a numerical attribute, b_l is its lower bound, b_u is its upper bound, $attr_{nom}$ is a nominal attribute, $value$ is its value, and $class$ is the value of the decision class indicated by the rule.

The pseudocode of the RILL algorithm is presented as Algorithm 1. It operates as follows. When a new learning example e_i is available, index of currently processed instance is incremented and the example e_i is added to the sliding window sw. Moreover, the distribution d of classes for learning examples in the window sw is updated according to the label of e_i (Alg. 1, line 2).

Next, RILL checks if e_i is covered by any positive rule with the same class label as e_i (Alg. 1, line 3). For every rule covering e_i, statistics like the number of covered positive examples in the window sw and the timestamp of its last usage are updated. Next, negative rules are checked. If they also cover e_i, their respective statistics are updated (Alg. 1, line 4).

If no positive rule covers e_i, the generalization procedure is fired (presented as Algorithm 2). First, this procedure looks for the nearest rule nr to the current

---

Algorithm 1. RILL (Rule-based IncrementaL Learner)

Input   : S: data stream of examples;
         w: maximum size of the sliding window;
         a: rule's maximum age
Output: RS: updated set of decision rules;
         sw: sliding window with learning examples;
         d: distribution of the learning examples in the sliding window

1 **foreach** *(learning example $e_i \in S$)* **do**
2   add example e_i to sw and update d;
3   *positiveCoverage* = `PositiveCoverage` (e_i);
4   *negativeCoverage* = `NegativeCoverage` (e_i);
5   **if** *(positiveCoverage = false)* **then**
6   *generalization* = `Generalization` (e_i, sw, d);
7   **if** *((positiveCoverage = false) and (generalization=false))* **then**
8   r = full description of the example e_i;
9         $RS_c \leftarrow RS_c \cup \{r\}$;
10   **if** *(swSize > w)* **then**
11   remove the oldest example from sw and update d;
12   RS = `DeleteOldRules` (RS, a);
13   RS = `DeleteImpureRules` $(RS, swSize, d)$;
14   RS = `DeleteErroneousRules` $(RS, swSize, d)$;

---

example e_i (Alg. 2, line 1), i.e. the rule with the smallest distance calculated using modified HVDM measure [12]. The value of this distance is defined as:

$$distance = \sum_{a \in attributes} \begin{cases} 0 \vee 1 & \text{if } a \text{ is nominal} \\ 0 \vee (b_l - value_a) \vee (value_a - b_u) & \text{if } a \text{ is numerical} \end{cases}.$$

This formula expresses the distance between the learning example and the decision rule as a sum of distances for each rule's conditional attribute. In case when rule's elementary condition for given attribute matches the example's value $(value_a)$, then the distance equals 0. When the rule does not match the example's value for given attribute, the distance depends on the type of the attribute. In case of nominal attributes, the distance equals 1. On the other hand, if the type of the attribute is numerical, the distance is calculated to the nearest bound of the elementary condition in the rule. Next, the nearest rule nr is generalized to cover e_i (Alg. 2, line 2). It is enhanced by dropping condition in case of nominal attributes or extending boundaries of numerical attributes $((b_l = value_a) \vee (b_u = value_a))$ to include attribute value describing the example e_i. After obtaining generalized rule gr, its statistics are updated and its purity is calculated (Alg. 2, line 4). If the obtained purity value, calculated as $\frac{\text{positive examples from the sliding window covered by the rule}}{\text{total number of covered examples from the sliding window}}$, is higher than the relative frequency of the rule's class calculated from the window sw (Alg. 2, line 5), then the procedure looks for the nearest negative example not covered by gr (Alg. 2, line 6). The motivation is to consider more general rule conditions. If such an example exist, the rule gr is extended on numerical attributes to the half of the distance to the negative example and the new rule er replaces the old rule nr in the current set of rules RS (Alg. 2, line 7—9). Otherwise, the rule nr is removed from the rules' set RS and the rule gr is added to the current set of rules RS (Alg. 2, line 11).

Algorithm 2. Generalization procedure

    Input   : e_i: current learning example;
                  sw: sliding window with number of recent learning examples;
                  d: distribution of learning examples in the sliding window
    Output: *generalization*: flag indicating whether generalization was performed

1   nr = find nearest rule to e_i;
2   gr = generalize nr to cover e_i;
3   **if** *(length of gr > 0)* **then**
4 | rule's gr $purity = \frac{positiveCoverage}{positiveCoverage+negativeCoverage}$;
5 | **if** *(purity$>=\frac{d(class)}{swSize}$)* **then**
6 | neg = find the nearest negative example not covered by gr;
7 | **if** *(neg $\neq$ null)* **then**
8 | | er = extend gr on numerical attributes to the half distance to neg;
9 | | $RS_c \leftarrow \{RS_c \setminus \{nr\}\} \cup \{er\}$;
10 | **else**
11 | | $RS_c \leftarrow \{RS_c \setminus \{nr\}\} \cup \{gr\}$;
12 | generalization = true;

In case when both procedures: finding positive coverage and generalization attempt fail, then full description of the learning example e_i is added to the set of rules RS as the most specific rule (Alg. 1, lines 7—9). Next, if the number of stored learning examples exceeds the maximum size of the sliding window sw, the oldest example from the sliding window sw is removed and the distribution d is updated (Alg. 1, lines 10—11).

The set of rules RS is pruned using three criteria. First, rules that are not used for more than a maximum age threshold are removed (Alg. 1, line 12). Secondly, rules with too low purity (below the relative frequency of its class) are deleted (Alg. 1, line 13). Finally, rules making too many prediction errors are removed (Alg. 1, line 14). This criterion uses the confidence intervals related to class probabilities, which are constructed for both, the rule classification accuracy and its class relative frequency observed over the sliding window. If the accuracy interval's higher endpoint is less than its class frequency interval's lower endpoint, then the rule is deleted [1].

Finally, the RILL's classification strategy assigns the new examples according to the class labeled of its nearest rule, which is consistent with the idea used in generalization process.

4 Experiments

Experimental Setup. The aims of this experiments are to evaluate the RILL algorithm and compare it with related classifiers. We consider the only available rule-based classifier in MOA—Very Fast Decision Rules (VFDR). Unfortunately, despite all our efforts, FACIL's code was inaccessible for the public use. Moreover, in order to be consistent with former experiments of VFDR [7], we chose other incremental tree classifier: Very Fast Decision Trees (VFDT) and Very Fast Decision Trees with Naïve Bayes leaves (VFDT-NB). Additionally, we also checked a single Naïve Bayes (NB). All classifiers, except RILL, are not adjusted to directly deal with concept drift, that is why we combined them with the sliding window of the same size as used in RILL. All algorithms are implemented in Java, including our implementation of RILL, and are embedded into the Massive Online Analysis framework for mining streams of data[1]. All classifiers were run with default values of their parameters. As for RILL, the size of the sliding window was set to 1000. We tested other sizes but results obtained for chosen value stated the best compromise between achieved accuracy of classification and computational demands. RILL's maximum age threshold was chosen experimentally and finally was set to 3000. Lower values caused removing rules, which were still up-to-date. On the other hand, higher values of age threshold resulted in decreasing value of classification accuracy.

Table 1. Characteristics of datasets

Dataset	#Examples	#Attributes	#Classes	Change type
AgrawalGradual	100000	9	2	gradual
CovType	581012	54	7	unknown
Crash[2]	999900	8	4	gradual
Electricity	45312	8	2	unknown
HyperplaneFaster	100000	10	4	gradual
HyperplaneSlow	100000	10	4	gradual
PAKDD09	50000	30	2	unknown
Poker	829201	11	10	unknown
Power	29928	2	24	unknown
RBFBlips	100000	20	4	blips
RBFGradualRecurring	100000	20	4	gradual
RBFNoDrift	100000	10	2	N/A
RBFSudden	100000	20	4	sudden
SEAGradual	100000	3	2	gradual
STAGGERGradual	100000	3	2	gradual
STAGGERSuddenFaster	100000	3	2	sudden

To estimate classification performance we used the Evaluate Prequential method from MOA [6]. It first uses each example in the stream to assess a classification accuracy and then this example can be used to update the classifier. Moreover, this method uses a sliding window or a fading factor as a forgetting mechanism. Besides the total classification accuracy, we also recorded values of accumulated processing time from the beginning of the learning phase and the size of current model (expressed by its used memory size).

---

[1] For more about MOA project see http://moa.cs.waikato.ac.nz/

[2] We would like to thank Radosław Ziembiński, who provided us this dataset.

We considered several datasets involving different types of changes, such as gradual drifts, sudden drifts, blips (representing rare events—outliers in a stable period, which a good classifier should not treat as real drifts), stability periods (no drifts for which a classifier should not be updated) and complex/mixed changes. To model precisely these drifts we used data stream generators available in the MOA framework to construct 11 synthetic datasets. To extend the study on more real world scenarios, we decided to additionally consider 5 publicly available real datasets previously used to test the related ensemble algorithms in several papers. However, for some of them there is no precise information about type of drifts. Detailed characteristics of these datasets are given in Table 1.

Experimental Results. Although we carried out more experiments, due to the page limits, we can present only the most representative results showing the general tendency. The accuracy values were averaged over recording time points (every 1000 examples, more frequent records did not influence the results). They are presented in Table 2, where the best results are denoted in bold.

Table 2. Average values of classification accuracy [%]

Dataset	VFDT	NB	VFDT-NB	VFDR	RILL
AgrawalGradual	68.09	**73.22**	71.71	66.11	67.91
CovType	60.06	79.41	78.91	61.39	**88.13**
Crash	50.97	53.23	50.97	30.91	**81.91**
Electricity	68.12	65.41	68.76	67.47	**76.02**
HyperplaneFaster	61.36	**86.35**	85.25	69.61	68.55
HyperplaneSlow	50.35	**88.78**	88.76	52.80	66.55
PAKDD09	**80.19**	53.75	53.31	80.10	68.99
Poker	59.25	56.67	57.22	58.42	**79.22**
Power	13.77	13.77	13.77	5.24	**14.08**
RBFBlips	39.52	76.19	76.02	51.59	**80.34**
RBFGradualRecurring	33.92	67.18	67.19	42.86	**78.34**
RBFNoDrift	49.99	71.24	71.24	70.33	**80.59**
RBFSudden	33.90	67.14	67.16	43.26	**78.81**
SEAGradual	64.22	**91.25**	**91.25**	87.96	87.88
STAGGERGradual	72.52	85.99	85.99	64.32	**92.15**
STAGGERSuddenFaster	59.82	78.32	78.32	60.85	**88.68**

To globally compare classifiers we carried out the ranked Friedman test with the significance level $\alpha = 0.05$. According to it ($p\text{-}value=0.00008$) we claim that there is a significant difference in the results of compared classifiers. Notice that RILL achieved the highest value of average ranks equal 4.19. The second and third were NB (3.47) and VFDT-NB (3.31), which were very near one another. The post-hoc analysis showed that the differences in average ranks between these methods are too small to conclude that they are significant. However, we can resume that each of the top three methods was significantly better than VFDR and VFDT. We additionally carried out the paired Wilcoxon tests for comparing RILL against NB or VFDT-NB. Results of this test were nearly at the significance level (in both cases *p-value* equals 0.07), so we were very close to confirm that RILL was significantly better. However, making win-loss analysis for single data, we noticed that RILL won 11 times, was third 2 times, and fourth—3 times.

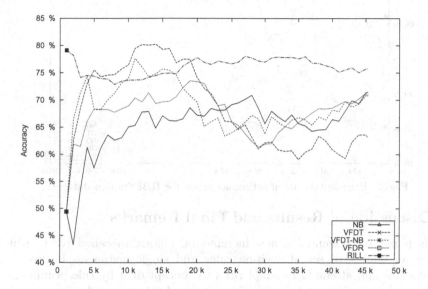

Fig. 1. Prequential classification accuracy for Electricity datasets

Better insight into dynamics of learning is available by studying figures of prequential classification accuracy with respect to processing every learning example. Again, due to the space limits, we present only the representative figures for real dataset Electricity and artificial data with sudden drift RBFSudden—see Figures 1, 2. For most of other datasets (except PAKDD09) RILL achieved also the best trend of classification accuracy. In case of artificial data with STAG-GER concepts, Radial Basis Function and crashes, RILL was definitely the best classifier in terms of the accuracy of classification. Moreover, results obtained by RILL are quite stable—without any drastic rises and falls. Only for Hyper-plane datasets and SEA functions, RILL was not able to outperform Very Fast Decision Tree with Naïve Bayes leaves and the single Naïve Bayes classifier.

In terms of memory consumption, the biggest amount of memory needs VFDR classifier. RILL have lower demands. However, the least memory consuming are VFDT, NB, and VFDT-NB. This tendency is also visible for most of the datasets and algorithms. Only for 2 datasets with STAGGER concepts, RILL uses more memory than VFDR classifier. In case of time, the fastest algorithms are VFDT, NB, and VFDT-NB. Other algorithms, RILL and VFDR, need more time. For majority of datasets VFDR is the slowest one. For some real problems, RILL needs more time however it becomes the most accurate classifier.

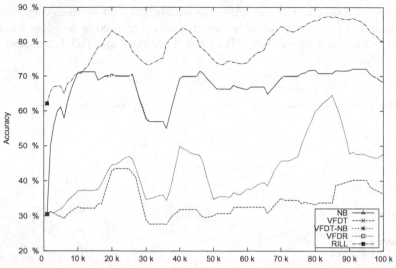

Fig. 2. Prequential classification accuracy for RBFSudden datasets

5   Discussion of Results and Final Remarks

In this paper we presented a new incremental algorithm called RILL, which induces an unordered set of decision rules and single instances. It processes data streams and attempts to adapt rules to concept drift by rule pruning and sliding window. Unlike related algorithms, it uses bottom-up rule generalization based on nearest rules, performs their intensive pruning and is integrated with the classification strategy also based on looking for the nearest rules. RILL was evaluated in a comparative study with available in MOA incremental classifiers on several data containing different types of concept drifts.

In terms of the total accuracy of classification, RILL was the best on most of the datasets. It won 11 times, was third 2 times, and fourth—3 times. With respect to more precise analysis we can say that it is competitive (according to statistical test) to the best decision trees VFDT-NB. Nevertheless, RILL outperformed the other rule-based classifier VFDR on 13 datasets and this difference is statistically significant. RILL left its opponents far behind especially on datasets with Radial Basis Function. The reason for this behavior is that these datasets represent the complex concept decomposed in many changing sub-concepts, which are not straightforward to approximate with a decision tree model. However, we have to admit, that RILL is not doing so well on datasets with moving hyperplane, SEA functions and loan functions introduced by Agrawal. We suspect that for the last two datasets, the problem lies in a definition of the concept as: (1) it is modeled as a quite difficult mathematical function with conditions over a subset of numerical attributes; (2) datasets also contain irrelevant numerical attributes. Notice that the current version of RILL generalization procedure does not allow to discard numerical attributes.

Furthermore, we experimentally showed that RILL is less computationally demanding, mainly from the memory usage, than VFDR rule classifiers.

Comparing RILL to FACIL algorithm, we were unable to do it experimentally. However, we could evaluate their differences with respect to computational costs. From the memory usage perspective, FACIL may be more demanding due to the fact that it stores examples for every rule. Moreover, its pruning is quite limited, so it still may keep many rules. On the other hand, RILL stores always the same number of examples in the global window and offers stronger pruning. As for processing time, RILL may operate longer during the rule generalization phase, because it calculates rule statistics from the complete window.

During experiments we also analyzed the number of rule generalizations performed by RILL. For most of the datasets almost half of the generalization attempts are successful. However, in case of the hyperplane datasets only 3% of generalizations are performed. Therefore in our future research we will consider modifications of these generalizations and dropping numerical conditions for such difficult datasets.

Acknowledgments. The research has been supported by internal grant no. 09/91/DSPB/0543.

References

1. Aha, D.W., Kibler, D.: Instance-based learning algorithms. Machine Learning 6, 37–66 (1991)
2. Deckert, M.: Incremental Rule-based Learners for Handling Concept Drift: An Overview. Foundations of Computing and Decision Sciences 38(1), 35–65 (2013)
3. Domingos, P., Hulten, G.: Mining high-speed data streams. In: Proceedings of the 6th ACM SIGKDD International Conference, KDD, pp. 71–80 (2000)
4. Ferrer-Troyano, F.J., Aguilar-Ruiz, J.A., Riquelme, J.C.: Data Streams Classification by Incremental Rule Learning with Parametrized Generalization. In: Proceedings of ACM Symposium on Applied Computing, SAC 2006, pp. 657–661. ACM (2006)
5. Fürnkranz, J., Gamberger, D., Lavrac, N.: Foundations of Rule Learning. Springer (2012)
6. Gama, J.: Knowledge Discovery from Data Streams. CRC Publishers (2010)
7. Gama, J., Kosina, P.: Learning Decision Rules from Data Streams. In: Proceedings of the 22th International Joint Conference on Artificial Intelligence, IJCAI 2011, vol. 2, pp. 1255–1260. AAAI Press (2011)
8. Kosina, P., Gama, J.: Handling time changing data with adaptive very fast decision rules. In: Proceedings of ECML/PKDD 2012, Bristol, United Kingdom, vol. 1, pp. 827–842 (2012)
9. Maloof, M.: Incremental Rule Learning with Partial Instance Memory for Changing Concepts. In: Proceedings of the International Joint Conference on Neural Networks, IJCNN 2003, vol. 4, pp. 2764–2769. IEEE Press (2003)
10. Napierala, K., Stefanowski, J.: BRACID: A comprehensive approach to learning rules from imbalanced data. Journal of Intelligent Information Systems 9(2), 335–373 (2012)
11. Widmer, G., Kubat, M.: Learning in the presence of concept drift and hidden contexts. Machine Learning 23, 69–101 (1996)
12. Wilson, D.R., Martinez, T.R.: Improved Heterogeneous Distance Functions. Journal of Artificial Intelligence Research 6(1), 1–34 (1997)

Community Detection for Multiplex Social Networks Based on Relational Bayesian Networks

Jiuchuan Jiang and Manfred Jaeger

Department of Computer Science, Aalborg University, Denmark
{jiuchuan,jaeger}@cs.aau.dk

Abstract. Many techniques have been proposed for community detection in social networks. Most of these techniques are only designed for networks defined by a single relation. However, many real networks are multiplex networks that contain multiple types of relations and different attributes on the nodes. In this paper we propose to use relational Bayesian networks for the specification of probabilistic network models, and develop inference techniques that solve the community detection problem based on these models. The use of relational Bayesian networks as a flexible high-level modeling framework enables us to express different models capturing different aspects of community detection in multiplex networks in a coherent manner, and to use a single inference mechanism for all models.

Keywords: Community detection; Multiplex networks; Relational Bayesian networks; Statistical relational learning.

1 Introduction

Social networks like Facebook, Twitter, Flickr, or Youtube have prospered in recent years. People in an online society here can communicate and interact with each other. Community structure is one of the most important characteristics for social networks [1]. Within a community, the connections between nodes are very dense but they are sparse in between communities. In a social network, a community can be a friend group which has close relations, a group of people with similar interests, a group of people in a same workplace, and so on. Community detection, therefore, has received significant attention in the research of social networks [2][3].

Most existing methods have been developed to analyze single relation networks, where there is only one type of relation between nodes. However, in the real world, social networks may often appear as multiplex networks, in which there exist different types of nodes, which are connected by different types of links [4][5]. For example, in Facebook, the relations between two users could be friends, common interests, alumni, and so on; and the users in Facebook could be humans, companies, or organizations, which makes the characteristics of users different from each other.

T. Andreasen et al. (Eds.): ISMIS 2014, LNAI 8502, pp. 30–39, 2014.
© Springer International Publishing Switzerland 2014

A few studies have investigated community detection for multiplex networks, but mostly these are characterised by strong simplifications that reduce community detection in multiplex networks to community detection in single relation networks. On the other hand, in Machine Learning the field of *statistical relational learning (SRL)* is specifically concerned with statistical models for multi-relational data. Since probabilistic models are a powerful tool for clustering in general, and community detection in networks in particular, it is natural to apply SRL techniques to the community detection problem in multiplex networks. In this paper we give an initial report on the application of the SRL modeling framework of *Relational Bayesian Networks (RBNs)* [6] to this task. The RBN framework provides an expressive and flexible representation language in which a variety of probabilistic community detection models can be specified. The different models are supported by a common set of generic inference algorithms. Such a general representation and inference framework provides a good basis to explore different aspects of communities in multiplex networks using different models, without the need to re-design inference methods for each model.

The rest of this paper is organized as follows. In Section 2, we compare our work with the related work on the subject; in Section 3, we present the problem description; in Section 4, we model the community detection in multiplex networks; in Section 5, we provide the experimental results; finally, we conclude our paper in Section 6.

2 Related Work

Community Detection in Single Relation Networks. Girvan and Newman [2] published the seminal paper on discovering the community structure in networks. After that, a lot of community detection methods for single networks has been developed. Typical methods include graph partitioning [7], hierarchical clustering [3], partitional clustering [8] and spectral methods [9]. A good review of community detection for single networks can be found in [10].

Community Detection in Multiplex Networks. Some researchers divide a multiplex network into single network layers and do community detection on these single networks, such as [4][5]. Yang et al. [11] extended traditional random walk algorithm to detect communities in signed networks which includes positive and negative relations. Breiger et al. [12] describe a hierarchical clustering algorithm that is based on an iterative transformation of incidence matrices into a block form, and which can be simultaneously applied to matrices representing multiple relations.

SRL Methods. An early paper that considered clustering in graphs as a possible application of SRL techniques is [13]. However, their model representation framework only allowed for the modeling of random attributes of nodes in a network, and not the modeling of random link structures, which are essential for natural probabilistic community detection models. *Markov Logic Networks* are a currently popular SRL framework, which in [14] also was applied to clustering in multi-relational data, though no application to a community detection problem was presented. Xu et al. [15] proposed to use the *Infinite Hidden Relational Model (IHRM)* for social network analysis. The resulting probabilistic model is quite similar to the basic RBN-based community detection model we will describe below. A main difference between [15] and our work lies in the fact that Xu et al. work with a fixed graphical model, whereas we are using a higher-level representation language that allows us to experiment with different clustering models without the need to re-design the update equations for inference in the underlying graphical model. On the other hand, a key concern of the IHRM is to support a nonparametric Bayesian approach for automatically finding the 'right' number of clusters or communities, whereas we are currently taking the simpler approach of requiring the number of communities to be a user-defined input.

3   Problem Description

In this paper, we will use probabilistic models to analyze community structure problems. Graphs will be used to represent social networks. The vertices and arcs in graphs correspond to nodes and relations in social networks, respectively. Considering only single relation networks first, we can give the problem description as follows.

Given a directed graph $G =< V, A >$, with vertices V and arcs A, we want to find the partition Γ with maximal probability

$$P_\theta(\Gamma|A) = \frac{P_\theta(A, \Gamma)}{P_\theta(A)} \tag{1}$$

Here P_θ is the underlying probabilistic model, which defines for the given set V a joint distribution over partitions and arcs. The model can depend on unknown parameters θ. Since $P_\theta(A)$ does not depend on Γ, the community detection problem therefore amounts to computing

$$\arg\max_\Gamma \max_\theta P_\theta(A, \Gamma) = \arg\max_\Gamma \max_\theta P_\theta(A|\Gamma)P_\theta(\Gamma) \tag{2}$$

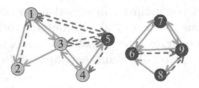

Fig. 1. A small example of multiplex network

We now turn to the generalization of this probabilistic community detection approach to multiplex networks. Multiplex networks are networks with more than one type of relations between nodes, and a node may have multiple attributes. Figure 1 gives an example of a multiplex network. This small network includes 9 nodes, two types of relations (green line and red dash), and two types of attributes (yellow and blue). The green line relation is assumed to be a positive relation and the red dash relation is a negative relation. Positive relations represent "attraction", such as "like", "friend of"; negative relations represent "repulsion", such as "dislike", "objector of". Two nodes tend to belong to a community if they are connected with positive relations, and belong to different communities if they are connected with negative relations. We can see that a reasonable community structure is given by two communities that consist of the left 4 nodes and right 5 nodes, respectively. However, node 5 would be in the left community if we only considered the green relation, but appears "closer"to the right community when also considering the red relation and the node attributes.

The information from all relations and attributes can be integrated into a probabilistic model; we only need to generalize (2) as follows.

• All types of relations should be considered. Therefore, if a network with p types of relations, $A = \{A_1, \ldots, A_p\}$, then (2) becomes

$$\arg \max_{\Gamma} \max_{\theta} P_\theta(A, \Gamma) = \arg \max_{\Gamma} \max_{\theta} P_\theta(A|\Gamma)P_\theta(\Gamma) \qquad (3)$$

• The attributes of the nodes can also play important roles in the community structure. The attributes of nodes within a same community should be as same as possible. Therefore, if a network with p types of relations and q types of attributes, $At = \{At_1, \ldots, At_p\}$, equation (3) should be extended as (4):

$$\arg \max_{\Gamma} \max_{\theta} P_\theta(A, At, \Gamma) = \arg \max_{\Gamma} \max_{\theta} P_\theta(A, At|\Gamma)P_\theta(\Gamma) \qquad (4)$$

4 RBN Models for Community Detection

4.1 Model Specification

We now briefly describe the elements of the RBN framework that we need for our community detection models. We begin by introducing a few basic concepts that are common for most SRL frameworks.

Given a set of relations A, a set of attributes, At, and a set of entities (vertices) V, we define following:

Definition 1. *A ground atom is an expression of the form* $A_i(v_{j_1}, v_{j_2})$ *or* $At_i(v_j)$, *where* A_i *is a relation,* At_i *an attribute, and the* v_j *are elements from* V. *In a non-ground atom, one can also have variables that range over all vertices as arguments for the* A_i, At_i.

A *probabilistic relational model* defines a joint probability distribution for all ground atoms $A_i(v_{j_1}, v_{j_k})$ or $At_i(v_j)$ as Boolean random variables [16]. Community membership will be represented by special attributes, so that also the joint distribution $P_\theta(A, At, \Gamma)$ is covered by these definitions. RBNs are a formal representation language for probabilistic relational models. It is based on *proba-bility formulas* that for each relation and each attribute specify the probability distributions for the ground atoms in a single declaration of the form

$$P(A_i(n_j, n_k) = true) \leftarrow ProbabilityFormula(n_j, n_k).$$

Here n_j, n_k are variables that range over vertices. The RBN language provides a small number of simple syntax rules with which complex probability formulas can be inductively constructed. The inductive definition is grounded by the base constructs of constants, parameters, and ground atoms. In this paper we will construct complex formulas only using the *convex combination* construct, which can be understood as a probabilistic if-then-else rule: if *PF1,PF2,PF3* are prob-ability formulas, then *(PF1:PF2,PF3)* is a new formula. This formula evaluates to a mixture of the values returned by *PF2,PF3*, with *PF1* defining the mixture weights. In the particular case that *PF1* is purely Boolean (i.e., evaluates to 0 or 1), this means that the formula evalutes to *PF2* if *PF1* returns *true*, and to *PF3* if *PF1* returns *false*. In this paper we will only make use of the convex combi-nation construct in this special form. Full definitions of syntax and semantics of RBNs can be found in [16].

We now turn to concrete encodings of community detection models using RBNs. For the time being, we only consider the case of two communities, and we use two special attributes $c1, c2$ to represent community membership. We then first define the prior distribution $P_\theta(\Gamma)$ using the two formulas

$$c1([Node]n) \leftarrow 0.5; \tag{5}$$

$$c2([Node]n) \leftarrow (c1(n) : 0, 1); \tag{6}$$

The first probability formula specifies that the probability of a node belongs to community $c1$ is 0.5; the second one specifies that a node with probability 0 belongs to cluster $c2$ if it belongs to community $c1$, and with probability 1 belongs to cluster $c2$ if it does not belong to community $c1$. This specification implies that the two communities form a partition of the nodes. By replacing the constants 0,1 in (6) with non-extreme values, the model will also allow over-lapping communities, and nodes belonging to no community.

The following are probability formulas that define the model $P_\theta(\mathbf{A}, \mathbf{At}|\Gamma)$ for the relations and attributes of the graph in Figure 1:

$$link\_green([Node]n1, [Node]n2) \leftarrow (c1(n1) : (c1(n2) : \theta_1, 0.01),$$
$$(c2(n1) : (c2(n2) : \theta_2, 0.01), 0.5)); \tag{7}$$

$$link\_red([Node]n1, [Node]n2) \leftarrow (c1(n1) : (c1(n2) : 0.01, \theta_3),$$
$$(c2(n1) : (c2(n2) : 0.01, \theta_4), 0.5)); \tag{8}$$

$$attribute\_yellow([Node]n) \leftarrow (c1(n) : \theta_5, \theta_6); \tag{9}$$

$$attribute\_blue([Node]n) \leftarrow (c2(n) : \theta_7, \theta_8); \tag{10}$$

The θ_i in these formulas are free parameters that are estimated in the optimization process. The nested if-then-else construct of formula (7) can be expanded as follows:

$$P(link\_green(n_1, n_2) = true) = \begin{cases} \theta_1 & if\ n_1 \in c_1 \land n_2 \in c_2 \\ 0.01 & if\ n_1 \in c_1 \land n_2 \notin c_2 \\ \theta_2 & if\ n_1 \notin c_1 \land n_1 \in c_2 \land n_2 \in c_2 \\ 0.01 & if\ n_1 \notin c_1 \land n_1 \in c_2 \land n_2 \notin c_2 \\ 0.5 & if\ n_1 \notin c_1 \land n_1 \notin c_2 \end{cases}$$

The placement of the constants 0.01 here encodes that $link\_green$ is a positive relation. Formula (8) is structurally similar, but parameterized differently in order to enforce that $link\_red$ is considered a negative relation. Formulas (9) and (10) represent in a generic manner the dependency of these attribute values on the community membership.

In the model given by (5)-(10) all attributes and relations are independent given the community membership. A dependency of the *green* on the *red* relation could be modeled by the formula

$$link\_green([Node]n1, [Node]n2) = (c1(n1) :$$
$$(c1(n2) : (link\_red(n1, n2) : \theta_9, \theta_{10}), \theta_{11}), \qquad (11)$$
$$(c2(n2) : (link\_red(n1, n2) : \theta_{12}, \theta_{13}), \theta_{14}));$$

4.2   Inference

To solve the inference problem (4) when P_θ is given by an RBN, we extend a datastructure and inference techniques that were introduced for RBN parameter learning in [17].

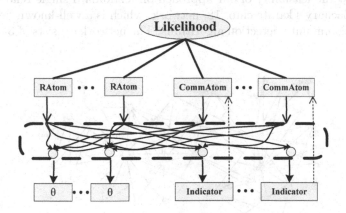

Fig. 2. The architecture of MAP-inference and parameter learning module

First, given the general RBN model and a concrete multiplex network, a representation of $P_\theta(A, At, \Gamma)$ in the form of a *likelihood graph* is constructed. Figure 2 illustrates the structure of this graph. Below the root *likelihood* node, the graph contains for each ground atom a node that represents the contribution of this ground atom to the overall likelihood. These nodes are inserted both for the atoms corresponding to observed relations and attributes in the network (denoted *RAtom* in Figure 2), and for atoms representing the unobserved community attributes (*CommuAtom*).

The leaves in the graph are the variables in the optimization problem (4): the free parameters θ of the model, and the truth settings for the community attributes, which jointly define Γ (*Indicator* nodes in Figure 2). Intermediate nodes in the graph (indicated by the dashed box in Figure 2) represent intermediate values that are obtained from probability sub-formulas in the recursive evaluation of the formulas associated with the ground atoms.

Using the likelihood graph as the common inference structure we compute (4) by alternating between maximization for θ and Γ. For maximizing θ we use the general parameter learning method of [17]. Maximization over Γ is a *Maximum A Posteriori Probability (MAP)* inference problem. This problem is intractable to solve optimally, and we use a local search procedure that combines greedy, random, and lookahead elements to obtain an approximate solution. In this process, the likelihood graph is used in two ways: first, it is used to compute the likelihood values of candidate solutions Γ, and second, the dependency structure of the likelihood function on the different atoms encoded in the graph structure is exploited to identify in the lookahead search candidate community membership atoms whose truth values might be changed to improve the likelihood.

5   Experiments

We first test the feasibility of our approach on a standard single relational network, the Zachary's karate club [18] network, which is a well-known benchmark for testing community detection algorithms. The network consists of 34 nodes as

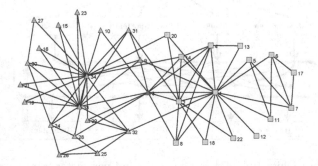

Fig. 3. The communities of Zacharys karate club network. Node shapes (and colors) indicate the community memberships of nodes.

members of the Karate club and 78 edges as friendships between members. We use probability formulas (5) and (6) to define $P_\theta(\Gamma)$. However, since the hard constraints on $c1$ and $c2$ are very difficult for MAP inference, we relax the model slightly by replacing 0 with 0.001, and 1 with 0.999. Since the relation in this network is positive, we model it using formula (7). The optimization terminates with parameter setting $\theta_1 = 0.259$,and $\theta_2 = 0.253$, and the communities shown in Fig.3. Yellow square and green triangle nodes represent nodes for which in the MAP solution for the community atoms exactly one of $c1$ and $c2$ was set to true. For node 3 (red triangle) in the middle both $c1$ and $c2$ were set to true in our solution. Apart from this ambiguous membership of node 3, our solution corresponds with the solution of [19].

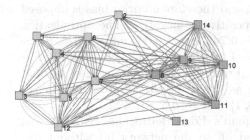

Fig. 4. The communities of bank wiring room network. Node colors indicate the community memberships of nodes.

Table 1. Parameters of the model for bank wiring dataset

	θ_1	θ_2	θ_3	θ_4
playing games together	0.500	0.563	-	-
friendship	0.276	0.219	-	-
helping	0.277	0.141	-	-
arguments	-	-	0.164	0.167
antagonism	-	-	0.272	0.271

We next conduct an experiment on the 'bank wiring room' multiplex network introduced in [12]. This network includes 14 employees work in a single wiring room. Relations between them are quite complex, and include *playing games together*(red arcs in Fig.4), *friendship*(yellow), *helping*(orange), *arguments about whether to open window*(blue) and *antagonism*(purple). The first three relations are positive and the latter two are negative. We define $P_\theta(\Gamma)$ as for the Zachary network. Each relation is modeled using the formulas of (7) and (8), depending of whether it is a positive or negative relation. The network and the computed communities are shown in Figure 4, and the learned parameters in Table 1. In this case, all nodes were assigned uniquely to one of the two communities (yellow and green squares in Figure 4), and the results coincide exactly with the structure suggested in [12].

In the preceding experiment the model encoded explicit information on which relations are positive, and which are nega tive. We next investigate the applicability of our approach when this distinction is not provided a-priori. For this, we use the network of Figure 5 (c), which contains two relations that are also separately drawn in Figure 5 (a) and (b). The network (a) shows an obvious community structure, whereas (c) shows no clear structure. We define a model with $P_\theta(\Gamma)$ as before, and for each of the two relations a formula of the form

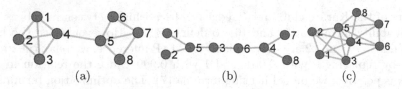

Fig. 5. A network which is a summation of two networks: (a) the network with obvious community structure; (b) the network with unobvious community structure; (c) summation of network (a) and (b).

$$link([Node]n1, [Node]n2) \leftarrow (c1(n1) : (c1(n2) : \theta_{14}, \theta_{15}),$$
$$(c2(n1) : (c2(n2) : \theta_{16}, \theta_{17}), 0.5)); \tag{12}$$

In these formulas all parameters are free, and therefore no prior bias is imposed on whether a relation is to be seen as positive or negative. The optimization terminates with parameters setting $\theta_{14} = 0.749$, $\theta_{15} = 0.063, \theta_{16} = 0.625$, $\theta_{17} = 0.063$ for the yellow relation (a), and $\theta_{14} = 1.2E-8$, $\theta_{15} = 0.312, \theta_{16} = 0.1.98E-6$, $\theta_{17} = 0.313$ for the green relation (b). From the parameters we can see that our model considers the yellow relation as positive and the green relation as negative. The results are clusters $\{1, 2, 3, 4\}$ and $\{5, 6, 7, 8\}$.

For comparison, we also apply Newman's Edge Betweenness(EB)[19] and Ruan's Spectral Clustering (SC) method [9] to the network (c) without distinguishing the two relations. We obtain the clusters $\{1, 2, 3, 4, 5\}, \{6, 7, 8\}$ from SC, and $\{1\}, \{2, 3, 4, 5, 6, 8, 8\}$ from EB. Neither of these two clusterings appear very meaningful, and it is clear that the clustering evidence provided by the relation (a) is not strong enough for single-relational methods to detect the two clusters, when it is masked by relation (b).

6 Conclusions

We addressed the problem of community detection in multiplex networks using RBNs as a high-level and flexible specification language for probabilistic models. The main benefit of this approach is that we can use a single coherent framework with uniform inference techniques to experiment with different models that can capture different aspects and objectives that arise in the context of multiplex networks. Even though our results are quite preliminary at this point, they already demonstrate that using this coherent framework we can easily reconstruct results that have previously been obtained using very different techniques (graph cut techniques for the Zachary network, and matrix permutation for the Wiring Room). Our current system can handle networks up to approximately 300 nodes with arbitrary link structure. Future work will be directed towards exploring additional aspects such as multiple and overlapping community detection, and the application to bigger datasets.

References

1. Newman, M.: Communities, Modules and Large-Scale Structure in Networks. Nature Physics 8(1), 25–31 (2011)
2. Girvan, M., Newman, M.E.: Community Structure in Social and Biological Networks. Proceedings of the National Academy of Sciences 99(12), 7821–7826 (2002)
3. Newman, M.E.: Detecting community structure in networks. The European Physical Journal B-Condensed Matter and Complex Systems 38(2), 321–330 (2004)
4. Mucha, P.J., Richardson, T., Macon, K., Porter, M.A., Onnela, J.P.: Community Structure in Time-Dependent, Multiscale, and Multiplex Networks. Science 328(5980), 876–878 (2010)
5. Cai, D., Shao, Z., He, X., Yan, X., Han, J.: Mining Hidden Community in Heterogeneous Social Networks. In: Proceedings of the 3rd International Workshop on Link Discovery, pp. 58–65. ACM (2005)
6. Jaeger, M.: Relational Bayesian Networks. In: Proceedings of the Thirteenth Conference on Uncertainty in Artificial Intelligence, pp. 266–273. Morgan Kaufmann Publishers Inc. (1997)
7. Flake, G.W., Lawrence, S., Giles, C.L., Coetzee, F.M.: Self-Organization and Identification of Web Communities. Computer 35(3), 66–70 (2002)
8. Rattigan, M.J., Maier, M., Jensen, D.: Graph Clustering with Network Structure Indices. In: Proceedings of the 24th International Conference on Machine Learning, pp. 783–790. ACM (2007)
9. Ruan, J., Zhang, W.: An Efficient Spectral Algorithm for Network Community Discovery and its Applications to Biological and Social Networks. In: Seventh IEEE International Conference on Data Mining, pp. 643–648. IEEE (2007)
10. Fortunato, S.: Community Detection in Graphs. Physics Reports 486(3), 75–174 (2010)
11. Yang, B., Cheung, W.K., Liu, J.: Community Mining from Signed Social Networks. IEEE Transactions on Knowledge and Data Engineering 19(10), 1333–1348 (2007)
12. Breiger, R.L., Boorman, S.A., Arabie, P.: An Algorithm for Clustering Relational Data with Applications to Social Network Analysis and Comparison with Multidimensional Scaling. Journal of Mathematical Psychology 12(3), 328–383 (1975)
13. Taskar, B., Segal, E., Koller, D.: Probabilistic Classification and Clustering in Relational Data. In: International Joint Conference on Artificial Intelligence, vol. 17, pp. 870–878 (2001)
14. Kok, S., Domingos, P.: Statistical Predicate Invention. In: Proceedings of the 24th International Conference on Machine Learning, pp. 433–440. ACM (2007)
15. Xu, Z., Tresp, V., Yu, S., Yu, K.: Nonparametric Relational Learning for Social Network Analysis. In: KDD 2008 Workshop on Social Network Mining and Analysis (2008)
16. Jaeger, M.: Complex Probabilistic Modeling with Recursive Relational Bayesian Networks. Annals of Mathematics and Artificial Intelligence 32(1-4), 179–220 (2001)
17. Jaeger, M.: Parameter Learning for Relational Bayesian Networks. In: Proceedings of the 24th International Conference on Machine Learning. pp. 369–376. ACM (2007)
18. Zachary, W.W.: An Information Flow Model for Conflict and Fission in Small Groups. Journal of Anthropological Research 33(4), 452–473 (1977)
19. Newman, M.E., Girvan, M.: Finding and Evaluating Community Structure in Networks. Physical Review E 69(2), 026113 (2004)

Mining Dense Regions from Vehicular Mobility in Streaming Setting

Corrado Loglisci and Donato Malerba

Dipartimento di Informatica, Università degli Studi di Bari "Aldo Moro"
via Orabona, 4 - 70126 Bari - Italy
{corrado.loglisci,donato.malerba}@uniba.it

Abstract. The detection of congested areas can play an important role in the development of systems of traffic management. Usually, the problem is investigated under two main perspectives which concern the representation of space and the shape of the dense regions respectively. However, the adoption of movement tracking technologies enables the generation of mobility data in a streaming style, which adds an aspect of complexity not yet addressed in the literature. We propose a computational solution to mine dense regions in the urban space from mobility data streams. Our proposal adopts a stream data mining strategy which enables the detection of two types of dense regions, one based on spatial closeness, the other one based on temporal proximity. We prove the viability of the approach on vehicular data streams in the urban space.

1 Introduction

The recent adoption of the mobile and ubiquitous technologies has enabled the tracking of moving objects. A challenging problem is the identification of regions with high density wherein the number of contained moving objects is above some threshold. The development of applications for monitoring fleets of vehicles makes the study of this kind of problems particularly prominent for designing intelligent traffic management systems and urban transportation planning technologies. Most of the existing works ([3,4]) follow a database-inspired approach based on density queries. Queries are efficiently answered by means of human-specified criteria or mechanisms of simplification of the query clauses. Common to these works there is the representation of the space in form of a grid, which forces the dense regions to be fixed-size cells. In real-world applications, dense regions could not have necessarily a pre-defined form, they can have a size and shape different from each other.

In this paper, we propose an approach to mine dense regions from mobility data in the urban vehicular traffic. Differently from the euclidean space, urban space takes the shape of a graph and presents some degrees of complexity due to the presence of buildings and infrastructures which constraint the movements of the vehicles. Strictly connected with the spatial dimension, we have to take into account the temporal component since the spatial location alone could not very useful. We propose to consider time both as dimension of analysis (movement

T. Andreasen et al. (Eds.): ISMIS 2014, LNAI 8502, pp. 40–49, 2014.
© Springer International Publishing Switzerland 2014

and speed of the vehicles are function of time) and as information associated to a dense region (some roads can be particularly dense in some hours of day only).

In the real-world applications, we cannot forget the streaming activity of the tracking technologies in recording the movements of the vehicles and producing unbounded sequences of trajectory data. This raises new challenges about the storage, acquisition and analysis that make most of the existing works on dense regions difficult to apply and ineffective. Based on these considerations, we argue that mining mobility data in streaming setting becomes necessary, especially in urban contexts where technologies able to process trajectories and detect congestions in real-time can be of support in several practical activities. To face the streaming activity several alternatives can be investigated [2], but we adopt the technique of the sliding windows which allows to combine the analysis of trajectories included into one window with the storage of new trajectories captured in the next windows. Trajectories are collected from tracking devices installed on the vehicles, which separately transmit their positions. Sliding windows appear appropriate to model a trajectory in its natural conception of sequence of positions and are particularly adequate to take into account time as physical measure (as velocity and space) in the discovery of dense regions. In this work, we propose a stream data mining solution to mine two types of dense regions, one based on the spatial closeness, the other one based on the temporal proximity. In particular, overlapping windows are used to *i)* acquire continuously coming trajectories and *ii)* monitor the number of the vehicles on the roads of the urban space. A rule-based mechanism is queried to mine dense regions.

The problem of mining dense regions was originally faced through a purely spatial dimension. Initially, dense regions were defined as the geographic areas which have particular spatial properties and which exhibit a particular concentration of objects. Wang et al. [6] study the problem through spatial and multidimensional domains by proposing a solution able to generate hierarchical statistical information from spatial data. A spatio-temporal perspective can be investigated when defining the dense regions on the basis of moving objects. Chen et al. [1] propose an effective clustering approach which, by means of a split-and-merge technique, first it identifies cluster units and then it creates different kinds of clusters based on the units. A quite similar method is implemented in [5] through an algorithm which discovers dense regions from cluster units. The latter are generated as groups of moving objects on the basis of the locations and patterns. A grid-based representation of dense regions is used in [3] in association with the density queries. The original space is modelled in the euclidean framework and partitioned into fixed-size cells to generate cells efficiently.

The remainder of the paper is organized as follows. The scientific problem is formalized in the Section 2, while the Section 3 describes the computational solution. Experiments are described in the section 4. Conclusions close the paper.

2   Basics and Problem Definition

Before formally defining the problem we intend to solve, some definitions are necessary. Consider $T : \{t_1, t_2, \ldots, t_k, \ldots, \}$ be the discrete time axis, where

$t_1, t_2, \ldots, t_k$ are time-points equally spaced, $M : \{o_1, o_2, \ldots, o_i, \ldots, o_m\}$ be the finite set of unique identifiers of the vehicles (or more generally, moving objects), $p_k^i \in \mathbb{R}^2$ is the latitude and longitude position of the ith vehicle at the time-point k^{th}. A trajectory stream S_i is the unbounded sequence of positions associated to the vehicle i^{th}, $S_i = \{p_1^i, p_2^i, \ldots, p_k^i, \ldots\}$. The time-points of recording of the positions are not necessarily equally spaced.

Urban space can be modelled as a directed graph $G(V, E)$, where V is the set of vertices representing road intersections and terminal points, and E is the set of edges representing roads each connecting two vertices. A road is defined by a unique identifier and the latitude and longitude positions of the terminal points p_s, p_e. For simplicity, we consider only one value of speed associated to a vehicle on a road. It can be obtained as the mean value of the speeds of the vehicle while transits though the road. Given the positions p_h^i, p_k^i ($t_k > t_h$), we can compute the speed of a vehicle on a road. While, given the speed, the position p_h^i and the length of a road, we can compute the time-point t_k ($t_k > t_h$) when the vehicle is expected to be out from the road (afterwards, t_{out}), where p_k^i in t_k denotes the position of the vehicle in correspondence to the terminal point of the road.

In this work, dense roads and regions are detected from the trajectory streams.

Definition 1 (Dense Road). *A road* $\sigma \in E$ *is dense if the following conditions hold:*

- *there exists a set of vehicles* $D \subseteq M$ *s.t.* $\forall o_i \in D$: $t_{out}^i < t_k^i$, t_{out}^i *is the time in which we expected the vehicle* o_i *is out from the road. This condition is true when* o_i *is still on the road at the time* t_k^i *and has position* p_k^i;
- $|D| > Dmin$, *Dmin a user-defined minimum density threshold*

A dense road σ *is identified by the tuple:* $\langle p_s, p_e, D, t_0, t_{max} \rangle$, *where* $t_{max} = \arg\max_{o_i \in D} t_{out}^i$ ($t_0 < t_{max}$), *so* $[t_0, t_{max}]$ *establishes the interval in which* σ *is dense.*

In this work, the meaning of dense road follows that illustrated in the works [3,5] and it depicts a congested area caused by slower speeds of the vehicles or increased queueing, which is related to time spent to transit through the road. This is different from the case of road with high intensity in which a high number of vehicles flow continuously, which is instead related to the traffic capacity.

A collection of dense roads can form a dense region on the basis of the spatial closeness. It would depict a traffic jam where neighbour roads exhibit density at the same time (the vehicles are fully stopped). More formally,

Definition 2 (Spatial Closeness-based Dense Region). *A collection of dense roads* Σ *forms a dense region of type SC if the following conditions hold:*

- $\forall \sigma', \sigma'' \in \Sigma : d(\sigma', \sigma'') \leq L$, L *is a user-defined parameter, d is a graph-based distance;*
- $\forall \sigma', \sigma'' \in \Sigma : t_0' = t_0''$ (t_0', t_0'' *are the start time-points of the time-intervals of the tuple of* σ' *and* σ'' *respectively);*
- $\forall \sigma', \sigma'' \in \Sigma : D' \cap D'' = \oslash$

A collection of dense roads can form a dense region on the basis of the spatial closeness and temporal proximity. It would depict a traffic congestion where neighbour roads exhibit density at consecutive times (the vehicles flow slowly). More formally,

Definition 3 (Spatial Closeness and Temporal Proximity-based Dense Region). *A collection of dense roads Σ is a dense region of type TP if the following conditions hold:*

- $\forall \sigma', \sigma'' \in \Sigma : t_0'' > t_0';$
- $\forall \sigma', \sigma'' \in \Sigma : t_0'' - t_0' < \Omega,$ Ω *user-defined parameter;*
- $\forall \sigma', \sigma'' \in \Sigma : D' \cap D'' \neq \oslash.$

The threshold Ω defines the temporal proximity within which the dense regions of type TP can be seek.

Given the set of vehicles M, the trajectory streams in the form of S_i, the urban space G, the problem at the hand is *To Detect* dense regions as defined in the Definitions 3 and 4. The parameters $Dmin$, L and Ω are used to filter out meaningless roads and regions.

3   Computation Solution

3.1   Overview

The notion of trajectory window is crucial in this work. Let $S_i, S_{i+1}, \ldots S_n$ be a stream of trajectories of vehicles $o_i, o_{i+1}, \ldots, o_n$. The trajectory window $[W]_w^u$ of width ω is a sub-sequence of the streams $S_i, S_{i+1}, \ldots S_n$ and comprises the positions observed in the time-interval $[t_u, t_w]$, where ω is a user-defined parameter.

The computational solution can be structured in two steps simultaneously performed. The first step collects data (about the positions and roads) continuously coming from the vehicles and fills partially overlapping windows. Each window is overlapped with (some of) the windows that both precede and follow it, so it contains part of the positions which are contained in the windows created before and after. For instance, the windows $[W]_w^u, [W]_s^r, [W]_n^m$ are being created in the order as they are written so that the orders $t_r < t_w$ and $t_m < t_s$ hold, while the order between t_w and t_m cannot be established a-priori since it depends from the rate (defined by the user) with which the windows are generated. The arrangement of the windows by partial overlaps is taken into account with the aim to share information of the vehicles among different windows and therefore it reduces the possibility of information loss when analyzing the movements.

The second step proceeds at the level of single windows and at the level of sequences of windows. More precisely, it processes one window at time and finds dense regions of type SC from that window: for each road, the positions buffered in one window are analyzed in order to determine whether that road is dense. The roads identified as dense in a window will be used to mine dense regions of type SC. In the meanwhile one window is processed, the next window is acquired.

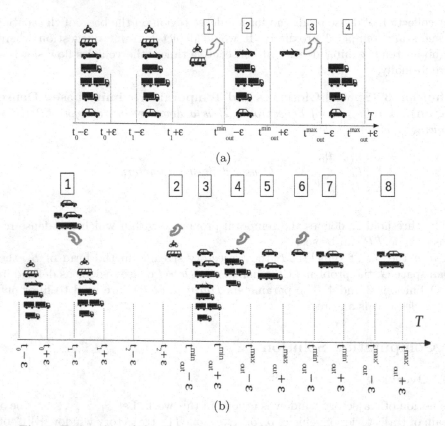

Fig. 1. Estimation of the times t_{out} and application of the check mechanism

When the algorithm turns to process the next window, it stores the *i)* dense roads discovered in the past window(s) and *ii)* expected times t_{out} computed in the past window(s). So, when a new window is processed, the algorithm has the results of the sequence of the windows since there analyzed. The expected times (computed before) are used to complete, in the current window, the search (started in the past window(s)) of dense roads. The discovery of dense regions of type TP is performed on the dense roads obtained from the sequences of windows. Only sequences of windows whose overall width does not exceed Ω are considered to mine dense regions of type TP.

3.2   Detection of Dense Roads

The method for detecting dense roads resorts to the queueing process according to which the delay of an object (with respect to the expected time) to leave the queue can be attributed to the increase of the number of the objects present in the queue or to a demand for resources greater than expected. We deem that, in absence of exceptional conditions of driving (e.g. accidents, icy roads), the increase of the vehicles is the most reasonable motivation of the delay on the roads of the urban space.

The method combines two steps. The first step aims at monitoring the roads by means of queries submitted in correspondence of the time-points in which it is expected that the vehicles will leave the road. Once a query has been completed, the second step checks whether the monitored vehicles are still on the road: if it is so, they will contribute to make that road dense. More precisely, the check mechanism adopts heuristics (modelled in the form of *if-then* rules) and performs a matching operation between (the information associated to) each road and the antecedent parts of the rules. The consequent parts indicate whether the road is dense or not. The possibility to query the trajectory streams only in correspondence of the expected time-points makes our approach different from the methods in which the queries have to be periodically submitted (e.g., [4]).

As anticipated in the section 2, the estimation of the times requires the speed of each vehicle, its position on the road and length of the road. In particular, the speed of the vehicle is determined when processing the current window by means of two queries (submitted in correspondence of two time-points t_0, t_1) that retrieve the vehicles and their positions. The length of the road is instead computed from the graph G. However, the vehicles can move with different velocities and can leave the road at different times. This may be determinant in the formation of dense roads. In order to take into account it, we consider both fast vehicles and slow vehicles when detecting dense roads. In this sense, we estimate two kinds of expected times, one, denoted as t_{out}^{min}, which indicates the time within which the fast vehicles should be out, the other one, denoted as t_{out}^{max}, which indicates the time within which the slow vehicles should be out. More precisely, t_{out}^{min} refers to the slowest vehicle out of the fastest vehicles, while t_{out}^{max} refers to the slowest vehicle in the set of monitored vehicles of that road. Intuitively, we expect that the fastest vehicles will be out before (within t_{out}^{min}), while the slowest ones will do it after (but within t_{out}^{max}). If this does not happen, we could have an high concentration of vehicles on that road. We encode this idea into *if-then* rules which work as follows. Once computed t_{out}^{min} and t_{out}^{max}, two queries are submitted, first in t_{out}^{min} and then in t_{out}^{max}, to check whether the fastest vehicles have left before than t_{out}^{min} and whether the slowest vehicles have left before than t_{out}^{max}: if the number of the vehicles in common ($|D|$) to the time t_{out}^{min} and to the time t_{out}^{max} is higher than the minimum threshold $Dmin$, then the road is dense.

Notice that, since it would be unrealistic that the positions of all vehicles are recorded at the same times, the queries cover time-interval values rather than punctual time-point values. More precisely, we consider time-intervals with fixed radius ϵ centred on the time-points t_{out}^{min}, t_{out}^{max}, t_0 and t_1. An illustration is reported in the Figure 1a. We see that the fast vehicles go out before than $t_{out}^{min} + \epsilon$ (square 1), while the other fast vehicles remain, although we expect their exit (square 2). Next, in correspondence of $t_{out}^{max} + \epsilon$ (square 3), we expect that the remaining fast vehicles and those slow go out: if, even after $t_{out}^{max} + \epsilon$, there are still vehicles (which totally exceed $Dmin$), then the road is dense (square 4).

During monitoring of some vehicles, other vehicles can appear in the urban space, indeed new vehicles can enter before than the time-interval centred on t_1 while monitoring the vehicles observed in the time-interval of t_0. In this case, the check mechanism can provide wrong results since the new vehicles are not included into the set of vehicles initially observed. This problem is solved by $i)$ extending the matching operation with *if-then* rules which are able to recognize the insertion of new vehicles and $ii)$ adapting the monitoring process in order to track and retrieve the new vehicles by means of the next queries. More specifically, an additional query, which covers the time-interval centred on t_2 $(t_1 < t_2)$, is submitted to compute the speed and expected time-points (afterwards, $t_{out}^{min'}$ and $t_{out}^{max'}$) for the new vehicles. Thus, the check mechanism is updated in order to monitor the new vehicles in correspondence of the $i)$ expected time-points $t_{out}^{min'}$ and $t_{out}^{max'}$, $ii)$ expected time-points t_{out}^{min} and t_{out}^{max} initially estimated. The values of the new time-points establish the new order of the queries to be submitted and, dependently on this order, different rules will be matched. The time-point t_0 and the longest estimated time-point $(t_{out}^{max}$ or $t_{out}^{max'})$ determine the time-interval $[t_0, t_{max}]$ of the tuple (Definition 1).

An illustration of dense road is reported for the case in which the time-points $t_{out}^{min'}$, $t_{out}^{max'}$ come after t_{out}^{min} and t_{out}^{max} (Figure 1b). As reported, the presence of new vehicles in t_1, with respect to t_0, forces to submit another query to complete the calculation of the speed for these new vehicles (square 1). In the meanwhile, the algorithm estimates t_{out}^{min} and t_{out}^{max} which will be the time-points that the check mechanism will use for the next queries. After the query at the time t_2, also $t_{out}^{min'}$ and $t_{out}^{max'}$ are estimated, so we have two additional queries to be submitted: the new order of the queries will comply with the order of the expected times $t_{out}^{min} < t_{out}^{max} < t_{out}^{min'} < t_{out}^{max'}$. Thus, the algorithm first checks that the new vehicles are present in t_{out}^{min} and t_{out}^{max} (squares 2-5), and then it checks that all have left before than $t_{out}^{max'}$ (square 8). In t_{out}^{min} and t_{out}^{max}, we observe that the initially monitored vehicles leave the road (squares 2,4) and that the other vehicles remain there (squares 3,5). The latter should leave before than $t_{out}^{max'}$, while instead the check mechanism detects that from $t_{out}^{min'}$ to $t_{out}^{max'}$ no vehicle has left (squares 6-8), so if the number of these vehicles exceeds $Dmin$, the road is considered dense.

3.3 Mining Dense Regions

Dense regions are mined with at least two dense roads which meet the Definitions 2 and 3. This provides some hints on the technique to use.

The algorithm to mine the regions of type SC is performed after that a window is processed and operates on the roads already detected and dense regions which are being generated during the process. Each road can be associated to only one region if $i)$ its distance from the roads (which have been already added to that region) is lower than the parameter L and $ii)$ it has no vehicle in common with those roads. If the examined road is added to none of the regions, it will be considered as seed for a new region. Finally, we have a set of roads which begin to be dense at the time t_0, which is common to all roads of one region.

The algorithm to mine the regions of type TP evaluates the dense roads obtained from a sequence of overlapping windows. In particular, when the values of t_{out}^{min} and t_{out}^{max} exceed the last time-point t_w of the window $[W]_w^u$, we need to adapt the check mechanism in order to submit the corresponding queries later, specifically when the window which contains t_{out}^{min} and t_{out}^{max} will be under analysis. The algorithm performs two selection operations after that a sequence of windows, whose total width is less than Ω, has been analyzed. The first operation selects one road from each window with the result to have dense roads which are chronologically ordered: for instance, $\sigma', \sigma'', \sigma'''$ are taken from the windows $[W]_w^u, [W]_s^r, [W]_n^m$ respectively under the condition $t_0' < t_0'' < t_0'''$ (t_0', t_0'', t_0''' are the start time-points of $\sigma', \sigma'', \sigma'''$ respectively). The second one refines the previous operation by selecting the roads which have vehicles in common. Searching dense roads which share vehicles over a finite sequence of windows introduces implicitly a spatial neighbourhood in which those vehicles move around. Such a neighbourhood defines the spatial closeness component of the Definition 3.

For both types of regions, the notion of the distance between two roads σ', σ'' corresponds to the shortest path computed in the graph-based urban space.

4 Experiments

We tested our proposal on a real-world dataset which comprised trajectories produced by taxies moving in the city of Bejing for one week (02/02/2008-08/02/2008) [1]. We have almost 89900 roads where the vehicle were recorded and 98600 vehicles, while the urban space of the city of Bejing [2] has totally 141,380 roads (edges) and 106,579 intersections and terminal points (vertices).

Experiments are performed to test the influence of the input threshold $Dmin$ on the final dense regions. We report results on the dense roads and dense regions mined in the peak hours for each day, namely 7:00-9:00pm, 12:00-14:00pm, 18:00-20:00pm. The thresholds and parameters are set as follows: $Dmin = 3, 5, 7$, $L = 1.5km$, $\Omega = 30mins$, $|t_1 - t_0| = 1min$, $|t_2 - t_1| = 1min$, $\epsilon = 10secs$, $\omega = 5mins$, the rate of generation of the windows equal to $2.5mins$.

From the results in the figure 2 we can drawn some considerations. As to the dense roads, a quite expected behaviour is that the number of dense roads decreases as the threshold $Dmin$ increases. This is common to all days for all hours. An analysis conducted on the basis of the peak hours may provide some indications of social nature: on the week-end, a greater flow can be observed only in the time-slot 18:00-20:00, while on the weekdays, the highest values of dense roads are produced in the time-slot 12:00-14:00.

As to the dense regions, the results basically follow the behaviour of the dense roads. In particular, by analyzing the time-slots 7:00-9:00 and 18:00-20:00,

---

[1] http://research.microsoft.com/apps/pubs/default.aspx?id=138035
[2] http://www.openstreetmap.org/

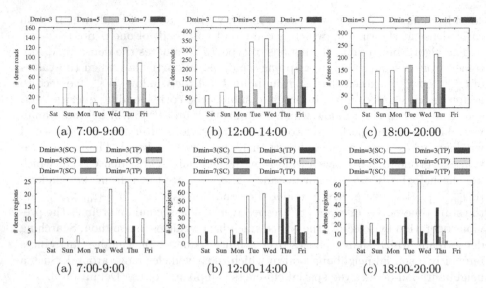

Fig. 2. Number of dense roads and dense regions at peak hours

numerous sets of regions of type SC are mined when we have more dense roads, while the greatest set of regions is generated in correspondence of the highest number of dense roads (12:00-14:00). Another expected result is the generation of great sets of dense regions of type SC as consequence of the generation of great sets of dense roads when $Dmin=3$. Indeed, the sets of mined dense roads can be so numerous that the roads can be closer to each other with the result to form many dense regions. Also for the regions of type TP, we have great sets when numerous sets of dense roads are generated, particularly in the time-slots 12:00-14:00 (on the weekdays) and 18:00-20:00 (on the week-end). This result can be explained with a slow movement of the vehicles.

We evaluated the final results through a quantitative measure which estimates the capacity of the approach to detect roads which have an high concentration of vehicles with respect to the neighbourhood. More formally, $\Theta(\sigma) = \frac{\sum_{j=1\dots m}(\sigma-\sigma_j)}{(m*1)+avg\sigma}$, where σ_j are non-dense roads close to σ in a diameter of 100m[3], m is the number of non-dense close roads, $avg\sigma$ is the average number of vehicles per road. The value of $\Theta(\sigma)$ would tend to 0 with an equal distribution of the vehicles among the road σ and its neighbours, while, when there is a strong concentration of the vehicles in the road σ, it tends to 1. Table 1 reports the mean of the values of Θ on all dense roads detected in each day: we observe the better performances in the days from Tuesday to Friday. This emphasizes the ability of the method to recognize correctly high concentrations of vehicles when the dense roads grow and maintain the robustness with respect to the noise.

---

[3] This value has been specifically computed on the graph of the city of Bejing by considering also that two parallel roads can be close within 100m.

Table 1. Evaluation on the dense roads by day

	Θ
02/02/2008 Saturday	0.52
03/02/2008 Sunday	0.67
04/02/2008 Monday	0.41
05/02/2008 Tuesday	0.54
06/02/2008 Wednesday	0.72
07/02/2008 Thursday	0.68
08/02/2008 Friday	0.64

5   Conclusions

In this paper we investigated the task of mining dense regions on vehicular mobility by taking into account two particular issues, namely the graph-based form of the urban space and regions, and the streaming style with which the mobility data are produced. The proposed method adopts a stream data mining strategy to detect dense roads and use them to form dense regions. It combines knowledge in the form of *if-then* rules and a querying mechanism to retrieve information about the roads from the mobility data stream. Experiments prove the applicability to real vehicular traffic data and remark a good accuracy of the method especially in highly congested roads at the peak hours. As future direction, we plan to extend the rule base with additional conditions and validate the results of the method with respect to the ground truth.

Acknowledgments.   In partial fulfilment of PON 02 00563 3470993 project "VINCENTE - A Virtual collective INtelligenCe ENvironment to develop sustainable Technology Entrepreneurship ecosystems" funded by the Italian Ministry of University and Research.

References

1. Chen, J., Lai, C., Meng, X., Xu, J., Hu, H.: Clustering moving objects in spatial networks. In: Kotagiri, R., Radha Krishna, P., Mohania, M., Nantajeewarawat, E. (eds.) DASFAA 2007. LNCS, vol. 4443, pp. 611–623. Springer, Heidelberg (2007)
2. Gama, J., Gaber, M.M.: Learning from Data Streams: Processing Techniques in Sensor Networks. Springer (November 2007)
3. Hadjieleftheriou, M., Kollios, G., Gunopulos, D., Tsotras, V.J.: On-line discovery of dense areas in spatio-temporal databases. In: Hadzilacos, T., Manolopoulos, Y., Roddick, J., Theodoridis, Y. (eds.) SSTD 2003. LNCS, vol. 2750, pp. 306–324. Springer, Heidelberg (2003)
4. Jensen, C.S., Lin, D., Ooi, B.C., Zhang, R.: Effective density queries on continuously moving objects. In: Liu, L., Reuter, A., Whang, K.-Y., Zhang, J. (eds.) ICDE, p. 71. IEEE Computer Society (2006)
5. Lai, C., Wang, L., Chen, J., Meng, X., Zeitouni, K.: Effective density queries for moving objects in road networks. In: Dong, G., Lin, X., Wang, W., Yang, Y., Yu, J.X. (eds.) APWeb/WAIM 2007. LNCS, vol. 4505, pp. 200–211. Springer, Heidelberg (2007)
6. Wang, W., Yang, J., Muntz, R.R.: Sting: A statistical information grid approach to spatial data mining. In: Jarke, M., Carey, M.J., Dittrich, K.R., Lochovsky, F.H., Loucopoulos, P., Jeusfeld, M.A. (eds.) VLDB, pp. 186–195 (1997)

Mining Temporal Evolution of Entities
in a Stream of Textual Documents

Gianvito Pio, Pasqua Fabiana Lanotte, Michelangelo Ceci, and Donato Malerba

University of Bari A. Moro
Dept. of Computer Science - Via Orabona, 4 - I-70125 Bari, Italy
{name.surname}@uniba.it

Abstract. One of the recently addressed research directions focuses on the problem of mining topic evolutions from textual documents. Following this main stream of research, in this paper we face the different, but related, problem of mining the topic evolution of entities (persons, companies, etc.) mentioned in the documents. To this aim, we incrementally analyze streams of time-stamped documents in order to identify clusters of similar entities and represent their evolution over time. The proposed solution is based on the concept of temporal profiles of entities extracted at periodic instants in time. Experiments performed both on synthetic and real world datasets prove that the proposed framework is a valuable tool to discover underlying evolutions of entities and results show significant improvements over the considered baseline methods.

1 Introduction

Topic Detection and Tracking (TDT) [3,22,5,10] is an important research area which applies data mining algorithms in order to find and follow topics in streams of news or, in general, in streams of textual documents. According to the classification suggested in [7], there are three main research lines in TDT: *i) Segmentation* - documents coming from a stream are clustered according to their topic. Each cluster represents the same topic across the time dimension. *ii) New topic detection* - new clusters are identified in the stream. *iii) Topic tracking* - evolutions of clusters are tracked. In this case, new documents can be associated to existing clusters, causing changes in clusters' properties.

By focusing our attention on topic tracking, in this paper, we argue that it is possible to use such techniques to discover evolutions of entities over time. We focus on entities (e.g. people, organizations) having particular roles (e.g. perpetrator, victim, in the risk identification and analysis domain) in particular types of domain-dependent relationships (e.g. kill, steal). These entities are considered as units of analysis. In this respect, the proposed framework identifies such entities and incrementally analyzes streams of documents in order to discovery clusters of "similar" entities and represent their evolution over time. To this aim, we apply a time-slice density estimation method [2] that allows us to represent the profile of each entity. Moreover, it allows us to analyze profiles evolution by measuring the rate of change of properties and peculiarities of entities activities'

T. Andreasen et al. (Eds.): ISMIS 2014, LNAI 8502, pp. 50–60, 2014.
© Springer International Publishing Switzerland 2014

over a given time horizon. At this purpose, we apply a time-slice density estimation method [2] that allows us to represent the profile of each entity. Moreover, it allows us to analyze profiles evolution by measuring the rate of change of properties and peculiarities of entities activities' over a given time horizon.

In the literature, several papers have faced the problem of mining evolutions in streams of documents and, in particular, the problem of tracking topics, ideas and "memes" [14,24]. However, most of the work considers single keywords or short phrases in the documents as units of analysis. On the contrary, we consider as units of analysis the entities that can be associated with (identified in) the documents. This means that we identify the evolution of entities by analyzing documents they are associated with. Evolutions are expressed according to relevant terms that allow us to represent and characterize entities. From the methodological viewpoint, we do not identify evolutions by evaluating whether a particular data mining model has become stale because of a change in the underlying data distribution [13,2], but we provide the user with an understanding of the changes, according to a content-based representation of the entities' profiles (entities are represented according to terms occurring in the textual documents).

The proposed framework could be profitably exploited in different application domains. For instance, in the analysis of papers belonging to the medical domain, it could support researchers in the identification of evolutions about the recognized role of biological entities (e.g. genes) over time. Another example is represented by the risk identification and analysis domain, which is considered in this paper. In this case, using publicly available news (e.g. daily police reports, public court records, legal instruments) about criminals, and assimilating the concept of *topic* with the concept of *crime typology* represented by a group of "similar" criminals, the proposed method can be considered as a valuable tool for law enforcement officers in risk and threat assessment.

The contribution of this paper is manifold: on the basis of entities identified in the documents, *i)* we define an unsupervised feature selection algorithm which overcomes limitations of existing unsupervised methods *ii)* we represent the entities' profile and on-line modify it according to more recent documents; *iii)* we generate clusters of similar entities and represent and analyze their evolution.

2   Related Work

In the literature, a variety of approaches to deal with evolving clusters from textual data streams can be found. For example, in [23] the authors propose an incremental and neural-network-based version of the "spherical" k-means which, according to an appropriate rule for updating the weights of the neural network, incrementally modifies the closest cluster center, given a new document. In [1] the authors cluster blogs by considering their labels and generating a "collective wisdom". In [15], stories, built from blogs, are clustered. After a set of initial clusters is built, a dynamic clustering algorithm incrementally updates clusters on the basis of the distance between new stories and clusters' stories.

Despite the clear relationship, there are differences between these researches and ours. In the former, clusters represent the same topic across the time dimension whereas we associate clusters to a single time interval. Consequently, we do not apply incremental clustering approaches, but we identify clusters for each time interval and compare them with those previously identified. Moreover, in topic tracking, clusters group documents on the same topic, i.e. the unit of analysis is the document, while in our case the unit of analysis is the entity. This means that we cluster entities on the basis of documents associated to them.

A similar approach to ours is proposed in [4], where clusters of keywords extracted from messages published in blogs are identified for each time interval. Clusters associated to consecutive time intervals are pairwise compared in order to identify pairs with the highest affinity. By combining affinity relationships over several time intervals, it is possible to identify the top-k paths that express the most significant evolutions of the initial clusters over time. The main difference with respect to our approach is that the considered unit of analysis is the "keyword", on the basis of the assumption that clusters of keywords characterize topics. Similarly, in [21] the authors propose an approach for defining and monitoring topics by clustering, for each time interval, blogs on the basis of their content. However, in this case, clustering is performed on the pairs (*class*, *similarity*) obtained by a centroid-based classifier. This means that clustering significantly depends on a preliminary supervised learning phase.

In [18], the authors propose the identification of evolutions of clusters over time, by considering the application of either batch or incremental clustering approaches for each time interval. Evolutions are represented through an *Evolution Graph*, where nodes are clusters and edges connect clusters belonging to adjacent time intervals, and are summarized through the so-called *fingerprints*. Also in this case, the units of analysis are the keywords of textual documents.

Finally, we mention the work presented in [12] where the authors propose to learn, from news, a generative model of terms which takes as input the topic and the mentioned entities. Although this work does not exploit the time dimension, it considers, similarly to our work, the possible correlation of news with other entities such as people, organizations, locations.

3   TB-EEDIS

In this section we define the framework **TB-EEDIS** *(Time-Based Entity Evolution DIScoverer)* that allows us to discover *evolutions* of entities from a stream of textual documents. In this respect, an *evolution* is defined as a relevant change of entities' properties in different time windows. All the necessary information are extracted from time-stamped textual documents which are implicitly associated to a single time window. In this work, *time windows* are defined as adjacent and disjunct time intervals, obtained by partitioning the entire period we intend to analyze into intervals of the same size. Evolutions are discovered by analyzing the changes identified among distinct time windows.

The textual content of each document d_j is represented according to the classical *Vector Space Model (VSM) with TF-IDF weighting*. Each entity is represented in the same space of terms used for representing documents (an example for news about criminals, that we will show in the experiments, is: [attack; fire; claim; suspect; report; injur; islam]), which better represents the profile of the entity in a given time window. Since the terms space can be very large, we select relevant features through an unsupervised feature selection algorithm.

Summarizing, the framework *TB-EEDIS* consists of the following phases: *i)* identification of entities from each document; *ii)* VSM representation of the documents (after applying classical pre-processing techniques), *iii)* feature selection, *iv)* identification of the position of each entity for each time window, *v)* clustering of entities for each time window and *vi)* evolution discovery and analysis. Entity identification is performed by applying two natural language processing techniques, that is, Named Entity Extraction [19] and Dependencies Analysis [16]. The adopted strategy considers the logical structure of the sentences and, starting from relationships, it identifies the involved entities. Since this task is not the main subject of the paper, for space constraints, we do not report further details about this phase. In the following, we explain the methods we use for selecting relevant features, representing entities and studying their evolution.

3.1 Feature Selection

We present two distinct unsupervised feature selection algorithms. The use of unsupervised approaches is motivated by the task we consider (i.e. clustering) and the consequent absence of any target (class) attribute to guide the selection.

Variance-based Feature Selection. The most straightforward way to perform feature selection is by computing the variance of the relative term-frequency of each term in the entire document collection and keeping the k terms with the highest variance. Intuitively, a term with high variance will better discriminate documents, whereas a term with low or zero variance will substantially describe the documents in the same way. This feature selection algorithm has the advantage of a linear time complexity, at the price of some disadvantages: *a)* It selects the terms which best discriminate between documents, disregarding their real similarity. In fact, it does not take into account the case in which similar documents share the same terms with similar relative term-frequency. This can lead to lose terms that characterize entire classes of documents. *b)* It does not consider the correlation between terms. In fact, two strongly correlated terms will be both selected if they are in the set of the top k terms with the highest variance. This leads to select redundant terms.

MIGRAL-CP. In [9], the authors propose to use the Laplacian Score to identify the features which better preserve samples similarity. However, the Laplacian Score rewards features for which similar samples show a small variation in the feature values, but does not reward those that show a large variation for dissimilar samples. Inspired by this work, we define a different method called *MInimum GRAph Loss with Correlation Penalty (MIGRAL-CP)*, which *i)* selects k terms to represent the whole collection of documents, showing both great variation for dissimilar documents and low variation for similar ones and *ii)* discards features correlated with already selected features.

Formally, given the set of documents $D = \{d_1, d_2, \ldots, d_n\}$ and the set of terms $T = \{t_1, t_2, \ldots, t_m\}$, we build the (fully-connected undirected) graph G, where each node represents a document and each edge between two documents d_i and d_j is labeled with their similarity computed as: $v_{i,j} = e^{-\|w_{d_i} - w_{d_j}\|^2}$, where w_{d_i} (w_{d_j}) is the relative term-frequency vector associated to the document d_i (d_j), defined according to the set of terms T. The similarity measure we use is defined in [9] but, obviously, any other similarity measure might be considered.

We define an iterative method to select a subset of k terms which satisfy the above requirements. The first term is selected in order to maximize the score:

$$Score_1(t_r) = \frac{1}{2}\left(1 - \frac{1}{n}\sum_{j=1}^{n} \rho(V_j, F_{r,j})\right) \tag{1}$$

where:

- $V_j = [v_{j,1}, v_{j,2}, \ldots, v_{j,n}]$ are the similarity values between the document j and all the other documents, using all the terms.
- $F_{r,j} = \left[(s_{r,j} - s_{r,1})^2, (s_{r,j} - s_{r,2})^2, \ldots, (s_{r,j} - s_{r,n})^2\right]$ are the dissimilarities between the document j and all the other documents, using the term t_r only. In this formula, $s_{r,j}$ is the relative term frequency of the term t_r in d_j.
- $\rho(\cdot, \cdot)$ is the Pearson correlation coefficient.

The first selected term ($\hat{t}_1 = \arg\max_{t_r \in T} Score_1(t_r)$) will be the one which determines the highest inverse linear correlation between the documents' similarities computed with all the terms and documents' dissimilarities computed with only that term. The remaining $k-1$ terms are selected according to:

$$Score_i(t_r) = Score_{i-1}(t_r) \times (1 - penalty(t_r, \hat{t}_{i-1})) \tag{2}$$

where, at the iteration i, $penalty(t_r, \hat{t}_{i-1})$ reduces the score of each term t_r according to its correlation to the term selected during the previous iteration, preventing the selection of redundant terms. Coherently with the correlation coefficient introduced before, we define $penalty(t_r, \hat{t}_{i-1}) = \max\left(0, |\rho(t_r, \hat{t}_{i-1})| - \gamma\right)$, where $0 \leq \gamma \leq 1$. The rationale of this choice is that a correlation value of $|\rho(t_r, \hat{t}_{i-1})| \leq \gamma$ is considered too small to result in a penalty.

3.2 Representing Entities

For each time window τ_z, the profile of each entity is represented in the same k-dimensional terms space identified in the feature selection phase. In the following we describe two possible alternatives.

Time-Weighted Centroid. In this case, the profile of the entity c in τ_z is:

$$X(c, \tau_z, h) = \frac{\sum_{<d_j, \tau_j> \in S_{c,\tau_z,h}} p_{\tau_z,\tau_j}(h) \times w_{d_j}}{\sum_{<d_j, \tau_j> \in S_{c,\tau_z,h}} p_{\tau_z,\tau_j}(h)}, \tag{3}$$

where $S_{c,\tau_z,h}$ is the set of documents associated to c and belonging to τ_z or to one of the previous $h-1$ time windows, and $p_{\tau_z,\tau_j}(h) = 1 - \frac{z-j+1}{h}$ is the time fading-factor which reduces the effect of the document d_j according to the distance between τ_z and the time window τ_j (i.e. the time window of d_j).

Max Density Point. In this solution, inspired to the work in [6], each document d_j is represented as a k-dimensional Gaussian function $d'_j(\cdot)\colon [0,1]^k \to \mathbb{R}^+$:

$$d'_j(x) = \prod_{i=1}^{k} \frac{1}{\sqrt{2\pi\sigma^2}} e^{-\frac{(x_i - s_{i,j})^2}{2\sigma^2}} \tag{4}$$

where $\sigma \in [0,1]$ is a parameter that defines the width of the Gaussian function. The position of c in τ_z is the point with the highest sum of contributions:

$$X(c,\tau_z,h) = \arg\max_{x \in [0,1]^k} \sum_{<d_j,\tau_j> \in S_{c,\tau_z,h}} p_{\tau_z,\tau_j}(h) \times d'_j(x) \tag{5}$$

where the time fading-factor $p_{\tau_z,\tau_j}(h)$ reduces the value of the Gaussian function.

For computational reasons, we search $X(c,\tau_z,h)$ in the discrete space Φ^k, where $\Phi = \left\{0, \frac{1}{\beta}, \ldots, \frac{\beta-1}{\beta}, 1\right\}$ and $\beta+1$ is the number of desired distinct values.

Moreover, we adopt two further optimizations: *i)* we limit the search to the areas interested by at least one document belonging to the time window τ_z, and to the position $X(c,\tau_{z-1},h)$, assumed in the previous time window (*incrementality*); *ii)* we adopt a greedy search that works only around the points for which the $d'_j(\cdot)$ functions, contributing to $X(c,\tau_z,h)$, reach the highest values. In particular, we focus (for each dimension) on a smaller area around the point for which $d'_j(\cdot)$ assumes the maximum value. Formally, let y be the value assumed on a given dimension by a document. Given the applied discretization, we analyze only $\beta\sigma$ values on both sides of y, leading to a total of $2\beta\sigma+1$ values[1], instead of $\beta+1$, thus covering all the available values in $[y-\sigma;y+\sigma]$ (see Figure 1).

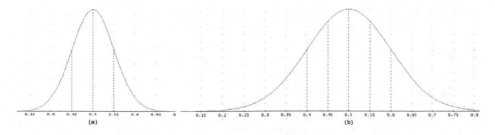

Fig. 1. Gaussian function defined on a single dimension with $y = 0.5$, $\beta = 20$, $\sigma = 0.05$ (a) and $\sigma = 0.10$ (b). In (a) it is enough to analyze only the values 0.45, 0.50 and 0.55, whereas in (b) it is necessary to analyze also the values 0.40 and 0.60.

---

[1] Reasonable values of σ are ≤ 0.1. In the experiments we use $\sigma \in \{0.05, 0.1\}$.

3.3 Clusters Evolution Discovery

The last necessary step, before analyzing the evolutions of entities, consists of searching for clusters of entities for each time window. Although we perform clustering for each time window independently, i.e. without considering the temporal neighborhood, it is noteworthy that the influence of documents belonging to previous time windows is caught by the proposed strategy for the identification of entities' profile, as already described in Section 3.2.

We use a variant of the K-means algorithm. Obviously, in TB-EEDIS, any clustering algorithm (also density based, e.g. DBSCAN [8]) can be plugged in.

Our improvement to the standard K-means algorithm consists in the automatic identification of the reasonable number of clusters to be extracted, which is necessary in the task at hand, since the number of clusters is not known a-apriori. The solution we adopt is that of exploiting the *Principal Component Analysis (PCA)*, which identifies a new (smaller) set of prototype features such that a given percentage of the variance in the data is explained [11]. By inverting the roles of features and examples, it is possible to identify a set of (orthogonal) prototype examples, according to which other examples can be aggregated. In our solution we use the number of prototype examples as an indication of the appropriate number of clusters, according to the underlying data distribution.

Once clustering is performed for each time window, it is possible to identify:

- the *position* of the cluster in the k-dimensional terms space. This can be identified by analyzing the terms with the highest weights in the cluster (e.g. of its centroid), which gives an idea about the entity typology it represents.
- a *matching* between clusters of different time windows by maximizing the similarity between the centroids of matched clusters, which are still represented in the same terms space.
- the *number of entities* which have evolved from the entity typology represented by C_i to that represented by C_j, where C_i and C_j are two generic clusters extracted for the time windows τ_{z-1} and τ_z, respectively (Figure 2).

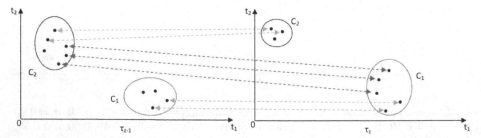

Fig. 2. An evolution: Three entities moved from C_2 in τ_{z-1} to C_1 in τ_z

4   Empirical Evaluation

The evaluation is performed on a set of synthetically generated datasets and on a real dataset. The synthetic datasets consist of documents about 50 entities, generated for 30 time windows. For each time window and entity, a set of 10 documents is generated by considering 7 specific vocabularies (representing different topics) and a generic English vocabulary, used to introduce "noise terms". For each time window and entity, the probability of changing the topic is set to 0.2. Three different datasets are generated, setting the number of time windows necessary to complete each evolution (which defines the "speed" of changes) to 4, 10 and 20. In order to reproduce realistic situations, when simulating an evolution from a topic A to a topic B, we gradually decrease the frequency of terms representing A and increase the frequency of terms representing B in the generated documents. During an evolution, no additional evolutions can start.

As real dataset, we consider the Global Terrorism Database (GTD)[2] for the risk identification and analysis domain. GTD consists of information on terrorist events (more than 104,000 cases) around the world from 1970 through 2011, including systematic data (such as the textual content of a news, involved criminals, publication date) on domestic as well as international terrorist incidents.

In the evaluation performed on the synthetic datasets, the feature selection is executed with both the proposed methods (variance-based, which is considered as baseline, and MIGRAL-CP), with $k = 10$ (number of features to select). For MIGRAL-CP, results are obtained with $\gamma = 0.5$ (γ is the minimum threshold on the Pearson coefficient for applying a penalty), which after preliminary experiments (not reported in this paper for space constraints), resulted in the best trade off between relevancy and allowed redundancy of selected terms. For MaxDensity, the discretization parameter β is set to 20 and σ of the Gaussian functions $d'_j(\cdot)$ is set to 0.05^3.

As regards GTD, we consider 13 annual time windows (from 1998 to 2010), for a total of 11,225 news concerning 82 criminals/criminal organizations. The feature selection is executed with both the proposed methods, with $k = 15$ and, for MIGRAL-CP, $\gamma = 0.5$. For MaxDensity, β is set to 20 and σ is set to $0.05^3$3.

Both synthetic and real datasets are analyzed with two different values for the PCA variance (90% and 95%) and with different values of h (2, 5 and 10). In particular, each synthetic dataset is analyzed with the corresponding value of h such that the number of time windows used to perform an evolution is $2h$. This solution is motivated by the fact that, in general, the system should be able to detect the change of the topic in the middle of the evolution. On the other hand, GTD is analyzed with all the considered values of h, since we do not know a priori the speed of evolutions in the dataset.

Results are collected in terms of running times (hh:mm) required to complete the whole evolution discovery process and in terms of a variant of the

---

[2] http://www.start.umd.edu/gtd/
[3] We also performed experiments with $\sigma = 0.10$. Since there was no significant difference in the results, for space constraints, we report only results with $\sigma = 0.05$.

Table 1. Results obtained on the synthetic datasets and on the GTD dataset. Italic indicates a better result with respect to the strategy for computing the entity position, while bold indicates a better result with respect to the feature selection strategy.

| | | | Synthetic Dataset | | | | | | GTD Dataset | | | |
| | | | Variance | | | MIGRAL-CP | | | Variance | | MIGRAL-CP | |
h	Position	Var	time	q-mod	NMI	time	q-mod	NMI	time	q-mod	time	q-mod
2	Centroid	90%	07:38	*0.581*	0.652	15:21	**0.613**	**0.757**	00:09	**0.294**	39:54	0.245
2	Centroid	95%	07:39	*0.606*	0.692	15:22	**0.659**	**0.799**	00:09	**0.319**	39:54	0.270
2	MaxDensity	90%	07:43	0.570	*0.689*	15:26	*0.672*	**0.770**	110:41	*0.322*	100:17	*0.447*
2	MaxDensity	95%	07:43	0.577	*0.698*	15:27	*0.710*	*0.800*	110:41	*0.509*	100:17	*0.479*
5	Centroid	90%	08:06	0.559	0.615	21:03	**0.616**	**0.726**	00:09	**0.297**	39:54	0.224
5	Centroid	95%	08:07	0.633	0.678	21:04	**0.634**	**0.739**	00:09	**0.316**	39:54	0.249
5	MaxDensity	90%	08:16	*0.614*	*0.679*	21:11	*0.654*	**0.773**	137:41	*0.325*	118:59	*0.454*
5	MaxDensity	95%	08:17	*0.665*	*0.713*	21:12	*0.690*	*0.797*	137:41	*0.521*	118:59	*0.487*
10	Centroid	90%	08:31	*0.534*	0.548	25:51	**0.547**	**0.698**	00:09	**0.304**	39:54	0.232
10	Centroid	95%	08:32	0.565	0.575	25:52	**0.587**	**0.731**	00:09	**0.322**	39:54	0.245
10	MaxDensity	90%	08:45	0.491	*0.603*	26:08	*0.607*	**0.724**	144:20	*0.400*	126:06	*0.452*
10	MaxDensity	95%	08:46	0.522	*0.623*	26:09	*0.660*	*0.762*	144:20	*0.524*	126:06	*0.479*

Q-Modularity measure [17], which allows us to evaluate the quality of the clustering with respect to a random clustering. This variant is described in the following. Let $e_{ij} = \frac{2}{r(r-1)} \sum_{c' \in C_i, c'' \in C_j} sim(X(c', \tau_z, h), X(c'', \tau_z, h))$ be a measure of the strength of the interconnections between entities in the cluster C_i and entities in the cluster C_j. In this formula, r represents the total number of entities and $sim(\cdot, \cdot) \in [0, 1]$ is the cosine similarity. Intuitively, we want clusters for which e_{ii} values are generally large and $e_{ij}(i \neq j)$ values are generally small. Formally: $Q = \sum_{i=1}^{k} \left(e_{ii} - a_i^2 \right)$, where $a_i = \sum_j e_{ij} = \sum_j e_{ji}$.

Moreover, for the synthetic datasets, we also evaluate the results in terms of the average Normalized Mutual Information (NMI) [20]. In particular, NMI is computed between the set of extracted clusters and the set of true clusters representing topics imposed during the generation of the datasets, in order evaluate the ability of TB-EEDIS to correctly catch the underlying evolutions.

As it can be observed in Table 1, for feature selection, the MIGRAL-CP algorithm always leads to better Q-Modularity and NMI results in the synthetic datasets, with respect to the variance-based method (which we consider as a baseline). The disadvantage is that better results are obtained at the price of significantly higher running times. These observations do not hold for the GTD dataset, where there is no clear advantage of MIGRAL-CP over the variance-based method in the case of MaxDensity (where we have better results). The motivation can be found in the fact that in the synthetic datasets we explicitly introduced redundancy in the text, while in GTD this phenomenon is not under control and the two methods almost equally perform.

As regards the method for computing the entities' position, the MaxDensity method always significantly outperforms the centroid-based method (which we consider as baseline) on GTD, at the price of a slightly higher running times, whereas on the synthetic datasets it shows betters results only in terms of NMI.

Observing the influence of the variance (for the PCA-based estimation of the number of clusters), we have that, for $Var = 95\%$, extracted clusters better adapt to the underlying topics models (see NMI in Table 1), without incurring in overfitting issues. This phenomenon is reflected on Q-Modularity values, also for GTD. However, it is noteworthy that the quality of results is less dependent on such parameter when the MIGRAL-CP feature selection method is adopted.

From a qualitative viewpoint, it is interesting to identify a description of the clusters, in order to deeply understand the evolutions in which they are involved. A possibility consists in the analysis of the terms describing the entities belonging to the cluster. For example, analyzing the centroid of a cluster identified from GTD (Var=90%, MaxDensity-MIGRAL-CP, $h = 5$), that is: [attack: 0.593; fire: 0.371; claim: 0.271; suspect: 0.1; report: 0.057; injur: 0.057; islam: 0.05] allows us to identify a specific type of crime (terrorist attack). For future work, we will investigate the possibility of performing an extensive qualitative analysis of the evolutions discovered from real datasets.

5   Conclusions

In this paper, we propose the framework TB-EEDIS to incrementally extract knowledge from time-stamped documents. In particular, it: identifies entities with domain-specific roles; represents documents by exploiting unsupervised feature selection algorithms; represents the entities' profile and identifies clusters of entities in order to represent and analyze their evolution.

Results show that the algorithms proposed for the unsupervised feature selection (MIGRAL-CP) and for the identification of the position of entities (Max Density-based) generally provide better results when compared to baseline approaches. Moreover, results obtained in terms of Normalized Mutual Information on synthetic datasets prove the ability of TB-EEDIS to catch the underlying evolutions of entities, making it applicable in additional domains (e.g. biological).

For future work, we intend to analytically identify the value of σ, with respect to h, such that the global optimum is guaranteed. Moreover, we will qualitatively evaluate the evolutions discovered on real datasets and we will analyze how different sizes of time windows can influence results.

Acknowledgements. We would like to acknowledge the support of the European Commission through the project MAESTRA - Learning from Massive, Incompletely annotated, and Structured Data (Grant number ICT-2013-612944).

References

1. Agarwal, N., Galan, M., Liu, H., Subramanya, S.: Wiscoll: Collective wisdom based blog clustering. Inf. Sci. 180, 39–61 (2010)
2. Aggarwal, C.C.: On change diagnosis in evolving data streams. IEEE Trans. Knowl. Data Eng. 17(5), 587–600 (2005)
3. Allan, J. (ed.): Topic Detection and Tracking: Event-based Information Organization. Kluwer International Series on Information Retrieval, Kluwer (2002)
4. Bansal, N., Chiang, F., Koudas, N., Tompa, F.W.: Seeking stable clusters in the blogosphere. In: VLDB, pp. 806–817. ACM (2007)

5. Brants, T., Chen, F., Farahat, A.: A system for new event detection. In: ACM SIGIR, pp. 330–337. SIGIR 2003. ACM (2003)
6. Ceci, M., Appice, A., Malerba, D.: Time-slice density estimation for semantic-based tourist destination suggestion. In: ECAI (2010)
7. Chung, S., McLeod, D.: Dynamic pattern mining: An incremental data clustering approach. In: Spaccapietra, S., Bertino, E., Jajodia, S., King, R., McLeod, D., Orlowska, M.E., Strous, L. (eds.) Journal on Data Semantics II. LNCS, vol. 3360, pp. 85–112. Springer, Heidelberg (2005)
8. Ester, M., Kriegel, H.P., Sander, J., Xu, X.: A density-based algorithm for discovering clusters in large spatial databases with noise. In: ICDM, pp. 226–231 (1996)
9. He, X., Cai, D., Niyogi, P.: Laplacian score for feature selection. In: NIPS (2005)
10. Jameel, S., Lam, W.: An n-gram topic model for time-stamped documents. In: Serdyukov, P., Braslavski, P., Kuznetsov, S.O., Kamps, J., Rüger, S., Agichtein, E., Segalovich, I., Yilmaz, E. (eds.) ECIR 2013. LNCS, vol. 7814, pp. 292–304. Springer, Heidelberg (2013)
11. Jolliffe, I.T.: Principal Component Analysis, 2nd edn. Springer (2002)
12. Kim, H., Sun, Y., Hockenmaier, J., Han, J.: Etm: Entity topic models for mining documents associated with entities. In: ICDM, pp. 349–358. IEEE (2012)
13. Kleinberg, J.: Bursty and hierarchical structure in streams. In: ACM SIGKDD, KDD 2002, pp. 91–101. ACM, New York (2002)
14. Leskovec, J., Backstrom, L., Kleinberg, J.: Meme-tracking and the dynamics of the news cycle. In: KDD 2009, pp. 497–506. ACM, New York (2009)
15. Li, X., Yan, J., Fan, W., Liu, N., Yan, S., Chen, Z.: An online blog reading system by topic clustering and personalized ranking. ACM Trans. Internet Technol. 9, 9:1–9:26 (2009)
16. de Marneffe, M.C., MacCartney, B., Manning, C.D.: Generating typed dependency parses from phrase structure trees. In: LREC (2006)
17. Newman, M.E.J.: Modularity and community structure in networks. Proceedings of the National Academy of Sciences 103(23), 8577–8582 (2006)
18. Ntoutsi, E., Spiliopoulou, M., Theodoridis, Y.: Fingerprint: Summarizing cluster evolution in dynamic environments. IJDWM 8(3), 27–44 (2012)
19. Sarawagi, S.: Information extraction. Foundations and Trends in Databases 1(3), 261–377 (2008)
20. Strehl, A., Ghosh, J.: Cluster ensembles — a knowledge reuse framework for combining multiple partitions. J. Mach. Learn. Res. 3, 583–617 (2003)
21. Varlamis, I., Vassalos, V., Palaios, A.: Monitoring the evolution of interests in the blogosphere. In: ICDEW, pp. 513–518 (2008)
22. Yang, Y., Carbonell, J., Brown, R., Pierce, T., Archibald, B., Liu, X.: Learning approaches for detecting and tracking news events. IEEE Intelligent Systems and their Applications 14(4), 32–43 (1999)
23. Zhong, S.: Efficient streaming text clustering. Neural Networks 18(5-6) (2005)
24. Zhu, Y., Shasha, D.: Efficient elastic burst detection in data streams. In: ACM SIGKDD, KDD 2003, pp. 336–345. ACM, New York (2003)

An Efficient Method for Community Detection Based on Formal Concept Analysis

Selmane Sid Ali[1], Fadila Bentayeb[1], Rokia Missaoui[2], and Omar Boussaid[1]

[1] Université Lyon 2 - Laboratoire ERIC
[2] Université du Québec en Outaouais

Abstract. This work aims at proposing an original approach based on formal concept analysis (FCA) for community detection in social networks (SN). Firstly, we study FCA methods which partially detect community in social networks. Secondly we propose a *GroupNode modularity* function whose goal is to improve a partial detection method taking into account all actors of the social network. Our approach is validated through different experiments based on real known social networks in the field and a synthetic benchmark networks. In addition, we adapted the *F-measure* function in the case of multi-class in order to evaluate the quality of a detected community.

1 Introduction

As far as social networks are concerned, the study of community structure of social networks has become a real challenge with many applications in multiple research areas. In Computer Science, social networks have been basically studied according to two different theories: the first one is based on graph theory [6] and the second one is based on the FCA [8]. Several approaches exist in FCA for community detection. However they imply a partial exploitation of social networks as they only consider a specific part of the actors in social networks who share certain properties.

The main objective of this work is to address FCA family methods for community detection [8] by proposing a new approach which refines the partial method proposed by [1]. Indeed, the author in [1] based his work on FCA in order to detect communities in social networks. Yet, as he considers a part of actors in social networks, we can consider his point of view as being subjective. In this paper, we propose to improve this method by considering all actors of the social network. Thus, we combine FCA tools [14] and concepts coming from graph theory so as to address in a comprehensive way the problem of community detection. Our approach has as input a formal context representing the social network ; and it allows us to determine partial communities. Then we assign the ignored nodes to the communities by maximizing *GroupNode modularity* function that we have proposed.

The rest of this paper is organized as follows. Section 2 sheds light upon a certain state of the art on community detection in social networks. Then in section

T. Andreasen et al. (Eds.): ISMIS 2014, LNAI 8502, pp. 61–72, 2014.
© Springer International Publishing Switzerland 2014

3, we study the community detection problem according to several approaches based on FCA namely [3] and [1]. Finally, section 4 details our approach and *GroupNode modularity* function. It also introduces our algorithm, the proposed community quality measure and the experimental results. At last, we will conclude this paper and provide future research directions in Section 5.

2   Communities in Networks

Social networks stand for groups of people or individuals (social entities) connected by several social ties. The different relationships between individuals depend on the context. Indeed we can find either friendly relationships in the case of friendship network or citation relations in a scientific publication network, physical links or logical connections in a computer network, etc..

The social network analysis covers a range of issues that mainly aim at identifying communities and their own evolution, studying the dynamics of a network, identifying the roles of nodes and communities that form a social network, studying links prediction and recommendation. These issues are applied in several areas: research and propagation of information [12], Biology [5], internet security etc.. Social graphs are characterized by specific areas. All the individuals of these ones have more links with each other than the rest of the social graph [2], [9]. This is what we call a community. Our own goal is to detect communities in networks, without knowing neither how many they are nor the size they have.

Other alternatives to graph theory underline the problem of community detection. In this work, we are particularly interested in the FCA [14]. It is a knowledge extraction technique based on the theory of concepts. [3] was the first to exploit FCA so as to describe a social network and to discover communities. Also, this method focuses on their key individuals by exploiting the idea of overlapping maximal cliques (MC) in a Galois lattice. His method was validated through small networks extracted from the real world. The drawback of this method is the removal of actors belonging to intermediate cliques in the detection process. [1] improves this approach and proposes to identify several groups at each level of the maximal cliques lattice taking into account all the actors belonging to maximal cliques without removing those belongings to intermediate cliques. However, this method does not consider the actors who belong to the social network graph and not belonging to the maximal cliques set. As far as social networks are concerned a large number of actors do not belong to the maximal cliques set. As we focus on this observation, our work proposes a new method to detect communities which would tend to improve *Falzon's* approach by considering all the actors of the social network. In other words, none of the actors of the social network is eliminated in the detection process we propose. We introduce *GroupNode modularity* function which is adapted from the Newman's modularity [9]. This function allows us to decide for each ignored actor what community he belongs to and refines the different communities we then obtain. We also define an adapted quality measure issued from F-measure [13]. It is used to evaluate the detected communities and to compare our method to *Falzon's*.

3 Community in FCA Based Approaches

Formal concept analysis [4] is a formalism for representation and knowledge extraction based on the notions of concepts and concept lattice (Galois lattice). FCA has been used successfully in several areas [10] such as computer software engineering, databases and data warehouses, mining and knowledge management and in several real-world applications such as medicine, psychology, linguistics and sociology. In this section we present a formalization of the problem of identifying community in the FCA based approaches.

3.1 Community Detection

In this subsection we introduce the different concepts necessary to understand the process of identifying a community in social networks using FCA.

Basic Definitions. Let $G = (V, E)$ be a graph representing a social network where V would be a set of actors (nodes) $\{x_i\}_{i=1}^n$ of the social network and $n = |V|$ would be the number of nodes in G ; E is the set of social links between actors and $m = |E|$ would be the number of edges in G.

A clique in an undirected graph $G = (V, E)$ is a subset of the vertex set $c \subseteq V$, such that for every two vertices in c, there exists an edge that connects the two. A maximal clique (MC) would be an unextendable one including one more adjacent vertex, that is to say, a clique which does not exist exclusively within the vertex set of a larger clique.

Now let $F = (V, C, I)$. The formal context associating the set C of maximal cliques with all actors V who belong to these maximal cliques. I is the binary relationship that binds V and C together. As a consequence, if an actor x_i belongs to a maximal clique C_j, $I(x_i, C_j) = 1$ else $I(x_i, C_j) = 0$.

Principle of Community Detection in Social Networks According to FCA. FCA methods exploit the Galois lattice based on the context $F = (V, C, I)$ with a view to identifying communities in social networks. In the following paragraph, we illustrate the definitions with an example of a social network represented by a graph $G(V, E)$ with 15 actors $V = \{1, 2, 3, ..., 15\}$ and 32 edges from this graph we enumerated 4 maximal cliques, $C = \{a, b, c, d\}$. Then we draw the formal context $\mathbb{K} := (V, C, I)$ (Figure 1). This example means to explain the formalization of the problem.

A formal context is a triple $\mathbb{K} := (G, M, I)$ where G is a set of objects, M a set of attributes and I a binary relation that links the sets G and M, $I \subseteq G \times M$. For $A \subseteq G$ and $B \subseteq M$, two subsets $A' \subseteq M$ and $B' \subseteq G$ are respectively defined as a set of attributes common to the objects in A and the set of objects that share all attributes in B. Formally, the derivation is denoted $'$ and is defined as follows: $A' := \{a \in M \mid o I a \ \forall o \in A\}$ (intension) and $B' := \{o \in G \mid o I a \ \forall a \in B\}$ (extension). This proposal defines a pair of correspondence $(',')$ among all the subsets of G and all subsets of M, which is a

Galois correspondence. The operators of the induced closure (in G and M) are denoted $''$. In our example, all the actors in V stand for the set of objects G and all maximal cliques in C represent the set of attributes.

Example: Let $A_1 = \{6,7\}, A_1 \subset G \Rightarrow (A_1)' = \{b,d\}$ and let $B_1 = \{c,d\}$, $M_1 \subset M \Rightarrow (B_1)' = \{4,7\}$

A *formal concept* (closed, rectangle) cf is a pair (A,B) with $A \subseteq G$, $B \subseteq M$, $A = B'$ and $B = A'$. The set A, which we will denote $\text{Ext}(cf)$, is called *extension* of cf while B is his *intention*, noted $\text{Int}(cf)$. A formal concept corresponds to a maximal rectangle in a formal context.

The set $\mathfrak{B}(\mathbb{K})$ of all concepts is formed as follows:

$$\mathfrak{B}(\mathbb{K}) = \{(A,B) \in (G', M') \smallsetminus A = B' \quad \text{and} \quad B = A'\}$$

Example: Let $A_1 = \{6,7\}, A_1 \subset G \Rightarrow (G_1)' = \{b,d\} \Rightarrow ((A_1)')' = (\{b,d\})'$ and $(\{b,d\})' = \{6,7,9\}$ then the set $A_1 \notin \mathfrak{B}(\mathbb{K})$ since the definition is not satisfied.

$\mathbb{K}$	a	b	c	d
1	1	0	0	0
2	1	1	0	0
3	1	0	0	1
4	1	0	1	1
5	0	0	1	0
6	0	1	0	1
7	0	1	1	1
8	0	0	0	1
9	0	1	0	1
10	0	0	1	0
11	0	0	0	0
12	0	0	0	0
13	0	0	0	0
14	0	0	0	0
15	0	0	0	0

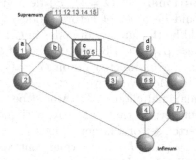

Fig. 1. Formal context $\mathbb{K} := (V, C, I)$

Fig. 2. $\mathfrak{B}(\mathbb{K})$ Galois lattice of the context $\mathbb{K}$

The set $\mathfrak{B}(\mathbb{K})$ of all concepts of the context $\mathbb{K}$ is partially ordered by: $(X_1, Y_1) \leq (X_2, Y_2) \Leftrightarrow X_1 \subseteq X_2$ constitutes a complete lattice, called the concept lattice of $\mathbb{K}$ and denoted $\mathfrak{B}(\mathbb{K})$. Figure 2 sheds the concept lattice $\mathfrak{B}(\mathbb{K})$ obtained from the formal context $\mathbb{K}$ of figure 1. The labelling of the lattice figure 2 is reduced at attributes so that the intent of a concept (node) **n** is given by the union of the attributes appearing in the node **n** as well as those appearing in concepts are smaller than **n**. For example, the node marked in red with the label $c - 10, 5$ represents the concept $(\{10, 4, 5, 7\}, \{c\})$ (Fig. 2).

The top of the lattice (supremum) $(11, 12, 13, 14, 15)$ shows that none of these actors belong to a maximal clique in [3] and [1] the authors did not consider them so they cannot be assigned to any group after the process of community identification, the bottom of the lattice (infimum) is empty which means that no actor belongs to all maximal cliques. Actors 2, 4 and 7 who are parents of the infimum will be considered as central actors in this network [3], [1].

The overlap of maximal cliques in a Galois lattice means that two maximal cliques in the lattice share at least one actor in common, i.e., if the maximal cliques meet only in the infimum they do not overlap). The overlap is reflexive, transitive and symmetric. Consequently, the overlapping is an equivalence relation. If we focus on the graphic, we can notice an overlapping between two nodes of the level k if they meet at level $k + 1$ (Fig. 2).

3.2   Freeman and Falzon Methods for Community Detection

The method Proposed by [1] is based on the same theoretical basis of social networks representation as proposed by [3], which combines the formal concept of maximal cliques with the formal concept of Galois lattice to determine communities.

In order to detect communities, [3] is mainly based on the notion of maximal cliques overlapping in Galois lattice. Considering the Galois lattice $\mathfrak{B}(\mathbb{K})$, [3] determines a set of maximal cliques which at least two paths from their current position in the lattice to the infimum have a different length (Figure 3 dashed paths.) He calls this type of maximal cliques "intermediate cliques". Then he removes edges starting from these nodes to get disjoint groups (Fig. 4).

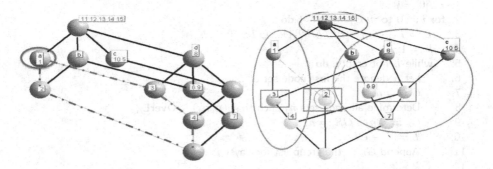

Fig. 3. Intermediate cliques   **Fig. 4.** Groups obtained by edges-removals

In large lattice nodes belong to several intermediate cliques and are eliminated with the process of detection proposed by *Freeman*[3]. Starting from this limit, [1] proposes a new approach which does not eliminate this type of node and justifies it through the observation of a property of the maximal cliques lattice. [1] suggests that the group structure evolves in line with the overlapping model in the lattice layers below the first one which represents the set of maximal cliques. For example in the lattice of figure 4, we notice that the nodes in layer 2 can be perfectly divided into three disjoint groups (represented by rectangles). Thus, we essentially get the same structure as obtained by removing the edges of intermediate cliques as proposed by [3]. In large networks, one would expect a more refined group structure whenever one descends into the lattice layers. The complexity of the Galois lattice construction is known as being exponential. However, [1] proposes a new construction method that calculates only in part the latter and reduces this complexity. In this lattice it also represents all actors belonging to maximal cliques. For its construction, we consider as an input all the maximal cliques noted L_1 then builds L_k sets for each level (k) which are obtained from the intersections of pairs of nodes at level $k - 1$. Then, creates lists $LS[k]$ of nodes sets for each level of layer (k). These lists contain several nodes in which actors do not appear in higher levels than k. Finally, each pair of adjacent layers in the lattice is compared so as to determine whether or not they have common nodes sets, the common nodes of the upper layer are removed. Algorithm 1 proposed by [1] of the groups structure calls repeatedly **OverL** (the overlapping node function). This function is an extension of the overlapping maximal cliques definition to nodes, its call is performed once for each layer in the network during the detection process.

Creating G_k groups is based on the generation of groups at level 1. These groups are formed by applying the overlap function on the nodes of level 1 and then the following groups (with $k > 1$) are formed by calling the overlap function on the nodes of level k and the nodes contained in the lists $LS[1]$ to $LS[k-1]$.

Algorithm 1. Group structure algorithm by *Falzon* method [1]

1: $LS[0] = \emptyset$
2: **for** $i := 0$ to $maxplayer - 1$ **do**
3:      $L := L_{i+1} \cup LS[1] \cup LS[2] \cup ... \cup LS[i]$;
4:      $k := 1$;
5:      **while** L not empty; **do**
6:          Let n denote the first node set of L;
7:          $GS := n$;
8:          Determine all node sets n_j such that $(n, n_j) \in$ **OverL** ;
9:          $G_k := \cup n_j$; $GS := n_j$;
10:          $L := L - GS$;
11:          Append G_k to the group list for Layer i;
12:          $k + +$
13:      $i := i + 1$;

---

[1] asserts that *Freeman* method for community detection is able to detect communities for small contexts whose associated maximal cliques lattice do not exceed three levels. Beyond three levels the number of actors belonging to intermediate cliques gets even more important. Therefore they are removed from the set of detected groups. [1] answers to the loss of this relevant information by providing a method which takes into account actors belonging to intermediate cliques. This approach also determines groups at each level of the lattice, but always from the same maximal cliques lattice which *Freeman* proposes to construct first. However in social networks graphs many actors do not belong to maximal cliques and these missed actors are not assigned to a community by the [1] method.

4   Our Approach

We can assert that several approaches based on FCA for community detection allow us to address the problem from a partial point of view. Indeed, those approaches consider only a certain part of the social network over which they apply their community identification approaches. Based on this observation, the idea of searching a new approach which takes into account all actors of the social network is relevant. Indeed, in the field of community detection, some actors can play important roles even though they do not maintain many links in the social network itself (i.e., they do not belong to maximal cliques). As instance, in a criminal organization, people who carry out attacks have few contacts with the others in the same group. They are just waiting for an order from an isolated person who assigns to them a mission. Thus, only one contact is necessary to the so-called criminal to commit his crime. Modeling this organization by a social network will represent this contact by a single link and thus the actor does not belong to any maximal clique. According to the approaches described in Section 3.2, these actors do not belong to any maximal clique and therefore they will not be taken into account during the detection process. On the contrary, the elements in charge of logistics, the preparation of this organization have many contacts with each other and belong to maximal cliques while these elements will not act concretely. Consequently, it seems necessary to underline the importance of considering these missed actors in the identification process, as it is possible not only to dismantle criminal networks but also to prevent criminals attacks from being successful. The action to lead is to connect these elements maintaining few contacts with the detected communities. In what follows, we introduce the approach we propose illustrated by an example all based on the formal definitions in Section 3.1.

4.1   Principle

The identification of communities is a difficult task as the number and the size (the number of nodes they contain) of communities are not basically known. Based on the definitions in Section 3.2, our approach can provide a solution that combines the approach [1] for community detection with a notion of graph theory which is the adaptation of the modularity function [9]. The process of identifying communities that we propose consists of two steps.

In the first one we determine the levels groups G_k by applying *Falzon* algorithm [1] on a part of the social network actors(the actors belonging to maximal cliques). These groups will constitute the core of the final communities. Then in the second step we make our context $K^* = (G(V, E), G_k, L_1)$ consisting in a social network graph $G(V, E)$, levels groups G_k and L_1 all nodes representing the maximal cliques at the first level of the lattice. Then, we determine the set of missed nodes N_a. They constitute the supremum of the lattice, i.e., actors who belong to the graph $G(V, E)$ and who do not belong to maximal cliques set L_1. Considering that an actor belongs to a group with which there is one link with one of the actors in the group at least. For each node n_i in the set N_a, we generate containers $B(n_i)$ from the set of edges E and the set of groups G_k. These ones are formed of pairs (n_j, G_k) representing the nodes and the groups with which a node n_i has links. Once these containers are made, we check if the node n_i has links only with nodes belonging to the same group or with nodes belonging to different groups. In the first case, the n_i node will be assigned to the group to which it has one or more links. In the second case we calculate the adapted modularity function in relation to different groups to which node n_i has links, i.e. by considering the node n_i belonging to each group G_k with whom he shares a link we calculate and compare the different values of *GroupNode modularity* denoted $Q(G_k, n_i)$ in order to get the maximal value, it is given by:

$$Q(G_k, n_i) = \sum_j (e_{jj}(G_k) - a_j^2(G_k^*)) \tag{1}$$

Where $e_{jj}(G_k)$ is the proportion of edges inside groups (the number of edges in group G_k, taking into account the node n_i divided by the total number of edges in the graph), $a_j(G_k^*) = \sum_j (e_{jj}(G_k^*))$ is the proportion of expected edges in the graph G by assigning node n_i in G_k^* group. The node n_i will be assigned to the group where $Q(G_k, n_i)$ is maximal.

4.2   F-measure for Evaluating Community Partitions

So as to evaluate the accuracy of partitioning we get through our approach. We propose to adapt the $F - measure$ function, an outcome of information retrieval[13], to the identification of community context. In our context, each actor of the social network is to be assigned (classified) into one, and only one, of l non-overlapping communities. The $F - measure$ is calculated as a function of Macro *Recall* and Macro *Precision* measures in the case of multi-class. We chose Macro-averaging because it treats all the class equally [11]. Communities are considered as classes and actors are elements to classify.

Let $P = (G_1, G_2, ..., G_k)$, $\bigcup_{(i=1..k)} G_i = V$ and $\forall i \neq j, G_i \cap G_j = \emptyset$ a partition of the set of actors V into k communities. Initially each node is assigned to a single community G_i, this allocation is determined through several types of studies (ethnographic, sociological, political, etc..) in the case of social networks dealing with real world, or obtained by the generation model for synthetic social networks. Through our method, we obtain a partitioning P' for each level of the lattice that we compare to the original partition P, $P' = (G_1', G_2', ..., G_l')$, $\bigcup_{(i=1..l)} G_i' = V$ and $\forall i \neq j, G_i' \cap G_j' = \emptyset$.

The $F - measure$ of partitioning is given by:

$$F(P, P') = \frac{2 \cdot (\mathbb{P}(P, P') \cdot \mathbb{R}(P, P')}{\mathbb{P}(P, P') + \mathbb{R}(P, P')} \tag{2}$$

The Precision $\mathbb{P}(P, P')$ and the Recall $\mathbb{R}(P, P')$ of partitioning are given by:

$$\mathbb{P}(P, P') = \frac{\sum_{i=1}^{l} \mathbb{P}\{G_i'\}}{l} \quad (3) \quad \mathbb{R}(P, P') = \frac{\sum_{i=1}^{l} \mathbb{R}\{G_i'\}}{l} \quad (4)$$

The Precision $\mathbb{P}\{G_i'\}$ is the number of actors who are correctly grouped together in a community G_i' compared with the number of individuals initially in the community G_i and the recall $\mathbb{R}\{G_i'\}$ is the number of actors correctly grouped together in a community G_i' compared to the total number of individuals in the community G_i'. These measures are given by:

$$\mathbb{P}\{G_i'\} = \frac{|G_i' \cap G_i|}{|G_i'|} \quad (5) \qquad \mathbb{R}\{G_i'\} = \frac{|G_i' \cap G_i|}{|G_i|} \quad (6)$$

4.3 Algorithm and Illustrative Example

The approach we propose is translated through the pseudo code of our algorithm 2. It takes as input the set of actors, the set of level groups and the set of maximal cliques. As an output we have refined groups that include all the actors of the social network. The first step of the algorithm (line 5) aims at determining ignored actors, (lines 6-9) making containers of nodes. These ones are obtained from ignored nodes and groups with which they have one link at least. Finally (lines 10-12) calculating the adapted modularity function $Q(G_k, n_i)$ for each container containing different groups, we compare these values and assign the ignored nodes to the group which maximize $Q(G_k, n_i)$ value.

Algorithm 2. Algorithm of refined group -Pseudo code

1: **Procedure** GROUPS $(G(V, E), G_k, L_1)$
2: **In:** $G(V, E)$ SN graph, G_k Partial group obtained from maximal cliques lattice and L_1 is the maximal cliques set
3: **Out:** G_K^* Set of refined Groups.
4: $G_K^* = G_k$
5: Construct the set $N_a = \{ n_i \mid n_i \in G(V, E) \wedge n_i \notin L_1 \}$
6: **for all** $n_i \in |N_a|$ **do**
7:     Construct containers $B(n_i)$ from G_k groups and n_i neighbors linked to G_k.
8:     **if** all nodes in $B(n_i)$ are in the same group G_i **then**
9:         $G_i^* = G_i \cup n_i$
10:     **else**
11:         Calculate $Q(G_k, n_i)$ for each element in $B(n_i)$
12:         Compare $Q(G_k, n_i)$ value for each group G_i then assign the node n_i to the group to which $Q(G_k, n_i)$ is maximal. $G_i^* = G_i \cup n_i$.
           In the case where the values are equals, we will assign n_i to the last group for which we calculate the modularity.
13: **Out:** G_i^*

To unroll our algorithm, we have focused on a simple social network known in the literature, *Zachary's* Karate Club of [15]. This social network is formed from 78 members of a karate club of the University of San Francisco in the United States. *Zachary* concentrated his work on the club members who have friendly relationships out of the club. Among all the members of the club (78 people), only 34 of them have maintain relations of friendship. The friendship network obtained from this club is shown in Figure 5. It consists of 34 nodes representing the club members and 78 links representing friendships between members. This social network contains two real communities, instructors and administrators of the club separated by a vertical line (Figure 5).

Enumerating maximal cliques is known to be a NP-hard problem, but we used [7] algorithm which runs in $O(M(n))$ time delay, where $M(n)$ denotes the time needed to multiply two $n \times n$ matrices, it can computes about 100.000 maximal cliques per second. We identified the set L_1 of 25 maximal cliques which represent the first layer of the lattice. This set contains only 32 of the 34 actors of the social network, the node 12 and node 10 do not belong to any maximal clique. The set containing these ignored nodes is denoted $N_a = \{10, 12\}$. We subsequently make the context $\mathbb{F}(Zachary) = (V, C, I)$ formed by the set of actors, the set of maximal cliques and the membership relationship.

From this context we get the lattice $\mathfrak{B}(\mathbb{F}(Zachary))$ of maximal cliques represented by the sets of portions which form its layers wherein the first one enumerates all maximal cliques and the following layers show the different intersections with the elimination of common nodes as proposed in [1].

With the application of Algorithm 1 on the lattice $\mathfrak{B}(\mathbb{F}(Zachary))$, we determine the partial groups at each level, partial groups at levels 3 and 4 in the Zachary Karate Club are shown in Figure 5.

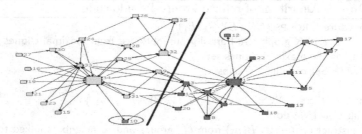

Fig. 5. Partial groups at levels 3 and 4 in the *Zachary* Karate Club

We finally got three distinct groups. Different groups obtained are colored in red, yellow and green. Actors 10 and 12 (in the blue circle) do not belong to maximal cliques so they will be assigned to one of three groups. The actor 12 has direct link with the group 3 in red will be assigned to this group. The actor 10 having two links, one with group 1 in yellow and the other related group 3 red then we calculate *GroupNode modularity* function $Q(G_k, n_i)$ to be able to assign it (Figure 5), for the yellow group *GroupNode modularity* is maximal so 12 will be part of this community.

5   Experiment and Discussion

In this section, we present an empirical evaluation of our method compared to that given by Falzon. We considered several social networks. First we tested on the examples quotes in [3] and [1] the results obtained were more refined, then on known social networks from the real world to show that even for midsize social networks many actors are eliminated and are not assigned to any community. In the example of Dolphin social network (Figure 6), we notice that 16 actors from the 62 actors in the social network, do not belong to maximal cliques which represent 26% of the social network. These 16 actors are not assigned to any community by [3] and [1] approaches but through our approach we identify which community they belong to.

SN	$|V|$	$|E|$	$|MC|$	$|Na|$	F_o	F_n
Zachary	34	78	25	2	0.77	0.80
Dolphin	62	159	46	16	0.83	0.96
Football	115	613	185	10	0.80	0.88
Polbooks	105	441	181	22	0.66	0.74
L1000n16c	1000	15168	1902	130	0.78	0.96
L5000n32c	5000	75946	9723	679	0.74	0.91

Fig. 6. F-measure for our approach F_n and Falzon approach F_o

Figure 6 summarizes the tests we carried out we took four social networks from the real world and two synthetic social networks generated from [6] model. The largest one contains 5000 nodes with 75946 edges. Most of the detection methods available in the literature are tested on these social networks to corroborate their performance in terms of execution time and precision. The purpose of these tests is only to validate the approach and show its accuracy, testing scalability and performance (execution time) by varying different parameters will be the object of a future work. We notice that our approach is more accurate than the one proposed by *Falzon* through all the social networks tested, the precision of the two approaches are equal but we have higher recall with our approach. The result is that we have been able to detect the correct communities relative to those already exist in the social network, even in non-obvious cases. We got almost identical results to the ethnographic study conducted by *Wayne Zachary* on karate club social network (Figure 5). The first group is the same as the one found in the *Zachary's* study by against we obtained in addition another group. This one gathered actors 6, 7 and 17, that could be intuitively merged with the group in red. We then obtain the same communities as *Zachary* got in his study, or consider it as a new group showing strong internal links.

The communities that we obtained by our method are closer to reality than other methods, several studies assign the actors in this example to communities whose number varies between 4 and 7 communities. Sometimes an approach is able to detect a different number of communities by varying parameters.

6   Conclusion

In this paper, we proposed a new method for community detection in social networks which improves *Falzon* method. Our approach consists in considering all the actors in a social network during the identification community process all based on FCA theoretical foundations. Through the process and the algorithm developed above, we have shown how to identify communities with accuracy. This approach has been evaluated by different tests led on real and synthetic social networks. They have shown that the accuracy of detection is good as well for social networks from the real world as for synthetic ones. This work broadens many opportunities for research. We plan in the short term to conduct an experimental study on synthetic social networks considering different test parameters: the number of nodes, the degrees of nodes and their density.

References

1. Falzon, L.: Determining groups from the clique structure in large social networks. Social Networks 22(2), 159–172 (2000)
2. Fortunato, S., Barthélemy, M.: Resolution limit in community detection. Proceedings of the National Academy of Sciences 104(1), 36 (2007)
3. Freeman, L.C.: Cliques, galois lattices, and the structure of human social groups. Social Networks 18(3), 173–187 (1996)
4. Ganter, B., Wille, R.: Formal Concept Analysis: Mathematical Foundations. Springer-Verlag New York, Inc. (1999); Translator-C. Franzke
5. Girvan, M., Newman, M.E.J.: Community structure in social and biological networks. PNAS 99(12), 7821–7826 (2002)
6. Lancichinetti, A., Fortunato, S.: Community detection algorithms: A comparative analysis. Physical Review E 80(5), 056117 (2009)
7. Makino, K., Uno, T.: New algorithms for enumerating all maximal cliques. In: Hagerup, T., Katajainen, J. (eds.) SWAT 2004. LNCS, vol. 3111, pp. 260–272. Springer, Heidelberg (2004)
8. Missaoui, R.: Analyse de réseaux sociaux par l'analyse formelle de concepts. In: EGC, pp. 3–4 (2013)
9. Newman, M.E.: Detecting community structure in networks. The European Physical Journal B-Condensed Matter and Complex Systems 38(2), 321–330 (2004)
10. Poelmans, J., Ignatov, D.I., Kuznetsov, S.O., Dedene, G.: Formal concept analysis in knowledge processing: A survey on applications. Expert Systems with Applications (2013)
11. Sokolova, M., Lapalme, G.: A systematic analysis of performance measures for classification tasks. Information Processing & Management 45(4), 427–437 (2009)
12. Tummarello, G., Morbidoni, C.: The dbin platform: A complete environment for semanticweb communities. Web Semantics: Science, Services and Agents on the World Wide Web 6(4) (2008)
13. Van Rijsbergen, C.: Information retrieval. dept. of computer science, university of glasgow (1979), citeseer.ist.psu.edu/vanrijsbergen79information.html
14. Wille, R.: Restructuring lattice theory: An approach based on hierarchies of concepts. In: Ferré, S., Rudolph, S. (eds.) ICFCA 2009. LNCS, vol. 5548, pp. 314–339. Springer, Heidelberg (2009)
15. Zachary, W.W.: An information flow model for conflict and fission in small groups. Journal of Anthropological Research, 452–473 (1977)

On Interpreting Three-Way Decisions through Two-Way Decisions

Xiaofei Deng, Yiyu Yao, and JingTao Yao

Department of Computer Science, University of Regina
Regina, Saskatchewan, Canada S4S 0A2
{deng200x,yyao,jtyao}@cs.uregina.ca

Abstract. Three-way decisions for classification consist of the actions of acceptance, rejection and non-commitment (i.e., neither acceptance nor rejection) in deciding whether an object is in a class. A difficulty with three-way decisions is that one must consider costs of three actions simultaneously. On the other hand, for two-way decisions, one simply considers costs of two actions. The main objective of this paper is to take advantage of the simplicity of two-way decisions by interpreting three-way decisions as a combination of a pair of two-way decision models. One consists of acceptance and non-acceptance and the other consists of rejection and non-rejection. The non-commitment of the three-way decision model is viewed as non-acceptance and non-rejection of the pair of two-way decision models.

1 Introduction

In concept learning, concept formation and classification, one typically uses a strategy of binary, two-way decisions. That is, an object is either accepted or rejected as being an instance of a concept or a class. One makes a decision with minimum errors or costs. An advantage of a two-way decision strategy is its simplicity. One only needs to consider two actions. On the other hand, when one is forced to make either an acceptance or a rejection decision, it is impossible to arrive at both a low level of incorrect acceptance error and a low level of incorrect rejection error at the same time [1]. To avoid this difficulty, three-way decisions are widely used in many fields and disciplines [2–11] as an alternative effective strategy. In contrast to two-way decisions, a third option called non-commitment, namely, neither acceptance nor rejection, is added. Three-way decisions enable us to reduce incorrect acceptance error and incorrect rejection error simultaneously at the expense of non-commitment for some objects.

In earlier formulations of three-way decisions with rough sets [12], three actions are considered and compared simultaneously. The consideration of six types of costs of three actions makes three-way decisions more complicated than two-way decisions, as the latter only need to consider four types of costs. A recent study by Yao [13] on three-way decisions based on two evaluations suggests that three-way decisions can in fact be interpreted through two-way decisions. In this

T. Andreasen et al. (Eds.): ISMIS 2014, LNAI 8502, pp. 73–82, 2014.
© Springer International Publishing Switzerland 2014

paper, we revisit the two-evaluation-based three-way decision model and explicitly show that three-way decisions can be formulated as a combination of a pair of two-way decision models, namely, an acceptance model and a rejection model. To achieve this goal, we slightly modify the interpretation of a two-way decision model. In the standard interpretation of a two-way decision model, acceptance and rejection are dual actions. That is, failing to accept is the same as rejecting, and vice versa. However, in our acceptance model, we have acceptance and non-acceptance decisions. Failing to accept is non-acceptance, rather than rejecting. Similarly, in the rejection model, we have rejection and non-rejection. Failing to rejection is non-rejection, rather than accepting. By combining the two models together, we have three actions of acceptance, rejection, and non-commitment, where non-commitment is interpreted as non-acceptance and non-rejection.

The rest of the paper are organized as follows. Section 2 gives a brief review of two-evaluation-based three-way decisions proposed by Yao [13]. Section 3 proposes an interpretation of three-way decisions based on a pair of two-way decisions. Section 4 calculates the acceptance and rejection thresholds using a pair of two-way decision models.

2   An Overview of Three-Way Decisions

The purpose of three-way decisions is to divide objects in an universe U, according to a criteria, into three pair-wise disjoint regions, namely, positive, negative and boundary regions, denoted by POS, NEG and BND. A criteria can be considered to be a set of conditions. Due to a lack of information, the true state of satisfiability of the criteria usually is unknown. The degree to which the criteria is satisfied must be estimated according to the partial information. We use a pair of evaluation functions to estimate whether the criteria is satisfied or not.

Definition 1. *Suppose U is a finite non-empty set of objects. Based on a pair of posets $(L_a, \preceq_a)$ and $(L_r, \preceq_r)$, where $\preceq_a$ and $\preceq_r$ are two partial orderings, an acceptance evaluation and a rejection evaluation can be defined as a pair of mappings: $v_a : U \longrightarrow L_a$ and $v_r : U \longrightarrow L_r$. For each object $x \in U$, $v_a(x) \in L_a$ and $v_r(x) \in L_r$ represent the acceptance-evaluation value and rejection-evaluation value, respectively.*

For two objects $x, y \in U$, $v_a(x) \preceq_a v_a(y)$ indicates that x is not more acceptable than y. Similarly, $v_r(x) \preceq_a v_r(y)$ indicates that x is not more rejectable than y. Notions, such as benefits, risks, costs or confidence, can be used to interpret partial orderings $\preceq_a$ and $\preceq_r$.

For an object, a decision is made if its evaluation value reaches a certain level. We introduce the notion of *designated values* as sets of values for acceptance and rejection.

Definition 2. *Suppose L_a^+ is the set of designated values for acceptance, satisfying condition $\emptyset \neq L_a^+ \subseteq L_a$, and L_r^- is the set of designated values for rejection,* satisfying condition $\emptyset \neq L_r^- \subseteq L_r$. The positive, negative and boundary regions of three-way decisions are defined by:

$$\text{POS}_{(L_a^+, L_r^-)}(v_a, v_r) = \{x \in U \mid v_a(x) \in L_a^+ \land v_r(x) \notin L_r^-\},$$

$$\text{NEG}_{(L_a^+, L_r^-)}(v_a, v_r) = \{x \in U \mid v_a(x) \notin L_a^+ \land v_r(x) \in L_r^-\},$$

$$\text{BND}_{(L_a^+, L_r^-)}(v_a, v_r) = (\text{POS}_{(L_a^+, L_r^-)}(v_a, v_r) \cup \text{NEG}_{(L_a^+, L_r^-)}(v_a, v_r))^c$$

$$= \{x \in U \mid (v_a(x) \notin L_a^+ \land v_r(x) \notin L_r^-) \lor$$

$$(v_a(x) \in L_a^+ \land v_r(x) \in L_r^-)\}, \tag{1}$$

where $(\cdot)^c$ denotes the complement of a set.

The three decision regions are pair-wise disjoint and some of them may be empty. They do not necessarily form a partition of the universe.

3   Three-Way Decisions as a Combination of a Pair of Two-Way Decisions

Two models of three-way decisions are interpreted based on two-way decisions. A general model uses two evaluations and a specific model uses a single evaluation.

3.1   Two-Evaluation-Based Model

Let $v_a : U \longrightarrow L_a$ denote the acceptance-evaluation function, $(A, \bar{A})$ denote the two-way decisions for acceptance, and $L_a^+ \subseteq L_a$ denote the designated values for acceptance. The two decision regions of an acceptance model, i.e., the $(A, \bar{A})$-model, are defined by:

$$\text{POS}_{L_a^+}(v_a) = \{x \in U \mid v_a(x) \in L_a^+\},$$

$$\text{NPOS}_{L_a^+}(v_a) = (\text{POS}_{L_a^+}(v_a))^c = \{x \in U \mid v_a(x) \notin L_a^+\}, \tag{2}$$

where $\text{POS}_{L_a^+}(v_a)$ is called the acceptance region and $\text{NPOS}_{L_a^+}(v_a)$ is called the non-acceptance region. For an object $x \in U$, we can make two-way decisions:

(A) If $v_a(x) \in L_a^+$, then take an acceptance action, i.e., $x \in \text{POS}_{L_a^+}(v_a)$;

($\bar{A}$) If $v_a(x) \notin L_a^+$, then take a non-acceptance action, i.e., $x \in \text{NPOS}_{L_a^+}(v_a)$.

The acceptance rule (A) classifies objects into an acceptance region. The non-acceptance rule ($\bar{A}$) classifies objects into the non-acceptance region. The two regions are disjoint and their union is the universe U.

Similarly, let v_r denote the rejection-evaluation function, $(R, \bar{R})$ denote the two-way decisions for rejection, and $L_r^- \subseteq L_r$ denote the designated values for rejection. The two-way decision regions of a rejection model, i.e., $(R, \bar{R})$-model, are defined by:

$$\text{NEG}_{L_r^-}(v_r) = \{x \in U \mid v_r(x) \in L_r^-\},$$

$$\text{NNEG}_{L_r^-}(v_r) = (\text{NEG}_{L_r^-}(v_r))^c = \{x \in U \mid v_r(x) \notin L_r^-\}, \tag{3}$$

where $\mathrm{NEG}_{L_r^-}(v_r)$ is called the rejection region and $\mathrm{NNEG}_{L_r^-}(v_r)$ is called the non-rejection region. For an object $x \in U$, we can make two-way decisions:

(R) If $v_r(x) \in L_r^-$, then take a rejection action, *i.e.*, $x \in \mathrm{NEG}_{L_r^-}(v_r)$;

($\bar{\mathrm{R}}$) If $v_r(x) \notin L_r^-$, then take a non-rejection action, *i.e.*, $x \in \mathrm{NNEG}_{L_r^-}(v_r)$.

The rejection rule (R) classifies objects into the rejection region. The non-rejection rule ($\bar{\mathrm{R}}$) classifies objects into the non-rejection region. The two regions are disjoint and their union is the universe U.

By combining decision rules of the pair of two-way decision models for acceptance and for rejection, we have three-way decision rules: for each object $x \in U$,

(P) If $v_a(x) \in L_a^+ \wedge v_r(x) \notin L_r^-$, then take an acceptance action,

   i.e., $x \in \mathrm{POS}_{(L_a^+, L_r^-)}(v_a, v_r)$;

(R) If $v_r(x) \in L_r^- \wedge v_a(x) \notin L_a^+$, then take a rejection action,

   i.e., $x \in \mathrm{NEG}_{(L_a^+, L_r^-)}(v_a, v_r)$;

(B) If $(v_a(x) \in L_a^+ \wedge v_r(x) \in L_r^-)$ or $(v_a(x) \notin L_a^+ \wedge v_r(x) \notin L_r^-)$,

   then take a non-commitment action, *i.e.*, $x \in \mathrm{BND}_{(L_a^+, L_r^-)}(v_a, v_r)$.

They in fact define the three regions, $\mathrm{POS}_{(L_a^+, L_r^-)}(v_a, v_r)$, $\mathrm{NEG}_{(L_a^+, L_r^-)}(v_a, v_r)$ and $\mathrm{BND}_{(L_a^+, L_r^-)}(v_a, v_r)$ of three-way decisions given in Equation (1). Table 1 shows the connection between two-way decisions and three-way decisions. An acceptance decision is interpreted as a combination of acceptance and non-rejection, i.e.,

$$\mathrm{POS}_{(L_a^+, L_r^-)}(v_a, v_r) = \mathrm{POS}_{L_a^+}(v_a) \cap \mathrm{NNEG}_{L_r^-}(v_r). \tag{4}$$

Combining rejection and non-acceptance decisions forms the rejection decision, i.e.,

$$\mathrm{NEG}_{(L_a^+, L_r^-)}(v_a, v_r) = \mathrm{NEG}_{L_r^-}(v_r) \cap \mathrm{NPOS}_{L_a^+}(v_a). \tag{5}$$

Making both an acceptance and a rejection decision is a contradiction of a pair of two-way decisions, which results in a non-commitment decision. Neither making an acceptance nor making a rejection decision leads to a different type of non-commitment decision. The union of the two sets forms the boundary region of three-way decisions:

$$\mathrm{BND}_{(L_a^+, L_r^-)}(v_a, v_r) = (\mathrm{POS}_{L_a^+}(v_a) \cap \mathrm{NEG}_{L_r^-}(v_r))$$
$$\cup (\mathrm{NPOS}_{L_a^+}(v_a) \cap \mathrm{NNEG}_{L_r^-}(v_r)). \tag{6}$$

In many applications, we tend to avoid making a contradiction during decision-makings by imposing the following condition:

$$\mathrm{POS}_{(L_a^+)}(v_a) \cap \mathrm{NEG}_{(L_r^-)}(v_r) = \emptyset. \tag{7}$$

As a result, we have $\mathrm{BND}_{(L_a^+, L_r^-)}(v_a, v_r) = \mathrm{NPOS}_{L_a^+}(v_a) \cap \mathrm{NNEG}_{L_r^-}(v_r)$, and the non-commitment decision are interpreted as neither acceptance nor rejection.

Table 1. Interpretation of three-way decisions based on two-way decisions

(R, R̄)-model — (A, Ā)-model	rejection	non-rejection
acceptance	non-commitment (contradiction)	acceptance
non-acceptance	rejection	non-commitment

3.2 Single-Evaluation-based Model

Suppose $v : U \longrightarrow L$ is an evaluation function defined on a totally ordered set $(L, \preceq)$, where $\preceq$ is a total ordering, and let $L^+ \subseteq L$ and $L^- \subseteq L$ denote the sets of designated values for acceptance and rejection, respectively. When we use v in the $(A, \bar{A})$-model, we define the two regions of two-way decisions for acceptance by:

$$\mathrm{POS}_{L^+}(v) = \{x \in U \mid v(x) \in L^+\},$$
$$\mathrm{NPOS}_{L^+}(v) = (\mathrm{POS}_{L^+}(v))^c = \{x \in U \mid v(x) \notin L^+\}, \tag{8}$$

where $\mathrm{POS}_{L^+}(v)$ is the acceptance region and $\mathrm{NPOS}_{L^+}(v)$ is the non-acceptance region. Similarly, when we use v in the $(R, \bar{R})$-model, we define the two regions of two-way decisions for rejection by:

$$\mathrm{NEG}_{L^-}(v) = \{x \in U \mid v(x) \in L^-\},$$
$$\mathrm{NNEG}_{L^-}(v) = (\mathrm{NEG}_{L^-}(v))^c = \{x \in U \mid v(x) \notin L^-\}, \tag{9}$$

where $\mathrm{NEG}_{L^-}(v)$ is the rejection region, and $\mathrm{NNEG}_{L^-}(v)$ is the non-rejection region.

As shown by Figure 1, combining the pair of two-way decisions forms three-way decision regions.

In order to ensure that the three-way decision regions are pair-wise disjoint, we assume that the following property holds:

$$\text{(t)} \quad L^+ \cap L^- = \emptyset.$$

According to condition (t), we can define three-way decision regions using a single evaluation by:

$$\mathrm{POS}_{(L^+, L^-)}(v) = \{x \in U \mid v(x) \in L^+\},$$
$$\mathrm{NEG}_{(L^+, L^-)}(v) = \{x \in U \mid v(x) \in L^-\},$$
$$\mathrm{BND}_{(L^+, L^-)}(v) = \{x \in U \mid v(x) \notin L^+ \wedge v(x) \notin L^-\}. \tag{10}$$

That is, a single-evaluation-based three-way decision model can be interpreted based on a pair of two-way decision models.

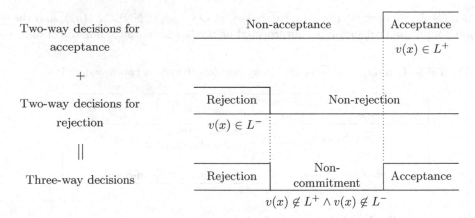

Two-way decisions for acceptance	Non-acceptance	Acceptance
		$v(x) \in L^+$

$+$

Two-way decisions for rejection	Rejection	Non-rejection
	$v(x) \in L^-$	

$\|$

Three-way decisions	Rejection	Non-commitment	Acceptance
		$v(x) \notin L^+ \wedge v(x) \notin L^-$	

Fig. 1. Interpretation of a single-evaluation-based three-way decisions

4   Probabilistic Three-Way Decisions

In order to apply three-way decisions, we need to investigate the following fundamental issues [13]:

 a) construction and interpretation of evaluation functions,
 b) construction and interpretation of the sets of designated values of acceptance and rejection, respectively, and
 c) analysis of the cost or risk of two-way decisions and three-way decisions, as well as their relationships.

We discuss these issues in the context of probabilistic three-way decisions, which is a generalization of decision-theoretic rough sets [12] and has received much attention recently [14–16].

4.1   Main Results of Probabilistic Three-Way Classifications

Suppose U is a universe of objects. An object in U is represented by a vector $\boldsymbol{x} = (x_1, x_2, \ldots, x_d)$ in a d-dimensional space, where x_i, $1 \leq i \leq d$, is the object's value on the i-th attribute. Using the terminology of rough set theory, a vector $\boldsymbol{x}$ is in fact the representation of the set of objects with the same description. In the rest of this paper, we simply consider U as a set of vectors. Let C denote a concept or a class of interest, C^c denote its complement, and $Pr(C|\boldsymbol{x})$ denote the conditional probability that an object is in C given that the object is described by $\boldsymbol{x}$. The main task of probabilistic classifications is to decide, according to $Pr(C|\boldsymbol{x})$, whether an object with description $\boldsymbol{x}$ is an instance of C.

The conditional probability $Pr(C|\boldsymbol{x})$ is considered to be an evaluation function for building a three-way decision model. The unit interval $[0, 1]$ is the set of evaluation status values, that is, $L = [0, 1]$. Given a pair of thresholds (α, β)

with $0 \leq \beta < \alpha \leq 1$, we construct the sets of designated values for acceptance and rejection as follows:

$$L^+ = \{a \in [0,1] \mid a \succeq \alpha\}, \qquad L^- = \{b \in [0,1] \mid b \preceq \beta\}. \tag{11}$$

By inserting L^+ and L^- into Equation (10), we immediately obtain the main results of probabilistic three-way classifications [17]:

$$\mathrm{POS}_{(\alpha,\cdot)}(C) = \{x \in U \mid Pr(C|x) \geq \alpha\},$$
$$\mathrm{NEG}_{(\cdot,\beta)}(C) = \{x \in U \mid Pr(C|x) \leq \beta\},$$
$$\mathrm{BND}_{(\alpha,\beta)}(C) = \{x \in U \mid \beta < Pr(C|x) < \alpha\}. \tag{12}$$

The threshold α is called the acceptance threshold and β is called the rejection threshold. A crucial issue is how to interpret and determine the pair of thresholds [1]. We present a solution by using a pair of two-way decision models based on Bayesian decision theory [18].

4.2 Calculating the Acceptance Threshold

To build a two-way decision model, i.e., $(A, \bar{A})$-model, we need to introduce an acceptance threshold $0 < \alpha \leq 1$. The two regions of probabilistic two-way decisions for acceptance are given by:

$$\mathrm{POS}_\alpha(C) = \{x \in U \mid Pr(C|x) \geq \alpha\},$$
$$\mathrm{NPOS}_\alpha(C) = \{x \in U \mid Pr(C|x) < \alpha\}. \tag{13}$$

Based on Bayesian decision theory [18], we calculate the optimal acceptance threshold α.

Let $\Omega = \{C, C^c\}$ denote the set of states and $Actions_A = \{a_A, a_{\bar{A}}\}$ denote the set of two decision actions, namely, an acceptance action a_A and a non-acceptance action $a_{\bar{A}}$. We assume each action is associated with certain cost, loss or risk. Such a loss function is given by a 2×2 matrix:

Action	$x \in C$ (Positive instance)	$x \in C^c$ (Negative instance)
a_A	$\lambda_{AP} = \lambda(a_A|C)$	$\lambda_{AN} = \lambda(a_A|C^c)$
$a_{\bar{A}}$	$\lambda_{\bar{A}P} = \lambda(a_{\bar{A}}|C)$	$\lambda_{\bar{A}N} = \lambda(a_{\bar{A}}|C^c)$

Each cell represents the loss or cost of taking an action $a \in Actions$ when the state of an object is $\omega \in \Omega$, namely, $\lambda(a|\omega)$. For example, $\lambda_{\bar{A}P} = \lambda(a_{\bar{A}}|C)$ represents the risk of taking action $a_{\bar{A}}$ given that $x \in C$. The conditional risks of taking actions a_A and $a_{\bar{A}}$ for $x \in U$ are given by:

$$R(a_A|x) = \lambda(a_A|C)Pr(C|x) + \lambda(a_A|C^c)Pr(C^c|x),$$
$$R(a_{\bar{A}}|x) = \lambda(a_{\bar{A}}|C)Pr(C|x) + \lambda(a_{\bar{A}}|C^c)Pr(C^c|x). \tag{14}$$

The overall risk of the acceptance and non-acceptance decisions for all objects can be expressed by:

$$R(\alpha) = R_{\mathrm{POS}}(\alpha) + R_{\mathrm{NPOS}}(\alpha),$$
$$= \sum_{x \in \mathrm{POS}_\alpha(C)} R(a_A|x) + \sum_{x \in \mathrm{NPOS}_\alpha(C)} R(a_{\bar{A}}|x). \tag{15}$$

By assuming the following two conditions,

$$(c0) \qquad \lambda_{AP} < \lambda_{\bar{A}P}, \quad \lambda_{\bar{A}N} < \lambda_{AN}$$

we obtain the optimal acceptance threshold α that minimizing R:

$$\alpha = \frac{(\lambda_{AN} - \lambda_{\bar{A}N})}{(\lambda_{AN} - \lambda_{\bar{A}N}) + (\lambda_{\bar{A}P} - \lambda_{AP})}$$

$$= \left(1 + \frac{\lambda_{\bar{A}P} - \lambda_{AP}}{\lambda_{AN} - \lambda_{\bar{A}N}}\right)^{-1} \qquad (16)$$

It can be verified that $0 < \alpha \leq 1$. The detailed procedure for deriving α can be found in [12].

4.3   Calculating the Rejection Threshold

To build a two-way decision model for rejection, i.e., $(R, \bar{R})$-model, we need a rejection threshold $0 \leq \beta < 1$. The two regions of probabilistic two-way decisions for rejection are given by:

$$\text{NEG}_\beta(C) = \{\boldsymbol{x} \in U \mid Pr(C|\boldsymbol{x}) \leq \beta\},$$
$$\text{NNEG}_\beta(C) = \{\boldsymbol{x} \in U \mid Pr(C|\boldsymbol{x}) > \beta\}. \qquad (17)$$

The rejection threshold β can be computed by using Bayesian decision theory.

Let $\Omega = \{C, C^c\}$ denote the set of states, and $Actions_R = \{a_R, a_{\bar{R}}\}$ denote the set of two decision actions, namely, a rejection action a_R and a non-rejection action $a_{\bar{R}}$. A loss function is given by a 2×2 matrix:

Action	$x \in C$ (Positive instance)	$x \in C^c$ (Negative instance)
a_R	$\lambda_{RP} = \lambda(a_R|C)$	$\lambda_{RN} = \lambda(a_R|C^c)$
$a_{\bar{R}}$	$\lambda_{\bar{R}P} = \lambda(a_{\bar{R}}|C)$	$\lambda_{\bar{R}N} = \lambda(a_{\bar{R}}|C^c)$

The overall risk of the rejection and non-rejection decisions for all objects is expressed by:

$$R'(\beta) = \sum_{\boldsymbol{x} \in \text{NEG}_\beta(C)} R(a_R|\boldsymbol{x}) + \sum_{\boldsymbol{x} \in \text{NNEG}_\beta(C)} R(a_{\bar{R}}|\boldsymbol{x}), \qquad (18)$$

where $R(a_R|\boldsymbol{x})$ and $R(a_{\bar{R}}|\boldsymbol{x})$ are conditional risks of taking actions a_R and $a_{\bar{R}}$ for $\boldsymbol{x} \in U$, respectively. By assuming the following conditions,

$$(c0') \qquad \lambda_{\bar{R}P} < \lambda_{RP}, \quad \lambda_{RN} < \lambda_{\bar{R}N}.$$

we obtain the optimal rejection threshold β that minimizing R':

$$\beta = \frac{(\lambda_{\bar{R}N} - \lambda_{RN})}{(\lambda_{\bar{R}N} - \lambda_{RN}) + (\lambda_{RP} - \lambda_{\bar{R}P})}$$

$$= \left(1 + \frac{(\lambda_{RP} - \lambda_{\bar{R}P})}{(\lambda_{\bar{R}N} - \lambda_{RN})}\right)^{-1} \qquad (19)$$

It can be verified that $0 \leq \beta < 1$.

4.4   Combing Results of a Pair of Two-Way Decision Models

To build a three-way classification model, we combine a pair of two-way classification models introduced in the last two subsections. According to condition (t), we require $0 \leq \beta < \alpha \leq 1$, that is,

$$(c1) \quad \alpha > \beta \iff (1 + \frac{(\lambda_{\bar{A}P} - \lambda_{AP})}{(\lambda_{AN} - \lambda_{\bar{A}N})})^{-1} > (1 + \frac{(\lambda_{RP} - \lambda_{\bar{R}P})}{(\lambda_{\bar{R}N} - \lambda_{RN})})^{-1}.$$

Using this condition, we immediate obtain three-way probabilistic regions in Equation (12).

To establish a connection to the existing formulation of probabilistic three-way classification [12], we further assume:

$$(c2) \quad \lambda_{NP} = \lambda_{\bar{A}P} = \lambda_{\bar{R}P}, \quad \lambda_{NN} = \lambda_{\bar{A}N} = \lambda_{\bar{R}N},$$

where λ_{NP} and λ_{NN} denote the costs of decisions of non-commitment. That is, the cost of non-acceptance is the same as the cost of non-rejection, and both of them are the same as the cost of non-commitment. In this case, for three-way decisions, we have a set of three actions. Suppose $Actions = \{a_A, a_R, a_N\}$ is the set of actions for acceptance, rejection and non-commitment. The loss function is given by a 3×2 matrix:

Action	$x \in C$ (Positive instance)	$x \in C^c$ (Negative instance)
a_A	$\lambda_{AP} = \lambda(a_A\|C)$	$\lambda_{AN} = \lambda(a_A\|C^c)$
a_R	$\lambda_{RP} = \lambda(a_R\|C)$	$\lambda_{RN} = \lambda(a_R\|C^c)$
a_N	$\lambda_{NP} = \lambda(a_N\|C)$	$\lambda_{NN} = \lambda(a_N\|C^c)$

It can be proved that the pair of thresholds (α, β) computed from Equations (16) and (19) in fact minimizes the overall risk of three-way classifications:

$$R(\alpha, \beta) = R_{\text{POS}}(\alpha) + R_{\text{NEG}}(\beta) + R_{\text{BND}}(\alpha, \beta), \tag{20}$$

where the risks of the three regions are defined similarly as earlier. In this way, we obtain three-way decisions classifications by combining a pair of two-way decision models.

5   Conclusion

A pair of two-way decision models, i.e., one for acceptance and non-acceptance, the other for rejection and non-rejection, is used to derive and interpret three-way decisions. An advantage of this interpretation is the simplicity derived from a consideration of only two actions, rather than three actions, simultaneously. That is, we investigate separately two 2×2 loss matrices of two-way decisions and combine the results into a 3×2 loss matrix of three-way decisions. We compute an acceptance threshold and a rejection threshold independently in a pair of two-way decision models. By combine the two thresholds together, we obtain a three-way classification model. Our analysis clearly shows the relative independence and the connection of the two thresholds in probabilistic three-way classification.

Acknowledgements. This work is partially supported by Discovery Grants from NSERC Canada.

References

1. Deng, X.F., Yao, Y.Y.: A multifaceted analysis of probabilistic three-way decisions. Fundamenta Informaticae (to appear, 2014)
2. Azam, N., Yao, J.T.: Analyzing uncertainties of probabilistic rough set regions with game-theoretic rough sets. International Journal of Approximate Reasoning 55, 142–155 (2014)
3. Grzymala-Busse, J.W., Clark, P.G., Kuehnhausen, M.: Generalized probabilistic approximations of incomplete data. International Journal of Approximate Reasoning 55, 180–196 (2014)
4. Grzymala-Busse, J.W., Yao, Y.Y.: Probabilistic rule induction with the lers data mining system. International Journal of Intelligent Systems 26, 518–539 (2011)
5. Jia, X.Y., Tang, Z.M., Liao, W.H., Shang, L.: On an optimization representation of decision-theoretic rough set model. International Journal of Approximate Reasoning 55, 156–166 (2014)
6. Li, H.X., Zhou, X.Z., Zhao, J.B., Liu, D.: Non-monotonic attribute reduction in decision-theoretic rough sets. Fundamenta Informaticae 126, 415–432 (2013)
7. Liang, D.C., Liu, D., Pedrycz, W., Hu, P.: Triangular fuzzy decision-theoretic rough sets. International Journal of Approximate Reasoning 54, 1087–1106 (2013)
8. Liu, D., Li, T.R., Liang, D.C.: Incorporating logistic regression to decision-theoretic rough sets for classifications. International Journal of Approximate Reasoning 55, 197–210 (2014)
9. Min, F., Hu, Q.H., Zhu, W.: Feature selection with test cost constraint. International Journal of Approximate Reasoning 55, 167–179 (2014)
10. Yu, H., Liu, Z.G., Wang, G.Y.: An automatic method to determine the number of clusters using decision-theoretic rough set. International Journal of Approximate Reasoning 55, 101–115 (2014)
11. Zhou, B.: Multi-class decision-theoretic rough sets. International Journal of Approximate Reasoning 55, 211–224 (2014)
12. Yao, Y.Y.: The superiority of three-way decisions in probabilistic rough set models. Information Sciences 181, 1080–1096 (2011)
13. Yao, Y.: An outline of a theory of three-way decisions. In: Yao, J., Yang, Y., Słowiński, R., Greco, S., Li, H., Mitra, S., Polkowski, L. (eds.) RSCTC 2012. LNCS, vol. 7413, pp. 1–17. Springer, Heidelberg (2012)
14. Liang, D.C., Liu, D.: Systematic studies on three-way decisions with interval-valued decision-theoretic rough sets. Information Sciences (2014), doi:10.1016/j.ins.2014.02.054
15. Ma, X., Wang, G.Y., Yu, H., Li, T.R.: Decision region distribution preservation reduction in decision-theoretic rough set model. Information Sciences (2014), doi:10.1016/j.ins.2014.03.078
16. Zhang, X.Y., Miao, D.Q.: Reduction target structure-based hierarchical attribute reduction for two-category decision-theoretic rough sets. Information Sciences (2014), doi:10.1016/j.ins.2014.02.160
17. Yao, Y.Y.: Three-way decisions with probabilistic rough sets. Information Sciences 180, 341–353 (2010)
18. Duda, R., Hart, P.: Pattern Classification and Scene Analysis. Wiley, New York (1973)

FHM: Faster High-Utility Itemset Mining Using Estimated Utility Co-occurrence Pruning

Philippe Fournier-Viger[1], Cheng-Wei Wu[2],
Souleymane Zida[1], and Vincent S. Tseng[2]

[1] Dept. of Computer Science, University of Moncton, Canada
[2] Dept. of Computer Science and Information Engineering,
National Cheng Kung University, Taiwan
{philippe.fournier-viger,esz2233}@umoncton.ca, silvemoonfox@gmail.com,
tseng@mail.ncku.edu.tw

Abstract. High utility itemset mining is a challenging task in frequent pattern mining, which has wide applications. The state-of-the-art algorithm is HUI-Miner. It adopts a vertical representation and performs a depth-first search to discover patterns and calculate their utility without performing costly database scans. Although, this approach is effective, mining high-utility itemsets remains computationally expensive because HUI-Miner has to perform a costly join operation for each pattern that is generated by its search procedure. In this paper, we address this issue by proposing a novel strategy based on the analysis of item co-occurrences to reduce the number of join operations that need to be performed. An extensive experimental study with four real-life datasets shows that the resulting algorithm named FHM (Fast High-Utility Miner) reduces the number of join operations by up to 95 % and is up to six times faster than the state-of-the-art algorithm HUI-Miner.

Keywords: Frequent pattern mining, high-utility itemset mining, co-occurrence pruning, transaction database.

1 Introduction

Frequent Itemset Mining (FIM) [1] is a popular data mining task that is essential to a wide range of applications. Given a transaction database, FIM consists of discovering frequent itemsets. i.e. groups of items (itemsets) appearing frequently in transactions [1]. However, an important limitation of FIM is that it assumes that each item cannot appear more than once in each transaction and that all items have the same importance (weight, unit profit or value). These assumptions often do not hold in real applications. For example, consider a database of customer transactions containing information about the quantities of items in each transaction and the unit profit of each item. FIM mining algorithms would discard this information and may thus discover many frequent itemsets generating a low profit and fail to discover less frequent itemsets that generate a high profit.

T. Andreasen et al. (Eds.): ISMIS 2014, LNAI 8502, pp. 83–92, 2014.
© Springer International Publishing Switzerland 2014

To address this issue, the problem of FIM has been redefined as *High-Utility Itemset Mining* (HUIM) to consider the case where items can appear more than once in each transaction and where each item has a weight (e.g. unit profit). The goal of HUIM is to discover itemsets having a high utility (e.g. generating a high profit). HUIM has a wide range of applications such as website click stream analysis, cross-marketing in retail stores and biomedical applications [2, 7, 10]. HUIM has also inspired several important data mining tasks such as high-utility sequential pattern mining [11] and high-utility stream mining [9].

The problem of HUIM is widely recognized as more difficult than the problem of FIM. In FIM, the *downward-closure property* states that the support of an itemset is anti-monotonic, that is the supersets of an infrequent itemset are infrequent and subsets of a frequent itemset are frequent. This property is very powerful to prune the search space. In HUIM, the utility of an itemset is neither monotonic or anti-monotonic, that is a high utility itemset may have a superset or subset with lower, equal or higher utility [1]. Thus techniques to prune the search space developed in FIM cannot be directly applied in HUIM.

Many studies have been carried to develop efficient HUIM algorithms [2, 6–8, 10]. A popular approach to HUIM is to discover high-utility itemsets in two phases using the Transaction-Weigthed-Downward closure model [2, 8, 10]. This approach is adopted by algorithms such as Two-Phase [8], IHUP [2] and UP-Growth [10]. These algorithms first generate a set of candidate high-utility itemsets by overestimating their utility in Phase 1. Then, in Phase 2, the algorithms perform a database scan to calculate the exact utility of candidates and filter low-utility itemsets. Recently, a more efficient approach was proposed in the HUI-Miner algorithm [7] to mine high-utility itemsets directly using a single phase. HUI-Miner was shown to outperform previous algorithms and is thus the current best algorithm for HUIM [7]. However, the task of high-utility itemset mining remains very costly in terms of execution time. Therefore, it remains an important challenge to design more efficient algorithms for this task.

In this paper, we address this challenge. Our proposal is based on the observation that although HUI-Miner performs a single phase and thus do not generate candidates as per the definition of the two-phase model, HUI-Miner explores the search space of itemsets by generating itemsets and a costly join operation has to be performed to evaluate the utility of each itemset. To reduce the number of joins that are performed, we propose a novel pruning strategy named EUCP (Estimated Utility Cooccurrence Pruning) that can prune itemsets without having to perform joins. This strategy is easy to implement and very effective. We name the proposed algorithm incorporating this strategy FHM (Fast High-utility Miner). We compare the performance of FHM and HUI-Miner on four real-life datasets. Results show that FHM performs up to 95 % less join operations than HUI-Miner and is up to six times faster than HUI-Miner

The rest of this paper is organized as follows. Section 2, 3, 4 and 5 respectively presents the problem definition and related work, the FHM algorithm, the experimental evaluation and the conclusion.

2   Problem Definition and Related Work

We first introduce important preliminary definitions.

Definition 1 (Transaction Database). Let I be a set of items (symbols). A *transaction database* is a set of transactions $D = \{T_1, T_2, ..., T_n\}$ such that for each transaction T_c, $T_c \in I$ and T_c has a unique identifier c called its Tid. Each item $i \in I$ is associated with a positive number $p(i)$, called its external utility (e.g. unit profit). For each transaction T_c such that $i \in T_c$, a positive number $q(i, T_c)$ is called the internal utility of i (e.g. purchase quantity).

Example 1. Consider the database of Fig. 1 (left), which will be used as our running example. This database contains five transactions $(T_1, T_2...T_5)$. Transaction T_2 indicates that items a, c, e and g appear in this transaction with an internal utility of respectively 2, 6, 2 and 5. Fig. 1 (right) indicates that the external utility of these items are respectively 5, 1, 3 and 1.

Definition 2 (Utility of an Item/Itemset in a Transaction). The utility of an item i in a transaction T_c is denoted as $u(i, T_c)$ and defined as $p(i) \times q(i, T_c)$. The utility of an itemset X (a group of items $X \subseteq I$) in a transaction T_c is denoted as $u(X, T_c)$ and defined as $u(X, T_c) = \sum_{i \in X} u(i, T_c)$.

Example 2. The utility of item a in T_2 is $u(a, T_2) = 5 \times 2 = 10$. The utility of the itemset $\{a, c\}$ in T_2 is $u(\{a, c\}, T_2) = u(a, T_2) + u(c, T_2) = 5 \times 2 + 1 \times 6 = 16$.

Definition 3 (Utility of an Itemset in a Database). The utility of an itemset X is denoted as $u(X)$ and defined as $u(X) = \sum_{T_c \in g(X)} u(X, T_c)$, where $g(X)$ is the set of transactions containing X.

Example 3. The utility of the itemset $\{a, c\}$ is $u(\{a, c\}) = u(a) + u(c) = u(a, T_1) + u(a, T_2) + u(a, T_3) + u(c, T_1) + u(c, T_2) + u(c, T_3) = 5 + 10 + 5 + 1 + 6 + 1 = 28$.

Definition 4 (Problem Definition). The *problem of high-utility itemset mining* is to discover all high-utility itemsets. An itemset X is a *high-utility itemset* if its utility $u(X)$ is no less than a user-specified minimum utility threshold *minutil* given by the user. Otherwise, X is a *low-utility itemset*.

Example 4. If *minutil* $= 30$, the high-utility itemsets in the database of our running example are $\{b, d\}$, $\{a, c, e\}$, $\{b, c, d\}$, $\{b, c, e\}$, $\{b, d, e\}$, $\{b, c, d, e\}$ with respectively a utility of 30, 31, 34, 31, 36, 40 and 30.

It can be demonstrated that the utility measure is not monotonic or anti-monotonic. In other words, an itemset may have a utility lower, equal or higher than the utility of its subsets. Therefore, the strategies that are used in FIM to prune the search space based on the anti-monotonicity of the support cannot be directly applied to discover high-utility itemsets. Several HUIM algorithms circumvent this problem by overestimating the utility of itemsets using a measure called the Transaction-Weighted Utilization (TWU) [2, 8, 10], which is anti-monotonic. The TWU measure is defined as follows.

Definition 5 (Transaction Utility). The *transaction utility* (TU) of a transaction T_c is the sum of the utility of the items from T_c in T_c. i.e. $TU(T_c) = \sum_{x \in T_c} u(x, T_c)$.

Example 5. Fig. 2 (left) shows the TU of transactions T_1, T_2, T_3, T_4, T_5 from our running example.

Definition 6 (Transaction Weighted Utilization). The *transaction-weighted utilization* (TWU) of an itemset X is defined as the sum of the transaction utility of transactions containing X, i.e. $TWU(X) = \sum_{T_c \in g(X)} TU(T_c)$.

Example 6. Fig. 2 (center) shows the TWU of single items a, b,c, d, e, f, g. Consider item a. $TWU(A) = TU(T_1) + TU(T_2) + TU(T_3) = 8 + 27 + 30 = 65$

The TWU measure has three important properties that are used to prune the search space.

Property 1 (Overestimation). The TWU of an itemset X is higher than or equal to its utility, i.e. $TWU(X) \geq u(X)$ [8].

Property 2 (Antimonotonicity). The TWU measure is anti-monotonic. Let X and Y be two itemsets. If $X \subset Y$, then $TWU(X) \geq TWU(Y)$ [8].

Property 3 (Pruning). Let X be an itemset. If $TWU(X) < minutil$, then the itemset X is a low-utility itemset as well as all its supersets. Proof. This directly follows from Property 1 and Property 2.

Algorithms such as Two-Phase [8], IHUP [2] and UPGrowth [10] utilizes the aforementionned properties to prune the search space. They operate in two phases. In Phase 1, they identify candidate high-utility itemsets by calculating their TWU. In Phase 2, they scan the database to calculate the exact utility of all candidates found in Phase 1 to eliminate low-utility itemsets. Recently, an alternative approach was proposed in the HUI-Miner algorithm [7] to mine high-utility itemsets directly using a single phase. HUI-Miner was shown to outperform previous algorithms and is thus the current best algorithm for HUIM [7]. HUI-Miner utilizes a depth-first search to explore the search space of itemsets. HUI-Miner associate a structure named *utility-list* [7] to each pattern. Utility-lists allow calculating the utility of a pattern quickly by making join operations with utility-lists of smaller patterns. Utility-lists are defined as follows.

Definition 7 (Utility-list). Let $\succ$ be any total order on items from I. The *utility-list* of an itemset X in a database D is a set of tuples such that there is a tuple $(tid, iutil, rutil)$ for each transaction T_{tid} containing X. The $iutil$ element of a tuple is the utility of X in T_{tid}. i.e $u(X, T_{tid})$. The $rutil$ element of a tuple is defined as $\sum_{i \in T_{tid} \wedge i \not\in X} U(i, T_{tid})$.

Example 7. The utility-list of $\{a\}$ is $\{(T_1, 5, 3)(T_2, 10, 17)(T_3, 5, 25)\}$. The utility-list of $\{e\}$ is $\{(T_2, 6, 5)(T_3, 3, 5)(T_4, 3, 0)\}$. The utility-list of $\{a, e\}$ is $\{(T_2, 16, 5), (T_3, 8, 5)\}$.

To discover high-utility itemsets, HUI-Miner perform a single database scan to create utility-lists of patterns containing single items. Then, larger patterns are obtained by performing the join operation of utility-lists of smaller patterns. Pruning the search space is done using the two following properties.

Property 4 (Sum of Iutils). Let X be an itemset. If the sum of *iutil* values in the utility-list of x is higher than or equal to *minutil*, then X is a high-utility itemset. Otherwise, it is a low-utility itemset [7].

Property 5 (Sum of Iutils and Rutils). Let X be an itemset. Let the *extensions* of X be the itemsets that can be obtained by appending an item y to X such that $y \succ i$ for all item i in X. If the sum of *iutil* and *rutil* values in the utility-list of x is less than *minutil*, all extensions of X and their transitive extensions are low-utility itemsets [7].

HUI-Miner is a very efficient algorithm. However, a drawback is that the join operation to calculate the utility-list of an itemset is very costly. In the next section, we introduce our novel algorithm, which improves upon HUI-Miner by being able to eliminate low-utility itemsets without performing join operations.

Tid	Transactions
T_1	(a,1)(c,1)(d,1)
T_2	(a,2)(c,6)(e,2)(g,5)
T_3	(a,1)(b,2)(c,1)(d,6),(e,1),(f,5)
T_4	(b,4)(c,3)(d,3)(e,1)
T_5	(b,2)(c,2)(e,1)(g,2)

Item	a	b	c	d	e	f	g
Profit	5	2	1	2	3	1	1

Fig. 1. A transaction database (left) and external utility values (right)

TID	TU
T_1	8
T_2	27
T_3	30
T_4	20
T_5	11

Item	TWU
a	65
b	61
c	96
d	58
e	88
f	30
g	38

Item	a	b	c	d	e	f
b	30					
c	65	61				
d	38	50	58			
e	57	61	77	50		
f	30	30	30	30	30	
g	27	38	38	0	38	0

Fig. 2. Transaction utilities (left), TWU values (center) and EUCS (right)

3   The FHM Algorithm

In this section, we present our proposal, the FHM algorithm. The main procedure (Algorithm 1) takes as input a transaction database with utility values and the *minutil* threshold. The algorithm first scans the database to calculate the TWU of each item. Then, the algorithm identifies the set I^* of all items having a

TWU no less than *minutil* (other items are ignored since they cannot be part of a high-utility itemsets by Property 3). The TWU values of items are then used to establish a total order $\succ$ on items, which is the order of ascending TWU values (as suggested in [7]). A second database scan is then performed. During this database scan, items in transactions are reordered according to the total order $\succ$, the utility-list of each item $i \in I^*$ is built and our novel structure named EUCS (Estimated Utility Co-Occurrence Structure) is built. This latter structure is defined as a set of triples of the form $(a, b, c) \in I^* \times I^* \times \mathbb{R}$. A triple (a,b,c) indicates that TWU($\{a, b\}$) = c. The EUCS can be implemented as a triangular matrix as shown in Fig. 2 (right) or as a hashmap of hashmaps where only tuples of the form (a, b, c) such that $c \neq 0$ are kept. In our implementation, we have used this latter representation to be more memory efficient because we have observed that few items co-occurs with other items. Building the EUCS is very fast (it is performed with a single database scan) and occupies a small amount of memory, bounded by $|I^*| \times |I^*|$, although in practice the size is much smaller because a limited number of pairs of items co-occurs in transactions (cf. section 5). After the construction of the EUCS, the depth-first search exploration of itemsets starts by calling the recursive procedure *Search* with the empty itemset $\emptyset$, the set of single items I^*, *minutil* and the EUCS structure.

Algorithm 1. The FHM algorithm

> **input**   : D: a transaction database, *minutil*: a user-specified threshold
> **output**: the set of high-utility itemsets

1 Scan D to calculate the TWU of single items;
2 $I^* \leftarrow$ each item i such that TWU(i) < *minutil*;
3 Let $\succ$ be the total order of TWU ascending values on I^*;
4 Scan D to built the utility-list of each item $i \in I^*$ and build the *EUCS* structure;
5 **Search** ($\emptyset$, I^*, *minutil*, *EUCS*);

The *Search* procedure (Algorithm 2) takes as input (1) an itemset P, (2) extensions of P having the form Pz meaning that Pz was previously obtained by appending an item z to P, (3) *minutil* and (4) the EUCS. The search procedure operates as follows. For each extension Px of P, if the sum of the *iutil* values of the utility-list of Px is no less than *minutil*, then Px is a high-utility itemset and it is output (cf. Property 4). Then, if the sum of *iutil* and *rutil* values in the utility-list of Px are no less than *minutil*, it means that extensions of Px should be explored (cf.). This is performed by merging Px with all extensions Py of P such that $y \succ x$ to form extensions of the form Pxy containing $|Px| + 1$ items. The utility-list of Pxy is then constructed as in HUI-Miner by calling the *Construct* procedure (cf. Algorithm 3) to join the utility-lists of P, Px and Py. This latter procedure is the same as in HUI-Miner [7] and is thus not detailed here. Then, a recursive call to the *Search* procedure with Pxy is done to calculate its utility and explore its extension(s). Since the *Search* procedure starts from single items, it recursively explore the search space of itemsets by appending single items and it only prunes the search space based on Property 5. It can be easily seen based on Property 4 and 5 that this procedure is correct and complete to discover all high-utility itemsets.

Co-occurrence-based Pruning. The main novelty in FHM is a novel pruning mechanism named EUCP (Estimated Utility Co-occurrence Pruning), which relies on a new structure, the EUCS. EUCP is based on the observation that one of the most costly operation in HUI-Miner is the join operation. EUCP is a pruning strategy to directly eliminate a low-utility extension Pxy and all its transitive extensions without constructing their utility-list. This is done on line 8 of the *Search* procedure. The pruning condition is that if there is no tuple (x, y, c) in EUCS such that $c \geq minutil$, then Pxy and all its supersets are low-utility itemsets and do not need to be explored.

This strategy is correct (only prune low-utility itemsets). The proof is that by Property 3, if an itemset X contains another itemset Y such that $TWU(Y) < minutil$, then X and its supersets are low-utility itemsets.

An important question about the EUCP strategy is: should we not only check the condition for x, y in each call to *Search* but also check the condition for all pairs of distinct items $a, b \in Pxy$? The answer is no because the *Search* procedure is recursive and therefore all other pairs of items in Pxy have already been checked in previous recursions of the *Search* procedure leading to Pxy. For example, consider an itemset $Z = \{a_1, a_2, a_3, a_4\}$. To generate this itemset, the search procedure had to combine $\{a_1, a_2 a_3\}$ and $\{a_1, a_2, a_4\}$, obtained by combining $\{a_1, a_2\}$ and $\{a_1, a_3\}$, and $\{a_1, a_2\}$ and $\{a_1, a_4\}$, obtained by combining single items. It can be easily observed that when generating Z all pairs of items in Z have been checked by EUCP except $\{a_3, a_4\}$.

4   Experimental Study

We performed experiments to assess the performance of the proposed algorithm. Experiments were performed on a computer with a third generation 64 bit Core i5 processor running Windows 7 and 5 GB of free RAM. We compared the performance of FHM with the state-of-the-art algorithm HUI-Miner for high-utility itemset mining. All memory measurements were done using the Java API. Experiments were carried on four real-life datasets having varied characteristics. The *Chainstore* dataset contains 1,112,949 transactions with 46,086 distinct items and an average transaction length of 7.26 items. The *BMS* dataset contains 59,601 transactions with 497 distinct items and an average transaction length of 4.85 items. The *Kosarak* dataset contains 990,000 transactions with 41,270 distinct items and an average transaction length of 8.09 items. The *Retail* dataset contains 88,162 transactions with 16,470 distinct items and an average transaction length of 10,30 items. The Chainstore dataset already contain unit profit information and purchase quantities. For other datasets, external utilities for items are generated between 1 and 1,000 by using a log-normal distribution and quantities of items are generated randomly between 1 and 5, as the settings of

---

Algorithm 2. The *Search* procedure

input : P: an itemset, *ExtensionsOfP*: a set of extensions of P, the *minutil* threshold, the *EUCS* structure
output: the set of high-utility itemsets

```
1  foreach  itemset Px ∈ ExtensionsOfP do
2  |   if SUM(Px.utilitylist.iutils) ≥ minutil then
3  |   |   output Px;
4  |   end
5  |   if SUM(Px.utilitylist.iutils)+SUM(Px.utilitylist.rutils) ≥ minutil then
6  |   |   ExtensionsOfPx ← ∅;
7  |   |   foreach  itemset Py ∈ ExtensionsOfP such that y ≻ x do
8  |   |   |   if ∃(x,y,c) ∈ EUCS such that c ≥ minutil) then
9  |   |   |   |   Pxy ← Px ∪ Py;
10 |   |   |   |   Pxy.utilitylist ← Construct (P, Px, Py);
11 |   |   |   |   ExtensionsOfPx ← ExtensionsOfPx ∪ Pxy;
12 |   |   |   end
13 |   |   end
14 |   |   Search (Px, ExtensionsOfPx, minutil);
15 |   end
16 end
```

[2, 7, 10]. The source code of all algorithms and datasets can be downloaded from http://goo.gl/hDtdt.

Execution Time. We first ran the FHM and HUI-Miner algorithms on each dataset while decreasing the *minutil* threshold until algorithms became too long to execute, ran out of memory or a clear winner was observed. For each dataset, we recorded the execution time, the percentage of candidate pruned by the FHM algorithm and the total size of the EUCS. The comparison of execution times is shown in Fig. 3. For Chainstore, BMS, Kosarak and Retail, FHM was respectively up to 6.12 times faster, 6 times faster, 4.33 times faster and 2.3 times faster than HUI-Miner.

Pruning Effectiveness. The percentage of candidates pruned by the FHM algorithm was 18% to 91%, 87%, 87 % and 31% to 95% for the Chainstore, BMS, Kosarak and retail datasets. These results show that candidate pruning can be very effective by pruning up to 95 % of candidates. As expected, when more pruning was done, the performance gap between FHM and HUI-Miner became larger.

Memory Overhead. We also studied the memory overhead of using the EUCS structure. We found that for the Chainstore, BMS, Kosarak and Retail datasets, the memory footprint of EUCS was respectively 10.3 MB, 4.18 MB, 1.19 MB and 410 MB. We therefore conclude that the cost of using the EUCP strategy in terms of memory is low.

Algorithm 3. The Construct procedure

input : P: an itemset, Px: the extension of P with an item x, Py: the
extension of P with an item y
output: the utility-list of Pxy

1 $UtilityListOfPxy \leftarrow \emptyset$;
2 **foreach** $tuple\ ex \in Px.utilitylist$ **do**
3   **if** $\exists ey \in Py.utilitylist\ and\ ex.tid = exy.tid$ **then**
4   **if** $P.utilitylist \neq \emptyset$ **then**
5   Search element $e \in P.utilitylist$ such that $e.tid = ex.tid.$;
6     $exy \leftarrow (ex.tid, ex.iutil + ey.iutil - e.iutil, ey.rutil)$;
7   **end**
8   **else**
9     $exy \leftarrow (ex.tid, ex.iutil + ey.iutil, ey.rutil)$;
10   **end**
11    $UtilityListOfPxy \leftarrow UtilityListOfPxy \cup \{exy\}$;
12   **end**
13 **end**
14 **return** $UtilityListPxy$;

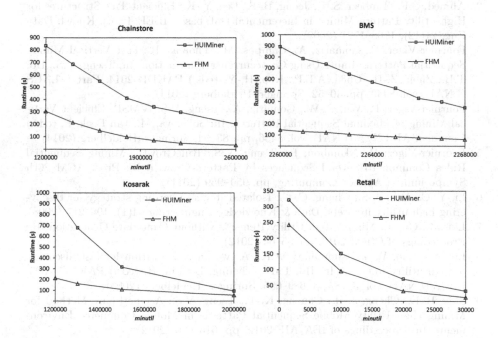

Fig. 3. Execution times

5   Conclusion

In this paper, we have presented a novel algorithm for high-utility itemset mining named FHM (Fast High-Utility Miner). This algorithm integrates a novel strategy named EUCP (Estimated Utility Cooccurrence Pruning) to reduce the number of joins operations when mining high-utility itemsets using the utility-list data structure. We have performed an extensive experimental study on four real-life datasets to compare the performance of FHM with the state-of-the-art algorithm HUI-Miner. Results show that the pruning strategy reduces the search space by up to 95 % and that FHM is up to 6 times faster than HUI-Miner. The source code of all algorithms and datasets used in our experiments can be downloaded from http://goo.gl/hDtdt, as part of the SPMF data mining library.

For future work, we are interested in exploring other optimizations for itemset mining, sequential pattern mining [3, 4] and sequential rule mining [5].

Acknowledgement. This work is financed by a National Science and Engineering Research Council (NSERC) of Canada research grant.

References

1. Agrawal, R., Srikant, R.: Fast algorithms for mining association rules in large databases. In: Proc. Int. Conf. Very Large Databases, pp. 487–499 (1994)
2. Ahmed, C.F., Tanbeer, S.K., Jeong, B.-S., Lee, Y.-K.: Efficient Tree Structures for High-utility Pattern Mining in Incremental Databases. IEEE Trans. Knowl. Data Eng. 21(12), 1708–1721 (2009)
3. Fournier-Viger, P., Gomariz, A., Campos, M., Thomas, R.: Fast Vertical Mining Sequential Pattern Mining Using Co-occurrence Information. In: Tseng, V.S., Ho, T.B., Zhou, Z.-H., Chen, A.L.P., Kao, H.-Y. (eds.) PAKDD 2014, Part I. LNCS (LNAI), vol. 8443, pp. 40–52. Springer, Heidelberg (2014)
4. Fournier-Viger, P., Wu, C.-W., Gomariz, A., Tseng, V.S.: VMSP: Efficient Vertical Mining of Maximal Sequential Patterns. In: Sokolova, M., van Beek, P. (eds.) Canadian AI. LNCS (LNAI), vol. 8436, pp. 83–94. Springer, Heidelberg (2014)
5. Fournier-Viger, P., Nkambou, R., Tseng, V.S.: RuleGrowth: Mining Sequential Rules Common to Several Sequences by Pattern-Growth. In: Proc. ACM 26th Symposium on Applied Computing, pp. 954–959 (2011)
6. Li, Y.-C., Yeh, J.-S., Chang, C.-C.: Isolated items discarding strategy for discovering high utility itemsets. Data & Knowledge Engineering 64(1), 198–217 (2008)
7. Liu, M., Qu, J.: Mining High Utility Itemsets without Candidate Generation. In: Proceedings of CIKM 2012, pp. 55–64 (2012)
8. Liu, Y., Liao, W.-k., Choudhary, A.K.: A two-phase algorithm for fast discovery of high utility itemsets. In: Ho, T.-B., Cheung, D., Liu, H. (eds.) PAKDD 2005. LNCS (LNAI), vol. 3518, pp. 689–695. Springer, Heidelberg (2005)
9. Shie, B.-E., Cheng, J.-H., Chuang, K.-T., Tseng, V.S.: A One-Phase Method for Mining High Utility Mobile Sequential Patterns in Mobile Commerce Environments. In: Proceedings of IEA/AIE 2012, pp. 616–626 (2012)
10. Tseng, V.S., Shie, B.-E., Wu, C.-W., Yu, P.S.: Efficient Algorithms for Mining High Utility Itemsets from Transactional Databases. IEEE Trans. Knowl. Data Eng. 25(8), 1772–1786 (2013)
11. Yin, J., Zheng, Z., Cao, L.: USpan: An Efficient Algorithm for Mining High Utility Sequential Patterns. In: Proceedings of ACM SIG KDD 2012, pp. 660–668 (2012)

Automatic Subclasses Estimation for a Better Classification with HNNP

Ruth Janning, Carlotta Schatten, and Lars Schmidt-Thieme

Information Systems and Machine Learning Lab (ISMLL), University of Hildesheim,
Marienburger Platz 22, 31141 Hildesheim, Germany
{janning,schatten,schmidt-thieme}@ismll.uni-hildesheim.de
http://www.ismll.uni-hildesheim.de

Abstract. Although nowadays many artificial intelligence and especially machine learning research concerns big data, there are still a lot of real world problems for which only small and noisy data sets exist. Applying learning models to those data may not lead to desirable results. Hence, in a former work we proposed a hybrid neural network plait (HNNP) for improving the classification performance on those data. To address the high intraclass variance in the investigated data we used manually estimated subclasses for the HNNP approach. In this paper we investigate on the one hand the impact of using those subclasses instead of the main classes for HNNP and on the other hand an approach for an automatic subclasses estimation for HNNP to overcome the expensive and time consuming manual labeling. The results of the experiments with two different real data sets show that using automatically estimated subclasses for HNNP delivers the best classification performance and outperforms also single state-of-the-art neural networks as well as ensemble methods.

Keywords: Image classification, subclasses, convolutional neural network, multilayer perceptron, hybrid neural network, small noisy data.

1 Introduction

Although nowadays most problems treated in artificial intelligence and especially in machine learning concern big data, there are still also many real world problems delivering less data than desired for machine learning approaches or delivering such noisy data that classification models deliver undesirable bad results. The first problem type addresses for instance problems in the field of radar like detecting reflections of buried objects in ground penetrating radar images ([5], [6], [7]) or recognizing craters on synthetic aperture radar images of the surface of planets ([10], [11]). The reason for small data sets in this field is the often very expensive and time consuming recording of the images as well as the need of manual labeling. Additionally, most of those data are not publicly available. Furthermore, in many cases it is unavoidable that the recording method or the ambiance induce noise into the data. The second problem type corresponds

T. Andreasen et al. (Eds.): ISMIS 2014, LNAI 8502, pp. 93–102, 2014.
© Springer International Publishing Switzerland 2014

to challenges like phoneme recognition ([1]). The phonemes spoken by different humans or even by the same human may be very different and on the other hand different phonemes may be very similar. This leads to high intra class variances as well as to small inter class variances. Furthermore, the phoneme extraction process is not exact and induces noise. Hence, the phoneme classification on phoneme data sets like for instance the TIMIT data set ([15]) is challenging. In [8] we presented a plait of hybrid neural networks (HNNP) for improving the classification performance in data sets from those problem types. We observed in [8] large intra class variance and small inter class variance in the investigated radar data (see also fig. 4) and hence we labeled the data for the training and testing within the HNNP structure not only with the main classes but also with subclasses to reduce the intra class variance. This splitting into subclasses was done manually by assigning similar looking examples of one main class to the same subclass. However, the question arises, if this approach really leads to better results and if yes, if this splitting into subclasses can be done also automatically, as the manual labeling is expensive and time consuming or even not possible for data like phonemes. Both questions are answered in this paper. We investigated on the one hand the difference in classification performance of the HNNP approach with only main classes and with more subclasses. On the other hand we propose an approach for automatic subclasses estimation for HNNP and compare the classification performance of HNNP with different numbers of automatically estimated subclasses per main class. The results of our experiments with two different real data sets show that using several subclasses per main class within the HNNP structure leads to better classification results than using just the original main classes. Furthermore, they show that an automatic estimation of the subclasses is even more effective. The main contributions of this paper are: (1) experiments showing the advantage of using subclasses (instead of using only main classes) for HNNP, (2) presentation of an approach of HNNP combined with automatic estimation of subclasses, (3) experiments showing the possibility and effectiveness of using automatically estimated subclasses for HNNP.

2 Related Work

Convolutional neural networks are typically used for image classification tasks like handwritten digit classification but also for problems like phoneme recognition which can be also considered as an image classification problem as in [1]. The well known LeNet-5 convolutional neural network for digits recognition is presented in [9]. A more recent convolutional neural network for digit recognition with a simpler and shallower architecture, which we used for our experiments on radar data, is proposed in [13]. In [1] a convolutional neural network for phoneme recognition is described. We used a version of this network for the experiments with the phoneme data. The combination of neural networks to ensembles is investigated e.g. in [12] and an investigation of stacking, or stacked generalization, can be found e.g. in [14]. A combination of more than one convolutional neural network is proposed in [3], where a committee of 7 convolutional neural networks is used and the outputs are averaged. Different kinds of feature sets are used in [4], where 6 feature sets are trained by multilayer perceptrons and the outputs are merged by using another multilayer perceptron. Our HNNP approach ([8], fig. 1)

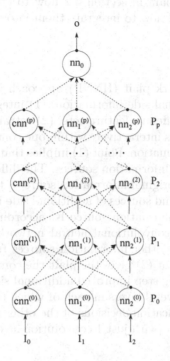

Fig. 1. Architecture of the hybrid neural network plait (HNNP). The plait is composed of $p+1$ layers $P_0, P_1, \ldots, P_p$. Each layer contains $k+1$ different neural networks (here $k = 2$) according to the $k+1$ different information sources $I_0, I_1, \ldots, I_k$ for the input. I_0 corresponds to the input image and the appropriate learning model is a convolutional neural network (cnn). The learning models for the other information sources $I_1, \ldots, I_k$ are multilayer perceptrons ($nn_1, nn_2, \ldots, nn_k$). In every layer from P_1 on the neural networks are retrained with additional input from the former layer. After the last layer P_p a further multilayer perceptron (nn_0) is attached to achieve one common output vector **o** delivering the final classification result.

combines the above mentioned approaches by using different feature sets and a committee of a convolutional neural network and multilayer perceptrons. However, for the HNNP approach different kinds of neural networks with adapted architecture are retrained interactively within a plait structure using additional side information gained before and during the retraining for a further improvement. Our approach for automatically estimating subclasses for HNNP is similar to the approach in [16] based on local clustering, however the focus of [16] lies on imbalanced class distributions, whereas we address the problem of high intra class variances and small inter class variances.

3   HNNP with Automatically Estimated Subclasses

In the following sections we will present the HNNP approach with automatically estimated subclasses. First we will introduce HNNP in section 3.1 and subsequently we will explain in section 3.2 how to estimate automatically the subclasses for HNNP and how to integrate them into the HNNP architecture.

3.1   HNNP

The hybrid neural network plait (HNNP) approach is based on two methods for incorporating additional side information: (1) integrating different information sources delivering different feature sets, (2) retraining the neural networks applied to this feature sets interactively within one common structure with additional improved side information. Point (1) implies that different neural networks are used for the different information sources. The different information sources are on the one hand the original information source – the pixel values of the image – and on the other hand sources of additional side information. The learning model for the original information source is, according to state-of-the-art approaches (see sec. 2), a convolutional neural network (cnn). For the training of the feature sets of the other information sources fully connected multilayer perceptrons are used. Point (2) gives the plait the complex structure and incorporates in every retraining step improved additional side information. The plait (fig. 1) is composed of several layers in each of which the networks are retrained by considering the classification decisions of the other networks from the former layer. A simplified version (with just 1 convolution layer instead of 2) of the cnn

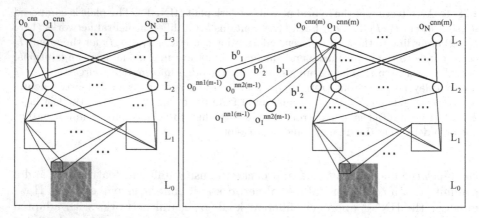

Fig. 2. Left: Architecture of a simplified cnn. It is composed of 4 layers L_0, L_1, L_2, L_3, where L_0 is the input layer fed with the image, L_1 is a convolution layer followed by a fully connected layer (L_2) and the output layer L_3. Right: Architecture of the cnn adapted to the plait structure with 3 information sources. In contrast to left the output neurons of this cnn get additional input from the other networks of the plait layer before.

used for HNNP is pictured in fig. 2. In these kind of networks the convolution layers and the subsampling serve as a local feature extractor and the fully connected layers as a trainable classifier. The output layer delivers an output vector $\hat{\mathbf{o}}^{\mathbf{cnn}} := (\hat{o}_0^{cnn}, \ldots, \hat{o}_N^{cnn}), -1 \leq \hat{o}_i^{cnn} \leq 1, i = 0, \ldots, N$, where the actual output $\hat{o}_i{}^{cnn}$ of the ith neuron in output layer n corresponds to

$$\hat{o}_i{}^{cnn} = \tanh(\sum_{l=0}^{N_{n-1}} w_n^{il} x_{n-1}^l) , \qquad (1)$$

with N_{n-1} as the number of neurons in layer $n-1$, w_n^{il} as the weight of the lth connection between neuron i and the neurons in layer $n-1$, x_{n-1}^l as the output of the lth neuron in layer $n-1$ and tanh (hyperbolic tangent) as activation function with an output within the interval $[-1, 1]$. Each neuron of the output layer is assigned to one class in $\{C_0, \ldots, C_N\}$. The classification result $C_{\text{argmax}_i \hat{o}_i{}^{cnn}}$ corresponds to the class C_i assigned to the output neuron with maximum output value $\hat{o}_i{}^{cnn}$. As usual in such architectures, the error in the last layer n for pattern $\mathcal{P}$ to be minimized is

$$E_n^{\mathcal{P}} := \frac{1}{2} \sum_{i=0}^{N} (\hat{o}_i{}^{cnn} - o_i^{cnn})^2 , \qquad (2)$$

with $\hat{o}_i{}^{cnn}$ as actual output (eq. (1)) and o_i^{cnn} as target output. In opposite to the cnn, fully connected multilayer perceptrons fulfill only a role as classifier (without local feature extraction) but use the same formulas for $\hat{\mathbf{o}}^{\mathbf{nn}_i}, E_n^{\mathcal{P}}$. cnns and multilayer perceptrons are interweaved in the HNNP structure like in a plait (fig. 1), which is enabled by adapting their architectures. In these new architectures (fig. 2) – using the example of the cnn – every neuron $o_i^{cnn^{(m)}}$ of the output layer L_n of a cnn in plait layer P_m is additionally connected to the outputs $\hat{o}_i^{nn_1^{(m-1)}}, \hat{o}_i^{nn_2^{(m-1)}}, \ldots, \hat{o}_i^{nn_k^{(m-1)}}$ of the k other networks $nn_1, nn_2, \ldots, nn_k$ from the previous plait layer P_{m-1}. This leads to new output formulas for the neural networks within the plait. In the case of k information sources for additional side information, the new output formulas for the $k+1$ adapted networks cnn and $nn_x, x = 1, \ldots, k$ are as follows:

$$\hat{o}_i^{cnn^{(m)}} = \tanh(\sum_{l=0}^{N_{n-1}} w_n^{il} x_{n-1}^l + \sum_{j=1}^{k} b_j^i \hat{o}_i^{nn_j^{(m-1)}}) , \qquad (3)$$

$$\hat{o}_i^{nn_x^{(m)}} = \tanh(\sum_{l=0}^{N_{n-1}} w_n^{il} x_{n-1}^l + b_0^i \hat{o}_i^{cnn^{(m-1)}} + \sum_{j=1, j \neq x}^{k} b_j^i \hat{o}_i^{nn_j^{(m-1)}}) . \qquad (4)$$

In this way each neural network learns to which degree it should consider the classification decisions of all other networks from the previous plait layer. The number $p+1$ of plait layers is a hyper parameter. The final component of the HNNP is an additional fully connected multilayer perceptron (nn_0 in fig. 1). nn_0 is fed with the outputs of every neural network in the last plait layer P_p and delivers one common final classification decision.

3.2 Automatic Subclasses Estimation for HNNP

Originally, the number N of output neurons in the last layer of the neural networks within the HNNP structure refers to the number M of main classes. However, as mentioned before, for our experiments in [8] we split the main classes into subclasses, which were estimated manually in a time consuming process. Hence, we propose an approach for an automatic estimation of subclasses for HNNP. Our approach applies a K-means clustering on the histograms to the set of examples of each main class. Each of the K clusters found is then assigned to one of K subclasses. K is, besides p, a further hyper parameter of our method. This approach leads for each neural network within the HNNP to a different number N of output neurons, namely $N := M \cdot K$ – instead of $N := M$ if just main classes are considered – as there is one output neuron per subclass. By using more output neurons within the HNNP structure a more fine granulated improved side information is passed through the architecture. Only nn_0 maps

---

Input:
Data sets $\mathcal{D}^{train} := \{d_0^{train}, \ldots, d_{|\mathcal{D}^{train}|}^{train}\}$
$\qquad = \bigcup_{i=0}^{M} \mathcal{D}_{main_i} = \bigcup_{i=0}^{M} \{d_0^{main_i}, \ldots, d_{|\mathcal{D}_{main_i}|}^{main_i}\}$
with main class labels $(main(d_0^{train}), \ldots, main(d_{|\mathcal{D}^{train}|}^{train}))$ and
$\mathcal{D}^{test} := \{d_0^{test}, \ldots, d_{|\mathcal{D}^{test}|}^{test}\}$,
number K of subclasses per main class, number p of plait layers.

Variables:
$(\mathcal{D}_{main_i}^{0}, \ldots, \mathcal{D}_{main_i}^{K})$,   // list of clusters of subclasses of main class i
$(sub(d_0^{main_i}), \ldots, sub(d_{|\mathcal{D}_{main_i}|}^{main_i}))$ // list of subclass labels of main class i
$\mathcal{D}^{train'}, \mathcal{L}^{train'}$ // train data and label set with subclasses
$(main(d_0^{test}), \ldots, main(d_{|\mathcal{D}^{test}|}^{test}))$ // list of predicted main class labels

1. **for** $i = 0$ to M **do**
    $((\mathcal{D}_{main_i}^{0}, \ldots, \mathcal{D}_{main_i}^{K}), (sub(d_0^{main_i}), \ldots, sub(d_{|\mathcal{D}_{main_i}|}^{main_i})))$
         $= K\text{-means}(\text{Histograms}(\mathcal{D}_{main_i}), K)$;
   end for

2. $\mathcal{D}^{train'} := \bigcup_{i=0}^{M} \bigcup_{j=0}^{K} \mathcal{D}_{main_i}^{j}$; $\mathcal{L}^{train'} := \bigcup_{i=0}^{M} (sub(d_0^{main_i}), \ldots, sub(d_{|\mathcal{D}_{main_i}|}^{main_i}))$

3. $\text{trainHNNP}(\mathcal{D}^{train'}, \mathcal{L}^{train'}, p, K)$;

4. $(main(d_0^{test}), \ldots, main(d_{|\mathcal{D}^{test}|}^{test})) = \text{applyHNNP}(\mathcal{D}^{test}, p, K)$;

5. **return** $(main(d_0^{test}), \ldots, main(d_{|\mathcal{D}^{test}|}^{test}))$;

Fig. 3. HNNP approach with automatically estimated subclasses

its input finally to the number M of main classes, i.e. for nn_0 holds $N := M$. The whole approach is shown in fig. 3. The input to the approach is on the one hand the train set $\mathcal{D}^{train} := \{d_0^{train}, \ldots, d_{|\mathcal{D}^{train}|}^{train}\}$ with a main class label for every example and on the other hand a test set $\mathcal{D}^{test} := \{d_0^{test}, \ldots, d_{|\mathcal{D}^{test}|}^{test}\}$. The examples $d_l^{train}, l = 0 \ldots |\mathcal{D}^{train}|$ and $d_l^{test}, l = 0 \ldots |\mathcal{D}^{test}|$ consist of as many feature vectors as there are information sources $(k + 1)$. A further input to the approach are the hyper parameters K and p. In step 1 a K-means algorithm is applied to the histograms of train set examples of each main class. The K-means algorithm finds K clusters in each main class set and assigns the appropriate subclass label to every example. The HNNP is trained in step 3 with these $M \cdot K$ gained subclasses. To predict the labels of test examples in step 4 the HNNP is applied to the test set. The output of the whole approach in step 5 is a predicted main class label for every test example. The described approach is proven by experiments presented in the next section 4.

4 Experiments

In the experiments we investigated three different questions: (1) Does HNNP outperform the single networks as well as the ensemble methods? (2) How does the classification performance behave if only main classes are used or if several subclasses are used? (3) How does the classification performance behave if automatically estimated subclasses are used? We focus in this paper on the two last questions, as the first one was already answered also for other data sets in [8]. We applied HNNP to 2 different real data sets (sec. 4.1). The first data set comes from a set of synthetic aperture radar (SAR) images of the surface of Venus, available at the UCI Machine Learning Repository [2]. The data were collected by the Magellan spacecraft ([10], [11]). The classification task in this data is the identification of volcanoes. To this SAR data set we applied HNNP on the one hand with 2 main classes (volcano, noise) and on the other hand with 10 manually estimated subclasses as well as with 10 automatically estimated subclasses. The second data set consists of examples of phonemes, or of the *Mel Frequency Cepstral Coefficients* (MFCC) vectors of speech signals interpreted as images by combining several subsequent vectors respectively (see [1]). The data are gained from the TIMIT data set ([15]) by choosing two different phonemes ('iy','n'). To this small TIMIT data set we applied HNNP with 2 main classes according to the 2 chosen phonemes and with 6, 10 and 20 automatically estimated subclasses. For both data sets we used the same experimental settings: the HNNP approach is compared to the five baselines *cnn, nn₁, nn₂, majority ensemble* and *stacking ensemble*. *cnn* is a single convolutional neural network (sec. 3.1) fed with the normalized pixel values of the images to classify. nn_1, nn_2 are single fully connected multilayer perceptrons (sec. 3.1) with input feature sets coming from the appropriate additional side information described in section 4.1. *Majority ensemble* classifies according to the majority vote of *cnn, nn₁* and *nn₂*. *Stacking ensemble* learns to combine the classification decisions of *cnn, nn₁* and *nn₂* by using the subsequent multilayer perceptron nn_0. We used in the experiments $k + 1 = 3$ information sources, $p + 1 = 3$ plait layers, $N = 2$ or 10 for the SAR data set and $N = 2, 6, 10$ or 20 (sub)classes for the small TIMIT data set and we conducted a 5-fold cross validation for every data set.

4.1 Data Sets

For the SAR data set we have chosen 5 images of the Magellan data. We extracted 898 examples (451 volcano, 447 noise). The examples are 29×29 pixel images (fig. 4), which build the input for the cnn. The cnn used for the experiments with the SAR data set is a 5-layer cnn with 2 convolution layers (1 with 10 maps and 1 with 50 maps) with integrated subsampling, based on the architecture described in [13]. In the fully connected layer there are 50 neurons and in the output layer, according to the value of N, 2 or 10 neurons. The input feature sets for nn_1 and nn_2 come from statistical information. nn_1 is fed with 4 histograms of 16 gray values, each of which represents one quarter of the input image. Accordingly, nn_1 has $4 \cdot 16$ neurons in its input layer, 64 neurons in the hidden layer and 2 or 10 neurons in the output layer. nn_2 is fed with the number of pixels with a light gray value (a value within the upper quarter of all gray values of the image) per area, where an area is one of 100 areas of the image (partitioned by a 10×10 grid). nn_2 has 100 neurons in the input layer, 100 neurons in the hidden layer and 2 or 10 neurons in the output layer.

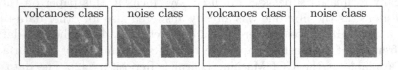

Fig. 4. Examples of 4 different automatically estimated subclasses of the SAR data

For the small TIMIT data set we have chosen 400 examples of the class of phonem 'iy' and 400 examples of the class of phonem 'n'. The examples are 43×19 pixel images consisting of MFCC vectors, which build the input for the cnn. The cnn used for the experiments with the small TIMIT data set is a 5-layer cnn with 1 convolution layer and 1 max pooling layer, based on the architecture described in [1]. In the convolution layer each 8×8 window of the input image is mapped to one value in each of the 6 maps. In the max pooling layer the maximum value of each 2nd 6×6 window of a convolution map is inserted in a pooling map. In the fully connected layer there are 100 neurons and in the output layer, according to the value of N, 2, 6, 10 or 20 neurons. The input feature sets for nn_1 and nn_2 are equivalent to the ones for the SAR data set.

4.2 Results

The results of the 5-fold cross validation experiments are shown in table 1. The first observation and answer to question (1) is that HNNP outperforms the

Table 1. Results of the 5-fold cross validation for the SAR and small TIMIT data set: Classification test errors (%) of the 5 baselines *cnn*, nn_1, nn_2, *majority ensemble* and *stacking ensemble* and the HNNP (standard deviations in brackets) with different numbers of classes

data	# classes	cnn	nn_1	nn_2	majority	stacking	HNNP
SAR	2	23.72	22.38	21.04	18.48	18.71	**17.61**
		(3.77)	(3.46)	(2.81)	(3.93)	(4.12)	(0.89)
SAR	10	23.84	23.50	19.82	17.60	16.15	**14.03**
	manual	(4.07)	(4.39)	(3.13)	(3.41)	(1.74)	(1.30)
SAR	10	26.50	18.92	20.05	16.62	14.94	**12.03**
	automatic	(1.95)	(2.27)	(2.77)	(3.21)	(2.01)	(2.56)
TIMIT	2	26.75	36.13	30.00	26.00	24.63	**16.38**
		(2.14)	(3.55)	(1.47)	(2.71)	(2.52)	(2.39)
TIMIT	6	29.88	35.88	27.75	25.38	21.25	**13.75**
	automatic	(2.18)	(5.28)	(1.22)	(2,45)	(1.71)	(2.54)
TIMIT	10	27.88	35.75	30.88	26.38	22.13	**12.50**
	automatic	(4.32)	(4.04)	(1.86)	(1.90)	(1.69)	(0.99)
TIMIT	20	25.38	34.88	32.88	26.00	20.75	**12.50**
	automatic	(3.24)	(3.17)	(3.18)	(2.19)	(1.90)	(3.03)

single networks as well as the ensemble methods. But in this work we focus on the answers of both other questions. Hence, for the SAR data set we investigated the difference in the classification performance of HNNP with only 2 main classes and HNNP with 10 manually or automatically estimated subclasses. Table 1 shows that using subclasses causes no big difference in the classification performance for the single neural networks but for *stacking* and HNNP the classification performance is improved by using subclasses. In the experiments with the small TIMIT data set we compared the classification performances of HNNP with different numbers of automatically estimated subclasses (as manually subclasses estimation is not possible). Already the use of 6 subclasses leads to a classification performance improvement of HNNP. Using 10 subclasses instead still improves the classification performance, whereas using 20 subclasses does not lead to a further improvement.

5   Conclusion

In this work we investigated the impact of using subclasses for HNNP and we proposed an approach of estimating subclasses for HNNP automatically. The experiments show that using (automatically estimated) subclasses within HNNP is able to improve the classification performance significantly. Next steps would be to apply our approach to multiclass problems like e.g. full phoneme recognition. We expect also for such problems a significant performance improvement by using HNNP with automatically estimated subclasses.

Acknowledgments. This work is co-funded by the EU project iTalk2Learn.

References

1. Abdel-Hamid, O., Mohamed, A., Jiang, H., Penn, G.: Applying Convolutional Neural Networks concepts to hybrid NN-HMM model for speech recognition. In: IEEE International Conference on Acoustics, Speech and Signal Processing (ICASSP), pp. 4277–4280 (2012)
2. Bache, K., Lichman, M.: UCI Machine Learning Repository. University of California, School of Information and Computer Science, Irvine, CA (2013), http://archive.ics.uci.edu/ml
3. Ciresan, D.C., Meier, U., Gambardella, L.M., Schmidhuber, J.: Convolutional Neural Network Committees For Handwritten Character Classification. In: International Conference on Document Analysis and Recognition (2011)
4. Cruz, R.M.O., Cavalcanti, G.D.C., Ren, T.I.: Handwritten Digit Recognition Using Multiple Feature Extraction Techniques and Classifier Ensemble. In: 17th International Conference on Systems, Signals and Image Processing (2010)
5. Janning, R., Horváth, T., Busche, A., Schmidt-Thieme, L.: GamRec: A Clustering Method Using Geometrical Background Knowledge for GPR Data Preprocessing. In: Iliadis, L., Maglogiannis, I., Papadopoulos, H. (eds.) AIAI 2012. IFIP AICT, vol. 381, pp. 347–356. Springer, Heidelberg (2012)
6. Janning, R., Horváth, T., Busche, A., Schmidt-Thieme, L.: Pipe Localization by Apex Detection. In: Proceedings of the IET International Conference on Radar Systems (Radar 2012), Glasgow, Scotland (2012)
7. Janning, R., Busche, A., Horváth, T., Schmidt-Thieme, L.: Buried Pipe Localization Using an Iterative Geometric Clustering on GPR Data. In: Artificial Intelligence Review. Springer (2013), doi:10.1007/s10462-013-9410-2
8. Janning, R., Schatten, C., Schmidt-Thieme, L.: HNNP – A Hybrid Neural Network Plait for Improving Image Classification with Additional Side Information. In: Proceedings of the IEEE International Conference on Tools With Artificial Intelligence (ICTAI) 2013, Washington DC, USA, pp. 24–29 (2013)
9. LeCun, Y., Bottou, L., Bengio, Y., Haffner, P.: Gradient-Based Learning Applied to Document Recognition. Proceedings of the IEEE 86(11), 2278–2324 (1998)
10. Pettengill, G.H., Ford, P.G., Johnson, W.T.K., Raney, R.K., Soderblom, L.A.: Magellan: Radar Performance and Data Products. Science 252, 260–265 (1991)
11. Saunders, R.S., Spear, A.J., Allin, P.C., Austin, R.S., Berman, A.L., Chandlee, R.C., Clark, J., Decharon, A.V., Dejong, E.M.: Magellan Mission Summary. Journal of Geophysical Research Planets 97(E8), 13067–13090 (1992)
12. Sharkey, A.J.C., Sharkey, N.E.: Combining diverse neural nets. The Knowledge Engineering Review 12(3), 231–247 (1997)
13. Simard, P.Y., Steinkraus, D., Platt, J.: Best Practices for Convolutional Neural Networks Applied to Visual Document Analysis. In: External Link International Conference on Document Analysis and Recognition (ICDAR), pp. 958–962. IEEE Computer Society, Los Alamitos (2003)
14. Ting, K.M., Witten, I.H.: Issues in stacked generalization. Journal of Artificial Intelligence Research 10, 271–289 (1999)
15. TIMIT Acoustic-Phonetic Continuous Speech Corpus, http://www.ldc.upenn.edu/Catalog/CatalogEntry.jsp?catalogId=LDC93S1
16. Wu, J., Xiong, H., Chen, J.: COG: local decomposition for rare class analysis. Data Mining and Knowledge Discovery 20, 191–220 (2010)

A Large-Scale, Hybrid Approach for Recommending Pages Based on Previous User Click Pattern and Content

Mohammad Amir Sharif and Vijay V. Raghavan

The Center for Advanced Computer Studies
University of Louisiana at Lafayette
Lafayette, LA 70503, USA
{mas4108,vijay@cacs.louisiana.edu}

Abstract. In a large-scale recommendation setting, item-based collaborative filtering is preferable due to the availability of huge number of users' preference information and relative stability in item-item similarity. Item-based collaborative filtering only uses users' items preference information to predict recommendation for targeted users. This process may not always be effective, if the amount of preference information available is very small. For this kind of problem, item-content based similarity plays important role in addition to item co-occurrence-based similarity. In this paper we propose and evaluate a Map-Reduce based, large-scale, hybrid collaborative algorithm to incorporate both the content similarity and co-occurrence similarity. To generate recommendation for users having more or less preference information the relative weights of the item-item content-based and co-occurrence-based similarities are user-dependently tuned. Our experimental results on Yahoo! Front Page "Today Module User Click Log" dataset shows that we are able to get significant average precision improvement using the proposed method for user-dependent parametric incorporation of the two similarity metrics compared to other recent cited work.

Keywords: Recommender Systems, Item-based Collaborative Filtering, Map-Reduce, Item-Item content-based similarity, Item-Item co-occurrence-based similarity, Mahout.

1 Introduction

In this age of Internet, the amount of information we come across is overwhelming. It is really difficult to find relevant information useful for a person. There are numerous research studies in this branch of research called information filtering [1, 2, 3, 4]. Information filtering system assists users by filtering the data source and delivers relevant information to the users. When the delivered information comes in the form of suggestions, an information filtering system is called a recommender system.

Recommender systems are mainly classified as content-based and collaborative filtering-based. In content-based recommendation, user's profile-vector is matched with item's profile vector to generate recommendation [2]. Content-based system depends on well structured attributes and reasonable distribution of attributes across

T. Andreasen et al. (Eds.): ISMIS 2014, LNAI 8502, pp. 103–112, 2014.
© Springer International Publishing Switzerland 2014

items. A content-based system is unlikely to find surprising connection. Rather, such a system aims to find a substituting item. In many cases, getting common attributes is not easy and complimentary items are preferred, rather than simple substitution. So, content-based systems are not preferred in many cases. In order to overcome these problems, collaborative filtering approach is introduced, which depends mainly on the users' item preferences information. It does not depend on the content information.

Collaborative filtering is classified as either user-based or item-based [3]. In the user-based collaborative filtering, recommendation is generated based on weighted average rating of similar or neighboring users' ratings. But as the number of users increases many users will not have sufficient rating and the user-item rating matrix will be very sparse. This can lead to finding no or dissimilar users as neighbors of a targeted user, making the recommendation task difficult. To overcome this sparsity problem, item-item collaborative recommendation is introduced, where item-item similarity is used to generate recommendation. Because it is more probable that each item will be rated by many users, finding effective neighbors of an item will be much easier.

In user-based method overcoming sparsity to get sufficient neighbor is acceptable, but in item-based collaborative algorithm, finding effective similar items is not enough, because some neighboring item may not be rated by the targeted users to contribute to the prediction calculation. The prediction is calculated by making weighted average of targeted user's rating to the neighboring items of the preferred item [4]. So if all or sufficient number of neighboring items can't contribute to recommendation calculation, then the recommendation will not be accurate. So, if a targeted user has a small number of preferred items, the recommendation generation needs to be modified.

Moreover, if the number of users and items increases to a massive scale, the traditional method of deploying an algorithm in a single machine does not work. So, scalability is a very common problem in current recommendation settings.

In this work, we propose to use a hybrid item-similarity score by combining item-item co-occurrence similarity and item-item content-based similarity. We use a parametric incorporation so that the parameter can have different values for different target users based on the length of their preference list. We test whether targeted users having a short preference lists get better recommendation if content-based similarity is given more weight for recommendation generation. Our experimental results validate the claim.

Moreover, we believe that the Map-Reduce based implementation of our proposed hybrid method is highly scalable to handle many recommendations in parallel. The experimental results on "Yahoo front page today module" dataset shows significant speed up in recommendation generation.

2   Literature Survey and Background

2.1   Sparsity Problem

In most e-commerce recommendation systems, the numbers of users and items are very large and many of the users don't rate or their actual preference is not obtainable.

In addition, even many popular items are not rated by many users. These factors cause user sparsity and item sparsity. Table 1 shows a user-item preference matrix where we can see that the matrix is not filled. In this case, the similarity calculation based on item cooccurrences will not be very much effective. Selecting either user-based or item-based collaborative approach automatically is also a method to handle sparsity [5]. Predictions corresponding to a user-based and item-based technique are calculated separately and, then, decisions are combined to make an integrated prediction in [5], which is infeasible for a large-scale implementation. Sometimes users sparsity is resolved by applying item-based filtering. But, in order to resolve item-sparsity, some studies have combined content-based item-similarity with the co-occurrence-based item similarity in item-based recommender systems [6, 7]. A weighted combination of content similarity and collaborative similarity is proposed in [6]. Table 2 shows an item-feature matrix. It is possible to calculate pair-wise similarity between items using the features of items. So, if the feature-based similarity and co-occurrence-based similarity are incorporated properly, better prediction is expected. Since the customization of weighting for the whole data set is practically infeasible, [7] proposes an incorporation technique giving the same weight for both the similarity values in multiplicative form. The performance of developed system in [7] is poor when the preference list size is small. Small preference list size hurts co-occurrence-based collaborative approach.

Table 1. User-Item preference matrix

Item User	I_1	I_2	I_3	I_4
U_1	$P_{1,1}$	$P_{1,2}$	$P_{1,3}$	?
U_2	$P_{2,1}$	$P_{2,2}$	$P_{2,3}$	$P_{2,4}$
U_3	$P_{3,1}$	$P_{3,2}$	?	$P_{3,4}$
U_4	?	$P_{4,2}$	?	?

So, if the active user does not have sufficient preference information, the resulting collaborative approach gives poor recommendation even though better item similarity is available. In this work we propose a user-dependent weighting technique so that, for the targeted users having short preference lists, more weight can be given on content-based similarity during prediction. On the other hand, users having longer preference lists will get more weight on co-occurrence based similarity prediction.

Table 2. Item-Feature matrix

Feature Item	F_1	F_2	F_3	F_4
I_1	$F_{1,1}$	$F_{1,2}$	$F_{1,3}$	$F_{1,4}$
I_2	$F_{2,1}$	$F_{2,2}$	$F_{2,3}$	$F_{2,4}$
I_3	$F_{3,1}$	$F_{3,2}$	$F_{3,3}$	$F_{3,4}$
I_4	$F_{4,1}$	$F_{4,2}$	$F_{4,3}$	$F_{4,4}$

2.2   Scalability Problem

As the number of users and items increases the computational complexity increases drastically in any recommendation setting.

In most of the recommender systems, the recommendation generation needs to be real time. There are already some works to do large-scale recommendation [8, 9]. While some of them have bottleneck in accuracy, others lack in the quality of performance. In this work, we propose an adaptive recommendation system that adjusts weight parameters according to the profile list lengths on top of the Apache Mahout's large-scale item-based recommendation systems. Mahout's Map-Reduce based recommendation systems run on top of Hadoop [10].

Map-Reduce. Map-reduce is a large-scale parallel computing framework developed by Google. A Map-reduce job mainly has two types of functions called *map ()* and *reduce()*, taking input from distributed file system (DFS).

The *Map* and *Reduce* functions of *Map-Reduce* are both defined with respect to data structured in (key, value) pairs. *Map* takes one pair of data with a type in one data domain, and returns a list of pairs in a different domain: Map(k1,v1)→ list(k2,v2).

The *Map* function is applied in parallel to every pair in the input dataset. This produces a list of pairs for each call. After that, the Map-Reduce framework collects all pairs with the same key from all lists and groups them together, creating one group for each key.

The *Reduce* function is then applied in parallel to each group, which in turn produces a collection of values in the same domain: Reduce(k2, list(v2)) →list(k3, v3).

Mahout's Large Scale Recommendation. In Mahout's algorithm the item similarity is calculated based on co-occurrences. Their distributed algorithm is based on parallel implementation of matrix multiplication. They run four Map-Reduce jobs to implement the recommendation.

Our content similarity-oriented, large-scale item based collaborative recommender system is built on top of an existing large-scale recommender system by apache Mahout [11]. We extended page co-occurrence to include content-based item-item similarity within the same Map-Reduce process. The structure is given in Fig. 1.

This algorithm runs four Map-Reduce jobs. The **first** Map-Reduce job builds user vector. Mapper takes file position as key and line of text containing user id, item id and preference as value; and outputs user id as key and (item id, preference) as value. The Reducer outputs user id as key and mahout's vector representation of items as value.

Second map-reduce job creates a co-occurrence matrix. The mapper takes user id as key and user vector as value; and outputs item id as key and other item id as value. The output from the mapper is itemId as key; and another-item with corresponding similarity as value. The reducer here outputs item id as key and a column vector of co-occurrence matrix as value.

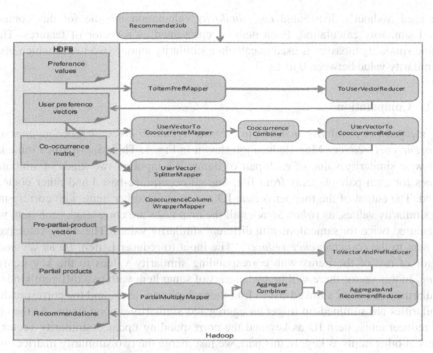

Fig. 1. Architecture of Mahout's Recommendation Engine

Inside-out technique is used for map-reduce based matrix multiplication. In **third** map-reduce job one mapper called *PartialMultiply1* takes the output of the first map-reduce job as input where user id is key and user's preference vector is the value. The output of this map job is item id as key and (user id, preference value) as value. Another mapper called *Partial Multiply2* inputs and outputs item id as key and co-occurrence matrix column Vector as value. The *Partial multiply* reducer inputs item id as key; and many (user ID, preference) pairs with a co-occurrence matrix column vector as value, which are value parts from previous two mapper's output. The reducer outputs item id as key and column vector with (user ID, preference) pair as value.

In the **fourth** map-reduce job a mapper called *aggregate mapper* takes the key value pair as input from the output of previous map-reduce job i.e. item id as key and column vector and (user ID, preference) pair as value. And the mapper outputs user id as key and column vector times preference as value. *Aggregate reducer* receives user id as key and vectors from previous mapper output for corresponding user id. Reducer sums to make recommendation vector and finds top n values for recommendations. For the top n values, the reducer outputs user id as key and (item ID, value) as value.

3   Our Approach

3.1   Content-Based Item-Item Similarity Calculation

As the content-based similarity is stable, we calculated the item-item content-based similarity off-line, and then incorporated it with the co-occurrence similarity later.

We used Mahout's distributed *row similarity* calculation module for this content-based similarity calculation. Each item is represented as a vector of features. Then cosine similarity measure is used to calculate similarity among two items, which gives a similarity value between 0 to 1.

3.2   Computation

We merged two similarity matrices in the *user vector to co-occurrence-mapper* and *co-occurrence-reducer* Map-Reduce job shown in Fig. 1. This Map-Reduce job takes pair wise similarity value of each pair of items. We inputted two kinds of similarity values for each pair of items from file; one co-occurrence-based and other content-based. The output of the mapper is item ID as *key* and another item, with corresponding similarity values, as *value*. So after all the map tasks are completed, each item will be emitted twice for same item with different similarity values. These key-value pairs are sent to the *co-occurrence-reducer*. The input to reducer is item ID as *key* and a vector of rest of the items with corresponding similarity values to the key item as *value*. In this vector there will be two entries of same item with two different kinds of similarities. We used a linked-list to find those duplicate items, and the corresponding similarities are summed-up to get an aggregated similarity among items. At the end, the reducer emits item ID as key and the corresponding updated similarity vector in terms of other items as key. In this part, we just merge the two similarity matrices into one. As the co-occurrence similarity ranges from [0, 185] and content similarity ranges from (0, 1], it is very easy to separate these similarity components later and give different weights of importance for combining the two similarity values based on different user preference list length, during recommendation generation.

In *aggregateAndReccomendReducer* module, we apply separate weights to the different similarity components based on the users' preference list length. In order to do so, we separate the whole and fractional part of the similarity value. The whole number is co-occurrence similarity and normalized to [0, 1] scale, the fractional part is content-similarity. After normalization we compute the weighted sum of the two different similarities for a user, based on her preference list length. In this way, the two different similarities are merged to generate the prediction.

3.3   Incorporating Content-Based and Co-occurrence-based Similarity

In our similarity calculation method, we used the following equation to calculate similarity, $Sim(ItemA, ItemB, U)$ among two items for a user U,

$$Sim(ItemA, ItemB, U)$$
$$= F(U) \times Content_{Sim(ItemA, ItemB)} + (1 - F(U))$$
$$\times Cooccurrence_{sim(ItemA, ItemB)}$$

where, $F(U)$: Content similarity weight to be considered for user U
$1 - F(U)$: Cooccurrence similarity weight to be considered for user U
$Content_{Sim(ItemA, ItemB)}$: Content similarity among items ItemA and ItemB
$Cooccurrence_{sim(ItemA, ItemB)}$: Cooccurrence similarity of ItemA and ItemB

4   Experimental Setup

We used the Yahoo! Front Page Today Module User Click Log Dataset for the experiments. The dataset is a fraction of user click log for news articles displayed in the Featured Tab of the Today Module on Yahoo! Front Page (http://www.yahoo.com) during the first ten days in May 2009. The dataset contains 45,811,883 user visits to the Today Module. For each visit, both the user and each of the candidate articles are associated with a feature vector of dimension 6. These features were represented by some numerical numbers without any identification. We preprocessed the data set to extract article click information for each user and the article vectors for each article for this experiment. There were 271 articles that were displayed in Yahoo! front page in those ten days.

4.1   Preference List Length Based Evaluation

We choose several different sets of users based on different preference list lengths for our evaluation. In this case, the preference means clicked pages. Each selected user set had users having preference list length 6 to 10, 11 to 20, 21 to 35, 36 to 55, more than 35 and more than 55. Although we had around 1.3 million users, very few of them had prefernce lists of size greater than 5. There were 560 users having preference list length 6 to 10, 333 users with length 11 to 20, 142 users with length 21 to 35, 87 users with length 36 to 55 and 214 users having prefernce length more than 55. For these sets, we trained the recommender system by 70% of the preferences of the users in the current user set along with 100% of other users' preferences and kept the remaining 30% preference list in the current user set for testing i.e. testing preferences. For these users, we explore the recommendations generated by the system for different weight combinations of content similarity and co-occurrence similarity. For each selected user set, we ran the experiment by increasing the content similarity weight; that is, by taking the content-based similarity and co-occurrence-based similarity weight pair respectively as (0, 1), (0.1, 0.9), (0.2, 0.8), (0.3, 0.7), (0.4, 0.6), (0.6, 0.4), (0.7, 0.3), (0.8, 0.2), (1, 0). Considering the rank order of first 10 recommendations, we calculated average precision for an active user, for each combination of weights, based on ranking of pages in the testing preferences. Then we found the mean average precision (MAP) over all the users within those selected user sets [12].

Fig. 2, 3 and 4 show how the recommender's performance changes with the increase of content similarity weight or decrease of co-occurrence similarity weight for users having preference list length respectively 6 to 10, 11 to 20, 21 to 35 and 36 to 55, more than 35 and more than 55.

From Fig. 2 and 3, we can see that for users having prefernce list length 6 to 10, 11 to 20 and 21 to 35 the best performance is obtained for content similarity weight 1.0 and cooccurence similarity weight 0.0. For users with prefence list length 36 to 55, we get better performance with content similarity weight 0.4 and cooccurence similarity weight 0.6, which shows the imporatnce of co-occurrence as the length grows. Fig. 4 says that for users having prefernce list length greater than 36 or 55, we get better

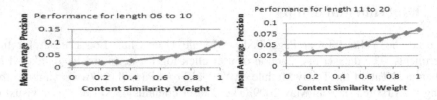

Fig. 2. Change of performance with increasing content similarity weight or decreasing co-occurrence similarity weight for users having preference list length 06 to 10 and 11 to 20

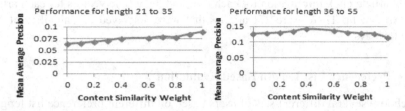

Fig. 3. Change of performance with increasing content similarity weight or decreasing co-occurrence similarity weight for users having preference list length of 21 to 35 and 36 to 55

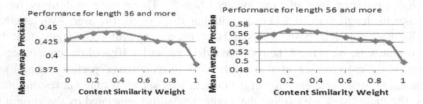

Fig. 4. Change of performance with increasing content similarity weight or decreasing co-occurrence similarity weight for users having preference length greater than 35 and 55

performance at content similarity weight of 0.4 and 0.3 respetcively. It also shows the importance of co-occurrence weight for users having long prefernce list. So, for shorter preference list length, less than 36, we give more weight on content.

Observing the results in Fig. 2, 3 and 4, we selected content similarity weight 1.0 for users having length less than 36; content similarity weight 0.4 for length 36 to 55 and content similarity weight 0.3 for length greater than 55. We used this weighting mechanism to do our full-scale testing. We tested our system with all the 1336 users having prefernce length greater than 5. Again, for each user in this set of 1336 users, 70% of papges in the preference list are used for training and the rest for testing. Table 3 shows the performance comparison of our user-dependent method with systems having only co-occurrence similarity, content similarity; and system described by Puntheeranurak, S., where the two similarities are just multiplied without using any parameter [7]. We can see from Table 3 that, we got 22% better performance than just using a simple merge process without parameter tuning.

Table 3. Comparative performance of our user-dependent parametric method with alternative approaches for all the users having preference list length greater than 5

	Co-occurrence similarity only	Content similarity only	Method in reference[7]	User-dependent adjustment
MAP	0.1174517	0.1578064	0.1253729	0.16152156

4.2   Scalability for Online Recommendations to Growing Number of Users

We ran our recommendation algorithm both in single node Hadoop ecosystem in a virtual machine and in a Cloudera 4.5 cluster with 3 nodes with one master node and two slave nodes. Each machine was with quad-core CPU with speed 2.4GHz, 4GB memory, running on Centos 6.2 OS. Table 4 shows the comparative elapsed time requirement for recommendation generation to varying number of users. The time required to generate a similarity matrix, which is used in recommendation generation is approximately 11 minutes for single machine and 5.5 minutes for the cluster environment. In table 4, we report the required time to generate recommendation from already created similarity matrix for different number of users.

Table 4. Comparison of recommendation time for different number of users

No. of users	Single Machine(time)	Cluster(time)
100K	3m29sec	1m32sec
200K	4m15sec	1m44sec
300K	5m1sec	1m57sec
400K	6m28sec	2m5sec
500K	7m52sec	2m26sec

From Table 4, we can see that for 100,000 users to 500,000 users the running time increases linearly with the number of targeted users for the single node recommendation generation. In the distributed environment, on a Hadoop cluster, the running time is also linear, but increases at a lower rate. Definitely a cluster with many nodes could make the running time nearly constant.

5   Conclusion

In this paper we have presented a Map-Reduce based, scalable recommender system that uses a unique, preference list length based weighting technique to incorporate content-based and co-occurrence-based item-item similarities to help make better recommendations in the situation where preference list sizes are small. Our length-based weighting technique gives 22% better performance compared to a recent work described in [7]. As we selected only users having preference list lengths greater than five for each test set, we could not show full potential of length-based approach for determining the weight parameters. But, in future, we plan to use the whole dataset to determine the length-based parameters applicable to a more global context. We will also use a bigger compute cluster to demonstrate better speed-up.

References

1. Hanani, U., Shapira, B., Shoval, P.: Information Filtering: Overview of Issues, Research and Systems. User Modeling and User-Adapted Interaction 11, 203–259 (2001)
2. Delgado, J., Ishii, N., Ura, T.: Content-based Collaborative Information Filtering: Actively Learning to Classify and Recommend Documents. In: Klusch, M., Weiss, G. (eds.) CIA 1998. LNCS (LNAI), vol. 1435, pp. 206–215. Springer, Heidelberg (1998)
3. Adomavicius, G., Tuzhilin, A.: Toward the Next Generation of Recommender Systems: A Survey of the State-of-the-art and Possible Extensions. IEEE Transactions on Knowledge and Data Engineering 11(6), 734–749 (2005)
4. Su, X., Khoshgoftaar, T.M.: A survey of collaborative filtering techniques. Journal of Advances in Artificial Intelligence (2009)
5. Hu, R., Lu, Y.: A Hybrid User and Item-based Collaborative Filtering with Smoothing on Sparse Data. In: Proceedings of the 16th International Conference on Artificial Reality and Telexistence–Workshops (2006)
6. Gong, S.J., Ye, H.W., Shi, X.Y.: A Collaborative Recommender Combining Item Rating Similarity and Item Attribute Similarity. International Seminar on Business and Information Management (2008)
7. Puntheeranurak, S., Chaiwitooanukool, T.: An Item-based Collaborative Filtering Method using Item-based Hybrid Similarity. In: 2nd International Conference on Software Engineering and Service Science (2011)
8. Jiang, J., Lu, J., Zhang, G., Long, G.: Scaling-up Item-based Collaborative Filtering Recommendation Algorithm based on Hadoop. IEEE World Congress on Services (2011)
9. Chen, Y., Pavlov, Y.: Large Scale Behavioral Targeting. In: 15th ACM SIGKDD International Conference on Knowledge Discovery and Data Mining (2009)
10. Apache Hadoop, http://hadoop.apache.org/
11. Apache mahout, https://mahout.apache.org/
12. Text retrieval quality, http://www.oracle.com/technetwork/database/enterprise-edition/imt-quality-092464.html

EverMiner Prototype
Using LISp-Miner Control Language

Milan Šimůnek and Jan Rauch

Faculty of Informatics and Statistics, University of Economics, Prague
nám W. Churchilla 4, 130 67 Prague 3, Czech Republic
{simunek,rauch}@vse.cz

Abstract. The goal of the *EverMiner* project is to run automatic data mining process starting with several items of initial domain knowledge and leading to new knowledge being inferred. A formal description of items of domain knowledge as well as of all particular steps of the process is used. The *EverMiner* project is based on the *LISp-Miner* software system which involves several data mining tools. There are experiments with the proposed approach realized by manual chaining of tools of the *LISp-Miner*. The paper describes experiences with the *LISp-Miner Control Language* which allows to transform a formal description of data mining process into an executable program.

1 Introduction

The *EverMiner* project is introduced in [8,14]. Its idea is to automate data mining process with help of several items of initial domain knowledge. All items of domain knowledge are formalized as well as particular steps of the process [7]. GUHA procedures [1,3,10] are used as core analytical tools. Input of each GUHA procedure consists of an analysed data and several parameters defining a large set of relevant patterns, output is a set of relevant patterns true in the analysed data.

The *EverMiner* project is based on the *LISp-Miner* system [12,9] which involves several GUHA procedures as well as modules for data preprocessing and dealing with items of domain knowledge. There are several experiments with the proposed approach realized by manual chaining of the procedure mining for generalized association rules and modules of *LISp-Miner* system [11]. Process of data mining with association rules is described by FOFRADAR – a formal frame for data mining with association rules [7] based on observational calculi. The goal of this paper is to describe experiences with a scripting language *LISp-Miner Control Language* (LMCL) which allows to transform a formal description of data mining process into an executable program.

Main features of the *EverMiner* project are outlined in Section 2. An example of a simple and manually implemented concept is described in Section 3. An application of LMCL to run an enhancement of this example is introduced in Section 4. The example concerns medical data. However, the goal of this example is not to get new medical knowledge, but to introduce the LMCL language.

T. Andreasen et al. (Eds.): ISMIS 2014, LNAI 8502, pp. 113–122, 2014.
© Springer International Publishing Switzerland 2014

There are various approaches to describe and automate data mining process [2,5,6,15]. Their detailed comparison with the presented approach is out of the scope of this paper and is left as a further work. Let us mention that they do not use the formalization of a data mining process based on observational calculi.

2   EverMiner Principles

The concept of the *EverMiner* is based on two loops – an outer loop (see the left part of Fig. 1) and an inner loop (see the right part of Fig. 1) in the phase (4) of answering analytical questions formulated in step (3) in the outer loop. A domain expert (1) is necessary to supervise the automated process. But his role is limited mainly to approval or disapproval of newly inferred knowledge. He or she doesn't intervene directly into the outer and inner loops so they could be really automated. Domain knowledge (2) contains both the initial domain knowledge prepared by the domain expert and the newly inferred domain knowledge which is clearly flagged as *data specific* and needs an approval from the domain expert before it could become part of accepted domain knowledge. Nevertheless, newly inferred knowledge could be used immediately to formulate new analytical questions so the whole automated process could carry on even if the expert is busy and not available. He or she could return back later and approve or disapprove the whole bunch of newly inferred knowledge afterwards.

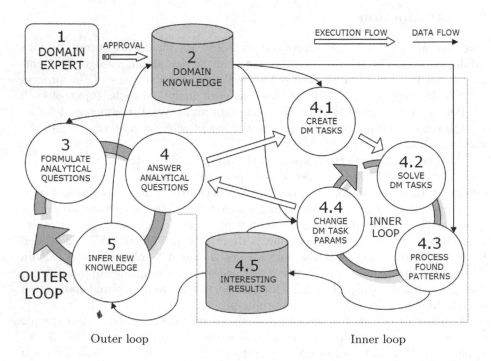

Fig. 1. *EverMiner* seen as two loops

Domain knowledge is formalized in a way both understandable to domain experts and suitable to be used in an automated data mining process. The outer loop starts with currently available domain knowledge (2). There are several ways to formulate reasonable analytical questions based on the current knowledge in step (3). Shortly, we can try (a) to verify that domain knowledge is valid in analysed data, (b) to found new patterns not yet covered by current knowledge or (c) to found exceptions to known patterns.

By answering analytical questions in step (4) we mean to create and solve several data mining tasks, results of which can contribute to a solution of the analytical question formulated in step (3). We plan to start with data mining tasks which can be solved by the GUHA procedures implemented in the *LISp-Miner*. An example of formulation of analytical questions and application of the GUHA procedure *4ft-Miner* dealing with association rules to solve one of the formulated analytical questions is in Section 3. Usually, several applications of GUHA procedures are necessary to solve one analytical question.

3   Mechanical Turk Proof of Concept

3.1   Data Set STULONG

As demo data we use the data matrix *Entry* belonging to the dataset STULONG concerning *Longitudinal Study of Atherosclerosis Risk Factors*[1].The data matrix concerns 1.417 male patients, each row describes one patient. The data matrix has 64 columns corresponding to particular attributes of patients. We use four groups of attributes – *Personal, Diet, Measures*, and *Examinations*.

Group *Personal* has three attributes: *Marital_Status* (4 categories i.e. possible values), *Education* (4 categories), and *Responsibility* in a job (4 categories). Group *Diet* has two attributes: *Beer* (3 categories) and *Coffee* (3 categories). Group *Measures* has only one attribute *BMI* (i.e. Body Mass Index) with 17 categories $- < 21, \langle 21; 22 \rangle, \langle 22; 23 \rangle, \dots, \langle 35; 36 \rangle, \geq 36$. Group *Examinations* has three attributes: *Diastolic* blood pressure (7 categories), *Systolic* blood pressure (9 categories) and *Cholesterol* in mg% (10 categories). Fig. 4 presents the selected attributes and categories.

3.2   Domain Knowledge and Analytical Questions

Three types of domain knowledge related to the STULONG data are managed by the *LISp-Miner* system [11]: *groups of attributes, information on particular attributes* and a *simple influence between attributes*. There are 11 basic groups of attributes defined at http://euromise.vse.cz/challenge2004/data/entry/ and four additional groups of attributes defined in the previous section. Groups of attributes are used to define reasonable analytical questions.

Information on particular attributes include information on types of attributes (nominal/ordinal/cardinal). An example of a simple influence between attributes

---

[1]   see http://euromise.vse.cz/challenge2004.

is an SI-formula *BMI ↑↑ Diastolic* saying that if BMI of a patient increases then patient's diastolic blood pressure increases too [11].

We outline how these items of knowledge can be used to define reasonable analytical questions, see outer loop in Fig. 1. We use groups of attributes *Personal, Diet, Measures,* and *Examinations* and SI-formula *BMI ↑↑ Diastolic.* An example of an analytical question is: *In the data matrix* Entry, *are there any interesting relations between combinations of attributes from* Personal, Diet, *and* Measures *on one side and attributes from* Examinations *on the other side?* We denote this analytical question as $\mathcal{AQ}_1$, symbolically

$\mathcal{AQ}_1$:   [*Entry: Personal, Diet, Measures* $\approx^?$ *Examinations*].

This question can be enhanced: *In the data matrix* Entry, *are there any interesting relations between combinations of attributes from* Personal, Diet, *and* Measures *on one side and attributes from* Examinations *on the other side? However, we are not interested in consequences of the known fact that if BMI increases, then diastolic blood pressure increases too.* Symbolically we can write

$\mathcal{AQ}_{1E}$:   [*Entry: BMI ↑↑ Diastolic ↛ Personal, Diet, Measures* $\approx^?$ *Examinations*].

3.3   EverMiner Analytical Questions

Similarly, additional analytical questions can be

$\mathcal{AQ}_2$:   [*Entry: Diet, Measures, Examinations* $\approx^?$ *Personal*]
$\mathcal{AQ}_3$:   [*Entry: Measures, Examinations, Personal* $\approx^?$ *Diet*]
$\mathcal{AQ}_4$:   [*Entry: Examinations, Personal, Diet* $\approx^?$ *Measures*].

We assume here that the only item of knowledge of the type *simple influence between attributes* is *BMI ↑↑ Diastolic.* It makes no sense to use this item in the introduced additional analytical questions. We assume here that the analytical questions $\mathcal{AQ}_1$, $\mathcal{AQ}_2$, $\mathcal{AQ}_3$, $\mathcal{AQ}_4$ are the only analytical questions to be solved. However, actually hundreds or thousands of similar analytical questions can be formulated using available groups of attributes and SI-formulas.

3.4   Solving a GUHA Task – An Example

The *EverMiner* project is based on applications of GUHA procedures implemented in the *LISp-Miner* system, there are nine GUHA procedures mining for various types of patterns [9,12]. Each analytical question formulated in the outer loop, see Fig. 1 is transformed into several data mining tasks. We outline how the GUHA procedure *4ft-Miner* can be used to solve the analytical question $\mathcal{AQ}_1$: [*Entry: Personal, Diet, BMI* $\approx^?$ *Examinations*].

The procedure *4ft-Miner* deals with association rules $\varphi \approx \psi$ where φ and ψ are Boolean attributes derived from columns of an analysed data matrix. Boolean attribute φ is called *antecedent* and ψ is called *succedent*. The symbol $\approx$ is a *4ft-quantifier*, it corresponds to a criterion concerning quadruples $\langle a, b, c, d \rangle$ of non-negative integers a, b, c, d such that $a + b + c + d > 0$. The association rule $\varphi \approx \psi$ is true in the data matrix $\mathcal{M}$ if the criterion corresponding to $\approx$ is satisfied

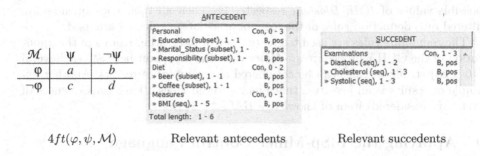

$4ft(\varphi, \psi, \mathcal{M})$ Relevant antecedents   Relevant succedents

Fig. 2. $4ft(\varphi, \psi, \mathcal{M})$ and definitions of sets of relevant antecedents and succedents

for a contingency table $4ft(\varphi, \psi, \mathcal{M})$ of φ and ψ in data matrix $\mathcal{M}$, see Fig. 2. Here a is the number of rows of $\mathcal{M}$ satisfying both φ and ψ, b is the number of rows of $\mathcal{M}$ satisfying φ and not satisfying ψ, etc.

Input of the *4ft-Miner* consists of an analysed data matrix, definitions of sets of relevant antecedents and succedents and of a 4ft-quantifier $\approx$. Output is a set of all rules $\varphi \approx \psi$ true in the analysed data matrix where φ is a relevant antecedent and ψ is a relevant succedent.

A definition of a set of relevant antecedents used to solve the analytical question $\mathcal{AQ}_1$ is in Fig. 2. An antecedent is a conjunction $\varphi_P \wedge \varphi_D \wedge \varphi_B$ where φ_P is a Boolean characteristics of the group *Personal*, similarly for φ_D, φ_B and the groups *Diet*, *BMI* respectively. The attribute φ_P is a conjunction of 0 - 3 Boolean attributes created from attributes *Education*, *Marital_Status*, and *Responsibility*. Four Boolean attributes *Education(basic)*, ..., *Education(university)* are automatically created from the attribute *Education*, similarly for *Marital_Status*, and *Responsibility*. The attributes φ_D are created analogously. The attributes φ_B are in a form $BMI(\alpha)$ where α is a sequence of 1–5 consecutive categories of BMI (see expression BMI(seq), 1 - 5 in the definition of relevant antecedents in Fig. 2). This way 75 Boolean characteristics of BMI are defined, $BMI(< 21, \langle 21; 22 \rangle, \langle 22; 23 \rangle)$ i.e. $BMI(< 23)$ being an example. The expressions B, pos means that all Boolean attributes are equally important [10].

The set of relevant succedents is defined similarly. Sequences of categories are again used, see right part of Fig. 2. We used the 4ft-quantifier $\Rightarrow_{p,B}$ of founded implication defined by the criterion $\frac{a}{a+b} \geq p \wedge a \geq B$. The rule $\varphi \Rightarrow_{p,B} \psi$ means that at least $100p$ per cent of rows of $\mathcal{M}$ satisfying φ satisfy also ψ and that there are at least B rows of $\mathcal{M}$ satisfying both φ and ψ. There are about 20 additional 4ft-quantifiers implemented in the *4ft-Miner*.

We started with the quantifier $\Rightarrow_{0.95,50}$. More than $4.5 * 10^6$ of association rules were verified in 150 sec. (PC with 4GB RAM and Intel i5-3320M processor at 2.6 GHz) and no true rule was found. After about 10 automatically computed modifications of parameters we get 98 true rule for 4ft-quantifier $\Rightarrow_{0.8,30}$. We used the way introduced in [11] to filter out 32 consequences of $BMI \uparrow\uparrow Diastolic$. This is based on transforming the SI-formula to a set of its atomic consequences i.e. all suitable simple rules $BMI(\alpha) \Rightarrow_{p,B} Diastolic(\beta)$ where α, β are subsets of possible values of *BMI*, *Diastolic* respectively. Then all their consequences are filtered out, deduction rules of the logic of association rules [9] are used.

The remaining rules are used to create a set of interesting results of the application of the GUHA procedure *4ft-Miner* with the 4ft-quantifier $\Rightarrow_{0.8,30}$. Several (10 – 20) strongest rules can be considered as examples of results. Another example of results is an assertion that among found rules 10 are consequences of (yet not considered) item of knowledge $BMI \uparrow\uparrow Systolic$.

4   Applying the LISp-Miner Control Language

Although theoretically proved valid, a manual implementation of the *EverMiner* steps is not feasible. The number of tasks to be solved could easily grow above hundreds and even thousands. Therefore an automated approach is necessary and here the *LISp-Miner Control Language* (LMCL) steps in. LMCL is a scripting language based on Lua and its syntax [4]. The main purpose of LMCL is to provide programmable means to automate all the main phases of data mining i.e. to import data, to pre-process them, to formulate reasonable analytical tasks, to process those tasks and finally to digest found patterns and to report only the interesting ones to the user or to infer directly new items of domain knowledge. In this sense, the language is a necessary prerequisite for automation of data mining process. But, it could serve other purposes too [13].

4.1   EverMiner Stulong Simple Algorithm

The *EverMiner Stulong Simple* demo presented here is really a simplified version of the *EverMiner* concept, in line with the proof of concept described in Section 3. Its main purpose is to proof that the LMCL is able to automate data mining process and to solve many data mining tasks in parallel with speed that would be never possible to achieve through standard user interface.

Few user-defined parameters guide the whole process which is fully automated. Apart from the rather technical ones (e.g. a connection string to database with analysed data), there is one piece of domain knowledge in form of groups of attributes and association of attributes with them. The second important input parameters are the minimal and maximal number of patterns to mine. There are several ways how to reduce (or enlarge) task search space to influence the number of found patterns, but they are out of scope of this paper. Just a very simple heuristic has been implemented for now. Nevertheless, it is successfully exploited in the step (4) of Fig. 1 to ensure the number of found patterns is within given range, see Fig. 3.

After a desired number of hypotheses was reached for each task (or a specified maximal limit of iterations was reached), a summary of task results could be prepared in form of an analytical report. An example of such a report (manually shortened) is in Fig. 4.

There is no space to discuss an automated data preprocessing in this paper (an example is in [13]). The first example where LMCL simplifies and speeds-up the data mining process is in the analytical task formulation phase. There

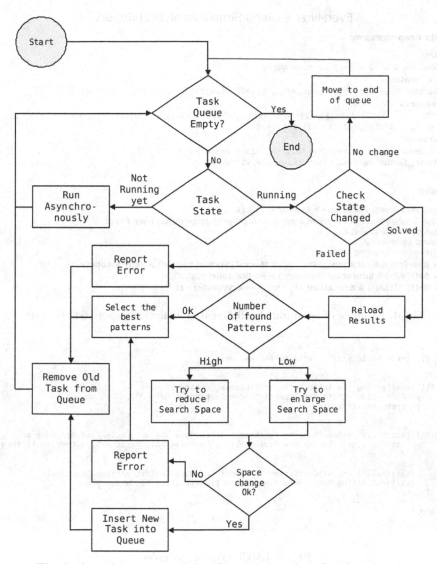

Fig. 3. Asynchronous parallel task processing algorithm description

is a data mining task automatically created for each analytical question. Task parameters are set to pre-defined initial values so the fine-tuning could turn in any direction based on number of hypotheses found. There is a simple heuristic implemented to take a special care of ordinal and cardinal values and to set coefficients of type *sequence* up to length of one third of number categories in corresponding attribute, see an example of LMCL syntax in Fig. 5.

EverMiner Stulong Simple Analytical Report

Data Preprocessing

Diet
Beer: no, 1 or 2 mugs, 3 or more mugs
Examinations
Diastolic: <50;70), <70;80), <80;90), <90;100), <100;110), <110;120), <120;150)
Measures
BMI: <21, <21;22), <22;23), <23;24), <24;25), <25;26), <26;27), <27;28), <28;29), <29;30),
<30;31), <31;32), <32;33), <33;34), >=34
Personal
Education: basic, apprentice, full secondary, university
Marital_Status: married, divorced, single, widower
...

Tasks

Diet, Measures, Personal -> Examinations (08)
Task finished succesfully and an acceptable number of patterns has been found
Number of iterations: 8
Found patterns: 20
The most interesting ones:
- **Beer**(*no*) & **Education**(*apprentice*) & **Marital_Status**(*married*) >÷< **Diastolic**(*<80;100*))
- **BMI**(*<24*) & **Education**(*university*) >÷< **Diastolic**(*<70;90*))
- **BMI**(*<21;24*)) & **Education**(*university*) >÷< **Systolic**(*<110;140*))

Fig. 4. Example of an automatically created analytical report (shortened)

```
for j, attribute in ipairs( attributeArray) do

    -- add literal for each attribute
    ftliteralSetting= lm.tasks.settings.FTLiteralSetting({
        pFTPartialCedentSetting= ftpartialCedentSetting,
        pAttribute= attribute
    });

    if ( (attribute.getDataCharacterTypeCode() == lm.codes.DataCharacterType.Ordinal) or
         (attribute.getDataCharacterTypeCode() == lm.codes.DataCharacterType.Cardinal)) then
    -- sequence up to 1/3 of number of categories

        ftliteralSetting.setCoefficientTypeCode( lm.codes.CoefficientType.Sequence);
        ftliteralSetting.MaxLen= math.max( math.floor( attribute.CategoryCount / 3), 1);
    else
    -- default (subset 1-1)
    end;
end;
```

Fig. 5. LMCL syntax example

The most important phase of the *EverMiner Stulong Simple* example is an asynchronous parallel processing of data mining tasks till the desired number of hypotheses is reached, see Fig. 3. It implements the *EverMiner* inner cycle using a task queue. The queue is initially filled with analytical tasks constructed in the step (4.1) of Fig. 1. The inner loop is processed till the queue is empty. The first step pops the top-most task in the queue and checks its state. If the task is not computed yet, it starts an asynchronous generation and verification of patterns by calling the *LISp-Miner ProcPooler* module for background computation of data mining tasks on multiple processor cores of a hosting computer. (Alternatively, the *LM GridPooler* module could be used to utilize distributed grid of computers.)

If the task has already been started, a query is made for the task state update (possibly made meanwhile by the *LM ProcPooler* module in another thread). If the state has not changed yet, the task is moved to the end of the queue and the first step is repeated for another task. If the task has finished with some error, its state is changed to *failed* and the task is removed from further processing. If the task state has been solved successfully, found patterns are loaded and the execution forks based on the number of found patterns. If it is within the defined acceptable range, the *most interesting patterns* (in this simplified version just the *first three*) are marked as *final results* to be included in the analytical report. If the number of found patters is outside the given range, the concerned task settings are changed to enlarge (respectively to reduce) the solution-space searched. Changes to the task settings are limited for now to a change of parameters p, B of the 4ft-quantifier $\Rightarrow_{p,B}$, see Section 3.4. In the above mentioned application 28 runs of the *4ft-Miner* procedure was used which required 2 minutes and 50 seconds.

4.2 LMCL Performance

LMCL is implemented by the *LM Exec* module of the *LISp-Miner* system using Lua script interpreter library of version 5.2. The used Lua interpreter is really lightweight and proved to be fast, so far tested with up to medium-sized scripts (thousands of code-lines). Script parsing and execution overhead costs are insignificant compared to data mining task solution times or to data transfers from database. Performance of LMCL scripts therefore depends solely on ability of the *LISp-Miner* system modules to compute data mining tasks. It has been proved already (see e.g. [10]) that the algorithms and optimizations techniques implemented in the *LISp-Miner* system lead to solution times linearly dependant on number of rows (objects) in analysed data.

5   Conclusions

We have demonstrated the first experience with the *LISp-Miner Control Language* which allows to transform a formal description of a data mining process into an executable program. It was shown that it is possible to use this language to automate data mining processes with association rules. In the next steps we assume to use theoretical results [7] and considerations [8,14] to enhance the described experiments by deeper application of additional formalized items of domain knowledge and other types of patterns.

Also, a deeper comparison of the presented approach with additional approaches, see e.g. [2,5,6,15], is necessary.

References

1. Hájek, P., Havránek, T.: Mechanising Hypothesis Formation - Mathematical Foundations for a General Theory. Springer, Heidelberg (1978)
2. Hájek, P., Havránek, T.: GUHA 80: An Application of Artificial Intelligence to Data Analysis. Computers and Artificial Intelligence 1, 107–134 (1982)

3. Hájek, P., Holeňa, M., Rauch, J.: The GUHA method and its meaning for data mining. J. Comput. Syst. Sci. 76, 34–48 (2010)
4. Ierusalimschy, R., Figueiredo, L.H., de Celes, W.: Lua an extensible extension language. Software: Practice & Experience 26, 635–652 (1996)
5. Mansingh, G., Osei-Bryson, K.-M., Reichgelt, H.: Using ontologies to facilitate post-processing of association rules by domain experts. Information Sciences 181, 419–434 (2011)
6. Phillips, J., Buchanan, B.G.: Ontology guided knowledge discovery in databases. In: Proc. First International Conference on Knowledge Capture, pp. 123–130. ACM, Victoria (2001)
7. Rauch, J.: Formalizing Data Mining with Association Rules. In: Proceedings of 2012 IEEE International Conference on Granular Computing (GRC 2012), pp. 406–411. IEEE Computer Society, Los Alamitos (2012)
8. Rauch, J.: EverMiner: consideration on knowledge driven permanent data mining process. International Journal of Data Mining, Modelling and Management 4(3), 224–243 (2012)
9. Rauch, J. (ed.): Observational Calculi and Association Rules. SCI, vol. 469. Springer, Berlin (2013)
10. Rauch, J., Šimůnek, M.: An Alternative Approach to Mining Association Rules. In: Lin, T.Y., Liau, C.-J., Ohsuga, S., Hu, X., Tsumoto, S. (eds.) Foundations of Data Mining and knowledge Discovery. SCI, vol. 6, pp. 211–231. Springer, Heidelberg (2005)
11. Rauch, J., Šimůnek, M.: Applying Domain Knowledge in AssociationRules Mining Process - First Experience. In: Kryszkiewicz, M., Rybinski, H., Skowron, A., Raś, Z.W. (eds.) ISMIS 2011. LNCS (LNAI), vol. 6804, pp. 113–122. Springer, Heidelberg (2011)
12. Šimůnek, M.: Academic KDD Project LISp-Miner. In: Abraham, A., Franke, K., Köppen, M. (eds.) Intelligent Systems Design and Applications. ASC, vol. 23, pp. 263–272. Springer, Tulsa (2003)
13. Šimůnek, M.: LISp-Miner Control Language – description of scripting language implementation. Submitted for publication in Journal of System Integration, http://www.si-journal.org ISSN: 1804-2724
14. Šimůnek, M., Rauch, J.: EverMiner – Towards Fully Automated KDD Process. In: Funatsu, K., Hasegava, K. (eds.) New Fundamental Technologies in Data Mining, pp. 221–240. InTech, Rijeka (2011)
15. Sharma, S., Osei-Bryson, K.-M.: Toward an integrated knowledge discovery and data mining process model. The Knowledge Engineering Review 25 49–67 (2010)
16. Tan, P.-N., Kumar, V., Srivastava, J.: Selecting the right objective measure for association analysis. Information Systems 29, 293–313 (2004)

Local Characteristics of Minority Examples in Pre-processing of Imbalanced Data

Jerzy Stefanowski, Krystyna Napierała, and Małgorzata Trzcielińska

Institute of Computing Science, Poznań University of Technology,
60-965 Poznań, Poland

Abstract. Informed pre-processing methods for improving classifiers learned from class-imbalanced data are considered. We discuss different ways of analyzing the characteristics of local distributions of examples in such data. Then, we experimentally compare main informed pre-processing methods and show that identifying types of minority examples depending on their k nearest neighbourhood may help in explaining differences in performance of these methods. Finally, we exploit the information about the local neighbourhood to modify the oversampling ratio in a SMOTE–related method.

1 Introduction

Many real-world applications have revealed difficulties in learning from imbalanced data, where at least one of the target classes contains a much smaller number of examples than the other classes. Most learning algorithms and classifiers fail to sufficiently recognize examples from the minority class. The specialized methods for imbalanced data proposed in the last years (see e.g. [3] for a review) are usually divided into two groups, applied either on the *data level* or on the *algorithmic level*. The first group includes classifier-independent *pre-processing methods* that rely on transforming the original data and modifying the balance between classes, e.g. by re-sampling. The other group of methods involves the modification of either the learning phase of the algorithm, its classification strategies or its adaptation to cost-sensitive learning.

Although implementing modifications on the algorithmic level could potentially further improve the given algorithm, pre-processing methods are still a dominating approach in the literature and are used with many algorithms to learn better classifiers. On the other hand, the popular simple random undersampling and oversampling have been shown to be less effective at improving the recognition of the minority class. Their drawbacks are solved by more sophisticated *informed methods* that use the characteristics of the *local data distribution* to remove only some difficult majority class examples or to add new minority class examples. The representatives of such methods are OSS, NCR, SMOTE or SPIDER [2,5,6,11]. However, there is still insufficient research on their general comparison and discussion of the competence area of each method.

In this study we want to analyze the performance of these informed pre-processing methods with respect to their different way of considering the local

T. Andreasen et al. (Eds.): ISMIS 2014, LNAI 8502, pp. 123–132, 2014.
© Springer International Publishing Switzerland 2014

characteristics of datasets. To have a common framework for experimentally comparing them on different datasets, we follow our earlier research [10], where we have claimed that data difficulty factors could be at least partly approximated by analyzing the *local characteristics* of learning examples from the minority class. Depending on the distribution of examples from the majority class in the neighbourhood of minority examples, we can evaluate how difficult it could be to learn the given data.

The main aim of this paper is to show that the differences in performance of the pre-processing methods depend on the type of the local characteristics in the given dataset. Moreover, we will show that the local neighbourhood of an example can be used for tuning the oversampling ratio in generalizations of the SMOTE method.

2   Related Works on Pre-processing Methods

The most popular re-sampling methods are random *over-sampling* which replicates examples from the minority class and random *under-sampling* which randomly eliminates examples from the majority classes until a required degree of balance between class cardinalities is reached. However, random under-sampling may potentially remove some important examples and over-sampling may lead to overfitting. Thus, recent research focuses on particular examples, taking into account information about their distribution in the attribute space [3].

The oldest informed method is *one-side-sampling* (OSS), which filters the majority classes in a focused way [5]. Its authors distinguish different types of learning examples: *safe* examples (located in homogeneous regions populated by examples from one class only), *borderline* (i.e., lying on a border between decision classes) and *noisy* examples. They propose to use Tomek links (two nearest examples having different labels) to identify and delete the borderline and noisy examples from majority classes.

Neighbourhood Cleaning Rule (NCR) represents another approach to the focused removal of examples from the majority class [6]. It deals directly with the local data characteristics by applying the edited nearest neighbour rule (ENNR) to the majority classes [13]. ENNR first looks for a specific number of $k-nearest$ *neighbours* ([6] recommends 3) of the "seed" example and uses their labels to predict the class label of the seed. In case of wrong prediction, the neighbours from the majority class are removed from the learning set.

The Synthetic Minority Over-sampling Technique ($SMOTE$) is also based on the k nearest neighbourhood, however it exploits it to selectively over-sample the minority class by creating *new synthetic examples* [2]. It considers each minority class example as a "seed" and finds its k-nearest neighbours also from the minority class. Then, according to the user-defined oversampling ratio $- o_r$, SMOTE randomly selects o_r of these k neighbours and randomly introduces new examples along the lines connecting the seed example with these selected neighbours.

SPIDER is a hybrid method that selectively filters out harmful examples from the majority class and amplifies the difficult minority examples [11]. In the first stage it applies ENNR to distinguish between safe and unsafe examples (depending how k neighbours reclassify the given "seed" example). For the majority class - outliers or the neighbours which misclassify the seed minority example are either removed or relabeled. The remaining unsafe minority examples are additionally replicated depending on the number of majority neighbours.

In all these methods k nearest neighbourhood is calculated with the HVDM metric (*Heterogeneous Value Difference Metric*) [12]. Recall that it aggregates normalized distances for both continuous and qualitative attributes, however it uses the Stanfil and Valtz value difference metric for qualitative attributes.

3   Modeling k-Nearest Neighbourhood for Data Difficulty

Studying local distributions of minority examples can also be applied to explain why some datasets are more difficult than others. Some researchers have already shown that class imbalance ratio is not the main source of difficulties [4]. The degradation of classification performance is also related to other data factors, such as the *decomposition* of the minority class into many sub-concepts with very few examples, the effect of *overlapping* between classes and the presence of single minority examples inside the regions of the majority classes.

We follow [10], where the data factors are linked to different types of examples creating the minority class distribution. Besides *safe* and *borderline* examples (as in [5]), other *unsafe* examples are also categorized into *rare cases* (isolated groups of few examples located deeper inside the opposite class), or *outliers*. The analysis of the class distribution inside a *local neighbourhood* of the considered example has been applied to identify these types of examples in the dataset.

In [10] such a neighbourhood has been modeled with *k-nearest* examples. Depending on the number of majority class examples in the k-neighbourhood of the given minority example, we can evaluate how difficult it could be to recognize the minority class examples in this region.

Following the study [10], for simplicity we consider $k = 5$. In this case the type of example x is defined as: if 5 or 4 of its neighbours belong to the same class as x, it is treated as a safe example; if the numbers of neighbours from both classes are similar (proportions 3:2 or 2:3) – it is a borderline example; if it has only one neighbour with the same label (1:4) it is a rare case; finally if all neighbours come from the opposite class (0:5) – it is an outlier.

Although this categorization is based on intuitive thresholding, its results are consistent with a probabilistic analysis of the neighbourhood, modeled with kernel functions, as shown in [8]. Moreover, the experiments with higher values of k have also led to a similar categorization of distribution of minority class examples [8]. We stay with $k = 5$ as it widely refers to more local data characteristics and is frequently considered in related pre-processing methods.

The experiments with UCI imbalanced datasets [10,8] have also demonstrated that most real-world data do not include many safe minority class examples.

Most studied data sets contain all types of examples, but in different proportions. On the other hand, most of majority class examples have been identified as safe ones. Depending on the dominating type of identified minority examples, the considered datasets have been labeled as: safe, border, rare or outlier. As a large number of borderline examples often occurred in many datasets, some of these datasets have been assigned both to border and more difficult categories. Moreover, comparing different learning algorithms has shown that the classifier performance depends on the category of data – for more details on the competence of each studied classifier see [8].

4   Extensions of SMOTE

Although SMOTE is often used in many experimental studies, as e.g. [1,3,9] some of the assumptions behind this method could still be questioned. First, using the same oversampling ratio to all minority examples may be doubtful for some data. Several researchers claim that unsafe examples are more liable to be misclassified, while safe examples located inside the class regions are easier to be learned and do not require such a strong oversampling. What is more, SMOTE may over-generalize the minority class as it blindly oversamples the attribute space without considering the distribution of neighbours from the majority class.

Therefore, its generalizations have recently been proposed. The first solution is to integrate SMOTE with a post-processing phase including filtering the most harmful examples. For instance, using ENNR after SMOTE performs quite well with tree classifiers [1] and rules [8]. In this paper we are interested in other "internal" extensions which are more focused either on the local neighbourhood or on the types of the minority class examples (for their review, see [7]).

Borderline-SMOTE is based on the assumption that not all minority examples are equally important and only the borderline examples should be oversampled. They are identified by analyzing the class labels inside the k neighbourhood of the minority example x. Let $SN(x)$ denote the number of the majority class examples in this neighbourhood. The borderline examples are those that satisfy the formula $k/2 \leq SN(x) < k$ and are further fed to the SMOTE procedure for generating synthetic examples around them with the oversampling ratio *or*.

Safe-Level-SMOTE attempts to decrease the effect of blind SMOTE over-generalization by considering the local characteristics around the seed example and its selected nearest minority class neigbour. For each of them a *safe level* is defined, expressing the ratio of closest minority examples to majority ones in its k-neighbourhood. Depending on the ratio of both safe levels the range of the SMOTE line for generating a new synthetic example is modified to direct the position of the new example closer to the safer region.

5   STARSMOTE: Local Oversampling Profiles

We propose a new extension, called *STARSMOTE (Sample Type and Relabeling Synthetic Minority Oversampling TEchnique)*, which uses local data characteristics to modify the oversampling ratio depending on the type of the minority examples. First, $SN(x)$ - the number of the majority class examples in the k-neighbourhood of the minority class example x is calculated. Then, we follow the rule saying that an unsafe example should be amplified more than a safe one. It is expressed by the user-defined *oversampling profile*, where for the minority example x the value of its *amplification*[1] is determined with respect to its value of $SN(x)$. In general, different profiles could be proposed, however in this study we consider two forms which are graphically presented in Figure 1. In the first profile the amplification is exponentially increasing for the more unsafe examples. The intuition is that an exponential oversampling modification may be beneficial for data, where the minority class is scattered into many rare cases and outliers and the number of safe examples is limited. The trapezoid profile strengthens the role of borderline (and rare) examples and it is expected to work well for borderline data. According to [8], both categories of such distributions occur in real-world datasets. Moreover, majority class outliers are relabeled to strengthen the minority class.

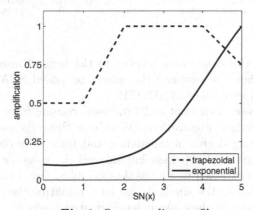

Fig. 1. Oversampling profiles

The other key issue is changing the SMOTE strategy of looking for k neighbours from the minority class only to finding really nearest neighbours, also from the majority class. We follow the arguments from [7] that it reduces the effect of SMOTE over-generalization. As a consequence, the new synthetic examples can also be generated between examples from different classes. However, in the case of majority class neighbours, we reduce the range of this random position to the half of the distance on the line between this neighbour and the minority seed example. The intuition is that we prefer to locate new examples closer to safer local regions of the minority examples, similarly to the hypothesis from [7].

6   Experiments

There are two aims of our experiments. First, we plan to carry out a comparative study of informed pre-processing methods, discussed in sect. 2, and verify the

---

[1] It influences the final oversampling of x as $or \times amplification$.

Table 1. Data characteristics

Dataset	# examples	# attributes	Minority class	IR
abdominal_pain	723	13	positive	2.58
acl	140	6	sick	2.5
new-thyroid	215	5	hyper	5.14
vehicle	846	18	van	3.25
car	1728	6	good	24.04
scrotal_pain	201	13	positive	2.41
ionosphere	351	34	bad	1.79
credit-g	1000	20	bad	2.33
ecoli	336	7	imU	8.60
hepatitis	155	19	die	3.84
haberman	306	4	died	2.78
breast-cancer	286	9	rec-events	2.36
cmc	1473	9	l-term	3.42
cleveland	303	13	pos	7.66
glass	214	9	v-float	11.58
hsv	122	11	bad	10
abalone	4177	8	0-4and16-19	11.56
postoperative	90	8	S	2.75
solar-flareF	1066	12	F	23.79
transfusion	748	4	1	3.20
yeast-ME2	1484	8	ME2	28.10
balance-scale	625	4	B	11.76

differences in their performance with respect to the type of minority examples occurring in data. Then, we evaluate the newly proposed STARSMOTE and compare it with other extensions of SMOTE.

Our experiments were carried out on 22 datasets coming from the UCI repository[2] or from our medical applications. We chose these datasets as they are characterized by varying degree of imbalance and they were often used in related experimental studies [1,10]. Less imbalanced data, as e.g breast cancer, was also chosen as it contained many unsafe examples. For datasets with more than two classes, we chose the smallest one as a minority class and combined the others into one majority class, like in some related works (e.g. [1], [6]). Their characteristics are presented in Table 1, where IR denotes the imbalance ratio.

First, we analyze the k nearest neighbours (using the method presented in sect. 3) to identify proportions of different types of minority class examples for each of the datasets. Due to page limits we skip the precise table (see [10] for similar results for most of our datasets). We identify the dominating types of examples in each dataset and we label 4 datasets as safe (ordered from abdominal pain to vehicle in Tab. 1), 3 datasets as safe and borderline (car, scrotal pain and ionosphere), 4 datasets (credit, ecoli, hepatitis, haberman) as mainly borderline and 7 datasets (ordered from haberman to post-operative in Tab. 1) as rare; as they also contain other types of examples, they partly overlap with the outlier category – data ordered from cleveland to balance-scale (e.g. balance-scale contains 90% outliers). Nevertheless, all datasets contain mostly safe majority class examples.

Then, we compare the pre-processing methods NCR, SMOTE (basic version), SPIDER with random over-sampling. They are applied with unpruned J48 decision

---

[2] UCI repository available at http://archive.ics.uci.edu/ml/.

tree[3], as this classifier is particularly sensitive to class imbalances and often used in related works [1,3,6,9]. We extend the comparison by reporting results for a single tree without preprocessing as a baseline. Oversampling ratios are tuned to balance both classes and k is equal to 5. The classification performance is evaluated with: *sensitivity* of the minority class (the minority class accuracy), its *specificity* (the accuracy of recognizing the majority class), their aggregation to the *geometric mean* (G-mean) and *F-measure*. For their definitions, see e.g., [3]. These measures are estimated with the stratified 10-fold cross-validation repeated 5 times to reduce the variance.

Table 2. Sensitivity values for the compared methods

Dataset	Baseline	NCR	Oversampling	SMOTE	SPIDER
abdominal-pain	0.6978	0.8278	0.7443	0.7071	0.8382
acl	0.8550	0.9200	0.8400	0.8450	0.8450
new-thyroid	0.9217	0.8733	0.9217	0.9200	0.8917
vehicle	0.8704	0.9095	0.8825	0.8925	0.9056
ionosphere	0.8271	0.8459	0.8351	0.8726	0.8303
car	0.7767	0.8486	0.8848	0.8086	0.8152
credit-g	0.4653	0.6900	0.5107	0.5473	0.6073
ecoli	0.5800	0.7583	0.5567	0.7367	0.7100
hepatitis	0.4317	0.6217	0.5583	0.5217	0.5633
scrotal-pain	0.5533	0.7280	0.6613	0.6820	0.7460
haberman	0.4103	0.6081	0.6069	0.6878	0.6592
breast-cancer	0.3867	0.6478	0.4683	0.4408	0.5611
cmc	0.3917	0.5784	0.4650	0.5000	0.5612
hsv	0.0000	0.0900	0.1000	0.0700	0.0400
abalone	0.3038	0.4077	0.3869	0.4693	0.4368
cleveland	0.2367	0.3983	0.2383	0.2517	0.3067
postoperative	0.0467	0.4267	0.2067	0.1633	0.3433
glass	0.3000	0.5500	0.3800	0.5400	0.3900
solar-flareF	0.2090	0.4410	0.4180	0.2900	0.4460
transfusion	0.4134	0.5539	0.5626	0.6865	0.7344
yeast-ME2	0.3087	0.3473	0.3633	0.4953	0.3553
balance-scale	0.0000	0.5300	0.0000	0.0200	0.0800

Due to space limit, we present in Table 2 results of Sensitivity only. We applied the Friedman ranked test to compare all methods over many datasets, finding that the differences in their performance are significant. The ranking of methods according to average ranks (the higher the better) is the following: NCR (4.07), SPIDER (3.65), SMOTE (3.25), oversampling (2.56) and baseline (1.36). Carrying out the post-hoc analysis and additionally the paired Wilcoxon test we can say that all informed pre-processing methods are significantly better than oversampling and baseline – however, the differences between the best methods are not significant. More interesting observations could be done if we split all datasets according to their categories. Considering only safe and borderline data (first 7 datasets) we still obtain the same ranking, however the average differences between best methods are smaller and SMOTE is performing practically as good as SPIDER (av. ranks 3.5 vs. 3.34). If we consider borderline and rare data only, NCR (av. rank 3.97) is still the best, followed by SMOTE (3.9) and

---

[3] All implementations were done in the WEKA framework.

SPIDER (3.6). Finally, for rare and outlier datasets the ranking is again different, as both SPIDER (3.83) and SMOTE (3.64) are slightly better than NCR (3.4). Considering F-measure and G-mean, the results are quite similar, however SMOTE is usually slightly better than SPIDER in their ranking. For specificity, we observe the best ranks for baseline (3.39), followed by oversampling (3.30), then NCR and SMOTE (both 3.05) and finally SPIDER (2.23). We have carried out a similar analysis using the PART classifier. Its results confirm the previous observations – NCR performs well for borderline data, while SMOTE and SPIDER work better for rare and outlier data.

Table 3. Sensitivity of SMOTE and its compared extensions

Dataset	Safe-Level SMOTE	Borderline SMOTE	SMOTE	STARSMOTE (exponential)	STARSMOTE (trapezoidal)
abdominal-pain	0.7623	0.7018	0.7071	0.7870	0.7704
acl	0.8500	0.8350	0.8450	0.8500	0.8500
new-thyroid	0.9250	0.9200	0.9200	0.9333	0.9433
vehicle	0.9057	0.9005	0.8925	0.8815	0.8895
car	0.7724	0.7533	0.8086	0.7419	0.7419
scrotal-pain	0.6873	0.6727	0.6820	0.6673	0.6393
ionosphere	0.8481	0.8395	0.8726	0.8482	0.8574
credit-g	0.5707	0.5260	0.5473	0.5480	0.5860
ecoli	0.7867	0.7100	0.7367	0.6467	0.6900
hepatitis	0.6167	0.4850	0.5217	0.5300	0.4900
haberman	0.6308	0.6672	0.6878	0.6553	0.6481
breast-cancer	0.5336	0.3908	0.4408	0.5244	0.5211
cmc	0.5657	0.4657	0.5000	0.5789	0.5742
cleveland	0.4033	0.2217	0.2517	0.3133	0.3100
glass	0.5000	0.4900	0.0054	0.6300	0.5800
hsv	0.1200	0.1200	0.0700	0.0600	0.0700
abalone	0.6110	0.4960	0.4693	0.4813	0.5196
postoperative	0.2833	0.1567	0.1633	0.2667	0.2933
solar-flareF	0.4280	0.2590	0.2900	0.3940	0.3980
transfusion	0.6441	0.5660	0.6865	0.6681	0.6375
yeast-ME2	0.5620	0.4847	0.4953	0.5367	0.5300
balance-scale	0.8980	0.0290	0.0200	0.6520	0.2310

Finally, we carried out a similar experimental analysis for the following extensions of SMOTE: Safe-Level-SMOTE (SLS), Borderline-SMOTE (BS) and two versions of the newly introduced STARSMOTE (exponential - SSEXP and trapezoidal - SST). The sensitivity values are presented in Table 3. Although according to the Friedman test the methods are not performing in a similar range ($p = 0.003$) over all datasets, their differences are not significant with the post-hoc analysis. The highest rank is achieved by SLS (av. rank 4.02), followed by SSEXP (3.2) and SST (3.1), then SMOTE (2.5) and BS (1.9). Again, the situation becomes clearer if we split the data into two categories. For safe and first 3 border datasets, the differences between the methods are insignificant with $p = 0.1$ and the best is Safe-Level-SMOTE. On the other hand, for remaining border, rare and outlier datasets, the differences are clearly visible and the ranking is SLS (av. rank 4.0) > SSEXP (3.6) > SST (3.3) > SMOTE (2.2) > BS (1.9). For G-mean the order of methods is nearly the same, although STARSMOTE is sligthly better than SLS (3.8 vs. 3.6) and BS (2.1) is also better than SMOTE

(1.8) for unsafe data. In the case of F-measure, the differences between methods are smaller, nevertheless STARSMOTE again performs better than SLS (3.4 vs. 3.25). Finally assuming that the best profile of STARSMOTE could be chosen depending on the dataset, a similar analysis shows that this method is much closer to SLS in the ranking of sensitivity and better for other measures (2.6 vs. 2.2 for F-measure).

7   Discussion of Results and Final Remarks

The recurring theme of this paper has been to incorporate the information about the local neighbourhood of a chosen minority class example in the process of constructing and analyzing informed pre-processing methods for imbalanced data. Following our earlier research in [10], we have evaluated the performance of different pre-processing methods depending on the dominating type of examples in the data and we have introduced a new method, called STARSMOTE, employing the local data distribution.

In the first stage of our experiments, we compared different pre-processing approaches, showing that the informed methods are significantly better than the baseline method and random oversampling. Our study has yielded the result that the competence area of each method depends on the data difficulty level, based on the types of minority class examples. In the case of safe data there are no significant differences between the compared methods. However, for borderline datasets NCR clearly performs best, which can be explained by its rather greedy removal of majority examples from the regions close to the decision border. On the other hand, SMOTE and SPIDER, which can add new examples to the data, seem to be more suitable for rare and outlier datasets.

A similar influence of different data types has been observed in the experiments with SMOTE and its extensions. For safe and partly for borderline datasets the differences between the compared methods are not signficant. In the case of rare and outlier datasets, although STARSMOTE is not as good as Safe-Level-SMOTE for sensitivity values, it is slightly better with respect to G-mean and F-measure. What is more, the choice of an oversampling profile may enable the user to adjust the method to the presence of difficult examples. The proposed trapezoidal profile tries to emphasize all types of unsafe examples, with particular stress on borderline ones, while the exponential profile aims to give priority to rare and outlier examples. Furthermore, considering a "closer" local neighbourhood of a seed example, without ignoring the majority class examples seems more appropriate than the original SMOTE solution, as the closest minority neighbours may be far away, especially in the case of difficult rare datasets. Notice that Safe-Level-SMOTE, also based on the analysis of the local distributions of examples (although in a different way than STARSMOTE), performs very good in experiments. It is consistent with our postulate on the role of the local characteristics of imbalanced data.

In our opinion, SMOTE and similar approaches depend too heavily on user-defined, global parameters, such as e.g. oversampling ratio. Thus, in our future research we would like to consider more intelligent techniques for locally tuning the appropriate values, depending on the data distribution. Moreover, the problem of classification of imbalanced data with multiple classes may be put into consideration.

This research is partially supported by grant 09/91/DSPB/0543.

References

1. Batista, G., Prati, R., Monard, M.: A study of the behavior of several methods for balancing machine learning training data. ACM SIGKDD Explorations Newsletter 6(1), 20–29 (2004)
2. Chawla, N., Bowyer, K., Hall, L., Kegelmeyer, W.: SMOTE: Synthetic Minority Over-sampling Technique. J. of Artificial Intelligence Research 16, 341–378 (2002)
3. He, H.: Yungian Ma: Imbalanced Learning. Foundations, Algorithms and Applications. IEEE - Wiley (2013)
4. Jo, T., Japkowicz, N.: Class Imbalances versus small disjuncts. SIGKDD Explorations 6(1), 40–49 (2004)
5. Kubat, M., Matwin, S.: Addresing the curse of imbalanced training sets: one-side selection. In: Proc. of the 14th Int. Conf. on Machine Learning, pp. 179–186 (1997)
6. Laurikkala, J.: Improving identification of difficult small classes by balancing class distribution. Tech. Report A-2001-2. University of Tampere (2001)
7. Maciejewski, T., Stefanowski, J.: Local neighbourhood extension of SMOTE for mining imbalanced data. In: Proc. IEEE Symp. on Computational Intelligence and Data Mining, pp. 104–111 (2011)
8. Napierala, K.: Improving rule classifiers for imbalanced data. Ph.D. Thesis. Poznan University of Technology (2013)
9. Napierała, K., Stefanowski, J., Wilk, S.: Learning from Imbalanced Data in Presence of Noisy and Borderline Examples. In: Szczuka, M., Kryszkiewicz, M., Ramanna, S., Jensen, R., Hu, Q. (eds.) RSCTC 2010. LNCS (LNAI), vol. 6086, pp. 158–167. Springer, Heidelberg (2010)
10. Napierala, K., Stefanowski, J.: Identification of different types of minority class examples in imbalanced data. In: Corchado, E., Snášel, V., Abraham, A., Woźniak, M., Graña, M., Cho, S.-B. (eds.) HAIS 2012, Part II. LNCS, vol. 7209, pp. 139–150. Springer, Heidelberg (2012)
11. Stefanowski, J., Wilk, S.: Selective pre-processing of imbalanced data for improving classification performance. In: Song, I.-Y., Eder, J., Nguyen, T.M. (eds.) DaWaK 2008. LNCS, vol. 5182, pp. 283–292. Springer, Heidelberg (2008)
12. Wilson, D.R., Martinez, T.R.: Improved heterogeneous distance functions. Journal of Artifical Intelligence Research 6, 1–34 (1997)
13. Wilson, D.R., Martinez, T.: Reduction techniques for instance-based learning algorithms. Machine Learning Journal 38, 257–286 (2000)

Visual-Based Detection
of Properties of Confirmation Measures

Robert Susmaga and Izabela Szczęch

Institute of Computing Science, Poznań Univesity of Technology,
Piotrowo 2, 60-965 Poznań, Poland

Abstract. The paper presents a visualization technique that facilitates
and eases analyses of interestingness measures with respect to their properties. Detection of properties possessed by these measures is especially
important when choosing a measure for KDD tasks. Our visual-based
approach is a useful alternative to often laborious and time consuming
theoretical studies, as it allows to promptly perceive properties of the visualized measures. Assuming a common, four-dimensional domain of the
measures, a synthetic dataset consisting of all possible contingency tables
with the same number of observations is generated. It is then visualized
in 3D using a tetrahedron-based barycentric coordinate system. Additional scalar function – an interestingness measure – is rendered using
colour. To demonstrate the capabilities of the proposed technique, we detect properties of a particular group of measures, known as confirmation
measures.

Keywords: Visualization, interestingness measures, confirmation measures, properties of measures.

1 Introduction

Within data mining tasks, a valid phase concentrates on the evaluation of induced patterns, in the form of e.g., "*if* premise, *then* conclusion" rules. The
number of rules induced even from relatively small datasets can be overwhelming for users/decision makers and needs to be limited, so that irrelevant or
misleading patterns can be discarded. Such an evaluation is commonly done by
means of interestingness measures, e.g., measures of support and confidence for
association rules or confirmation measures for decision rules. The variety of measures of interest proposed in the literature makes it, however, difficult to choose
a measure or a group of measures (in case of a multi-criteria evaluation) for a
particular application.

To help users make that choice, properties of measures are intensively studied
[4,7,9,10]. Theoretical analyses of measures with respect to their properties are
often time consuming and require laborious mathematical calculations. In this
paper we adopt (similarly to [11]) a visualization technique that aids the process
of analysing measure properties. It allows to conduct a preliminary detection of
properties satisfied by each visualized measure. In particular, it facilitates finding

T. Andreasen et al. (Eds.): ISMIS 2014, LNAI 8502, pp. 133–143, 2014.
© Springer International Publishing Switzerland 2014

counterexamples, i.e., examples that discard particular properties. Our visual-based property detection eases the analysis, not only for the measures already known in the literature, but also for newly developed ones (e.g., automatically generated).

In this paper, the visualization technique is adopted to the detection of properties of a particular group of measures, called *confirmation measures*. These measures are designed to quantify the strength of confirmation that a rule's premise gives to its conclusion, and are thus a commonly used tool for the evaluation of decision rules [3,6]. The particular properties that we will focus on in this paper are:

- property M, of monotonic dependency of the measure on the number of objects supporting or not the rule's premise or conclusion [6,12],
- property of *maximality/minimality*, indicating the necessary and sufficient conditions under which the measures should obtain their extreme values [5],
- property of hypothesis symmetry HS, stating that the measures should obtain the same value but of opposite sign for rules $E \to H$ and $E \to \neg H$ [1,2,5,8]. Hypothesis symmetry is actually one of several symmetry properties considered in the literature, but it is the only one that all the authors agree to find truly desirable.

Let us observe that the chosen properties also reflect many other properties, such as the well known Piatetsky-Shapiro's properties [10]. For example, the Piatetsky-Shapiro's requirement for a measure to obtain value 0 when the premise and conclusion are statistically independent corresponds to a part of the definition of confirmation (see Section 2). Similarly, given a rule $E \to H$, the requirement for a measure to monotonically increase with $P(E,H)$ when $P(E)$ (or $P(H)$) remains the same coincides with property M; analogously, the requirement that a measure monotonically decreases with $P(E)$ (or $P(H)$) when $P(E,H)$ and $P(H)$ (or $P(E)$) remain the same is included in property M(see Section 3). One can thus conclude that the property of confirmation and property M extend the requirements known as the Piatetsky-Shapiro's properties.

The rest of the paper is organized as follows. A selection of popular confirmation measures is described in Section 2, followed by their commonly studied properties in Section 3. Next, Section 4 demonstrates the proposed visualization technique, which is then employed to visual-based analyses of the selected measures with respect to those properties in Section 5. Final remarks are collected in Section 6.

2   Confirmation Measures

The confirmation measures, some of which are presented and analysed in this paper, constitute an important group among measures of interest. They evaluate rule patterns induced from a set of objects U and described by a set of attributes with respect to their relevance and utility [3,5,7,8]. Formally, for a rule $E \to H$, an interestingness measure $c(H, E)$ is a confirmation measure when it satisfies the following conditions:

$$c(H, E) \begin{cases} > 0 \text{ when } P(H|E) > P(H), \\ = 0 \text{ when } P(H|E) = P(H), \\ < 0 \text{ when } P(H|E) < P(H). \end{cases} \qquad (1)$$

Confirmation is, thus, regarded as an increase in the probability of the conclusion H provided by the premise E (similarly for neutrality and disconfirmation).

In practical applications, where frequentionist's approach is often employed, the relation between E and H is quantified by four non-negative numbers:

- a: the number of objects in U for which E and H hold,
- b: the number of objects in U for which $\neg E$ and H hold,
- c: the number of objects in U for which E and $\neg H$ hold,
- d: the number of objects in U for which $\neg E$ and $\neg H$ hold.

Naturally, a, b, c and d all sum up to n, being the cardinality of the dataset U.

Working with a, b, c and d has two advantages, namely they can be used to estimate probabilities e.g., the conditional probability of the conclusion given the premise is $P(H|E) = P(H \cap E)/P(E) = a/(a+c)$, and, secondly, they can be used to calculate the value of confirmation measures, as each of them is a scalar function of a, b, c and d. From a long list of alternative and ordinally non-equivalent confirmation measures, let us concentrate on six exemplary ones, defined in Table 1 (the selection inspired by [5,6]).

Table 1. Popular confirmation measures

$M(H, E) = P(E|H) - P(E) = \dfrac{a}{a+b} - \dfrac{a+c}{n}$
$S(H, E) = P(H|E) - P(H|\neg E) = \dfrac{a}{a+c} - \dfrac{b}{b+d}$
$F(H, E) = \dfrac{P(E|H) - P(E|\neg H)}{P(E|H) + P(E|\neg H)} = \dfrac{ad - bc}{ad + bc + 2ac}$
$FS(H, E) = \dfrac{1}{2}(F(H, E) + S(H, E))$
$\phi(H, E) = \dfrac{P(E, H) - P(E)P(H)}{\sqrt{P(E)P(H)P(\neg E)P(\neg H)}} = \dfrac{ad - bc}{\sqrt{(a+c)(a+b)(b+d)(c+d)}}$
$F\phi(H, E) = \dfrac{1}{2}(F(H, E) + \phi(H, E))$

The domains of the six selected measures range from -1 to $+1$, allowing us to translate them later on into a predefined colour map. Let us also note that the idea behind the formulation of measures $FS(H, E)$ and $F\phi(H, E)$ is that they benefit from properties of their constituent measures [5].

To better characterize the measures and to help choose the most suitable ones for the user, many properties have been proposed and compared in the literature

[2,4,5,7,8]. The definitions of the most commonly required ones are presented in the next section, followed by visual-based analyses of measures from Table 1 with respect to these properties.

3   Properties of Confirmation Measures

Property of Monotonicity M

Property of monotonicity M, introduced by Greco, Pawlak and Słowiński in [6], requires that a confirmation measure $c(H, E)$ is a function:

- non-decreasing with respect to a and d, and
- non-increasing with respect to b and c.

As a result, measures possessing M cannot decrease its value when a new positive example is introduced to the dataset (i.e., a increases). Similarly for objects not satisfying neither the rule's premise nor its conclusion (i.e., increase of d). At the same time, when new counterexamples are put into the dataset (i.e., increase of c), then a measure possessing property M cannot increase its value (analogously with b).

Property of Maximality/Minimality

The property of maximality/minimality has been introduced by Glass in [5] as a requirement stating necessary and sufficient conditions under which measures should obtain their extreme values. Formally, provided a confirmation measure $c(H, E)$ is defined, it enjoys property of maximality/minimality when:

- $c(H, E)$ is maximal if and only if $P(E, \neg H) = P(\neg E, H) = 0$, and
- $c(H, E)$ is minimal if and only if $P(E, H) = P(\neg E, \neg H) = 0$.

Equivalently, a measure satisfying the maximality/minimality requirement obtains its maximum iff $c = b = 0$, and obtains its minimum iff $a = d = 0$ [5]. Let us observe that the maximality/minimality property is closely related to properties Ex_1 [1] and weak Ex_1 [7,8], as well as to properties L [1,3] and weak L [7,8].

Property of Hypothesis Symmetry

In the literature there is an intensive discussion about a whole group of symmetry properties, where each symmetry considers how the value of a confirmation measure $c(H, E)$ relates to its value obtained for the situation in which the rule's premise and/or conclusion is negated and/or when the premise and conclusion switch positions. The symmetry properties have been analysed, among others, by Eells and Fitelson [2], Crupi, Tentori and Gonzalez [1], Greco, Słowiński and

Szczęch [8] as well as Glass [5]. They have different opinions as to which symmetries properties are desirable, but they all agree that a valuable confirmation measure should satisfy hypothesis symmetry HS stating that

$$c(H, E) = -c(\neg H, E).$$

This means that the value of a confirmation measure for rule $E \to H$ should be the same but of opposite sign as for a rule $E \to \neg H$. To present the intuition behind hypothesis symmetry, let us consider a rule "*if x is a square, then x is a rectangle*". Obviously, the strength with which the premise (x is a square) confirms the conclusion (x is a rectangle) is the same as the strength with which the premise disconfirms the negated conclusion (x is not a rectangle). Thus, it is natural to expect that $c(H, E) = -c(\neg H, E)$.

4   The Visualization Technique

The visual-based detection of properties of confirmation measures employs, first of all, synthetic data that consist of an exhaustive and non-redundant set of contingency tables. Given a constant $n > 0$ (the total number of observations), it is generated as the set of all possible $\left[\begin{smallmatrix} a & c \\ b & d \end{smallmatrix}\right]$ tables satisfying $a + b + c + d = n$. The set thus contains exactly one copy of each such table. We use $n = 128$ implying $t = 366145$ in all further visualizations.

Let us observe that using a synthetic dataset allows us to gain an insight into all areas that can be possibly occupied by the domain of a measure, including areas that could otherwise be omitted when using a real-life dataset. Thus, our approach reveals all possible behaviours and features of the considered measure. It also makes the results general, rather than application-specific.

The operational data set comprises t rows and 4 columns (representing a, b, c and d). Four columns correspond to four degrees of freedom, thus, visualization of such data would require four dimensions. However owing to the constraint $a + b + c + d = n$, the number of degrees of freedom is reduced to 3. Thus, it is possible to represent such data in three dimensions (3D) using tetrahedron-based barycentric coordinates. The tetrahedron, as used throughout the paper (and rendered in what will be referred to as the standard view), has its four vertices A, B, C and D coinciding with points of the following $[x, y, z]$ coordinates: A: $[1, 1, 1]$, B: $[-1, 1, -1]$, C: $[-1, -1, 1]$ and D: $[1, -1, -1]$.

The interpretation of the tetrahedron points is as follows: the vertex A corresponds to the (single) contingency table satisfying $a = n$ and $b = c = d = 0$, the edge AB corresponds to the (multiple) contingency tables satisfying $a + b = n$ and $c = d = 0$, etc. A standard view of a skeleton tetrahedron (only edges visible) is depicted in Figure 1.

For a more comprehensive visualization, the standard view will be accompanied by a rotated view, which depicts the DAB face of the tetrahedron (not visible in the standard view). The combination of these two views will be referred to as the 3D 2-view visualization of the tetrahedron.

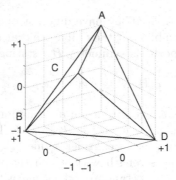

Fig. 1. A skeleton visualization of the tetrahedron

Let us now observe that, since the points of the tetrahedron may be displayed in colour, the proposed visualization technique can also visualize a function $f(a, b, c, d)$ of the four arguments, e.g., any interestingness measure. It is assumed that the value set of this function is a real interval $[r, s]$, with $r < s$, so that its values may be rendered using a pre-defined colour map. Non-numeric values, i.e., $+\infty$, NaN and $-\infty$, if generated by a particular function, may be rendered as colours not occurring in the map.

Notice that the 3D 2-view visualization of a 'solid' tetrahedron shows only extreme values of the arguments of the visualized function (external view). If areas located strictly inside the tetrahedron need to be additionally visualized, various variants of the visualization may be generated (internal views), however due to page limit, we shall not cover that topic in this paper.

5   Visual-Based Detection of Properties of Measures

The proposed visualization technique presents a coloured tetrahedron for each of the analysed confirmation measures. The values of the measures range from -1 to $+1$, thus a standard colour map[1] with shades of dark blue (minimal values) through pale green (neutral values) to dark brown (maximal values) shall be used. The NaN values are depicted with a special character ('*').

The following subsections explain how the visualization eases and speeds up the detection of particular properties of confirmation measures. It should be stressed, however, that essentially the same techniques may be applied to other measures, e.g. the group of measures that assess the performance of binary classification procedures (sensitivity, specificity, F-score, precision, accuracy, etc.). One of the fundamental differences between measures from these groups is that while the confirmation measures are bi-polar (i.e. their values range from -1

---

[1] Owing to the printing restrictions, the standard colour map for all the measures visualized in Figure 2 had to be substituted with a grey colour map, with black and white corresponding to -1 and $+1$, respectively. Thus, in the following subsections, read 'dark blue' as 'black', 'pale green' as 'light grey' and 'dark brown' as 'white'.

to +1), the performance measures are customarily uni-polar (i.e. their values range 0 to +1). Certainly, this does not prevent the performance measures from being visualized using the presented technique, however, the bi-polarity makes the confirmation measures and their analysis more difficult, and thus more interesting. In result, it is the confirmation measures that have been chosen to be discussed and visualized below.

5.1 Detection of Property M

The definition of property of monotonicity M expects a confirmation measure $c(H, E)$ to be a function non-decreasing with respect to a and d, and non-increasing with respect to b and c. Such requirements directly translate to particular colour changes that are allowed at the visualized tetrahedrons. Precisely, the "non-decreasing with a and d" condition should be reflected in the visualization as colours changing towards dark brown (increase of confirmation) around vertices A and D and the "non-increasing with b and c" condition should be reflected in the visualization as colours changing towards dark blue (increase of disconfirmation) around vertices B and C. The proposed visualization technique allows thus to promptly perceive cases (counterexamples) that are contrary to the colour patterns required by property M. In fact, if there occurs any colour change that does not follow the expected gradient towards dark brown (dark blue) around vertices A and D (B and C), we can conclude that the visualized measure does not possess the property M. Let us stress, however, that potential counterexamples can also be hidden inside the tetrahedron, thus a thorough analysis of property M must take into account the inner parts of the shape as well.

Among the measures considered in this paper, clearly, measure $M(H, E)$, depicted in Figure 2 (row 1), does not satisfy property M, as in the visualization the colour changes from dark brown at vertex D to pale green at vertex A, violating the demands the of the non-decrease with a.

On the other hand, there are no observable counterexamples to property M in the external visualizations of other considered measures (see e.g., measure $S(H, E)$ in Figure 2 (row 2)) which, together with additional analysis of the inside of the tetrahedrons, determines the possession of the property by those measures.

5.2 Detection of Maximality/Minimality Property

The maximality/minimality property requires that a confirmation measure obtains its maximal values if and only if $b = c = 0$. Additionally, minimal values are to be obtained if and only if $a = d = 0$. This translates into expectation that the dark brown (dark blue) colour must be found on the AD (BC) edge of the tetrahedron and cannot be found anywhere else. Let us observe that the

AD (BC) edge contains all points for which $b = c = 0$ ($a = d = 0$), i.e., the points most distant from the vertices B and C (A and D). Similarly to property M, it must be stressed that a thorough analysis with respect to maximality/minimality also requires an insight into the tetrahedron as potential counterexamples to this property may be located inside the shape.

Visual-based detection of maximality/minimality property among the measures in Table 1 reveals that measures $M(H, E)$ (see Figure 2 (row 1)) and $F(H, E)$ (see Figure 2 (row 3)) are the ones that do not satisfy this property. It is, among others, due to the fact that the points with maximal values of measure $M(H, E)$ are located only at vertex D of AD edge and thus do not cover the whole edge (i.e., there are too few of them), and the points with maximal values of measure $F(H, E)$ cover the whole ABD face (i.e., there are too many of them). The remaining measures do satisfy maximality/minimality (see e.g., measure $FS(H, E)$, presented in Figure 2 (row 4)). It is because the external views of the tetrahedra representing the corresponding measures do not depict counterexamples to this property and the additional insight into the shapes did not reveal any such cases either.

5.3   Detection of Property of Hypothesis Symmetry

By definition, the hypothesis symmetry demands that a measure $c(H, E)$ obtains the same values, but of the opposite sign, for rules $E \to H$ and $E \to \neg H$, i.e., $c(H, E) = -c(\neg H, E)$. Let us assume that the first rule is characterized by a contingency table with numbers a, b, c and d, and the latter by a table with numbers a', b', c' and d'. Then, the conditions for hypothesis symmetry translate into: $c(H, E) = f(a, b, c, d) = -c(\neg H, E) = -f(a', b', c', d') = -f(c, d, a, b)$, reflecting the exchange of columns in the contingency tables ($a = c'$, $b = d'$, $c = a'$ and $d = b'$). In the context of our visualization technique, hypothesis symmetry detection boils down to the determination if the two presented views of a considered measure have the same gradient profile (i.e., if the left view is just like the right one, provided the colour map is reversed). If the 'recoloured' views are not the same, then the visualized measure does not possess the hypothesis symmetry. Again, a thorough analysis of HS requires an insight into the tetrahedron, as potential counterexamples can also lie inside the shape.

Figure 2 (row 1), shows counterexamples to hypothesis symmetry of measure $M(H, E)$, since e.g., the BCD face has a gradient profile that is characterized by straight lines running parallel to edge BD, while the DAB face has a profile that is characterized by curved lines coinciding with vertex D. Simple change of colours will, thus, not result in unifying faces BCD and ABD, which implies that measure $M(H, E)$ does not satisfy the hypothesis symmetry. On the other hand, there are no observable counterexamples to this property in other considered measures (see e.g., measure $\phi(H, E)$ in Figure 2 (row 5) and measure $F\phi(H, E)$ in Figure 2 (row 6)).

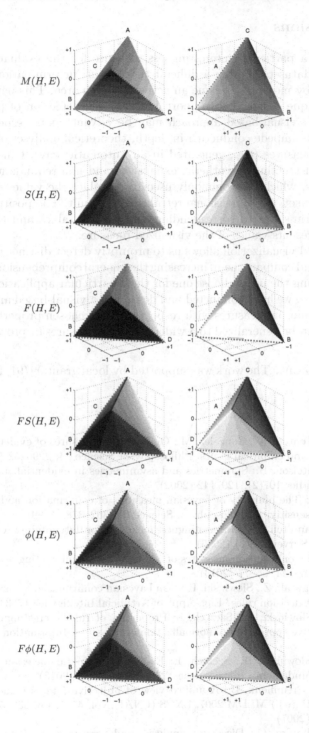

Fig. 2. A 3D 2-view visualization of the selected confirmation measures

6   Conclusions

The choice of a particular interestingness measure for the evaluation of rules induced from datasets is often a difficult task. Thus, determination of properties possessed by measures became an active research area. This paper presents a visual technique designed to support and ease the detection of properties of measures. Such visual-based approach may be advantageous, especially when time constraints impede conducting in-depth, theoretical analyses of large numbers of such measures (e.g., generated in an automatic way). Our proposition starts with constructing a synthetic, exhaustive and non-redundant set of contingency tables, which are commonly used to calculate the values of measures. Using such dataset, a 3-dimensional tetrahedron is built. The position of points in the shape translates to corresponding contingency tables and the colour of the points represents values of the visualized measure.

The proposed visualization allows us to promptly detect distinct properties of the measures and compare them, increasing the general comprehension of the measures and helping the users choose one for their particular application. For illustrative purposes, we have conducted and described a visual-based analysis of six popular confirmation measures with respect to three chosen properties. Clearly, the analyses can be generalized to a wider range of measures or properties.

Acknowledgment. The work was supported by local grant 09/91/DSPB/0543.

References

1. Crupi, V., Tentori, K., Gonzalez, M.: On bayesian measures of evidential support: Theoretical and empirical issues. Philosophy of Science 74, 229–252 (2007)
2. Eells, E., Fitelson, B.: Symmetries and asymmetries in evidential support. Philosophical Studies 107(2), 129–142 (2002)
3. Fitelson, B.: The plurality of bayesian measures of confirmation and the problem of measure sensitivity. Philosophy of Science 66, 362–378 (1999)
4. Geng, L., Hamilton, H.J.: Interestingness measures for data mining: A survey. ACM Computing Surveys 38(3) (2006)
5. Glass, D.H.: Confirmation measures of association rule interestingness. Knowlegde Based Systems 44, 65–77 (2013)
6. Greco, S., Pawlak, Z., Słowiński, R.: Can bayesian confirmation measures be useful for rough set decision rules? Eng. Appl. of Artifficial Intelligence 17, 345–361 (2004)
7. Greco, S., Słowiński, R., Szczęch, I.: Properties of rule interestingness measures and alternative approaches to normalization of measures. Information Sciences 216, 1–16 (2012)
8. Greco, S., Słowiński, R., Szczęch, I.: Finding meaningful bayesian confirmation measures. Fundamenta Informaticae 127(1-4), 161–176 (2013)
9. Hébert, C., Crémilleux, B.: A unified view of objective interestingness measures. In: Perner, P. (ed.) MLDM 2007. LNCS (LNAI), vol. 4571, pp. 533–547. Springer, Heidelberg (2007)
10. Piatetsky-Shapiro, G.: Discovery, analysis, and presentation of strong rules. In: Knowledge Discovery in Databases, pp. 229–248. AAAI/MIT Press (1991)

11. Susmaga, R., Szczęch, I.: Visualization of interestingness measures. In: Proceedings of the 6th Language & Technology Conference: Human Language Technologies as a Challenge for Computer Science and Linguistics, pp. 95–99 (2013)

12. Szczęch, I.: Multicriteria attractiveness evaluation of decision and association rules. In: Peters, J.F., Skowron, A., Wolski, M., Chakraborty, M.K., Wu, W.-Z. (eds.) Transactions on Rough Sets X. LNCS, vol. 5656, pp. 197–274. Springer, Heidelberg (2009)

A Recursive Algorithm
for Building Renovation in Smart Cities*

Andrés Felipe Barco**, Elise Vareilles,
Michel Aldanondo, and Paul Gaborit

Université de Toulouse, Mines d'Albi
Route de Teillet Campus Jarlard, 81013 Albi Cedex 09, France
`abarcosa@mines-albi.fr`

Abstract. Layout configuration algorithms in civil engineering have
two major strategies called *constructive* and *iterative improvement*. Both
strategies have been successfully applied within different facility scenarios such as room configurations and apartment layouts. Yet, most of the
work share two commonalities: They attack problems in which the reference plane is parallel to the Earth and, in most cases, the number of
activities are known in advance. This work aims to close that gap by developing a constructive-based algorithm for the layout configuration of
building facades in the context of a French project called CRIBA. The
project develops a smart-city support system for high-performance renovation of apartment buildings. Algorithm details are explained and one
example is presented to illustrate the kind of facades it can deal with.

1 Introduction

As pointed out by Liggett in [10], a layout configuration, commonly referred as
space planing or layout synthesis, "...is concerned with the allocation of activities to space such that a set of criteria (for example, area requirements) are
met and/or some objective optimized...". Layout configuration algorithms have
two major and often mixed strategies. The first strategy is called *constructive*:
Place one activity (e.g. room, office, panel) at a time. The *iterative improvement* strategy, on the other hand, is based on the improvement of an already
configured space. Both strategies have been applied, for instance, in room configurations [14], apartment layouts [9], activities within a business office [8] and
finding an optimal configuration for hospital departments [4]. Underlying models
used in these approaches include but are not limited to evolutionary computation (genetic algorithms [12]), graph theoretic models (adjacency graphs [7]) and
constraint satisfaction problems (filtering algorithms [2,6]) .

Yet, regardless the considerable body of literature, most of the work share two
commonalities. On the first hand, they attack problems in which the reference

---

* The authors wish to acknowledge the TBC Générateur d'Innovation company, the
Millet and SyBois companies and all partners in the CRIBA project, for their in-
volvement in the construction of the CSP model.
** Corresponding author.

T. Andreasen et al. (Eds.): ISMIS 2014, LNAI 8502, pp. 144–153, 2014.
© Springer International Publishing Switzerland 2014

plane is parallel to the Earth, meaning that they do not deal with gravity or other natural forces that will, potentially, affect the configuration. On the other hand, in most cases the number of activities are known in advance, given the possibility to use existing algorithms to tackle the problem [4,9,10,12,14] .

Our work aims to close that gap by developing two algorithms, one greedy and one constraint-based, for the layout configuration of facades as part of a decision support system for buildings renovation [5,13]. In this paper we focus our attention on the first algorithm. The decision support system, and hence the algorithm, uses the notion of Constraint Satisfaction Problems (CSPs) [11] to describe relations among components. It has been proved that CSPs modeling fits neatly in the constrained nature of layout synthesis [2,14]. The presented algorithm deals with the geometric of facades and the weight of panels by using the knowledge of constraints inherent to any facade and thus improves performance at the conception and implementation of the renovation.

The paper is structured as follows. We present the context of the project, called CRIBA, and the environment setup in Section 2. In Section 3 the constraint model describing the problem is introduced. Afterwards we present the first version of the layout configuration algorithm in Section 4. An example illustrating the algorithm is drawn in Section 5. Finally, some conclusions and future work are discussed in Section 6.

2 Preliminaries

The CRIBA project aims to industrialize high performance thermal renovation of apartment buildings [5,13]. This industrialization is based on an external new thermal envelope which wraps the whole building. The envelope is composed of prefabricated rectangular panels comprising insulation and cladding, and sometimes including in addition, doors, windows and solar modules. As a requirement for the renovation, facades have to be strong enough to support the weight added by the envelope. Within CRIBA several tools, needed to industrialize the renovation process, will be developed: a) a new method for three-dimensional building survey and modelling, b) a configuration system for the design of the buildings new thermal envelope (bill of material and assembly process), and c) a working site planning model with resource constraints. At the core of the renovation are:

Facades. Are compositions of apartments along with its doors, windows and so on. At the model level, they will be represented by a 2D coordinate plane which includes a set of rectangles defining frames, a set of supporting areas and rectangles defining zones out of configuration. For convenience, the origin of coordinates (0,0) is the bottom-left corner of the facade.

Configurable Components. At the current stage of the project, we only consider rectangular panels that are attached to the facade by means of fasteners. In addition, the panels may come with rectangular frames (windows, doors or solar modules). Panels are prefabricated in the factory when the user inputs the renovation profile.

A facade configuration will be made by one or more of these panels. Following the constructive approach, we establish what we consider a well-configured facade. A panel is well configured if it satisfies all its facade related constraints (presented in Section 3.2), i.e., posses the right dimensions, can be hang on the facade, is consistent with the facade frames, does not overlap with other panels and if it does not interfere with other panels placement. A facade is said to be well configured if all its composing panels are well configured and if they cover all facade area.

Consider the facade (a) in Figure 1 which represents a facade to be renovated. Horizontal and vertical lines represent the supporting areas in the facade. These are places in which we are allowed to attach weight-fasteners to supports panels. On panels, fasteners are attached in the edges (corners been mandatory). On the facade, fasteners will be aligned with the center of each supporting area in order to evenly distributed the panel's weight. As a constraint, distance between two fasteners is in $[0.9, 4]$ meters. Small rectangles in the facade are frames (e.g. windows and doors) that must be *completely covered* by one and only one panel. Two zones in the facade are out the configuration: The gable and the bottom part before the first horizontal supporting area. Those parts need specific panels design.

Now, facade (b) in Figure 1 presents three ill-configured panels. This is due to the impossibility to place another panel north to the already placed panel p_1, because there is not supporting areas at the corners of panel p_2 and, in the case of panel p_3, because it partially overlaps a frame. Non of these cases are allowed in a configuration solution. Finally, facades (c) and (d) in Figure 1 are well configured because they satisfy all criteria.

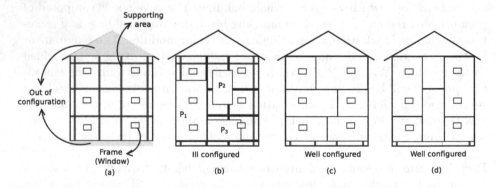

Fig. 1. Facade to renovate along with well and ill-configured panels

3   Constraint Model

This section introduces the constraint model describing the renovation. Recall that a CSP problem is described in terms of a tuple $\langle \mathcal{V}, \mathcal{D}, \mathcal{C} \rangle$, where $\mathcal{V}$ is a set of variables, $\mathcal{D}$ is a collection of potential values associated for each variable, also known as domains, and C is a set of relations over those variables, referred to as constraints. See [1,3] for further references.

3.1 Constraint Variables

Following the CSP model, we have identified 16 variables that allow us to represent the core of the layout configuration for a given building facade: The spatial positioning of panels. The list of constraint variables and their domains is presented in Table 1.

Table 1. 16 crucial constraint variables

Variable	Description	Domain
w_{fac}	Width of facade	$[2, 18]$ meters
h_{fac}	Height of facade	$[3, 21]$ meters (≤ 7 stories)
e_{fac}	Environmental property	$[easy, hard]$
(p_{x0}, p_{y0})	Origin (bottom-left) of panel p	$x0 \in [0, w_{fac}]$, $y0 \in [0, h_{fac}]$
(p_{x1}, p_{y1})	End (top-right) of panel p	$x1 \in [0, w_{fac}]$, $y1 \in [0, h_{fac}]$
w_p	Width of panel p	$[0.9, 13.5]$
h_p	Height of panel p	$[0.9, 13.5]$
(f_{x0}, f_{y0})	Origin (bottom-left) of frame f	$x0 \in [0, w_{fac}]$, $y0 \in [0, h_{fac}]$
(f_{x1}, f_{y1})	End (top-right) of frame f	$x1 \in [0, w_{fac}]$, $y1 \in [0, h_{fac}]$
fai_{load}	Maximum weight load of fastener at supporting area i	$[0, 500]$
X_{sa}	Collection of horizontal supporting areas of the form (w, h)	$\{\forall(w_i, h_i) \in X_{sa} :$ $w_i \in [0, w_{fac}] \wedge h_i \in [0, h_{fac}]\}$
Y_{sa}	Collection of vertical supporting areas of the form (w, h)	$\{\forall(w_i, h_i) \in Y_{sa} : \}$ $w_i \in [0, w_{fac}] \wedge h_i \in [0, h_{fac}]\}$

Given the description of origin and end coordinates of panels, bottom-left and top-right corners, we can deduce the first constraint: $p_{x0} < p_{x1}$ and $p_{y0} < p_{y1}$.

3.2 Components Relationship

In order to configure the layout of a given facade we use constraints to ensure relations over the variables representing components. In this section, we present the set of relevant constraints over panels and components w.r.t. the facade. The underlying CSP model we use is that of Disjunctive CSP [2]. Disjunctive CSP are boolean combination of atomic constrains (e.g. $<, \leq, >, \geq$). The canonical form of a disjunctive constraint is expressed as $C_i = (d_{i1} \vee d_{i2} \vee ... \vee d_{ik})$ where each d_i are atomic constraint connected by the **and** operator, $d_j = (c_{j1} \wedge c_{j2} \wedge ... \wedge c_{jk})$ [2]. Some of the constraint in our model, presented in Table 2, follow this approach.

Environmental. Impact on domains from environmental properties are expressed as inequalities. The width w_p and height h_p of panels may be constrained because accessibility difficulties to the facade (e.g. trees, water sources, high voltage lines, etc), transportation issues (e.g. only small trucks available) or even climatological aspects (e.g. wind speed more than a given threshold). Constraint C_1 express these constraints, where Γ and Θ represents the upper bound for panel dimensions, width and height respectively.

Dimension. The width w_p and height h_p of each panel is in the range $[0.9, 13.5]$. However, this is actually a combination of values. In other words, it is possible to have a panel with dimensions 0.9×13.5, 3×8.4 or 13.5×0.9, but it is not possible to have one with dimensions 13.5×13.5, this is due to fabrication and transportation constraints. In consequence, we constrain the combination of values for the width and height of panels using C_2.

Table 2. Atomic and disjunctive constraints

C_1 Environmental constraint
$\quad (w_p \leq \Gamma) \wedge (h_p \leq \Theta) \wedge (e_{fac} = Hard)$
C_2 Dimension constraint
$\quad ((w_p \in [0.9, 3.5] \wedge h_p \in [0.9, 13.5]) \vee (w_p \in [0.9, 13.5] \wedge h_p \in [0.9, 3.5]))$
C_3 Area constraint
$\quad w_{fc} \times h_{fc} = \sum_{i=1}^{N} (w_{pi} \times h_{pi})$
C_4 Non-overlap constraint
$\quad (p_{x1} < q_{x0}) \vee (q_{x1} < p_{x0}) \vee (p_{y0} < q_{y1}) \vee (q_{y0} < p_{y1})$
C_5 Weight Constraint
$\quad \sum_{j=1}^{|ATP_i|} computeWeight(\text{ATP}_i[j]) \leq \text{fai}_{load}$
C_6 Panels and frames constraints
$\quad ((p_{x1} + \Delta \leq f_{x0}) \vee (p_{x0} - \Delta \geq f_{x1}) \vee (p_{y1} + \Delta \leq f_{y0}) \vee (p_{y0} - \Delta \geq f_{y1})) \vee$
$\quad ((p_{y0} + \Delta \leq f_{y0}) \wedge (f_{y1} \leq p_{y1} - \Delta) \wedge (p_{x0} + \Delta \leq f_{x0}) \wedge (f_{x1} \leq p_{x1} - \Delta))$

Area. As a requirement, we have that the entire area of the facade must be renovated, provided it has the corresponding supporting areas. Thus, a constraint forcing the sum of panel areas ($w_p \times h_p$) to be equal to the facade area ($w_{fac} \times h_{fac}$) is posted. The constraint C_3 express this relation, where N is the number of panels covering the facade.

Non-Overlap. In addition, we must ensure that the panels do not overlap so we can have a valid solution. Thus, for each pair of panels p and q we define the non-overlap constraint using the disjunctive constraint C_4.

Weight. A given fastener in a supporting area is defined by its coordinates and its maximum weight load. Let ATP_i be the panels attached to the fastener fa_i and let $computeWeight(p)$ be a function[1] that returns the weight of panel p. Constraint over panels weight is defined by C_5.

Panel vs. Frames. We shorten the width or height of a given panel if there exists a frame near to it. Either the panel overlaps the frame or the panel is right, left, up or down to the frame. This is a typical case addressed by

---

[1] This function uses the next values to calculate the weight of a panel: dimensions of the panel, insulation type of the panel, weight of the frames within the panel (if any) and weight of any other component (e.g. solar modules).

disjunctive CSP. In any case, due to the internal structure of the panel, borders of frames and borders of panels must be separated by a minimum distance that we denote by Δ. This disjunctive constraint is modeled in C_6.

4 Greedy-Recursive Algorithm

Bearing in mind the above description, we proceed by developing an algorithm that solves the layout configuration in a greedy fashion. It makes local decisions for positioning panels following the constructive approach. It is worth noting that the algorithm uses the knowledge of constraints inherit by any building facade and thus reducing the search space. Also, it exploits recursion, simulating backtracking, when positioning a panel is not possible due to constraint conflicts. Moreover the algorithm is parameterized with an heuristic (soft constraint) that limits panel dimensions: Try to use, as much as possible, either vertical (i.e., left part of C_2) or horizontal panels (i.e., right part of C_2).

First, we present the Algorithm 1 that checks whether an initial origin point and end point for a panel, bottom-left and top-right corners respectively, violates the disjunctive constraint C_6. Its complexity is $\mathcal{O}(N_f)$ where N_f is the number of frames in the facade.

Algorithm 1. Panels versus frames validation

```
1  def panelVSframes(p_{x0},p_{y0},p_{x1},p_{y1},w_{fac},h_{fac},frames,Δ):
2      if (p_{y1} ≠ h_{fac}) then
3        ⌊ reduceDimensions(w_{fac}-p_{y1} ≥ 0.9);/* Dimension constraint         */
4      stack ← {f ∈ frames | p_{y0} ≤ f_{y0} ≤ p_{y1} + Δ };
5      while (stack ≠ ∅) do
6          f = pop(stack);
7          if (p_{y1} - f_{y1} < Δ) then
8              reduceDimensions(p_{y1} +Δ ≤ f_{y0});/* Panel vs Frames constraint    */
9          else
10            ⌊ mark(f); /* Frame successfully covered by panel in this axis...   */
11     Repeat from 2 to 10 with x-coordinate;
12     foreach f in frames do
           /* Discard frames overlaped in two axis...                            */
13         if (|marks(f)|==2) then discard(f);
14       ⌊ else unmark(f) ;
15   ⌊ return (p_{x1},p_{y1});
```

The first step of Algorithm 1 is to leave enough space for the next panel (lines 2-3). Then, it uses a stack to perform an ordered check of all frames covered by the panel in a given axis (lines 4-10). In the case there is a conflict, the algorithm proceeds by constraining one of the coordinates of the end point (line 8). If there is no conflict panel-frame, the algorithm marks the frame as good (line 10). Finally, the algorithm discards all frames successfully covered by the panel in order to avoid forthcoming checks (lines 12-16). The end point of the panel is returned: A point which is consistent with all frames.

The greedy-recursive algorithm, presented in Algorithm 2, works as follows. It begins by retrieving an available origin point and finding an end point given the heuristic (lines 3-4). It proceeds by generating a new valid point using the Algorithm 1. If dimensions of the panel violate dimensions constraints then it fails at positioning the panel (lines 6). After computing the weight of the panel (line 7) it checks whether it is possible to hang it using an horizontal (block at 8-14) or vertical supporting areas (block at 15-28). To hang the panel in an horizontal supporting area, it checks if the area is strong enough to support the weight of the panel (lines 9-11), in which case it propagates the weight to supporting areas (line 10). In the case it is not possible, it reduces the dimensions of the panel (line 13). To hang the panel in vertical supporting areas, it checks if the number of panels needed to hang the panel are less than or equal than

---

Algorithm 2. Greedy-recursive algorithm for layout synthesis.

```
1  def GreddyRecursive(w_fac,h_fac,frames,heu,op,solution):
2      if (op == ∅)then return True;
3      (p_x0,p_y0) ← getOriginPoint (op); /* Non-overlap constraint            */
4      (p_x1,p_y1) ← getEndPoint(p_x0,p_y0,heu); /* Non-overlap constraint     */
5      (p_x1,p_y1) ← panelVSframes(p_x0,p_y0,p_x1,p_y1,frames);
6      if (checkDimensions(p_x0,p_y0,p_x1,p_y1) == False)then  return False;
7      weight ← computeWeight(p_x0,p_y0,p_x1,p_y1);
8      if (p_y0 ∈ Y_sa)then
9          if (weight ≤ getNearestSA(p_y0)_load)then
10             getNearestSA(p_y1)_load ← getNearestSA(p_y1)_load−weight;
11             goto 29;
12         else
13             (p_x1,p_y1) ← reduceDimensions(p_x0,p_y0,p_x1,p_y1,heu); /* Weight constraint  */
14             goto 5;
15     else
16         n ← computeFasteners(p_x0,p_x1,weight);
17         m ← panelOverlaps(p_x0,p_y0,p_x1,p_y1);
18         if (n ≤ m)then
19             ssAreas ← selectedSA(p_x0,p_y0,p_x1,p_y1);
20             foreach (area ∈ ssAreas)do
21                 if ((weight/n) > area_load)then
22                     (p_x1,p_y1) ← reduceDimensions(p_x0,p_y0,p_x1,p_y1,heu); /* Weight
                           constraint                                              */
23                     goto 5;
24             foreach (area ∈ ssAreas)do
25                 area_load ← area_load − weight/n;
26         else
27             (p_x1,p_y1) ← reduceDimensions(p_x0,p_y0,p_x1,p_y1,heu);
28             goto 5;
       /* Place the panel in p_x0,p_y0,p_x1,p_y1. Do recursive call to place next panel!   */
29     newOp ← computePoints(p_x0,p_y0,p_x1,p_y1);
30     next ← GreddyRecursive(w_fac,h_fac,frames,newOp,solution);
31     if (next == False)then
32         (p_x1,p_y1) ← reduceDimensions(p_x0,p_y0,p_x1,p_y1,heu); /* Area constraint   */
33         goto 5;
34     else
35         solution.append(new Panel (p_x0,p_y0,p_x1,p_y1));
36         return True;
```

the actual vertical overlaped supporting areas (lines 16-18). If the panel can not be hang in those supporting areas given its weight, it proceeds by reducing the dimensions of the panel (block lines 19-23). Otherwise propagate the weight to supporting areas (lines 24-25). Finally, if the panel is well positioned, it proceeds by computing new origin points and adding the next panel recursively (lines 29-30). If the next panel can not be placed, dimensions for current panel are reduced and another check is run (lines 31-33). Otherwise add the solution to solution list and return (lines 35-36). The algorithm runs in $\mathcal{O}(r \times s(N_f + N_{sa}))$ in the best case (i.e., no failures in recursive calls) and $\mathcal{O}(r \times s(N_f + N_{sa})^{r \times s})$ in the worst case (i.e., no solution found), where N_f is the number of frames, N_{sa} the number of supporting areas and r and s are the maximum number of panels that can be fixed vertically and horizontally, respectively.

5   Example

In what follows, we present a behavior illustration for the greedy-recursive algorithm. Figure 2 shows a facade in the commune Saint Paul-lès-Dax in the department of Landes, France. This facade is part of a 5 block working site called *La Pince*. In our illustration the heuristic used to find a solution is vertical panels first, i.e., the algorithm tries to put a vertical panel as big as possible and resolves constraint conflicts. Additionally, the setup simulates a customization where the upper bound for panel's height has been set in 10 meters.

Due to paper-length constraints, Figure 2 shows the most representative states of the execution. State 1 is a failed attempt to position the first panel with dimensions 3.5×10; constraint C_6 (*Panels vs. frames*) is violated. Algorithm 2 changes p_{y1} to match an horizontal supporting area and thus reducing the panel dimensions. State 2 shows the final position of the first panel. The same occurs to the second panel in State 3, thus resulting the State 4. It is worth noting that the second panel is constrained in its width by the zone out of configuration. In State 5 the third panel is well configured because it does not enters in constraint conflict and because it allows a panel to be placed above it (it can be installed using its corners). The State 9 shows the result of placing the panels 4 and 5: Constraint conflicts in y-axis are solved. Nonetheless, the panel number 5 is not well configured because it does not allow another panel to be placed at it's right. Thus, the algorithm reduce its width resulting in the configuration of State 10. State 12 shows the correct placement of another panel, with valid dimensions, at the right edge of the facade. A panel is placed then above the zone out of configuration in the intermediate State 13. In State 14 is presented the correct configuration of two panels in the top-right of the facade. States 15 to 18 correspond to the correct placement of another two panels at the top of the facade. Finally, the algorithm stops at State 19 given that there is no more origin points for positioning panels.

Fig. 2. Configuration example using the greedy-recursive algorithm

6 Conclusions

In the present document we have shown a constructive-based algorithm for the layout synthesis of building facades. This work is part of a project that investigates the possibility of automated building renovation based on rectangular panels and supported by an intelligent system. Our problem is interesting and our results novel because it integrates a vertical oriented layout synthesis, a diverse set of constraints (e.g. geometrical, structural, global constraints) and user preferences. Conception and implementation for the renovation of buildings in smart cities are then improved. In addition, the algorithm presented in the paper contributes with the field of layout synthesis and civil engineering discipline. A constraint-based algorithm, implemented by means of global constraints and using a constraint solver, is totally valid to solve the problem but will be proposed in further communications. We acknowledge that the paper presents preliminary results that need to be improved. On this regard, the following objective is a strategic direction within the project.

Providing Structural Analysis. Intuitively, a human configuration takes advantages of the facade dimensions and positions of frames to find a solution.

Thus, it is adequated to add new constraints consequence of previous structural analysis of the facade. For instance, an analysis may look for symmetries in the facade, distances between windows and between supporting areas. Moreover, a structural analysis may throw different origin points or even determine which is the optimal number of panels given the facade structure. Consequences of this preprocessing are transparent to the supporting system given that constraint posting is a monotonic operation, i.e., it can only reduce the search space.

References

1. Barták, R.: Constraint Programming: In Pursuit of the Holy Grail. In: Proceedings of the Week of Doctoral Students WDS (June 1999)
2. Baykan, C.A., Fox, M.S.: Spatial synthesis by disjunctive constraint satisfaction. Artificial Intelligence for Engineering, Design, Analysis and Manufacturing 11, 245–262 (1997)
3. Brailsford, S.C., Potts, C.N., Smith, B.M.: Constraint satisfaction problems: Algorithms and applications. European Journal of Operational Research 119(3), 557–581 (1999)
4. Elshafei, A.N.: Hospital layout as a quadratic assignment problem. Operational Research Quarterly 28(1), 167–179 (1977)
5. Falcon, M., Fontanili, F.: Process modelling of industrialized thermal renovation of apartment buildings. In: eWork and eBusiness in Architecture, Engineering and Construction, European Conference on Product and Process Modelling (ECPPM 2010), pp. 363–368 (2010)
6. Flemming, U., Baykan, C., Coyne, R., Fox, M.: Hierarchical generate-and-test vs constraint-directed search. In: Gero, J., Sudweeks, F. (eds.) Artificial Intelligence in Design 1992, pp. 817–838. Springer, Netherlands (1992)
7. Goetschalckx, M.: An interactive layout heuristic based on hexagonal adjacency graphs. European Journal of Operational Research 63(2), 304–321 (1992)
8. Hassan, M.M.D., Hogg, G.L., Smith, D.R.: Shape: A construction algorithm for area placement evaluation. International Journal of Production Research 24(5), 1283–1295 (1986)
9. Lee, K.J., Kim, H.W., Lee, J.K., Kim, T.H.: Case-and constraint-based project planning for apartment construction. AI Magazine 19(1), 13–24 (1998)
10. Liggett, R.S.: Automated facilities layout: Past, present and future. Automation in Construction 9(2), 197–215 (2000)
11. Montanari, U.: Networks of constraints: Fundamental properties and applications to picture processing. Information Sciences 7(0), 95–132 (1974)
12. Tate, D.M., Smith, A.E.: A genetic approach to the quadratic assignment problem. Computers and Operations Research 22(1), 73–83 (1995)
13. Vareilles, E., Barco, A.F., Falcon, M., Aldanondo, M., Gaborit, P.: Configuration of high performance apartment buildings renovation: a constraint based approach. In: Conference of Industrial Engineering and Engineering Management (IEEM). IEEE (2013)
14. Zawidzki, M., Tateyama, K., Nishikawa, I.: The constraints satisfaction problem approach in the design of an architectural functional layout. Engineering Optimization 43(9), 943–966 (2011)

Spike Sorting Based upon PCA
over DWT Frequency Band Selection

Konrad Ciecierski[1], Zbigniew W. Raś[2,1], and Andrzej W. Przybyszewski[3]

[1] Warsaw Univ. of Technology, Institute of Comp. Science, 00-655 Warsaw, Poland
[2] Univ. of North Carolina, Dept. of Comp. Science, Charlotte, NC 28223, USA
[3] UMass Medical School, Dept. of Neurology, Worcester, MA 01655, USA
K.Ciecierski@ii.pw.edu.pl, ras@uncc.edu,
Andrzej.Przybyszewski@umassmed.edu

Abstract. When analyzing the neurobiological data many of its aspects
have to be carefully looked upon. Data coming from the MRI[1], EMG[2] or
microrecording all have its special properties that have to be extracted
during the process analysis. In case of recordings coming from the microrecording procedure i.e. from microelectrodes placed within the neu-
ronal tissue signal can be analyzed in at least two ways. First approach
focuses on the background noise present in such recordings. Second one,
looks upon the presence of the spikes - electrical signs of the bioelectrical
neurophysiological activity of neuron cells. In a given recording one may
often find many spikes with different shapes. For further analytical reasons it is frequently desired that spikes are to be grouped according to
their shape. Such grouping / shape clustering is called spike sorting and
there are many known approaches to that problem. Still, before spikes
are detected and sorted the raw recorded signal is almost always filtered
and altered in various DSP processes. This preliminary DSP operations
may significantly hamper the spike sorting efficiency. Analysis presented
in this paper provides answer as to which frequency bands are alone
sufficient for proper and successful spike sorting.

Keywords: Spike, Spike sorting, Wavelet, DWT decomposition, Band
filtering, PCA, Hierarchical clustering, Silhouette.

Introduction

During many neurosurgical procedures the microrecording is being used. In such
procedures a thin electrode is placed within brain tissue to monitor and record its
electrical activity. Electrical activity is produced by neuron cells. Neurons produce spikes – short (lasting about 1.5 ms) electrical impulses of high amplitude.
Typically the recording microelectrode can register electrical activity in a radius
of about 50 μm. In such volume, depending on the location within the brain, one

---

[1] Magnetic Resonance Imaging.
[2] Electromyography.

T. Andreasen et al. (Eds.): ISMIS 2014, LNAI 8502, pp. 154–163, 2014.
© Springer International Publishing Switzerland 2014

can find well over 100 neurons [4]. Many of them can produce spikes simultaneously and in a given recording there might be present spikes that originated in different neurons. It is also not uncommon that couple of neurons might produce spikes within few milliseconds of each other and that the recorded spikes would overlap. Single neuron cell measured at affixed point has an invariant shape of the spike. Shape derives from the physical structure of a given cell and it does not change by itself [3]. On Fig. 1a there are shown 721 aligned spikes that were found in 10 s long recording. It is evident that the shape of spikes is constant and unchanging, one can so infer that all 721 spikes were produced by a single neuron.

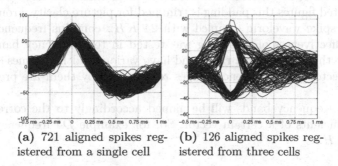

(a) 721 aligned spikes registered from a single cell   (b) 126 aligned spikes registered from three cells

Fig. 1. Spikes aligned to the time of their extremal value

In another recording (see Fig. 1b) only 126 spikes were found. Here, however the shapes are much more varied. Careful observation reveals that there are three classes of spike shapes, two with positive extreme value and one with negative one.

Discrimination of the spikes according to their shape is called Spike Sorting and there are many approaches to that subject [5][6][7]. It is therefore important to know which frequency band are essential to spike sorting and which can be safely filtered out or modified.

1 Spike Waveform

Each proper spike has a well defined point of maximal absolute amplitude. Clearly the whole spike takes place since some time before the time of the maximal amplitude and lasts for some time after. In this paper it is assumed that spikes last for 0.5 ms before reaching its maximal absolute amplitude and for 1.1 ms after it. Signals recorded by microelectrodes that are being analyzed in this paper have all been sampled with 24 KHz. It means that there are 24 samples for each millisecond of the recording. Spike is assumed to last 0.5 ms before and 1.1 ms after reaching its extreme amplitude. That means that amplitude vector for a spike contains $\lfloor 0.5 * 24 \rfloor + 1 + \lfloor 1.1 * 24 \rfloor = 39$ samples, 12 samples for pre-extremal part, one for extremum and 26 for post-extremal part of the spike.

1.1 Waveform DWT Decomposition

Basing on the properties of the Discrete Wavelet Transform (DWT) [1] one can obtain the image of the spike shape in certain frequency band through the proper wavelet coefficients. In this paper, the Daubechies D4 base function is used for DWT. D4 bears some similarity to the spike shape. As wavelet transform requires that the input vector has length being the power of two, the 39 samples of the spike are – prior to DWT transformation – right padded with twenty five zeros. In all presented figures this padding is trimmed for picture clarity. From Nyquist theorem [2] spike waveform sampled with 24 KHz contains frequencies up to 12 KHz. Effectiveness of clustering was tested in ten frequency bands, all of them fit into the $0 - 12KHz$ range and have various widths. Names of wavelet coefficient vectors and frequency ranges represented by them are presented in Table 1.

Later on, frequency bands will be named accordingly to the corresponding wavelet coefficients of a certain vector. So, for example $D4$ frequency band is $750 - 1500\ Hz$.

Table 1. Frequency bands

Wavelet coefficients vector name	Frequency range (Hz)		Wavelet coefficients vector name	Frequency range (Hz)	
$S1$	0	6000	$D1$	6000	12000
$S2$	0	3000	$D2$	3000	6000
$S3$	0	1500	$D3$	1500	3000
$S4$	0	750	$D4$	750	1500
$S5$	0	375	$D5$	375	750

Table 2 shows how the $0 - 12\ KHz$ range can be fully divided.

Table 2. Frequency band divisions

(a)	(b)	(c)	(d)	(e)
$S1$ $D1$	$S2$ $D2$ $D1$	$S3$ $D3$ $D2$ $D1$	$S4$ $D4$ $D3$ $D2$ $D1$	$S5$ $D5$ $D4$ $D3$ $D2$ $D1$

It is a reasonable hypothesis that some frequency bands residing below 12 KHz might be more essential to spike sorting than others. This can be readily seen on Fig. 2 where spikes seen on Fig. 1b are shown in two selected frequency bands. While on Fig. 2a we can still identify the shape classes, it becomes more difficult to distinguish them in the $1500 - 3000\ Hz$ band shown on Fig. 2b.

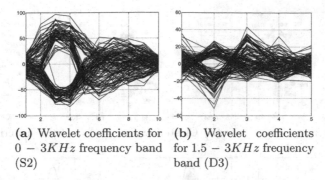

(a) Wavelet coefficients for $0 - 3KHz$ frequency band (S2)

(b) Wavelet coefficients for $1.5 - 3KHz$ frequency band (D3)

Fig. 2. Spikes representation in different frequency bands

2   Clusterings

All clusterings presented in this paper are based upon first two principal components obtained from either spike waveform or DWT coefficients in a given frequency band. Calculations using first three components were also tested but they did not show significant improvement in the quality of the obtained results. Clusterings were built using Hierarchical Clustering method with Euclidean distance and Ward's minimum variance approach [11]. Since no more than few hundred spikes are to be expected in a single $10\ s$ long recording, this clustering method can be safely applied regardless of its computational complexity. To achieve the optimal target number of clusters, the Silhouette function was applied. Clustering has been run a number of times with a target number of clusters ranging from 2 to 6, In each run the average value of the Silhouette was calculated. Finally, the clustering that produced maximal average Silhouette value was picked up.

Fig. 3a shows the result of a clustering found on the first two components of PCA based upon spikes padded waveforms. Such clustering is labeled as C_0. Fig. 3b shows the result of a clustering found on the first two components of PCA based upon $S2$ coefficients of spikes. Clusterings that are firstly found on a certain vector of wavelet coefficients are labeled accordingly to the name of that vector. So C_{S2} is the label assigned to that clustering. Fig. 3c shows the result of C_{D3} clustering.

Figures 3d, 3e and 3f show spikes that were assigned by C_{S2} clustering to clusters $\alpha_{C_{S2}}$, $\beta_{C_{S2}}$ and $\gamma_{C_{S2}}$. It is evident that shape classes were correctly separated in C_{S2} results.

From the Fig. 3c one can notice that C_{D3} clustering did not produce three clusters. Instead only two clusters were discovered. Figures 3g and 3h show spikes that were assigned by C_{D3} clustering to clusters $\alpha_{C_{D3}}$ and $\beta_{C_{D3}}$. The $\gamma_{C_{S2}}$ clearly corresponds to $\alpha_{C_{D3}}$. $\beta_{C_{D3}}$ seems to contain spikes from $\alpha_{C_{S2}}$ and $\beta_{C_{S2}}$. In the next section, a method showing how to obtain degree of agreement between clusters is provided. It will allow us to measure how big is the agreement between C_0 and C_{S2} and also between C_0 and C_{D3}.

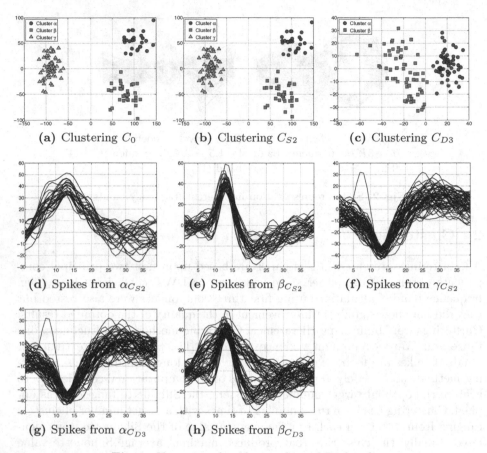

(a) Clustering C_0 **(b)** Clustering C_{S2} **(c)** Clustering C_{D3}

(d) Spikes from $\alpha_{C_{S2}}$ **(e)** Spikes from $\beta_{C_{S2}}$ **(f)** Spikes from $\gamma_{C_{S2}}$

(g) Spikes from $\alpha_{C_{D3}}$ **(h)** Spikes from $\beta_{C_{D3}}$

Fig. 3. Clusterings based upon S2 and D3 bands

2.1 Comparison of Clusterings

Let's assume that clustering C_0 produced m clusters $C_0 = \{c_{0,i}\}_{i=1..m}$ denoted as $\alpha_{C_0}, \beta_{C_0}, ...$ and clustering C_L produced n clusters $C_L = \{c_{L,i}\}_{i=1..n}$ denoted as $\alpha_{C_L}, \beta_{C_L}, ...$. There are $m \times n$ intersections between those resulting clusters. The cardinalities of such intersections form arrays like these shown in Tables 3a and 3b.

When mapping between clusters is exact, i.e. $\forall_{c_{0,i}} \exists_{c_{L,j}} c_{0,i} = c_{L,j}$ each row and column of such array must contain exactly one non–zero element (see Table 3a). When the mapping is not exact, it must be decided which cluster from C_0 should be 1–1 mapped to which one from C_L. If $\overline{\overline{C_0}} > \overline{\overline{C_L}}$, then at least one cluster from C_0 will not be mapped to a cluster from $c_{L,j}$. If $\overline{\overline{C_0}} < \overline{\overline{C_L}}$, then at least one cluster from C_L will not be mapped to a cluster from $c_{0,i}$. It is desirable to have the not mapped clusters as small as possible.

Table 3. Cluster intersections

(a)   (b)

$\cap$	α_{S2}	β_{S2}	γ_{S2}
α_0	28	0	0
β_0	0	35	0
γ_0	0	0	63

$\cap$	α_{D3}	β_{D3}
α_0	0	28
β_0	0	35
γ_0	63	0

Assuming that C_0 and C_L are the clustering results that are to be compared, the mapping can be obtained using simple greedy approach

1. Find largest not mapped $c_{0,i}$;   exit when not found
2. Find not mapped $c_{L,j}$ such that $\overline{c_{0,i} \cap c_{L,j}}$ is maximal and non zero
3. Exit when no $c_{L,j}$ has been found
4. Map $c_{0,i}$ to $c_{L,j}$
5. Go to 1

Results of the described mapping for Tables 3a and 3b are shown in Table 4a and Table 4b.

Table 4. Cluster mappings

(a)   (b)

$\cap$	α_{S2}	β_{S2}	γ_{S2}
α_0	28	0	0
β_0	0	35	0
γ_0	0	0	63

$\cap$	α_{D3}	β_{D3}
α_0	0	28
β_0	0	35
γ_0	63	0

As the C_{D3} contains only two clusters, one of the clusters from C_0 remains not mapped, it is the α_0 cluster – smallest cluster found in C_0.

Having all intersections calculated and the mapping affixed, one can define the observed probability P_o of agreement between clusterings C_0 and C_{S2} and between C_0 and C_{D3}. P_o is the proportion of the spikes that according to the mapping have been correctly placed in both clusterings to the total number of clustered spikes [8].

$$P_o(C_0, C_{S2}) = \frac{28 + 35 + 63}{126} = 1 \tag{1}$$

$$P_o(C_0, C_{D3}) = \frac{63 + 35}{126} = \frac{7}{9} \approx 0.778 \tag{2}$$

Second probability, denoted by P_e is the probability of random agreement between clusterings [8]. That is for C_0 and C_{S2} case, P_e is a probability that given element will be in α_0 and α_{S2} or that it will be in β_0 and β_{S2} or that it will be in γ_0 and γ_{S2}.

$$P_e(C_0, C_{S2}) = \frac{28}{126} * \frac{28}{126} + \frac{35}{126} * \frac{35}{126} + \frac{63}{126} * \frac{63}{126} = \frac{61}{162} \approx 0.377 \qquad (3)$$

For C_0 and C_{D3} case, P_e is a probability that given element will be in β_0 and β_{D3} or that it will be in γ_0 and α_{D3}.

$$P_e(C_0, C_{D3}) = \frac{35}{126} * \frac{28 + 35}{126} + \frac{63}{126} * \frac{63}{126} = \frac{7}{18} \approx 0.389 \qquad (4)$$

The measure of agreement between clusterings, the Cohen's kappa coefficient [9] is defined as

$$\kappa = \frac{P_o - P_e}{1 - P_e} \qquad (5)$$

Finally,

$$\kappa(C_0, C_{S2}) = 1 \qquad \kappa(C_0, C_{D3}) = \frac{7}{11} \approx 0.636 \qquad (6)$$

Only the $\kappa(C_0, C_{S2})$ exceeds the threshold of 0.75 that indicates very good clustering agreement [10][12]. Results of clusterings can also be compared using the Jaccard similarity measure.

3   Evaluation

For the analysis we used 2917 real biological recordings. Each recording is 10 s long and contains at least 100 spikes and at least two spike shape classes. All recordings were made during the Deep Brain Stimulation (DBS) surgeries in the Institute of Psychiatry and Neurology in Warsaw, Poland.

For each of these recordings, eleven clusterings were calculated. First of them, the referencing one, is made upon first two PCA components calculated for all spikes found in the recordings. This clustering is denoted as $C_0(rec)$.

Next, 10 DWT based clusterings are computed. Each of them is calculated basing on specific vector of wavelet coefficients (see Table 1). $C_{S1}(rec)$ is a clustering calculated basing upon the first two PCA components calculated for $S1$ wavelet coefficients of all spikes found in the recordings. $C_{D1}(rec)$, $C_{S2}(rec)$, ..., $C_{D5}(rec)$ are calculated accordingly.

Finally for each set of wavelet coefficients / frequency band, the Cohen's κ is calculated. We obtain ten values: $\kappa_{S1} = \kappa(C_0, C_{S1})$, $\kappa_{D1} = \kappa(C_0, C_{D1})$, ..., $\kappa_{D5} = \kappa(C_0, C_{D5})$. As a result of this, we have a measure that is telling us how closely each of DWT based clusterings matches the C_0. This in turn gives important information as to which frequency bands are by themself sufficient for spike sorting of quality similar to this of C_0.

In this paper it has been assumed that a given frequency band BND is deemed to be suitable for spike sorting only if for over 90% of recordings the $\kappa_{BND} >$

0.75. Kappa of value above 0.75 indicates very good agreement between clustering results [10][12]. This would ensure that with 90 % probability, clustering based on frequency band BND would produce results of very good agreement with C_0.

In the Table 5, for each band there are percentages of recordings for whom the value of κ was above 0.75 threshold.

Table 5. Percentages of recordings with κ above 0.75

S1	D1	S2	D2	S3	D3	S4	D4	S5	D5
98	63	97	82	93	56	88	61	3	87

One can see, that some bands are clearly worse than others. For example for $S5$ only 3 % reaches κ value of 0.75 or above. It clearly indicates that S5 band i.e. $0 - 375\ Hz$ is poorly suited for spike shape discrimination. Interesting is that for $D5$, 87 % of κ_{D5} values are greater than 0.75. All of it knowing that there are only two $D5$ coefficients for each spike. It might be expected that such two value based discrimination works well for two shape classes, when one of them has negative extremum and other one positive (see Fig. 3e and 3f).

If that is true, κ_{D5} should have much lower values when considering recordings with at least three shape classes (as seen on Fig. 1b). In the Table 6 there are percentages of such recordings having κ_{S1}, ..., κ_{D5} values above 0.75 threshold.

There are only 95 recordings having spikes of at least three shape classes. While this is as little as 3.3 % of the initially selected recordings, still there are brain regions for which such divergence is characteristic [4].

Table 6. Percentages of recordings with κ above 0.75

S1	D1	S2	D2	S3	D3	S4	D4	S5	D5
94	22	80	25	62	21	37	31	2	23

As expected, all of the percentages shown in Table 6 are lower than those seen in Table 5. Now only for the $S1$ band there are over 90 % of recordings having very good κ, i.e. over 0.75.

Still, the $S2$ band gives almost 80 % probability for recording to have κ over 0.75. The case shown in section 2 is an example of randomly chosen recording with three spike shape classes for which $\kappa_{S2} = 1$, i.e. there is ideal agreement between C_0 and C_{S2}.

4 Summary

In this paper various clusterings of spike shapes based upon DWT frequency band selection have been tested. The hierarchical clustering based upon first two principal components of spikes waveforms's PCA was used as the reference clustering. Test clusterings were based upon first two principal components calculated from a wavelet coefficients proper for each selected frequency band. In both reference and test clusterings, the target number of clusters was selected automatically as to maximize the average silhouette value.

For testing of agreement between clustering results the Cohen's kappa (κ) was used. Two cluster results were assumed to be in agreement if the calculated κ was larger than 0.75.

The ultimate goal of this research was to find frequency bands for whom the majority of the recordings exhibited agreement between the reference and test clustering. The spike sorting for such bands, based upon the whole spike waveform and its band's wavelet coefficients, would produce highly analogous spike sorting results.

Depending on the number of shape classes in the recording, different frequency bands can be successfully used for spike sorting. In the case of two or more distinct shape classes, sufficient information is contained in bands $S1$: $0 - 6000\ Hz$, $S2$: $0 - 3000\ Hz$, and $S3$: $0 - 1500\ Hz$. Only in the case of these bands for over 90 % of recordings, the κ was above 0.75.

In fact for $S1$ and $S2$, the κ were above 0.75 for more than 95 % of recordings. What must be noted is that $S3$ band is a frequency subset of $S2$ band which is in turn a frequency subset of $S1$ band. The narrower the band, the percentage of recordings with high κ becomes smaller.

The $S4$: $0 - 750\ Hz$, $D5$: $375 - 750\ Hz$, and $D2$: $3000 - 6000\ Hz$ bands gave $\kappa > 0.75$ for at least 80 % of recordings.

Out of all evaluated recordings containing at least two spike shape classes only 3.3 % of them actually do contain more than two spike shape classes.

For recordings with three or more spike shape classes (see Table 6) only the $S1$ – the most broad band containing low frequencies – does provide information complete enough to allow very good spike sorting. For this band the κ was above 0.75 for 94 % of recordings.

The $S1$ frequency subset – $S2$ gave κ above 0.75 for 80 % of recordings. While this percentage does not meet the 90 % limit it still can be successfully used for spike sorting as it is shown in section 2.

All recordings used for research described in this paper have been recorded during Deep Brain Stimulation (DBS) surgeries in the Institute of Psychiatry and Neurology in Warsaw, Poland. The recordings were not preselected in any way besides the requirement for the sufficient number of observed spikes and shape classes.

During those deeply situated surgeries, Parkinson Disease affected part of the brain (STN) must be identified [3][4]. As an aid in precise identification of that brain region, a specialized software has been developed by the authors of this paper. At the certain stage of analysis, this software performs spikes sorting procedures that are described in this paper.

This software is currently being used during deep brain stimulation surgeries for Parkinson Disease performed in the Institute of Psychiatry and Neurology in Warsaw, Poland.

References

1. Jensen, A., Cour-Harbo, A.I.: Ripples in Mathematics. Springer (2001)
2. Smith, S.W.: Digital Signal Processing. Elsevier (2003)
3. Nolte, J.: The Human Brain, An Introduction to Its Functional Anatomy (2009)
4. Israel, Z., Burchiel, K.J.: Microelectrode Recording in Movement Disorder Surgery. Thieme Medical Publishers (2004)
5. Archer, C., Hochstenbach, M.E., Hoede, C., et al.: Neural spike sorting with spatio-temporal features. In: In:Proceedings of the 63rd European Study Group Mathematics with Industry, January 28 - February 1 (2008)
6. Quian Quiroga, R., Nadasdy, Z., Ben-Shaul, Y.: Unsupervised Spike Detection and Sorting with Wavelets and Superparamagnetic Clustering. MIT Press (2004)
7. Lewicki, M.S.: A review of methods for spike sorting: The detection and classification of neural action potentials. 9(4), R53–R78 (1998)
8. Reilly, C., Wang, C., Rutherford, M.: A rapid method for the comparison of cluster analyses. Statistica Sinica 15, 19–33 (2005)
9. Cohen, J.: A Coefficient of Agreement for Nominal Scales. Educational and Psychological Measurement 20(1), 37–46 (1960)
10. Seigel, D.G., Podgo, M.J., Remaley, N.A.: Acceptable values of kappa for comparison of two groups. American Journal of Epidemiology 135(5), 571–578 (2012)
11. Kaneko, H., Suzuki, S.S., Okada, J., Akamatsu, M.: Multineuronal spike classification based on multisite electrode recording, whole-waveform analysis, and hierarchical clustering. IEEE Transactions on Biomedical Engineering 46(3) (1999)
12. Fleiss, J.L.: Measuring nominal scale agreement among many raters. Psychological Bulletin, Vol 76(5), 378–382 (1971)

Neural Network Implementation
of a Mesoscale Meteorological Model

Robert Firth and Jianhua Chen

Division of Computer Science and Engineering
School of Electrical Engineering and Computer Science
Louisiana State University
Baton Rouge, LA 70808
rfirth1@tigers.lsu.edu,
jianhua@csc.lsu.edu

Abstract. Numerical weather prediction is a computationally expensive task that requires not only the numerical solution to a complex set of non-linear partial differential equations, but also the creation of a parameterization scheme to estimate sub-grid scale phenomenon. This paper outlines an alternative approach to developing a mesoscale meteorological model – a modified recurrent neural network that learns to simulate the solution to these equations. Along with an appropriate time integration scheme and learning algorithm, this method can be used to create multi-day forecasts for a large region.

The learning method presented in this paper is an extended form of Backpropagation Through Time for a recurrent network with outputs that feed back through as inputs only after undergoing a fixed transformation.

Keywords: Recurrent neural networks, spatial-temporal, weather prediction, forecasting, temperature, wind.

1   Introduction

While huge accuracy gains have been made in weather forecasting, it still remains a challenging task. Many approaches have been developed including heuristics like persistence and trends, numerical weather prediction, and neural networks.

Numerical weather prediction is a computationally expensive task that requires not only the numerical solution to a complex set of non-linear partial differential equations (PDEs), but also the creation of a parameterization scheme to estimate sub-grid scale phenomenon [1].

This paper proposes a method to replace the primitive equations, the set of PDEs that govern atmospheric dynamics [2], with a set of recurrent neural networks.

2   Previous Work

Computational approaches that have been developed to forecast weather include heuristics like persistence and trends, numerical weather prediction, and neural networks.

T. Andreasen et al. (Eds.): ISMIS 2014, LNAI 8502, pp. 164–173, 2014.
© Springer International Publishing Switzerland 2014

Numerical weather prediction techniques focus on simulating the evolution of the atmosphere through numerical solutions to the simplified set of partial differential equations known as the primitive equations to find the time derivative of the variables and then applying forward time integration [2]. Models such as the NCEP's RAP (Rapid Refresh) model use this technique to generate forecasts.

Parameterization schemes are used to estimate sub-grid scales phenomena that can't be directly simulated [1]. Krasnopolsky et al. replaced the shortwave and long-wave atmospheric radiation parameterization schemes of the NCAR CAM-2 model with a neural network. The network proved to be a fast and accurate replacement and resulted in a 50-60 times faster computation of the radiation parameterization [6].

Zakerinia et al. developed a neural network to create a wind forecast for a single site using 3 inputs, 20 hidden nodes, and 1 output node that represented the 1 hour wind forecast [3]. Corne et al. also developed a neural network to forecast wind speed for a single site, but used 7 input variables (cloud cover, humidity, pressure, temperature, visibility, wind speed, and wind direction) for the single site. They tested using the 7 variables as inputs, the 7 variables plus 7 more from an hour before, and the 7 variables plus their 1 hour deltas [7].

Abdel-Aal et al. used abductive networks to create a 24 hour hourly temperature forecast. The inputs to the network were temperatures for the 24 previous hours, minimum and maximum temperature for the previous day, and the minimum and maximum forecasted temperature. The output is the temperature for a given hour on the following day [4].

Previous work in this direction has been focused mainly on either forecasting weather variables for a single location and learn using inputs from only that site, or focused on creating a hybrid dynamic climate model by applying machine learning to the parameterization scheme. The former ignores the important spatial component that is available and essential to a successful forecast, while the latter hybrid model only partially relies on machine learning. For this reason, the method proposed is a generalized recurrent neural network that utilizes both spatial and temporal information to generate a forecast for a wide region. Instead of developing a hybrid model, the method almost exclusively relies on learning with some domain knowledge.

The learning method presented in this paper is an extended form of Backpropagation Through Time for a recurrent network with outputs that feed back through as inputs only after undergoing a fixed transformation.

Backpropagation Through Time involves unfolding the network in time until all cycles are removed, then applying normal backpropagation [5].

3   Method

3.1   The Grid and the Forecast

The input data to forecast wind is the wind speed in the east-west (U) and north-south (V) directions, geopotential height, and latitude at 2744 locations across the southeastern United States. The input data to forecast temperature is temperature, wind, cloud cover, and solar angle. These observations are on a 56x49 grid as shown in fig. 3.

The goal of the forecast is to determine the future state of the gridded variables 1 hour in the future. To accomplish this, a 6 minute forecast is generated for every point on the grid. This 6 minute forecast can be further extended by using it as the input to the forecast system again to time step further and further into the future. This is done 10 times to generate a forecast 1 hour in the future.

3.2   Finite Differences

Many inputs are not fed directly into the network. Instead, they are fed in as partial derivatives with respect to east-west and north-south grid coordinates x and y. These partial derivatives are computed numerically using a centered finite difference scheme:

$$\frac{\partial T_{ij}}{\partial x} = \frac{T_{i+1,j} - T_{i-1,j}}{2}$$

$$\frac{\partial T_{ij}}{\partial y} = \frac{T_{i,j+1} - T_{i,j-1}}{2}$$

3.3   Time Integration

The learning task of our recurrent neural network is to learn to compute the partial derivative of each meteorological variable with respect to time. For temperature, this would be $\frac{\partial T_{ij}}{\partial t}$. Once this is known, we can time step forward to get the next value of T:

$$T_1 = T_0 + \Delta t \frac{\partial T_0}{\partial t}$$

$$T_{t+1} = T_{t-1} + 2 \times \Delta t \frac{\partial T_t}{\partial t}$$

The first formula is a forward integration technique, while the second is a centered-in-time technique, or leapfrog. While we could use the first one for every time step, errors quickly ruin the forecast unless a very small time step is used [10]. It was confirmed experimentally with a 15 second time step that the forward integration technique underperforms the leapfrog scheme using only a 5 minute time step. For this reason, we only use the forward scheme in the first time step to get the leapfrog scheme started.

This same process is used to forecast all meteorological variables.

3.4   The CFL Condition

We are using 20km resolution input data and 1 hour later target values. Ideally, we would take that input data and create a 1 hour forecast. However, it was discovered by

Courant, Friedrichs, and Lewy that forecast stability is a function of grid resolution, time step, and velocity [2].

$$C = \frac{u_{max}\Delta t}{\Delta x} \leq C_{max}$$

The ideal value of C_{max} depends on many factors, including the system solution method. For our purposes, we'll take it to be equal to 1. This means that with a 1 hour time step and 20 km grid spacing, the maximum wind velocity we can simulate without the simulation becoming unstable is approximately 5.5 m/s or 12 mph. This is much lower than the typical maximum wind speed, even at the surface. Jet streams and cyclones can have wind speeds that exceed 150 mph.

If we change our time step to 6 minutes and keep the same grid spacing, we can simulate wind speeds up to 55.5 m/s, or 124 mph. This necessarily smaller time step makes designing our system much more difficult because we don't have target values for only 6 minute later. It also means that the forecast system must run for 10 iterations in order to create a 1 hour forecast, increasing computation time by a factor of 10.

For reference, the RAP model uses a 1 minute time step. This allows for very high wind speeds in a very stable model.

3.5   Inputs and Outputs

The recurrent networks to forecast U, V, and T each require 5 inputs and generate 1 output. This is possible because of our pre-processing stage where we compute the spatial derivatives at the point we wish to forecast.

Alternately, instead of computing the partial derivatives for use as inputs to the network, we could train an autoencoder network. This could take the 6 nearest neighbor grid points as inputs and use unsupervised learning to learn a lower dimensional representation.

If we instead input every temperature value in the 1-region, every U and V component of wind, then this would be 12 more inputs – a total of 17 inputs to the network. Because training slows and the network becomes less able to capture the desired function with increased dimensionality, this is undesirable.

This paper describes how to represent and learn N1, N2 (by symmetry with N1), and N3.

3.6   Error and Error Attribution

Error is computed by comparing the output of the system to the 1 hour later initialization data for the RAP model.

However, because the network is not a normal recurrent network – the inputs do not directly connect to the inputs – backwards propagation of error is not straightforward. Two rules are therefore created to propagate error back to the output layer of the network (so that normal backpropagation can begin). First, the last step, time integration:

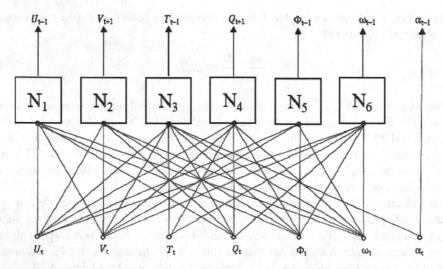

Fig. 1. Networks N1 through N6 are independent and can be trained separately. Each network computes the time derivative of a different meteorological observable variable.

$$T_{t+1} = T_t + \Delta t \frac{\partial T_t}{\partial t}$$

Error in T_{t+1} can be attributed to two sources, T_t and $\frac{\partial T_t}{\partial t}$. We can weight this error as:

$$E = (1 - \lambda)E + \lambda E$$

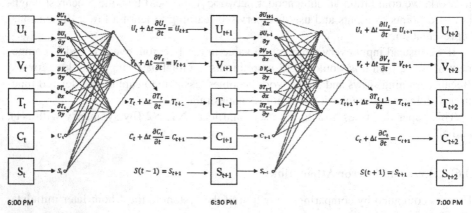

Fig. 2. Temperature forecasting recurrent network unfolded though time with time step Δt equal to 30 minutes. U and V are the east-west and north-south components of the wind speed, respectively. T is the temperature, C is the cloud cover, and S is the sun's altitude above the horizon. Outputs of the network do not directly re-enter as inputs. Instead, they must go through a pre-processing stage, which complicates backpropagation training.

Here E is the target – predicted, $(1-\lambda)E$ is the error attributed to T_1, and λE is the error attributed to an error in $\frac{\partial T_t}{\partial t}$. Also, λ is a parameter $0 \leq \lambda \leq 1$ and ideally decaying with time.

The portion attributed to T_1 is directly used to compute a grid of errors in T_1, which is passed backward in time to the previous time step. The portion of error attributed to $\frac{\partial T_t}{\partial t}$ is directly used as the error in the output layer of the network and is backpropagated normally using the backpropagation algorithm and updates the weights in the network.

4   Results

4.1   Implementation and Experimental Setup

The proposed method was implemented in Python and C++. All neural network code was written by the author specifically for this task. The experiment was run on a laptop with a 1.6GHz Intel Core 2 Duo U7600 processor and 4GB RAM. Training time was limited to 1 day, but could be allowed to run longer for reduced error. Because wind speed was forecast on 37 levels of the atmosphere, this required training 74 different networks – two for each level for the U and V components.

4.2   Data Sets

The network was trained using the hourly input data sets to the Rapid Refresh (RAP) model for days divisible by 3 in January 2014 and validated against days 3n+1 in that same month.

The RAP model is run hourly out to 18 hours on a 301x225 Lambert conformal projected grid with a 20km horizontal resolution and 37 vertical levels with pressure coordinates. This data can be downloaded from either the NCEP or NCDC ftp server [8,9].

Because the same learned network is applied to every grid point on a given pressure surface to create a forecast, and interaction with land/water at the surface therefore needs to be taken into account, this effect is reduced by selecting a 56x49 subgrid that covers the southeastern US and no ocean. This is a roughly homogenous region. This is only necessary at the lower levels of the atmosphere that are influenced by interaction with land, the planetary boundary layer. This region of the atmosphere is known as the planetary boundary layer.

Training and validation data was generated by computing input data for every grid point and generating target values for a 1 hour forecast by using the input files for the RAP model initialized 1 hour later.

4.3   Analysis

Error was measured as the Mean Absolute Error (MAE):

$$MAE = \frac{1}{n}\Sigma_{ij}|t_{ij} - y_{ij}|$$

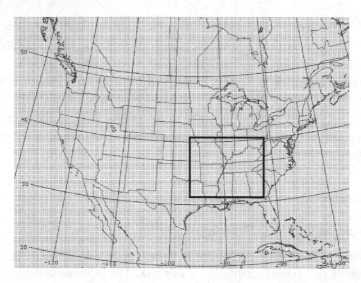

Fig. 3. NCEP Grid 252, a 301x225x37 grid of observations with approximately 20km horizontal resolution. Only the 56x49 grid over the southeastern US is used in this paper. This represents a roughly homogenous region with similar elevation and no ocean.

The forecasts generated by the proposed approach were compared to forecasts generated by the RAP model. The MAE of the 1 hour forecast generated by the RAP model is calculated for the same 56x49 sub-grid.

Table 1. Summary of results forecasting U and V for 10 levels of the atmosphere, where AIM3 represents the results of the proposed method. Error is MAE in m/s (meters per second).

Level	RAP U	RAP V	AIM3 U	AIM3 V
1000	0.4652	0.5219	0.4514	0.5807
900	0.7464	0.8519	1.0420	1.0868
800	0.7506	0.7588	1.0017	1.0764
700	0.7945	0.7715	1.0209	1.0714
600	0.8570	0.8695	1.1019	1.2007
500	1.0294	1.1761	1.2813	1.4829
400	1.3186	1.4488	1.8873	1.9040
300	1.5315	1.4577	2.0526	2.5082
200	1.2195	1.2825	1.5223	1.6046
100	0.7350	0.7379	0.9235	1.0091

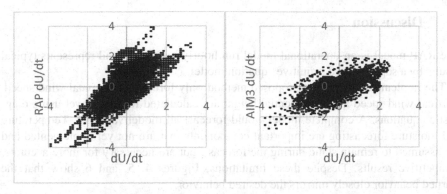

Fig. 4. Scatterplot for U Forecast on 1000mb Level

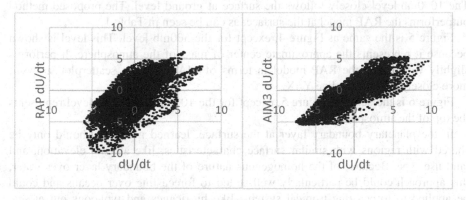

Fig. 5. Scatterplot for U Forecast on 500mb Level

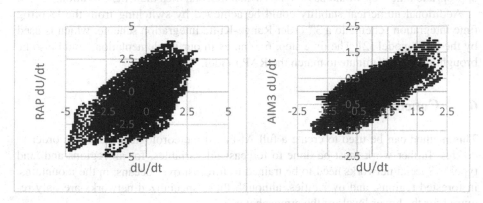

Fig. 6. Scatterplot for U Forecast on 100mb Level

5   Discussion

The RAP model is an operational model run hourly by NCEP, and represents typical results by a sophisticated primitive equation model.

The implementation of the proposed method only forecasts horizontal wind speed. Vertical wind speed and geopotential height are calculated for each level using diagnostic equations. A complete model would forecast all model variables. Temperature and moisture forecasting are important components that are not yet implemented and are assumed to remain static during the forecast, but are necessary for more accurate, competitive results. Despite these limitations, Figures 4, 5, and 6 show that the learned behavior closely mirrors the desired behavior.

Figure 4 is a scatterplot comparing the actual to the forecasted one hour change in wind speed for both the proposed method and the RAP model for the 1000mb level. The 1000mb level closely follows the surface at ground level. The proposed method outperforms the RAP model at the surface, as can be seen in Table 1.

Figure 5 is the same as Figure 4, except for the 500mb level. This level is shown because it represents the approximate center of mass of the atmosphere. It performs slightly worse than the RAP model in terms of MAE, but the scatterplot shows it more closely follows the line Y=X.

Figure 6 is the same as Figure 5, except for the 100mb level. This level represents the top of the atmosphere.

In the planetary boundary layer at the surface, learned networks should only be shared with regions with similar surface characteristics, like albedo, elevation, and land use type. Because of the homogenous nature of the boundary layer over water, this approach could be particularly well-suited to forecasting over oceans and could be applied to forecasting tropical systems like hurricanes and typhoons out at sea. However, special consideration would have to be made for landfalling systems and an appropriate time step would have to be chosen that satisfies the CFL condition.

Additional numerical stability could be achieved by switching from the leapfrog time integration scheme to a 3rd order Runge-Kutta integration scheme, which is used by the RAP model [2]. The time step, 6 minutes in our implementation, could also be brought down to 1 minute to match the RAP model.

6   Conclusion

This method can be used to create a full AI-based meteorological model. In order to do this, further work must be done to forecast all variables, for all regions and land types. Special networks need to be trained to forecast over oceans, in the mountains, in forested regions, and over cities, although these specialized networks are only required for the lower levels of the atmosphere.

Networks also need to be designed to forecast moisture transport, phase change, and latent heat. The remaining work there mainly involves selecting the appropriate inputs to consider. Although unimplemented, the network for forecasting temperature is given in Figure 2. With this, we would have a full forecast system.

Our implementation of the proposed approach has been successfully run out to 6 hours, but needs these additional components to generate competitive forecasts.

Acknowledgements. This work is partially supported by the Louisiana Board of Regents grant LEQSF-EPS(2013)-PFUND-307. We are grateful to Dr. Kevin Robbins from the NOAA Southern Regional Climate Center at LSU and Dr. Frederick Carr from the University of Oklahoma for many helpful discussions on topics related to this work.

References

1. Stensrud, D.J.: Parameterization Schemes: Keys to Understanding Numerical Weather Prediction Models, pp. 7–9. Cambridge, New York (2007)
2. Coiffier, J.: Fundamentals of Numerical Weather Prediction, vol. 4-6, pp. 15–16. Cambridge, New York (2011)
3. Zakerinia, M., Ghaderi, S.F.: Short Term Wind Power Forecasting Using Time Series Neural Networks. University of Tehran, Tehran (2011)
4. Abdel-Aal, R.E.: Hourly temperature forecasting using abductive networks. Eng. App. of Art. Intel. (2004)
5. Mitchell, T.M.: Machine Learning, pp. 119–121. McGraw-Hill, Singapore (1997)
6. Krasnopolsky, V.M., Michael, S.F., Dmitry, V.C.: New Approach to Calculation of Atmospheric Model Physics: Accurate and Fast Neural Network Emulation of Longwave Radiation in a Climate Model. Mon. Wea. Rev. 133, 1370–1383 (2005)
7. Corne, D., Reynolds, A., Galloway, S., Owens, E., Peacock, A.: Short term wind speed forecasting with evolved neural networks. In: Blum, C. (ed.) 15th Genetic and Evolutionary Computation Conference Companion (GECCO 2013 Companion), pp. 1521–1528. ACM, New York (2013)
8. National Centers for Environmental Prediction,
   ftp://ftp.ncep.noaa.gov/pub/data/nccf/com/rap/prod/
9. National Climatic Data Center,
   http://nomads.ncdc.noaa.gov/thredds/dodsC/rap252/
10. Warner, T.T.: Numerical Weather and Climate Prediction, pp. 456–459. Cambridge, New York (2011)

Spectral Machine Learning for Predicting Power Wheelchair Exercise Compliance

Robert Fisher[1], Reid Simmons[1], Cheng-Shiu Chung[2], Rory Cooper[2],
Garrett Grindle[2], Annmarie Kelleher[2], Hsinyi Liu[2], and Yu Kuang Wu[2]

[1] Carnegie Mellon University
5000 Forbes Ave, Pittsburgh PA 15213
[2] Human Engineering Research Laboratories
University of Pittsburgh
6425 Penn Avenue, Pittsburgh PA 15206

Abstract. Pressure ulcers are a common and devastating condition faced
by users of power wheelchairs. However, proper use of power wheelchair
tilt and recline functions can alleviate pressure and reduce the risk of ulcer occurrence. In this work, we show that when using data from a sensor
instrumented power wheelchair, we are able to predict with an average accuracy of 92% whether a subject will successfully complete a repositioning
exercise when prompted. We present two models of compliance prediction.
The first, a spectral Hidden Markov Model, uses fast, optimal optimization techniques to train a sequential classifier. The second, a decision tree
using information gain, is computationally efficient and produces an output that is easy for clinicians and wheelchair users to understand. These
prediction algorithms will be a key component in an intelligent reminding
system that will prompt users to complete a repositioning exercise only in
contexts in which the user is most likely to comply.

Keywords: Machine learning, spectral learning, HMMs, healthcare applications.

1 Introduction

More than 2.5 million Americans experience pressure ulcers every year[14]. One
particularly high risk group of individuals are those with severe physical disabilities that utilize power wheelchairs as their primary means of mobility. Power
wheelchairs can very effectively relieve pressure by raising the user to an re-
clined, titled position for several seconds. However, research indicates that less
than 40% of power wheelchair users correctly use their power seat functions to
relieve pressure and prevent deadly ulcers[12]. This leads many power wheelchair
users to be exposed to preventable pressure ulcer formation, which can often lead
to complicating infections or even death.

The Virtual Coach smart power wheelchair system was designed to track the
power wheelchair usage of users in order to help them better conform to the
pressure relief guidelines set forth for them by clinicians, as well as improve

T. Andreasen et al. (Eds.): ISMIS 2014, LNAI 8502, pp. 174–183, 2014.
© Springer International Publishing Switzerland 2014

their overall posture. By monitoring encoders in the power wheelchair's joints (chair tilt, leg rest elevation, seat elevation, etc), the system is able to determine when a user has successfully performed the repositioning exercise prescribed by the user's healthcare provider. Users are reminded on a periodic basis to conduct pressure relief, and the system tracks whether the user complied with the reminder or not. Initial results suggest that users that receive regular reminders from the system have higher rates of compliance with their exercise regime than those users not receiving instruction. However, there is a risk of the system becoming intrusive and annoying if users are given too many undesired reminders.

Instead of simply reminding users with a static periodicity, we would like to devise an intelligent system that selects the most appropriate contexts in which to issue reminders. In this paper, we present the results of applying machine learning algorithms to predict if a user is likely to comply with a reminder given information about the user's current context collected by several sensors onboard the Virtual Coach system. We show that when averaged across all users, we can attain a predictive accuracy of 92%. We present results with an expressive and statistically consistent spectral Hidden Markov Model, as well as a computationally lightweight decision tree model that creates an output which is easy for clinicians and users to understand. The goal for this work is to create an intelligent, decision theoretic reminding system that would select the best moments to issue reminders to users given the likelihood of compliance in the current context, the time since the last pressure relief was completed, and the recommendations given by the overseeing clinician. Predicting whether a user is likely to comply in a given context is the first step towards realizing these goals.

2   Related Work

Predicting pressure relief compliance using sensor data is related to the task of predicting user interruptability given contextual information. For instance, there has been some work predicting whether or not a user would prefer to have their smartphone ringer enabled in a given setting. Some of this work has leveraged decision theoretic models of user preferences[15], while other work has emphasized active sampling of user preferences coupled with features generated using sensor processing algorithms[10].

There has been some previous work utilizing machine learning to predict patient behavior in healthcare applications, but rarely with the fine level of granularity seen in the Virtual Coach system. For instance, one author used statistical machine learning techniques with patient data to predict which subjects suffering from coronary artery disease would be likely to comply with pharmaceutical guidelines for managing cholesterol levels[8]. These predictions were based primarily on demographic data, and did not give indications as to which contexts and circumstances would lead to non-compliance for a subject. There has also been quite a bit of work using machine learning to predict health care outcomes and complications, but this work has not been focused on understanding or altering patient behavior to improve outcomes[16,6].

In this paper, we present results using *spectral methods* for machine learning, which has become a very popular topic in machine learning research. Based on eigenvector decomposition, such as is used in Singular Value Decomposition (SVD) and Principle Component Analysis (PCA), spectral methods can give optimal results for many optimization tasks using a small fraction of the computation required by comparable algorithms. Spectral algorithms exist for learning latent-variable PCFG's[5], dynamical systems[3], and Hidden Markov Models[11]. From an applications standpoint, spectral methods have been used to estimate a student's aptitude for key tasks in a classroom setting[9], for dependency parsing of natural language text[7], and for image segmentation and classification tasks[13].

3   The Virtual Seating Coach System

The Virtual Seating Coach (VC) system is a smart power wheelchair outfitted with a variety of sensors and an onboard computer. The sensors installed on these chairs include encoders in each of the chair's joints and wheels, accelerometers in the base of the seat, a seat occupancy sensor, a thermometer, and a light sensor. Additionally, a subset of users in the clinical trails used chairs that were outfitted with GPS chips and microphones. A complete list of the sensors and the machine learning features that were computed is shown in Table 1.

Fig. 1. A researcher demonstrating the use of the Virtual Coach system

A tablet computer is attached to the arm of the chair, as seen in Figure 1. The software on this tablet is used to issue reminders to users, and to provide feedback while an exercise is being conducted. Additionally, users can set preferences that dictates if reminders with be given using audio queues, silent text, or if the reminders should be disabled completely. When a reminder is given, a user is given the choice of snoozing the reminder, which will result in another prompt being given in 5 minutes, or the reminder can be dismissed for 1 hour. If the user does not comply with the reminder, it is automatically snoozed for 5 minutes.

If the user successfully completes the exercise, the reminding system will be reset for a duration determined by a clinician.

Table 1. The features computed using the Virtual Coach sensors

Data Source	Features
Chair angle encoders	Mean and variance of encoder values over previous 10 seconds
Wheel Encoders	Mean and variance of rotational velocity over previous 10 seconds
Accelerometers	Average Fourier power in 500 Hertz bands over 0-4000 Hz
Thermometer	Average temperature over previous 30 seconds
Seat Occupancy	A 0/1 value indicating state of pressure sensor in seat
Clock	The current day of the week and hour of the day
Recent Behavior	Time since the last reminder
Recent Behavior	Outcome of previous reminder (Snooze, dismiss, compliance)
GPS Data[1]	The current encrypted longitude and latitude coordinates
Audio Data[1]	Average Fourier power in 100 Hertz bands over 0-1000 Hz

Several prototype versions of the system have been undergoing clinical trials, and more than 20 volunteers have participated in the trials using the VC system over a period of 6 weeks per user. Volunteers were either assigned to a control group, being given a sensor equipped chair without a reminding system, or they were given the full VC system with tablet computer. The system stores clinician pressure relief guidelines for the individual user. For instance, the recommendation may be for the user to receive 30 seconds of pressure relief once every 60 minutes. During clinical trials the system issued reminders on a fixed basis according to the prescription, and the user's response to the reminder was recorded.

4   Spectral Hidden Markov Models

In recent years, spectral methods have become increasingly popular tools for solving a wide variety of optimization problems in machine learning. Spectral algorithms utilize eigenvector decomposition of data matrices to estimate the parameters of a learning model. Spectral algorithms are statistically consistent, do not suffer from local optima, and can be orders of magnitude faster than comparable optimization techniques, such as Expectation Maximization (EM). Spectral algorithms have been particularly successful when learning latent variable models[11].

Hidden Markov Models operate on sequences of observed data and assume the existence of a latent state variable that allows for the Markovian assumption. Put another way, given a hidden state $h_i \in [1, m]$ and a sequence of observations $x_1, x_2, ...x_t$, $x_i \in \mathbb{R}^n$, the assumptions driving the HMM give us the following equality:

---

[1] Data only available for subset of users.

$$P(x_t|x_1, x_2, ...x_{t-1}) = P(x_t|h_i)$$

Because the hidden state is unobserved and latent, the current state is generally represented as a distribution over possible states. We can then marginalize over the latent state distribution to estimate the probability of observing a given outcome.

There are three sets of parameters that must be learned to describe an HMM. The state transition probability matrix, $T \in \mathbb{R}^{m \times m}$, defines the probability of transitioning from one hidden state to another. The observation matrix, $O \in \mathbb{R}^{n \times m}$, defines the probability of seeing a given observation given a state, where n is the dimensionality of our observation space. Finally, the initial state distribution, $\pi \in \mathbb{R}^m$, defines the probability of beginning in each of the m latent states. The mean squared error of estimating T and O goes to 0 with the number of observations. However, the mean squared error of the estimate of π goes to 0 with the number of observed sequences. Therefore, a model trained on a single sequence of many observations may have high predictive error because the estimate of π will be incorrect. We prefer, therefore, to use several short independent sequences of operations. For our application, we consider each day that a subject uses the Virtual Coach system to be a separate independent sequence. In practice it is not necessary to know m, the number of hidden states, because this value can be tuned using cross validation.

Historically, the parameters of an HMM have been trained using the EM algorithm or a similar approach. However, these optimization techniques tend to be very slow and prone to fall into local optima. This is in part due to the fact that learning HMM parameters exactly has been shown to be intractable under cryptographic assumptions[17]. Recently, spectral formulations of the HMM learning problem have been posed which allow an HMM to be learned quickly and optimally when the parameter matrices are of rank no greater than m[1,11].

The spectral formulation of an HMM utilizes subspace identification by creating an m-dimensional subspace that preserves the state transition dynamics of the original model. This type of subspace identification is closely related to the use of Predictive State Representations, which have been used quite successfully in a variety of dynamical systems applications[4]. To project into this subspace, we must compute an invertible subspace transformation matrix, $U \in \mathbb{R}^{n \times m}$, that allows us to transform data in the original $n \times n$ space into the $n \times m$ subspace. To accomplish this, we begin by computing the empirical probabilities of sequential cooccurence of all observation doubles and store the results in a matrix $\hat{P}_{2,1} \in \mathbb{R}^{n \times n}$. Put another way, this matrix represents the probability of seeing observation x_i followed immediately by observation x_j. The matrix U can then be recovered by performing Singular Value Decomposition (SVD) on the matrix $\hat{P}_{2,1}$. Now, rather than learning the HMM parameters in the original space, we learn the parameters in this reduced subspace. If the subspace dimensionality, m, is no less than the rank of the parameter matrices T and O, then this subspace embedding will act as a lossless compression, and the new parameters that we learn will have the same dynamics as the original system.

Of further interest, these new subspace parameters can be computed using only computationally efficient operations, such as empirical probability estimates and standard matrix operations. We denote the subspace parameters as $\hat{\pi}_U$, $\hat{T}_U$, and $\hat{O}_U$. X^+ represents the Moore-Penrose pseudo-inverse of a matrix. P_1, $P_{2,1}$, and $P_{3,x,1}$ respectively represent the empirical single, double, and triple cooccurrence probabilities of observation. It can then be proven that the following equalities hold:

$$\hat{\pi}_U = U^T P_1$$

$$\hat{T}_U = (P_{2,1}^T U) + P_1$$

$$\hat{O}_U = U^T P_{3,x,1}(U^T P_{2,1})^+ \forall x \in [n]$$

For the sake of brevity, we omit the proofs here, but they can be found in Hsu, et al 2012[11]. We see from the above equalities that the parameters of the spectral HMM can be recovered exactly using only empirical distribution estimates with matrix transposition, multiplication, and pseudo-inverse operations.

5   Decision Trees

Power wheelchairs are a principle component in the life of many people living with a physical disability, and research suggests that chair users prefer to remain in full control of their chair's functionality [2]. In light of this, we may wish to create a reminding system that is maximally transparent, so that a user can better understand how the intelligent system is deciding when to remind. The spectral HMM, while very powerful, relies on a latent variable distribution that often lacks a simple, relatable explanation.

In an effort to explore a model that would be easier for end users to comprehend, we turned to one of the oldest and most popular machine learning classifiers: decision trees. Decision trees conduct classification through a series of logical binary operations applied to the learning features. *Information gain* is most often used to determine which features to place near the root of the tree. Information gain is defined as the reduction in statistical entropy gained by learning the state of a random variable. If we denote $H(Y)$ to be the statistical entropy of a random variable, the information gain of X with regards to label Y is $H(Y) - H(Y|X = a)$.

Decision trees have been shown to be high bias classifiers, leading to a great deal of overfitting [18]. To combat this, decision trees are generally pruned near the leaves of the tree to reduce bias and improve the tree's capacity for generalization. Smaller trees will also be easier for users to view and comprehend.

An example of a heavily pruned decision tree trained on one user's Virtual Coach data is shown in Figure 2. In this figure, a lead node of 1 indicates that the user is expected to comply with a reminder, 0 indicates predicted non-compliance. The tree in this figure was trimmed to depth 3, leaving only encrypted location and audio features remaining. It is worth noting that the audio feature with the highest information gain is the Fourier signal power in the 300 to 400Hz range of the spectrum. This represents the lower end of the audio spectrum that the human voice inhabits, so the decision tree seems to indicate that this given user would prefer not to be reminded if there is evidence of a speech signal present in the environment. If a Virtual Coach user were to elect to use a decision tree to determine when the intelligent system should issue reminders, the user would be able to verify, and possibly modify, the behavior of the system. This would help to curb any confusion a user may have as to why the system is behaving in the way that it is.

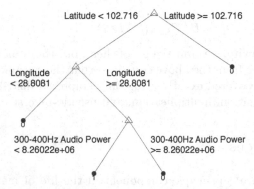

Fig. 2. A Decision Tree Showing the Compliance Predictions of a User

6   Results

When collecting empirical results, we used 15 fold cross validation to determine prediction accuracy of our models by randomly selecting one week of data to hold out for validation. Figure 3 shows the average prediction accuracy of the spectral HMM and decision tree models as a function of training data volume. We also considered several other classifiers, including Support Vector Machines, Nearest Neighbors, Naive Bayes' algorithm, and Conditional Random Fields. None of these other approaches performed as well as the spectral HMM, or produced output that was as easy to understand as the decision trees.

We see from these results that the spectral HMM performs particularly well with small amounts of data, with a very steep peak in the training curve after being supplied with only 20% of the data (roughly one week's worth of data). This result fits with intuition, because we expect accuracy to plateau after seeing one full instance of a user's weekly schedule. The spectral HMM peaks at around 92% classification accuracy, while the decision tree model peaks at 89%. It is not surprising that the spectral HMM performs the best, because the HMM reflects recent observations through the latent state distribution, while the decision tree performs classification using only the current sensor values. For comparison, a naive model that always predicts the most likely prior attains 69% classification accuracy.

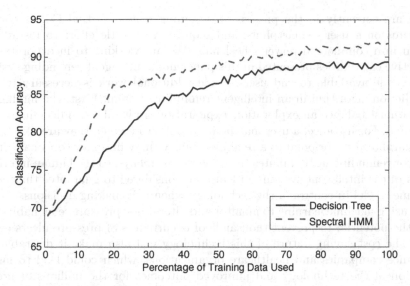

Fig. 3. Empirical Training Curve

We determined that both models performed best when supplied with training and testing data from only a single user. Models trained with data from multiple users resulted in many contradictory training examples, which is intuitive, because we would expect different users to react differently to a reminder under the same circumstances.

Prediction accuracy with users for whom audio and GPS data were available was higher than the rest of the population, showing 94% average accuracy with the HMM, and 92% average accuracy with the decision tree. We see a smaller differential between the two models in this case due to the fact that the GPS data shows very high information gain, often placing it near the root of the decision trees. For these users we can attain a predictive accuracy surpassing 85% using trees of only depth four that were built only using time of day, GPS coordinates, and the outcome of the most recent reminder. This suggests that simpler models are almost as effective for users that are willing to share personal information, such as location data, with the Virtual Coaching system.

7 Conclusions and Future Work

Initial results show that we can accurately predict repositioning exercise compliance given a set of fairly common sensors embedded in a power wheelchair. The spectral Hidden Markov Model leverages Markovian information about the past in a framework that has many desirable theoretical and practical qualities. On the other hand, the decision tree model is computationally very lightweight, and presents an easily understandable graphical representation of the model. The tradeoffs of these two methods means that each may be preferred to the other under certain conditions.

We are currently in the process of designing a new Virtual Coach system that runs on a user's smartphone and requires very little effort to install the system in a commercial power wheelchair. We are working to incorporate the predictive capabilities presented in this paper in an intelligent reminding system to be made available to end users. Some additional work is necessary to use a prediction algorithm in an intelligent reminding system. First, the intelligent system must balance an exploration/exploitation trade-off when deciding when to remind. For instance, a user may be in a context in which they are likely, but not guaranteed, to respond to a reminder, but we may find more optimal conditions for reminding in 10 minutes time. Questions related to the human interface design of the intelligent system must also be considered to guarantee that users continue to feel in control of the technology when it is making decisions.

By using machine learning to monitor and alter users' pressure relief habits, we have the potential to prevent thousands of occurrences of pressure ulcers every year. The cost saving nature of this technology will also make it desirable for insurance companies and healthcare organizations, which could lead to faster adoption of the technology and improved outcomes for the millions of power wheelchair users in the world.

Acknowledgements. This material is based upon work supported by the National Science Foundation under Cooperative Agreement EEC-0540865 as well as by a National Science Foundation Graduate Research Fellowship. We also acknowledge the Pittsburgh chapter of the American Rewards for College Scientists (ARCS) program for their generous support.

References

1. Bailly, R.: Quadratic weighted automata: Spectral algorithm and likelihood maximization. Journal of Machine Learning Research 20, 147–162 (2011)
2. Beach, S.R., Schulz, R., Matthews, J.T., Courtney, K., Dabbs, A.D.: Preferences for technology versus human assistance and control over technology in the performance of kitchen and personal care tasks in baby boomers and older adults. Disability and Rehabilitation: Assistive Technology, 1–13 (2013)
3. B. Boots, G.J. Gordon.: An online spectral learning algorithm for partially observable nonlinear dynamical systems. In: AAAI (2011)
4. Boots, B., Siddiqi, S.M., Gordon, G.J.: Closing the learning-planning loop with predictive state representations. The International Journal of Robotics Research 30(7), 954–966 (2011)
5. Cohen, S.B., Stratos, K., Collins, M., Foster, D.P., Ungar, L.: Spectral learning of latent-variable pcfgs. In: Proceedings of the 50th Annual Meeting of the Association for Computational Linguistics: Long Papers. Association for Computational Linguistics, vol. 1, pp. 223–231 (2012)
6. Cruz, J.A., Wishart, D.S.: Applications of machine learning in cancer prediction and prognosis. Cancer Informatics 2, 59 (2006)
7. Dhillon, P.S., Rodu, J., Collins, M., Foster, D.P., Ungar, L.H.: Spectral dependency parsing with latent variables. In: Proceedings of the 2012 Joint Conference on Empirical Methods in Natural Language Processing and Computational Natural Language Learning. Association for Computational Linguistics, pp. 205–213 (2012)

8. Dubey, A.K.: Using rough sets, neural networks, and logistic regression to predict compliance with cholesterol guidelines goals in patients with coronary artery disease. In: AMIA Annual Symposium Proceedings. American Medical Informatics Association, vol. 2003, p. 834 (2003)

9. Falakmasir, M.H., Pardos, Z.A., Gordon, G.J., Brusilovsky, P.: A spectral learning approach to knowledge tracing (2010)

10. Fisher, R., Simmons, R.: Smartphone interruptibility using density-weighted uncertainty sampling with reinforcement learning. In: 2011 10th International Conference on Machine Learning and Applications and Workshops (ICMLA), vol. 1, pp. 436–441. IEEE (2011)

11. Hsu, D., Kakade, S.M., Zhang, T.: A spectral algorithm for learning hidden markov models. Journal of Computer and System Sciences 78(5), 1460–1480 (2012)

12. Lacoste, M., Weiss-Lambrou, R., Allard, M., Dansereau, J.: Powered tilt/recline systems: why and how are they used? Assistive Technology 15(1), 58–68 (2003)

13. Minh, H.Q., Cristani, M., Perina, A., Murino, V.: A regularized spectral algorithm for hidden markov models with applications in computer vision. In: 2012 IEEE Conference on Computer Vision and Pattern Recognition (CVPR), pp. 2384–2391. IEEE (2012)

14. Reddy, M., Gill, S.S., Rochon, P.A.: Preventing pressure ulcers: A systematic review. JAMA 296(8), 974–984 (2006)

15. Rosenthal, S., Dey, A.K., Veloso, M.: Using decision-theoretic experience sampling to build personalized mobile phone interruption models. In: Lyons, K., Hightower, J., Huang, E.M. (eds.) Pervasive 2011. LNCS, vol. 6696, pp. 170–187. Springer, Heidelberg (2011)

16. Song, X., Mitnitski, A., Cox, J., Rockwood, K.: Comparison of machine learning techniques with classical statistical models in predicting health outcomes. Med. Info. 11(pt 1), 736–740 (2004)

17. Terwijn, S.A.: On the learnability of hidden markov models. In: Adriaans, P.W., Fernau, H., van Zaanen, M. (eds.) ICGI 2002. LNCS (LNAI), vol. 2484, pp. 261–268. Springer, Heidelberg (2002)

18. Allan, P.: White and Wei Zhong Liu. Technical note: Bias in information-based measures in decision tree induction. Machine Learning 15(3), 321–329 (1994)

Mood Tracking of Radio Station Broadcasts

Jacek Grekow

Faculty of Computer Science, Bialystok University of Technology,
Wiejska 45A, Bialystok 15-351, Poland
j.grekow@pb.edu.pl

Abstract. This paper presents an example of a system for the analysis of emotions contained within radio broadcasts. We prepared training data, did feature extraction, built classifiers for music/speech discrimination and for emotion detection in music. To study changes in emotions, we used recorded broadcasts from 4 selected European radio stations. The collected data allowed us to determine the dominant emotion in the radio broadcasts and construct maps visualizing the distribution of emotions in time. The obtained results provide a new interesting view of the emotional content of radio station broadcasts.

Keywords: Emotion detection, Mood tracking, Audio feature extraction, Music information retrieval, Radio broadcasts.

1 Introduction

The overwhelming number of media outlets is constantly growing. This also applies to radio stations available on the Internet, over satellite, and over the air. On the one hand, the number of opportunities to listen to various radio shows has grown, but on the other, choosing the right station has become more difficult. Music information retrieval helps those people who listen to the radio mainly for the music. This technology is able to make a general detection of the genre, artist, and even emotion.

Listening to music is particularly emotional. People need a variety of emotions, and music is perfectly suited to provide them. Listening to the radio station throughout the day, whether we want to it not, we are affected by the transmitted emotional content. In this paper, we focus on emotional analysis of the music presented by radio stations. During the course of a radio broadcast, these emotions can take on a variety of shades, change several times with varying intensity. This paper presents a method of tracking changing emotions during the course of a radio broadcast. The collected data allowed to determine the dominant emotion in the radio broadcast and construct maps visualizing the distribution of emotions in time.

Music emotion detection studies are mainly based on two popular approaches: categorical or dimensional. The categorical approach [1][2][3][4] describes emotions with a discrete number of classes - affective adjectives. In the dimensional approach [5][6], emotions are described as numerical values of valence and arousal.

T. Andreasen et al. (Eds.): ISMIS 2014, LNAI 8502, pp. 184–193, 2014.
© Springer International Publishing Switzerland 2014

In this way an emotion of a song is represented as a point on an emotion space. In this work, we use the categorical approach.

There are several other studies on the issue of mood tracking [7][8][9]. Lu et al. [3], apart from detecting emotions, tracked them, divided the music into several independent segments, each of which contained a homogeneous emotional expression. The use of mood tracking for indexing and searching multimedia databases has been used in the work of Grekow and Ras [10].

One wonders how long it takes a person to recognize emotion in musical compositions to which he/she is listening. Bachorik et al. [11] concluded that the majority of music listeners need 8 seconds to identify the emotion of a piece. This time is closely related to the length of a segment of music during emotion detection. Xiao et al. [12] found that the segment length should be no shorter than 4 sec and no longer than 16 sec. In various studies on the detection of emotion in music the segments are varying lengths. In [4][1], segment length is 30 sec. A 25 second segment was used by Yang et al. [5]. Fifteen second clips are used as ground truth data by Schmidt et al. [13]. The length of 1 sec segment was used by Schmidt and Kim in [6]. In this work, we use 6-second segments.

A comprehensive review of the methods that have been proposed for music emotion recognition was prepared by Yang et al. [14]. Another paper surveying state-of-the-art automatic emotion recognition was presented by Kim et al. in [15].

The issue of mood tracking is not only limited to music. The paper by Mohammad [16] is an interesting extension of the topic; the author investigated the development of emotions in literary texts. Yeh et al. [17] tracked the continuous changes of emotional expressions in Mandarin speech.

A method of profiling radio stations was described by Lidy and Rauber [18]. They used a technique of Self-Organizing Maps to organize the program coverage of radio stations on a two-dimensional map. This approach allows to profile the complete program of a radio station.

2   Music Data

To conduct the study of emotion detection of radio stations, we prepared two sets of data. One of them was used for music/speech discrimination, and the other for the detection of emotion in music.

A set of training data for music/speech discrimination consisted of 128 wav files, including 64 designated as speech and 64 marked as music. The training data were taken from the generally accessible data collection project MARSYAS (http://marsyas.info/download/data_sets).

The training data set for emotion detection consisted of 374 six-second fragments of different genres of music: classical, jazz, blues, country, disco, hip-hop, metal, pop, reggae, and rock. The tracks were all 22050Hz Mono 16-bit audio files in wav format.

In this research we use 4 emotion classes: energetic-positive, energetic-negative, calm-negative, calm-positive, presented with their abbreviation in Table 1. They cover the four quadrants of the 2 dimensional Thayer model of emotion [19]. They correspond to four basic emotion classes: happy, angry, sad and relaxed.

Table 1. Description of mood labels

Abbreviation	Description
e1	energetic-positive
e2	energetic-negative
e3	calm-negative
e4	calm-positive

Music samples were labeled by the author of this paper, a music expert with a university musical education. Six-second music samples were listened to and then labeled with one of the emotions (e1, e2, e3, e4). In the case when the music expert was not certain which emotion to assign, such a sample was rejected. In this way, the created labels were associated with only one emotion in the file. As a result, we obtained 4 sets of files: 101 files labeled e1, 107 files labeled e2, 78 files labeled e3, and 88 files labeled e4.

To study changes in emotions, we used recorded broadcasts from 4 selected European radio stations:

- Polish Radio Dwojka (Classical/Culture), recorded on 4.01.2014;
- Polish Radio Trojka (Pop/Rock), recorded on 2.01.2014;
- BBC Radio 3 (Classical), recorded on 25.12.2013;
- ORF OE1 (Information/Culture), recorded on 12.01.2014.

For each station we recorded 10 hours beginning at 10 A.M. The recorded broadcasts were segmented into 6-second fragments using sfplay.exe from MARSYAS software. For example, we obtained 6000 segments from one 10 h broadcast.

3   Features Extraction

For features extraction, we used the framework for audio analysis of MARSYAS software, written by George Tzanetakis [20]. MARSYAS is implemented in C++ and retains the ability to output feature extraction data to ARFF format [21]. With the tool bextract.exe, the following features can be extracted: Zero Crossings, Spectral Centroid, Spectral Flux, Spectral Rolloff, Mel-Frequency Cepstral Coefficients (MFCC), and chroma features - 31 features in total.

For each of these basic features, four statistic features were calculated: 1. *The mean of the mean* (calculate mean over the 20 frames, and then calculate the mean of this statistic over the entire segment); 2. *The mean of the standard deviation* (calculate the standard deviation of the feature over 20 frames, and then calculate the mean these standard deviations over the entire segment); 3. *The standard deviation of the mean* (calculate the mean of the feature over

20 frames, and then calculate the standard deviation of these values over the entire segment); 4. *The standard deviation of the standard deviation* (calculate the standard deviation of the feature over 20 frames, and then calculate the standard deviation of these values over the entire segment). In this way, we obtained 124 features.

The input data during features extraction were 6-second segments in wav format, sample rate 22050, channels: 1, Bits: 16.

An example of using `bextract.exe` from the MARSYAS v0.2 package to extract features:

```
bextract.exe -fe -sv colllection -w outputFile
```

where `collection` is file with list of input files and `outputFile` is name of output file in ARFF format.

For each 6-second file we obtained a representative single feature vector. The obtained vectors were used for building classifiers and for predicting new instances.

4 Classification

4.1 The Construction of Classifiers

We built two classifiers, one for music/speech discrimination and the second for emotion detection, using the WEKA package [21]. During the construction of the classifier for music/speech discrimination, we tested the following algorithms: J48, RandomForest, BayesNet, SMO [22]. The classification results were calculated using a cross validation evaluation CV-10. The best accuracy (98%) was achieved using SMO algorithm, which is an implementation of support vector machines (SVM) algorithm. The second best algorithm was Random-Forest (94% accuracy). During the construction of the classifier for emotion detection, we tested the following algorithms: J48, RandomForest, BayesNet, IBk (K-nn), SMO (SVM). The highest accuracy (55.61%) was obtained for SMO algorithm. SMO was trained using polynominal kernel. The classification results were calculated using a cross validation evaluation CV-10. After applying attribute selection (attribute evaluator: WrapperSubsetEval, search method BestFirst), classifier accuracy improved to 60.69%.

Table 2. Confusion matrix

classified as →	a	b	c	d
a = e1	65	26	5	5
b = e2	23	77	2	5
c = e3	13	4	37	24
d = e4	19	2	19	48

The confusion matrix (Table 2) obtained during classifier evaluation shows that the most recognized emotion was e2 (Precision 0.706, Recall 0.72, F-measure 0.713), and the next emotion was e1 (Precision 0.542, Recall 0.644, F-measure 0.588). We may notice a considerable amount of mistakes between the emotions of the left and right quadrants of the Thayer model, that is between e1 and e2, and analogously between e3 and e4. This is confirmed by the fact that detection on the arousal axis of Thayers model is easier. There are less mistakes made between the top and bottom quadrants. At the same time, recognition of emotions on the valence axis (positive-negative) is more difficult.

4.2   Analysis of Recordings

During the analysis of the recorded radio broadcasts, we conducted a two-phase classification. The recorded radio program was divided into 6-second segments. For each segment, we extracted a feature vector. This feature vector was first used to detect if the given segment is speech or music. If the current segment was music, then we used a second classifier to predict what type of emotion it contained. For features extraction, file segmentation, use of classifiers to predict new instances, and visualization of results, we wrote a Java application that connected different software products: MARSYAS, MATLAB and WEKA package.

5   Results of Mood Tracking in Radio Stations

The percentages of speech, music, and emotion in music obtained during the segment classification of 10-hour broadcasts of four radio stations are presented in Table 3. On the basis of these results, radio stations can be compared in two ways. The first way is to compare the amount of music and speech in the radio broadcasts, and the second is to compare the occurrence of individual emotions.

Table 3. Percentage of speech, music, and emotion in music in 10-hour broadcasts of four radio stations

	PR Dwojka	PR Trojka	BBC Radio 3	ORF OE1
speech	59.37%	73.35%	32.25%	69.10%
music	40.63%	26.65%	67.75%	30.90%
e1	4.78%	4.35%	2.43%	2.48%
e2	5.35%	14.43%	1.00%	0.92%
e3	20.27%	6.02%	56.19%	22.53%
e4	10.23%	1.85%	8.13%	4.97%
e1 in music	11.76%	16.32%	3.58%	8.02%
e2 in music	13.16%	54.14%	1.47%	2.98%
e3 in music	49.89%	22.59%	82.93%	72.91%
e4 in music	25.17%	6.94%	12.00%	16.08%

5.1   Comparison of Radio Stations

The dominant station in the amount of music presented was BBC Radio 3 (67.75%). We noted a similar ratio of speech to music in the broadcasts of PR Trojka and ORF OE1, in both of which speech dominated (73.35% and 69.10%, respectively). A more balanced amount of speech and music was noted on PR Dwojka (59.37% and 40.63%, respectively).

Comparing the content of emotions, we can see that PR Trojka clearly differs from the other radio stations, because the dominant emotion is e2 energetic-negative (54.14%) and e4 calm-positive occurs the least often (6.94%).

We noted a clear similarity between BBC Radio 3 and ORF OE1, where the dominant emotion was e3 calm-negative (82.93% and 72.91%, respectively). Also, the proportions of the other emotions (e1, e2, e4) were similar for these stations. We could say that emotionally these stations are similar, except that considering the speech to music ratio, BBC Radio 3 had much more music.

The dominant emotion for PR Dwojka was e3, which is somewhat similar to BBC Radio 3 and ORF OE1. Compared to the other stations, PR Dwojka had the most (25.17%) e4 calm-positive music.

5.2   Emotion Maps

The figures (Fig. 1, Fig. 2, Fig. 3, Fig. 4) present speech and emotion maps for each radio broadcast. Each point on the map is the value obtained from the

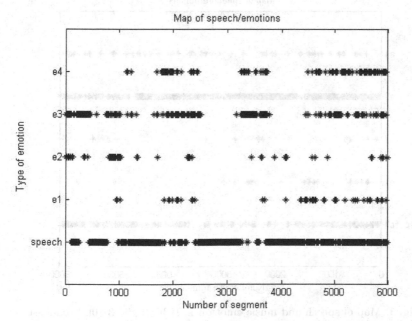

Fig. 1. Map of speech and music emotion in PR Dwojka 10h broadcast

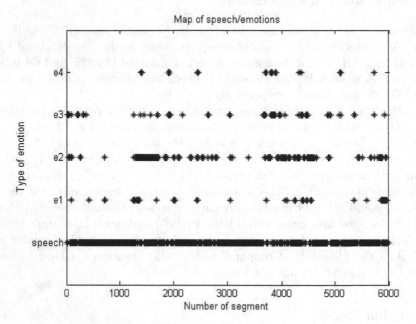

Fig. 2. Map of speech and music emotion in PR Trojka 10h broadcast

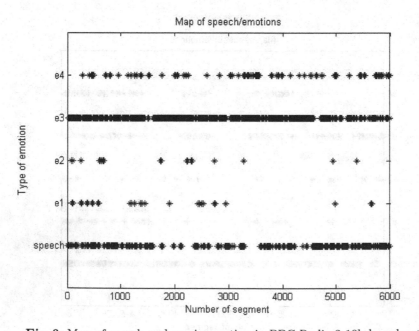

Fig. 3. Map of speech and music emotion in BBC Radio 3 10h broadcast

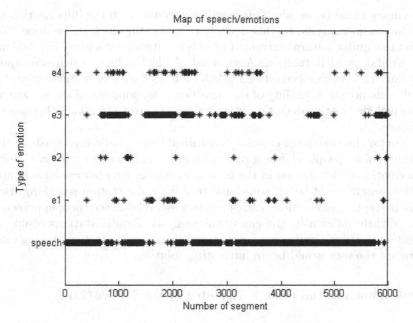

Fig. 4. Map of speech and music emotion in ORF OE1 10h broadcast

classification of a 6-second segment. These show which emotions occurred at given hours of the broadcasts.

For PR Dwojka (Fig. 1), there are clear musical segments (1500-2500, 2300-3900) during which e3 dominated. At the end of the day (4500-6000), emotion e2 occurs sporadically. It is interesting that e1 and e4 (from right half of the Thayer model) did not occur in the morning. For PR Trojka (Fig. 2), emotion e4 did not occur in the morning, and e2 and e3 dominated (segments 1200-2800 and 3700-6000). For BBC Radio 3 (Fig. 3), we observed almost a complete lack of energetic emotions (e1 and e2) in the afternoon (segments after 3200). For ORF OE1 (Fig. 4), e3 dominated up to segment 3600, and then broadcasts without music dominated.

The presented analyses of maps of emotions could be developed by examining the quantity of changes of emotions or the distribution of daily emotions.

6 Conclusions

This paper presents an example of a system for the analysis of emotions contained within radio broadcasts. The collected data allowed to determine the dominant emotion in the radio broadcast and present the amount of speech and music. The obtained results provide a new interesting view of the emotional content of radio stations.

The precision of the constructed maps visualizing the distribution of emotions in time obviously depends on the precision of the classifiers of emotion detection.

Their accuracy could be better. This is still associated with the imperfection of features for audio analysis. In this matter, there is still much to be done. We could also test audio features extracted by other software for feature extraction, such as jAudio or MIR toolbox. Also, musical file labeling, which are input data for learning classifiers, could be made by a bigger number of music experts; this would enhance the reliability of the classifiers. Development of the presented system to include emotion detection in speech also seems to be a logical prospect in the future.

A system for the analysis of emotions contained within radio broadcasts could be a helpful tool for people planning radio programs enabling them to consciously plan the emotional distribution in the broadcast music. Another example of applying this system could be an additional tool for radio station searching. Because the perception of emotions can be subjective and different people perceive emotions slightly differently, the emotional analysis of radio stations could be dependent on the user's preferences. Search profiling of radio stations taking into consideration the user would be an interesting solution.

Acknowledgments. This paper is supported by the S/WI/3/2013.

References

1. Li, T., Ogihara, M.: Detecting emotion in music. In: Proceedings of the Fifth International Symposium on Music Information Retrieval, pp. 239–240 (2003)
2. Grekow, J., Raś, Z.W.: Detecting emotions in classical music from MIDI files. In: Rauch, J., Raś, Z.W., Berka, P., Elomaa, T. (eds.) ISMIS 2009. LNCS (LNAI), vol. 5722, pp. 261–270. Springer, Heidelberg (2009)
3. Lu, L., Liu, D., Zhang, H.J.: Automatic mood detection and tracking of music audio signals. IEEE Transactions on Audio, Speech and Language Processing 14(1), 5–18 (2006)
4. Song, Y., Dixon, S., Pearce, M.: Evaluation of Musical Features for Emotion Classification. In: Proceedings of the 13th International Society for Music Information Retrieval Conference (2012)
5. Yang, Y.-H., Lin, Y.C., Su, Y.F., Chen, H.H.: A regression approach to music emotion recognition. IEEE Transactions on Audio, Speech, and Language Processing 16(2), 448–457 (2008)
6. Schmidt, E., Kim, Y.: Modeling Musical Emotion Dynamics with Conditional Random Fields. In: Proceedings of the 12th International Society for Music Information Retrieval Conference, pp. 777–782 (2011)
7. Schmidt, E.M., Turnbull, D., Kim, Y.E.: Feature Selection for Content-Based, Time-Varying Musical Emotion Regression. In: Proc. ACM SIGMM International Conference on Multimedia Information Retrieval, Philadelphia, PA (2010)
8. Schmidt, E.M., Kim, Y.E.: Prediction of time-varying musical mood distributions from audio. In: Proceedings of the 2010 International Society for Music Information Retrieval Conference, Utrecht, Netherlands (2010)
9. Grekow, J.: Mood tracking of musical compositions. In: Chen, L., Felfernig, A., Liu, J., Raś, Z.W. (eds.) ISMIS 2012. LNCS (LNAI), vol. 7661, pp. 228–233. Springer, Heidelberg (2012)

10. Grekow, J., Raś, Z.W.: Emotion based MIDI files retrieval system. In: Raś, Z.W., Wieczorkowska, A.A. (eds.) Advances in Music Information Retrieval. SCI, vol. 274, pp. 261–284. Springer, Heidelberg (2010)
11. Bachorik, J.P., Bangert, M., Loui, P., Larke, K., Berger, J., Rowe, R., Schlaug, G.: Emotion in motion: Investigating the time-course of emotional judgments of musical stimuli. Music Perception 26(4), 355–364 (2009)
12. Xiao, Z., Dellandrea, E., Dou, W., Chen, L.: What is the best segment duration for music mood analysis? In: International Workshop on Content-Based Multimedia Indexing (CBMI 2008), pp. 17–24 (2008)
13. Schmidt, E.M., Scott, J.J., Kim, Y.E.: Feature Learning in Dynamic Environments: Modeling the Acoustic Structure of Musical Emotion. In: Proceedings of the 12th International Society for Music Information Retrieval Conference, pp. 325–330 (2012)
14. Yang, Y.H., Homer, H., Chen, H.H.: Machine Recognition of Music Emotion: A Review. ACM Transactions on Intelligent Systems and Technology 3(6), Article No. 40 (2012)
15. Kim, Y., Schmidt, E., Migneco, R., Morton, B., Richardson, P., Scott, J., Speck, J., Turnbull, D.: State of the Art Report: Music Emotion Recognition: A State of the Art Review. In: Proceedings of the 11th International Society for Music Information Retrieval Conference, pp. 255–266 (2010)
16. Mohammad, S.: From Once Upon a Time to Happily Ever After: Tracking Emotions in Novels and Fairy Tales. In: Proceedings of the ACL 2011 Workshop on Language Technology for Cultural Heritage, Social Sciences, and Humanities, Portland, OR, USA, pp. 105–114 (2011)
17. Yeh, J.-H., Pao, T.-L., Pai, C.-Y., Cheng, Y.-M.: Tracking and Visualizing the Changes of Mandarin Emotional Expression. In: Huang, D.-S., Wunsch II, D.C., Levine, D.S., Jo, K.-H. (eds.) ICIC 2008. LNCS, vol. 5226, pp. 978–984. Springer, Heidelberg (2008)
18. Lidy, T., Rauber, A.: Visually Profiling Radio Stations. In: Proceedings of the 7th International Conference on Music Information Retrieval (2006)
19. Thayer, R.E.: The biopsychology arousal. Oxford University Press (1989)
20. Tzanetakis, G., Cook, P.: Marsyas: A framework for audio analysis. Organized Sound 10, 293–302 (2000)
21. Hall, M., Frank, E., Holmes, G., Pfahringer, B., Reutemann, P., Witten, I.H.: The WEKA Data Mining Software: An Update. SIGKDD Explorations, 11(1) (2009)
22. Witten, I.H., Frank, E.: Data Mining: Practical machine learning tools and techniques. Morgan Kaufmann, San Francisco (2005)

Evidential Combination Operators for Entrapment Prediction in Advanced Driver Assistance Systems

Alexander Karlsson[1], Anders Dahlbom[1], and Hui Zhong[2]

[1] Informatics Research Center, University of Skövde, Skövde, Sweden
{alexander.karlsson,anders.dahlbom}@his.se
[2] Advanced Technology & Research, Volvo Group Trucks Technology,
Gothenburg, Sweden
hui.zhong@volvo.com

Abstract. We propose the use of evidential combination operators for advanced driver assistance systems (ADAS) for vehicles. More specifically, we elaborate on how three different operators, one precise and two imprecise, can be used for the purpose of entrapment prediction, i.e., to estimate when the relative positions and speeds of the surrounding vehicles can potentially become dangerous. We motivate the use of the imprecise operators by their ability to model uncertainty in the underlying sensor information and we provide an example that demonstrates the differences between the operators.

Keywords: Evidential combination operators, advanced driver assistance systems, Bayesian theory, credal sets, Dempster-Shafer theory.

1 Introduction

Advanced driver assistance systems (ADAS) such as adaptive cruise control, forward collision warning, advanced emergency braking systems, or lane departure warning are now available not only for passenger cars but also for commercial vehicles on the market. Existing ADAS are typically based on focusing on a *single vehicle* which at present constitutes the greatest *risk* to the ego vehicle. Such a selection schema is based on real-time vehicle data, such as relative position, velocity, acceleration, and driving path of the predicted vehicle. Research on next generation ADAS [1] is aimed at recognizing more *complex traffic scenarios* and supporting the driver with both preventive information and active intervention to avoid accidents. We here take one step in this direction by exploring three different *evidential combination operators* that can be used for this purpose. More specifically, we will propose a way of modeling pieces of evidences based on *low-level filtered sensor information* in order to detect dangerous *entrapment situations* (see Fig. 1).

The paper is organized as follows: in Section 2, we elaborate on evidential combination operators. In Section 3, we describe how one can construct pieces of evidence and in Section 4, we provide an example of using the different operators. Lastly, in Section 5, we summarize the paper and present our conclusions.

T. Andreasen et al. (Eds.): ISMIS 2014, LNAI 8502, pp. 194–203, 2014.
© Springer International Publishing Switzerland 2014

Fig. 1. An example of a dangerous entrapment situation where the ego vehicle is the truck. The car to the left of the ego vehicle matches its speed while the cars in front and to the rights have lower speeds.

2 Preliminaries

We use the term *evidential combination operators* for a family of operators that combine pieces of independent evidence, e.g., *Dempster's combination operator* [2]. In the following sections we present the three evidential combination operators that we consider in the paper.

2.1 Bayesian Combination

Bayesian combination [3–5] relies on that evidences can be represented by *normalized likelihood functions*. Given a state space Ω_X regarding an unknown variable X and observations $y_i \in \Omega_{Y_i}$ from different sources $i \in \{1, 2\}$, e.g., sensor features, we can formulate a joint likelihood as:

$$p(y_1, y_2|X) = p(y_1|X)p(y_2|X) \tag{1}$$

by assuming conditional independence between Y_1 and Y_2 given X. This equation is essentially all we need in order to formulate a Bayesian combination operator. However, for convenience, we normalize the likelihoods so that the joint likelihood does not monotonically decrease with the number of combinations. We are now ready to formally define the Bayesian combination operator [3–5]:

Definition 1. *Bayesian combination is defined as:*

$$\Phi_\mathcal{B}(\hat{p}(y_1|X), \hat{p}(y_2|X))) \triangleq \frac{\hat{p}(y_1|X)\hat{p}(y_2|X)}{\displaystyle\sum_{x \in \Omega_X} \hat{p}(y_1|x)\hat{p}(y_2|x)}, \tag{2}$$

where $\hat{p}(y_i|X)$, $i \in \{1, 2\}$, are independent evidences in the form of normalized likelihood functions, i.e:

$$\hat{p}(y_i|X) \triangleq \frac{p(y_i|X)}{\displaystyle\sum_{x \in \Omega_X} p(y_i|X)}. \tag{3}$$

Note that $\sum_{x \in \Omega_X} \hat{p}(y_1|x)\hat{p}(y_2|x) = 0$ represents a situation where the sources mutually excludes all possibilities within the state space and, hence, Bayesian combination is undefined.

2.2 Credal Combination

Credal combination [6, 3–5] is an extension of Bayesian combination where *imprecision* in probabilities (or normalized likelihoods) is taken into account [7]. There could in principle be several different reasons to model imprecision in an application [8]. Perhaps the most well-known reason is to make the combination more *robust* as is the case in *robust Bayesian theory* [9] where the imprecision has a sensitivity interpretation, i.e., the imprecision stems from small perturbations.

Instead of using single normalized likelihood functions, as was the case in Bayesian combination, we are now allowed to use convex sets of such functions (polytopes), also known as *credal sets* [6] and denoted $\hat{\mathcal{P}}(y|X)$ instead. In order to consider two credal sets $\hat{\mathcal{P}}(y_i|X)$, $i \in \{1,2\}$, as independent, the extreme points of the corresponding joint credal set $\hat{\mathcal{P}}(y_1, y_2|X)$ must factorize [10], i.e., each extreme point of $\hat{\mathcal{P}}(y_1, y_2|X)$ must be equivalent to some product $\hat{p}(y_1|X)\hat{p}(y_2|X)$ where $\hat{p}(y_i|X) \in \hat{\mathcal{P}}(y_i|X)$ for $i \in \{1,2\}$. Now if we have the same situation as before, i.e., observations $y_i \in \Omega_{y_i}$, $i \in \{1,2\}$, and corresponding credal sets $\hat{\mathcal{P}}(y_i|X)$, we can combine them by using the *credal combination operator* [3–5].

Definition 2. *Credal combination is defined as:*

$$\Phi_C(\hat{\mathcal{P}}(y_1|X), \hat{\mathcal{P}}(y_2|X))) \triangleq$$
$$\mathcal{CH}\left(\left\{\Phi_B(\hat{p}(y_1|X), \hat{p}(y_2|X))) : \hat{p}(y_i|X) \in \hat{\mathcal{P}}(y_i|X), i \in \{1,2\}\right\}\right), \quad (4)$$

where $\mathcal{CH}$ is the convex hull operator and where $\hat{\mathcal{P}}(y_i|X)$, $i \in \{1,2\}$, are independent evidences in the form of credal sets of normalized likelihood functions.

2.3 Dempster-Shafer Theory

Dempster-Shafer theory [2, 11] is another way of modeling imprecision. The basic structure to model evidence in Dempster-Shafer theory is a *mass function*, formally defined as:

$$m(\emptyset) = 0 \quad (5)$$
$$\sum_{A \subseteq \Omega_X} m(A) = 1 . \quad (6)$$

The idea behind such a structure is that one allocate mass imprecisely over the state space. The independence assumption underlying the combination is not completely clear, but can according to Smets [12] be interpreted "in practice" as ordinary stochastic independence. In order to combine two mass function m_1 and m_2, one uses *Dempster's combination operator* [2], formally defined as:

Definition 3. *Dempster's combination operator is defined as:*

$$\Phi_{\mathcal{D}}(A, m_1, m_2) \triangleq \frac{\displaystyle\sum_{\substack{B \cap C = A \\ B, C \subseteq \Omega_X}} m_1(B) m_2(C)}{1 - \displaystyle\sum_{B \cap C = \emptyset} m_1(B) m_2(C)} , \tag{7}$$

where $A \subseteq \Omega_X$.

3 Evidential Combination for Entrapment Prediction

Assume that the ego vehicle is equipped with three different radar sensors, one in the front and one at each side of the vehicle. Based on these radars, we will for all samples at time t assume that we obtain estimates of positions of n vehicles in different lanes surrounding the ego vehicle, i.e., we have access to a set $\Theta_t \triangleq \{p_t^i\}_{i=1}^n$ where p_t are the positions relative to the ego vehicle. Furthermore, we will assume that these positions are given with some standard deviation σ in the direction of the ego vehicle's velocity vector (provided by the sensor vendor).

Let X be a random variable for future entrapment, denoted e, or no entrapment, denoted $\bar{e}$, i.e., the state space for our combination methods is $\Omega_X \triangleq \{e, \bar{e}\}$. The main idea is to construct two pieces of evidence with respect to future entrapment based on position and relative change of positions with respect to the surrounding vehicles and then combine these two pieces of evidence to a joint evidence which can be used to determine the likelihood of future entrapment. More formally, we need to define functions f that gives an evidence based on the surrounding vehicles' position's in relation to the ego vehicle and a function g that provides an evidence based on the relative change of those positions. The functions f and g could in principle be defined in many different ways and we do not claim that the definitions we use in this paper are optimal in any sense.

3.1 Bayesian Combination

We here suggest the use of *splines* as a flexible parametric way of defining the characteristics of f and g. In our coming definitions and examples, we will utilize a simple *Bézier* curve, but one could also use some of the more advanced curves if more advanced shapes of f and g are desirable. Let us first define f where the intuition behind the function is that it constructs evidence based on the maximum distance to a car in a lane in the neighborhood of the ego vehicle (if a lane is empty, we will assume an infinite distance):

$$f(p_t^*, \Theta_t, \{V_i\}_{i=1}^m) \triangleq \begin{cases} 1 - \epsilon & \text{if } d_t < \underline{\alpha} \\ \mathcal{B}(d_t, \{V_i\}_{i=1}^m) & \text{if } \underline{\alpha} \leq d \leq \bar{\alpha} , \\ \epsilon & \text{otherwise} \end{cases} \tag{8}$$

where $\mathcal{B}$ denotes a Bézier curve with the control points in the set $\{V_i\}_{i=1}^m$; p_t^* is the position of the ego vehicle at time t; Θ_t are the positions of the surrounding vehicles; the parameters $\underline{\alpha}$ respective $\overline{\alpha}$ constitute distance thresholds for when the evidence with respect to entrapment should be close to one respective zero[1]; ϵ is a small number; and d_t is the distance defined by:

$$d_t \triangleq \max_{\boldsymbol{p}_t \in \Theta_t} |\boldsymbol{p}_t^* - \boldsymbol{p}_t|, \tag{9}$$

where $|\cdot|$ denotes the Euclidean norm. The control points $\{\boldsymbol{V}_i\}_{i=1}^m$ for the Bézier curve can be set/calibrated to a desired risk level but they need to contain: $[\underline{\alpha}, 1 - \epsilon]^\mathrm{T}$ and $[\overline{\alpha}, \epsilon]^\mathrm{T}$ in order to avoid discontinuities.

We can now construct a Bayesian evidence, i.e., a probability function, based on positions by[2]:

$$\begin{aligned} p_1(e) &\triangleq f(\boldsymbol{p}_t^*, \Theta_t, \{\boldsymbol{V}_i\}_{i=1}^m) \\ p_1(\bar{e}) &\triangleq 1 - f(\boldsymbol{p}_t^*, \Theta_t, \{\boldsymbol{V}_i\}_{i=1}^m) \end{aligned}. \tag{10}$$

In order to construct evidence based on relative change in positions, i.e., the function g, we define:

$$g(\boldsymbol{p}_t^*, d_t', \{\boldsymbol{W}_i\}_{i=1}^m) \triangleq \begin{cases} 0.5 & \text{if } d_t' < \underline{\beta} \\ \mathcal{B}(d_t', \{\boldsymbol{W}_i\}_{i=1}^m) & \text{if } \underline{\beta} \le d_t' \le \overline{\beta} \\ 1 - \epsilon & \text{otherwise} \end{cases} \tag{11}$$

where d_t' is defined as:

$$d_t' \triangleq -\left(\frac{d_t - d_{t-1}}{t_s}\right) \tag{12}$$

where t_s denotes the sample time, and where the remaining variables and parameters are the same as for f. Based on g, we can define the corresponding evidence by:

$$\begin{aligned} p_2(e) &\triangleq g(\boldsymbol{p}_t^*, d_t', \{\boldsymbol{W}_i\}_{i=1}^m) \\ p_2(\bar{e}) &\triangleq 1 - g(\boldsymbol{p}_t^*, d_t', \{\boldsymbol{W}_i\}_{i=1}^m) \end{aligned}. \tag{13}$$

The reason for the lower bound of 0.5 of $\hat{p}_2(e)$ is that we do not want this evidence to be able to suppress evidence $\hat{p}_1(e)$ in the combination.

Finally, since we now have constructed pieces of evidence based on position and relative change in position between the ego vehicle and the surrounding vehicles, we can combine these to a single joint evidence by the Bayesian combination operator:

$$p_{1,2}(X) \triangleq \Phi_{\mathcal{B}}(p_1(x), p_2(x)) . \tag{14}$$

Note that in the above evidence schema, we have not utilized any information regarding the quality of the estimates by the sensors, i.e., the standard deviation σ. Such information can, however, be incorporated in the credal and Dempster-Shafer combination operators that we will describe in the two coming sections.

---

[1] Zero probabilities should generally be avoided when constructing evidence, see further Arnborg (2006) [4].

[2] We omit the more cumbersome notation for evidence that we used in Section 2 but it is important to remember that the evidence here should be interpreted as normalized likelihood functions.

3.2   Credal Combination

In contrast to Bayesian combination, credal combination enables one to model imprecision in probabilities. For this purpose, we will utilize the information regarding standard deviation σ.

Let the minimum and maximum distance, denoted $\underline{d}_t^i$ and $\overline{d}_t^i$, between the *line segment* $l_t^i \triangleq p_t^i \pm [0, 2\sigma]^T$, $p_t^i \in \Theta_t$, and the ego vehicle at position p_t^* be the smallest respectively greatest euclidean distance between any of the points on the line segment and p_t^* (similar to [13, Definition 3.6]). We can now formulate the sets:

$$\begin{aligned}\underline{D}_t &\triangleq \{\underline{d}_t^i : i \in \{1, \ldots, n\}\} \\ \overline{D}_t &\triangleq \{\overline{d}_t^i : i \in \{1, \ldots, n\}\}\end{aligned} \tag{15}$$

and use that in order to construct an evidence in the form of a credal set by:

$$\mathcal{P}_1(X) \triangleq \Big\{ p(X) : f(p_t^*, \max \overline{D}_t, \{V_i\}_{i=1}^m) \le p(e) \le$$
$$f(p_t^*, \max \underline{D}_t, \{V_i\}_{i=1}^m), \sum_{x \in \Omega_X} p(x) = 1 \Big\} \tag{16}$$

We can also obtain imprecise evidence by first defining[3]:

$$\underline{d}_t' \triangleq - \left(\frac{\max \underline{D}_t - \max \underline{D}_{t-1}}{t_s} \right) \tag{17}$$

$$\overline{d}_t' \triangleq - \left(\frac{\max \overline{D}_t - \max \underline{D}_{t-1}}{t_s} \right), \tag{18}$$

and then:

$$\mathcal{P}_2(X) \triangleq \Big\{ p(X) : g(p_t^*, \underline{d}_t', \{W_i\}_{i=1}^m) \le p(e) \le$$
$$g(p_t^*, \overline{d}_t', \{W_i\}_{i=1}^m), \sum_{x \in \Omega_X} p(x) = 1 \Big\} . \tag{19}$$

These two pieces of evidences in the form of credal sets can now be combined to a joint evidence using credal combination:

$$\mathcal{P}_{1,2}(X) \triangleq \Phi_C(\mathcal{P}_1(X), \mathcal{P}_2(X)) \tag{20}$$

Based on this joint evidence, one can construct lower and upper bounds by:

$$\underline{p}_{1,2}(e) \triangleq \min_{p_{1,2}(X) \in \mathcal{P}_{1,2}(X)} p_{1,2}(e) \tag{21}$$

$$\overline{p}_{1,2}(e) \triangleq \max_{p_{1,2}(X) \in \mathcal{P}_{1,2}(X)} p_{1,2}(e) . \tag{22}$$

The interval $[\underline{p}_{1,2}(e), \overline{p}_{1,2}(e)]$ can in this case be interpreted as a type of model uncertainty stemming from the quality of sensors and tracking methods.

---

[3] We could equally well have used $\max \overline{D}_t - \max \overline{D}_{t-1}$ in the nominator of Eq. (17).

3.3 Dempster-Shafer Combination

In the case of Dempster-Shafer theory [11], we will also model imprecision in terms of standard deviations. For the position we use f straightforwardly by:

$$m_1(e) \triangleq f(\boldsymbol{p}_t^*, \max \overline{D}_t, \{\boldsymbol{V}_i\}_{i=1}^m)$$
$$m_1(\bar{e}) \triangleq 1 - f(\boldsymbol{p}_t^*, \max \underline{D}_t, \{\boldsymbol{V}_i\}_{i=1}^m) \qquad\qquad , \qquad (23)$$
$$m_1(\Omega_X) \triangleq f(\boldsymbol{p}_t^*, \max \underline{D}_t, \{\boldsymbol{V}_i\}_{i=1}^m) - f(\boldsymbol{p}_t^*, \max \overline{D}_t, \{\boldsymbol{V}_i\}_{i=1}^m)$$

i.e., we interpret the imprecision as ignorance since it is equivalent to the mass on the state space Ω_X. For the relative change in position, we change g, to g', so that we obtain a mass of zero when the relative change is lower than the threshold $\underline{\beta}$:

$$\hat{g}(\boldsymbol{p}_t^*, d_t', \{\boldsymbol{W}_i\}_{i=1}^m) \triangleq \begin{cases} 0 & \text{if } d_t' < \underline{\beta} \\ \mathcal{B}(d_t'), \{\boldsymbol{W}_i\}_{i=1}^m) & \text{if } \underline{\beta} \leq d_t' \leq \overline{\beta} \\ 1 - \epsilon & \text{otherwise}, \end{cases} \qquad (24)$$

and use that in order to construct the following mass function:

$$m_2(e) \triangleq \hat{g}(\boldsymbol{p}_t^*, \underline{d}_t', \{\boldsymbol{W}_i\}_{i=1}^m)$$
$$m_2(\bar{e}) \triangleq 0 \qquad\qquad . \qquad (25)$$
$$m_2(\Omega_X) \triangleq 1 - \hat{g}(\boldsymbol{p}_t^*, \underline{d}_t', \{\boldsymbol{W}_i\}_{i=1}^m)$$

The reason for defining zero mass on the event "no entrapment", i.e., $\bar{e}$, is that we do not want this evidence to act as a counter evidence to entrapment, i.e., we do not want relative change to be able to suppress evidence based on position.

These two pieces of evidence can now be combined using Dempster-Shafer combination:

$$m_{1,2}(A) \triangleq \Phi_\mathcal{D}(A, m_1, m_2) \qquad (26)$$

where $A \subseteq \Omega_X$. In order to obtain lower and upper bounds on probabilities, similar to Eqs. (21) and (22), one can use what is known within Dempster-Shafer theory as *belief* and *plausibility*, defined as functions:

$$Bel(A) \triangleq \sum_{B \subseteq A} m(B)$$
$$Pl(A) \triangleq 1 - Bel(\bar{A}) = \sum_{B \cap A \neq \emptyset} m(B) \ . \qquad (27)$$

The interval $[Bel(e), Pl(e)]$ represents lower and upper bounds on a probability for the future entrapment e.

4 Example

In this section we exemplify entrapment prediction using the three different combination operators and parameter settings shown in Table 1. The scenario, which is simulated by using realistic sensor models, is seen in Fig. 2. Initially all vehicles match the speed of the ego vehicle, which is illustrated as part a in the figure. At time $t = t_2$, car A has increased its speed and has overtaken the ego vehicle. At $t = t_3$ car B has decreased its speed, meaning that it will come closer to the ego vehicle. At $t = t_4$, both A and B have decreased their speeds resulting in the ego vehicle closing in on them. Effectively, the ego vehicle has at $t = t_4$ been trapped between the three cars.

Table 1. The table describes the chosen parameters for the example scenario

Parameter(s)	Value(s)
σ	1 m
t_s	0.05 s
$(\underline{\alpha}, \overline{\alpha})$	$(5, 50)$ m
$(\underline{\beta}, \overline{\beta})$	$(0.05, 7)$ m/s
$\{v_i\}_{i=1}^m$	$[\underline{\alpha}, 1 - \epsilon]^{\mathrm{T}}, [\underline{\alpha} + (\overline{\alpha} - \underline{\alpha})/2, 0.2]^{\mathrm{T}}, [\overline{\alpha}, \epsilon]^{\mathrm{T}}$
$\{W_i\}_{i=1}^m$	$[\underline{\beta}, 0.5]^{\mathrm{T}}, [\underline{\beta} + (\overline{\beta} - \underline{\beta})/2, 0.5]^{\mathrm{T}}, [\overline{\beta}, 1 - \epsilon]^{\mathrm{T}}$

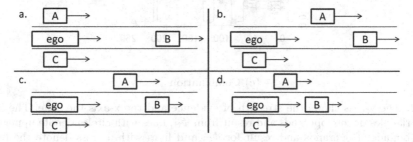

Fig. 2. Illustration of the example scenario at time $t = 0$ (a), $t = t_2$ (b), $t = t_3$ (c) and at $t = t_4$ (d). The lengths of the arrows show approximate speeds of the vehicles in relation to each other. A, B and C are cars and ego denotes the ego vehicle.

The results of modelling the pieces of evidences and the result of combining them is seen in Fig. 3. A first observations from the simulated data is that there exists gaps, i.e., blind spots of the sensor, which yields "dips" in evidence strength. This is a problem that we omit to handle in this paper but in reality one could use some filtering methods for smoothing away such dips. We see that the credal approach tends to highlight imprecision more in comparison to Dempster's operator since the difference between the upper and lower bounds are greater in the combined evidence in Fig. 3(c).

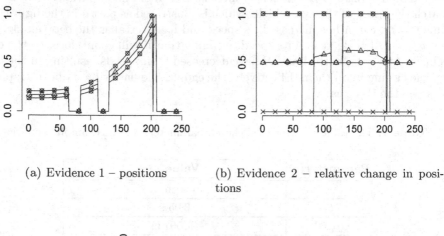

(a) Evidence 1 – positions

(b) Evidence 2 – relative change in positions

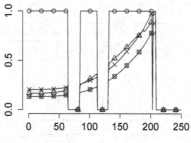

(c) Combination

Fig. 3. The y-axis shows the strength of evidence and the x-axis the time. The lines with triangles denote operands and result from $\Phi_\mathcal{B}$, lines with circles denote upper and lower bounds of operands and result for $\Phi_\mathcal{C}$, and lines with a cross denote the belief and plausibility of operands and result for the operator $\Phi_\mathcal{D}$.

5   Summary and Conclusions

We have elaborated on how entrapment prediction, as an example of ADAS functionality, can be implemented by using an evidential approach based on information stemming from sensors. By using information about positions and relative change of positions of surrounding vehicles we encode and combine evidence with respect to future entrapment. It should be noted that the combination schemas that we presented here are context dependent so one would need some high-level functionality to detect which situation the ego vehicle exists in. When using the credal and Dempster's operators, we were also able to take the sensor quality into account by modeling imprecision. The upper and lower bounds based on the sensor quality can be used as way to capture the *system uncertainty* which can be further utilized by other ADAS subsystems or conveyed to the driver in autonomous driving situations [14], e.g., when system uncertainty increases the driver needs to take control of the vehicle.

Acknowledgements. This work was supported by the Information Fusion Research Program (University of Skövde, Sweden), in partnership with the Swedish Knowledge Foundation under grant 2010-0320 (URL: http://www.infofusion. se, UMIF project).

References

1. Brannstrom, M., Sandblom, F., Hammarstrand, L.: A probabilistic framework for decision-making in collision avoidance systems. IEEE Transactions on Intelligent Transportation Systems 14(2), 637–648 (2013)
2. Dempster, A.P.: A generalization of bayesian inference. Journal of the Royal Statistical Society 30(2), 205–247 (1969)
3. Arnborg, S.: Robust Bayesianism: Imprecise and paradoxical reasoning. In: Proceedings of the 7th International Conference on Information fusion (2004)
4. Arnborg, S.: Robust Bayesianism: Relation to evidence theory. Journal of Advances in Information Fusion 1(1), 63–74 (2006)
5. Karlsson, A., Johansson, R., Andler, S.F.: Characterization and empirical evaluation of bayesian and credal combination operators. Journal of Advances in Information Fusion 6(2), 150–166 (2011)
6. Levi, I.: The Enterprise of Knowledge: An Essay on Knowledge, Credal Probability, and Chance. MIT Press (1983)
7. Walley, P.: Towards a unified theory of imprecise probability. International Journal of Approximate Reasoning 24, 125–148 (2000)
8. Walley, P.: Statistical Reasoning with Imprecise Probabilities. Chapman and Hall (1991)
9. Berger, J.O.: An overview of robust Bayesian analysis. Test 3, 5–124 (1994)
10. Couso, I., Moral, S., Walley, P.: A survey of concepts of independence for imprecise probabilities. Risk Decision and Policy 5, 165–181 (2000)
11. Shafer, G.: A Mathematical Theory of Evidence. Princeton University Press (1976)
12. Smets, P.: Analyzing the combination of conflicting belief functions. Information Fusion 8, 387–412 (2007)
13. Irpino, A., Tontodonato, V.: Cluster reduced interval data using Hausdorff distance. Computational Statistics 21, 241–288 (2006)
14. Helldin, T., Falkman, G., Riveiro, M., Davidsson, S.: Presenting system uncertainty in automotive uis for supporting trust calibration in autonomous driving. In: Proceedings of the 5th International Conference on Automotive User Interfaces and Interactive Vehicular Applications. AutomotiveUI 2013, pp. 210–217. ACM, New York (2013)

Influence of Feature Sets on Precision, Recall, and Accuracy of Identification of Musical Instruments in Audio Recordings

Elżbieta Kubera[1], Alicja A. Wieczorkowska[2], and Magdalena Skrzypiec[3]

[1] University of Life Sciences in Lublin, Akademicka 13, 20-950 Lublin, Poland
[2] Polish-Japanese Institute of Information Technology, Koszykowa 86, 02-008 Warsaw, Poland
[3] Maria Curie-Skłodowska University in Lublin, Pl. Marii Curie-Skłodowskiej 5, 20-031 Lublin, Poland
elzbieta.kubera@up.lublin.pl,
alicja@poljap.edu.pl,
mskrzypiec@hektor.umcs.lublin.pl

Abstract. In this paper we investigate how various feature sets influence precision, recall, and accuracy of identification of multiple instruments in polyphonic recordings. Our investigations were performed on classical music, and musical instruments typical for this music. Five feature sets were investigated. The results show that precision and recall change to a great extend, beyond the usual trade-off, whereas accuracy is relatively stable. Also, the results depend on the polyphony level of particular pieces of music. The investigated music varies in polyphony level, from 2-instrument duet (with piano) to symphonies.

1 Introduction

Identification of musical instruments in audio recordings is a challenging task, within the interest of Music Information Retrieval area of research. If successful, it could facilitate automatic transcription; after multi-pitch tracking, each note could be assigned to a particular instrument. However, automatic identification of instrument is difficult, especially in polyphonic environment. While the quality of the instrument recognition of single isolated sounds can even reach 100% (for a few classes; about 40% for 30 or more classes [11]), the identification of instruments in polyphonic audio excerpts is much more challenging. It is sometimes addressed through multi-pitch tracking [10], often supported with external provision of pitch data, or limited to the sound identification of a predominant instrument [2], [7]. In our research, we aim at identification of possibly all instruments playing in the investigated segment, without providing external data, without performing multi-pitch tracking, and without segmentation. Therefore, the sound analysis is performed on short-frame basis, and the results are shown for 0.5 second segments, which is an arbitrary set length and can be freely changed if needed.

T. Andreasen et al. (Eds.): ISMIS 2014, LNAI 8502, pp. 204–213, 2014.
© Springer International Publishing Switzerland 2014

1.1   Background

The research on identification of musical instruments in recordings is performed on various sets of data, so it is not feasible to compare all results directly. Still, we would like to sketch a general outlook of the state of the art in this area, and show the broad range of approaches and classification methods applied.

Because of tedious labeling of ground truth data, mixes and single sounds are commonly applied in such this research. The polyphony level (number of instrument sounds played together) varies from two sounds, i.e. duets [5], [13], [27], to symphonic, high level. Instrument recognition is difficult because the sound waves of instruments overlap, and also some harmonic spectral components (partials) overlap, too. Overlapping partials are sometimes omitted in the recognition process [5], yielding about 60% accuracy for duets from 5-instrument set using Gaussian Mixture Models. In [4], the approach applied was inspired by non-negative matrix factorization, with an explicit sparsity control. Direct spectrum/template matching was also sometimes performed for instrument identification, without feature extraction [13], [14]. In [14], this approach yielded 88% accuracy for the polyphony of 3 instruments (flute, violin and piano), when musical context was integrated into the system. LDA (Linear Discriminant Analysis) based approach in [16] resulted in 84.1% recognition rate for duets, 77.6% for trios, and 72.3% for quartets; in [1], LDA yielded a high recall of 86–100% and 60% average precision for instrument pairs (300 pairs, 25 instruments). SVM (Support Vector Machine), decision trees, and k-nearest neighbor classifiers were also used for instrument identification [6], [19]. In [6], SVMs yielded the average accuracy of 53%, for the polyphony of up to four jazz instruments. For the polyphony of up to 4 notes for 6 instruments [20] 46% recall and 56% precision was obtained, based on spectral clustering, and PCA (Principal Component Analysis); this research aimed at sound separation, which is one of the steps towards automatic music transcription. In [15], semi-automatic music transcription is addressed through shift-variant non-negative matrix deconvolution (svNMD) and k-means clustering; the accuracy obtained for 5 instruments in recordings dropped below 40%. Generally, the higher polyphony, the lower accuracy is usually obtained in automatic instrument recognition.

Motivation. In our previous research [18], we investigated identification of 10 instruments typical for classical music. Training was performed on 20,000 or 40,000 samples, but increasing the training set twice did not substantially change the results. The training sets were prepared in two versions: one based on single sounds of instruments, and mixes of these sounds, and the second one based on sound taken from real recordings (without segmentation). Testing was performed on real recordings, and the results showed that real recordings used for training improved recall, but decreased precision. The feature set was applied before [17] and fixed through all these experiments. This time we would like to investigate whether the use of different feature sets influences the results, and if so, in what way. The test audio data are the same, i.e. real recordings from RWC Classical music set [8]. Since using real recordings requires laborious labeling of audio data, we decided to use single sounds and their mixes for training, and see whether applying a different feature set can improve the results.

2   Data

Digital audio data in typical uncompressed formats represent amplitude changes vs. time. Such data are rarely used in raw form in classification. These data are parameterized, to obtain relatively compact and representative feature vector. There is no standard feature set used worldwide, but some features are commonly used. The audio data, originally in mono or stereo format, were used as mono input; mixes of the left and right channel (i.e. the average value of samples in both channels) were taken for stereo data. Our parametrization is performed for 40 ms frames, and Fourier transform is applied to calculate spectral features, with Hamming window. No segmentation or pitch (especially multi-pitch) extraction are required. The audio data are labeled indicating the instrument or instruments playing in a given frame. The sounds were recorded at 44.1 kHz sampling rate with 16-bit resolution, or converted to this format.

Identification of instruments is performed frame by frame, for consequent frames, with 10 ms hop size. Final classification result is calculated as an average of classifier output over 0.5 second segment of the recording, in order to avoid tedious labeling of ground-truth data over shorter frames.

2.1   Feature Sets

Audio data were parameterized in a few ways, and 4 resulting feature sets were investigated. The basic feature set consists of 46 features which we already applied in our previous research [17], with encouraging results. We also investigate other feature sets, to check how the feature vector influences the recognition results.

Basic Feature Set. The basic feature vector (*Basic*) consists of 46 features; most of them represent MPEG-7 low-level audio descriptors, often used in audio research [12]. These 46 features are (see also [17], [18]):

- *Audio Spectrum Flatness, $flat_1, \ldots, flat_{25}$* — 25 parameters describing the flatness property of the power spectrum within a frequency bin for selected bins; 25 out of 32 frequency bands were used;
- *Audio Spectrum Centroid* — the power weighted average of the frequency bins in the power spectrum. Coefficients were scaled to an octave scale anchored at 1 kHz [12];
- *Audio Spectrum Spread* — RMS (root mean square) of the deviation of the log frequency power spectrum wrt. *Audio Spectrum Centroid* [12];
- *Energy* — energy (in log scale) of the spectrum;

- *MFCC* — 13 mel frequency cepstral coefficients. The cepstrum was calculated as the logarithm of the magnitude of the spectral coefficients, and then transformed to the mel scale, reflecting properties of the human perception of frequency. 24 mel filters were applied, and the results were transformed to 12 coefficients. The 13th parameter is the 0-order coefficient of MFCC, corresponding to the logarithm of the energy [23];
- *Zero Crossing Rate* of the time-domain representation of the sound wave; a zero-crossing is a point where the sign of the function changes;
- *Roll Off* — the frequency below which 85% (experimentally chosen threshold) of the accumulated magnitudes of the spectrum is concentrated; parameter originating from speech recognition, where it is used for distinguishing between voiced and unvoiced speech;
- *NonMPEG7 - Audio Spectrum Centroid* — the linear scale version of *Audio Spectrum Centroid*;
- *NonMPEG7 - Audio Spectrum Spread* — the linear scale version of *Audio Spectrum Spread*;
- *Flux* — the sum of squared differences between the magnitudes of the DFT (Discrete Fourier Transform) points calculated for the starting and ending 30 ms sub-frames within the main 40 ms frame.

Other Feature Sets. Four other feature sets were also investigated:

- *Chroma* - 12-element chroma vector [22] of summed (through octaves) energy of pitch classes, corresponding to the equal-tempered scale, i.e. C, C#, D, D#, E, F, F#, G, G#, A, A#, and B. Chroma vector was calculated using Chroma Toolbox [21], after conversion of the sampling rate to 22.05 kHz.
- *Dif-10* - 91-element feature vector, consisting of the basic set, and changes of the basic features, with the exception of *Flux*. Changes measure the differences between the basic features calculated for the starting 30 ms subframe of the main frame, and the 30 ms subframe taken with 10 ms hop size.
- *Dif-40* - 91-element feature vector, constructed similarly, but for changes measured between consecutive 40 ms frames (with 40 ms hop size).
- *B+Chrom* - 58 features: basic feature set, and *Chroma*.

2.2   Audio Data

The instruments used in our research represent chordophones (stringed instruments) and aerophones (wind instruments), typical for classical music. Although these instruments produce sounds of definite pitch, we do not assume this information is provided nor calculate it, thus avoiding introducing possible errors in calculating pitch/multiple pitches. The sounds were played in various ways, i.e. with various articulation, including bowing vibrato and pizzicato, and recorded at 44.1 kHz sampling rate and 16-bit resolution (or converted to this format). Frames of the RMS level below 300 were treated as silence; this threshold was empirically set in our previous experiments. The investigated classes include:

- flute (fl),
- oboe (ob),
- bassoon (bn),
- clarinet (cl),
- French horn (fh),
- violin (vn),
- viola (va),
- cello (ce),
- double bass (db), and
- piano (pn).

For each class, a binary classifier was trained to indicate whether the target instrument is playing in a given sound segment. If all classifiers yield negative answers, we conclude that segment represents an unknown instrument or instruments, or silence. Also, silence segments were used as negative training examples during training of our classifiers.

Training Data. Training of the classifiers was performed on 40 ms sound frames, as this is the length of the analyzing frame applied in our feature extraction. The training frames were drawn from the audio recordings with no overlap between the frames, to have as diversified data as possible. The training set of 20,000 frames was based on single sounds of musical instruments, taken from RWC [9], MUMS [24], and IOWA [26] sets of single sounds of musical instruments, and on mixes of up to three instrument sounds. Altogether, 5,000 frames of single sounds representing positive examples for a target instrument were used, 5,000 frames of single sounds of other instruments (or silence) constituted negative examples, 5,000 frames of mixes (including the target instrument) constituted positive examples, and 5,000 frames of mixes (not containing the target instrument) constituted negative examples. The set of instruments in mixes was always typical for classical music, and the probability of instruments playing together in the mix reflected the probability of these instruments playing together in the RWC Classical Music Database.

Testing Data. No training data were used in our tests. Testing was performed for 5 pieces - on the first minute of each of the following recordings from RWC Classical Music Database recordings [8]:

- No. 1 (C01), F.J. Haydn, Symphony no.94 in G major, Hob.I-94 'The Surprise'. 1st mvmt., with the following instruments playing in the first minute: flute, oboe, bassoon, French horn, violin, viola, cello, double bass;
- No. 2 (C02), W.A. Mozart, Symphony no.40 in G minor, K.550. 1st mvmt.; instruments playing in the first minute: flute, oboe, bassoon, French horn, violin, viola, cello, double bass;
- No. 16 (C16), W.A. Mozart, Clarinet Quintet in A major, K.581. 1st mvmt.; instruments playing in the first minute: clarinet, violin, viola, cello;

- No. 18 (C18), J. Brahms, Horn Trio in Eb major, op.40. 2nd mvmt.; instruments playing in the first minute: piano, French horn, violin;
- No. 44 (C44), N. Rimsky-Korsakov, The Flight of the Bumble Bee; flute and piano.

These pieces pose diverse difficulties for the classifier, including short sounds, and multiple instruments playing at the same time. To avoid tedious labeling the ground truth data of very short segments and use the segment length easier for perception, labeling was done on 0.5 s segments. Testing was performed on 40 ms frames (with 10 ms hop size, i.e. with overlap, to marginalize errors on note edges), and for each instrument, the outputs of the classifier over all frames within the 0.5 s segment were averaged. If the result exceeded the 50% threshold, the output of the binary classifier was considered positive, meaning that the target instrument was playing in this 0.5 s segment.

3   Classification

In our research, we applied random forest (RF) classifiers, since their proved successful in our previous research [17], [18]. RF is a classifier based on a tree ensemble, constructed using procedure minimizing bias and correlations between individual trees. Each tree is built using a different N-element bootstrap sample of the N-element training set, obtained through drawing with replacement from the original N-element set. Roughly 1/3 of the training data are not used in the bootstrap sample for any given tree. For objects described by a vector of K attributes (features), k attributes are randomly selected ($k \ll K$, often $k = \sqrt{K}$) at each stage of tree building, i.e. for each node of any particular tree in RF. The best split on these k attributes is used to split the data in the node, and Gini impurity criterion is applied (minimized) to choose the split. This criterion is the measure of how often an element would be incorrectly labeled if labeled randomly, according to the distribution of labels in the subset. Each tree is grown without pruning to the largest possible extent. By repeating this randomized procedure M times a collection of M trees is obtained. Classification of each object is made by simple voting of all trees in such a random forest [3].

The classifier used in our research consists of a set of binary RFs, each one trained to identify the target instrument in the analyzed audio frame. The decision is based on standard majority voting. If the percentage of votes of the trees in RF is 50% or more, then the answer of the classifier is considered to be positive (target instrument present in this sound), otherwise – negative (absent).

4   Experiments and Results

The outcomes of our experiments are presented using the following measures, based on true positives (TP), true negatives (TN), false positives (FP) and false negatives (FN):

- precision pr calculated as [25]:

$$pr = \frac{TP + 1}{TP + FP + 1},$$

- recall rec calculated as [25]:

$$rec = \frac{TP + 1}{TP + FN + 1},$$

- f-measure f_{meas} calculated as:

$$f_{meas} = \frac{2 \cdot pr \cdot rec}{pr + rec},$$

- accuracy acc calculated as:

$$acc = \frac{TP + TN}{TP + TN + FP + FN}.$$

Such pr and rec are never equal to 0, even if the classifier indicates no TP. However, for huge amount of analyzed sound frames this is not likely to happen. Still, adding unities in the denominator allows avoiding division by 0, and this was often the case when a particular music pieces was analyzed, because not all target instruments played there. Confusion matrices cannot be produced in our research, as we do not know which instrument is mistaken with which one when multiple answers are obtained (and are true) for each input frame.

The amount of TP, FP, FN and FP for *Basic* feature set for 5 drawings of training data are shown in Table 1; the results are shown as sums after these 5 runs. Flute, violin, oboe and clarinet are rarely mistakenly indicated (few FP); French horn, viola and violin are often omitted (many FN). Comparison of pr, rec, f_{meas} and acc for all 5 investigated feature sets is shown in Table 2.

The classifier based on *Chroma* often indicated all instruments as playing in the analyzed frame, so recall was high, but precision low in this case. *Basic* classifier yielded much better accuracy and f-measure. Adding more features decreases pr and increases rec, but the change is not proportional to the number of features in the set. The best recall is obtained for *Chroma*, and it is improved comparing to *Basic* for all instruments, even though it is not the largest feature set. It is also interesting that results for *Dif-10* only slightly differ from those

Table 1. Results of musical instruments recognition in the selected RWC Classical recordings, for 5 drawings of the training data; results summarized after these 5 runs

Result	bn	ob	cl	fl	fh	pn	ce	va	vn	db
TP	177	150	49	422	95	570	880	533	1308	549
FP	264	36	44	7	444	494	746	328	17	413
FN	403	355	161	408	715	260	225	627	562	211
TN	2156	2459	2746	2163	1746	1676	1149	1512	1113	1827

Table 2. Results of the recognition of musical instruments in the selected RWC Classical recordings, for 5 investigated feature sets. Average values after 5 runs of training and testing of RFs are shown.

Features	Measure	bn	ob	cl	fl	fh	pn	ce	va	vn	db	Average
Basic	pr	40%	81%	53%	98%	18%	54%	54%	62%	99%	57%	62%
	rec	31%	30%	24%	51%	12%	69%	80%	46%	70%	72%	48%
	f_{meas}	35%	44%	33%	67%	14%	60%	64%	53%	82%	64%	52%
	acc	78%	87%	93%	86%	61%	75%	68%	68%	81%	79%	78%
Chroma	pr	17%	14%	7%	28%	25%	31%	40%	42%	66%	28%	30%
	rec	68%	74%	83%	90%	77%	85%	96%	95%	94%	94%	86%
	f_{meas}	28%	24%	12%	42%	38%	46%	57%	58%	78%	43%	43%
	acc	31%	19%	16%	32%	33%	44%	46%	47%	66%	36%	37%
Dif-10	pr	38%	79%	49%	98%	18%	48%	52%	56%	99%	56%	59%
	rec	32%	31%	26%	53%	13%	59%	83%	45%	68%	71%	48%
	f_{meas}	35%	44%	34%	68%	15%	53%	64%	50%	81%	63%	51%
	acc	77%	87%	93%	87%	61%	71%	66%	65%	80%	79%	76%
Dif-40	pr	35%	71%	27%	98%	21%	36%	52%	56%	97%	56%	55%
	rec	38%	32%	55%	56%	22%	45%	91%	67%	75%	75%	55%
	f_{meas}	37%	44%	36%	71%	21%	40%	66%	61%	85%	64%	52%
	acc	75%	86%	87%	87%	56%	62%	66%	67%	83%	79%	75%
B+Chrom	pr	31%	59%	11%	94%	27%	36%	46%	53%	95%	54%	51%
	rec	57%	30%	90%	56%	55%	95%	96%	86%	81%	82%	73%
	f_{meas}	40%	40%	19%	70%	37%	52%	62%	66%	87%	65%	54%
	acc	67%	85%	47%	87%	49%	52%	57%	65%	85%	78%	67%

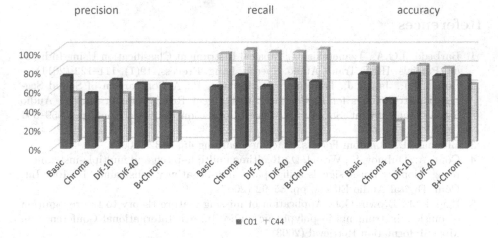

Fig. 1. Precision, recall, and accuracy for RWC Classic No. 1 (C01) and No. 44 (C44), for the investigated feature sets

for *Basic*, whereas for *Dif-40* we see the usual trade-off between *pr* and *rec*, i.e. decreased *pr* and increased *rec*. It is also worth mentioning that the results depend on the music piece. The best precision was obtained for the piece C01, and lowered gradually for consequent pieces, with definitely worst precision for C44. Recall was highest for *Chroma* (96%) and *Dif-40* (87%) for C16, but for C44 they were similar (97% and 89% respectively), with higher recall (averaged through feature sets) for C44. Therefore, we present the results for pieces C01 and C44 for illustration. C01 is a highly polyphonic symphony, whereas C44 is a duet. For low polyphony we have high recall, and precision much lower than recall. For higher polyphony, precision and recall are relatively similar.

5   Summary and Conclusions

In this research, we compared identification of musical instruments in polyphonic recordings for 5 features sets, using binary RFs to identify target instruments. The results show that using chroma features alone or with the basic feature set improves recall, more than having a larger feature set, describing changes between neighboring or overlapping frames. The usual trade-off between precision and recall is also observed, and dependency of these measures on the polyphony level can be seen, too. We believe that the results can be improved when domain knowledge is applied to clean the RF results for overlapping and consecutive frames, for removing spurious breaks and misidentifications.

Acknowledgment. This work was partially supported by the Research Center of PJIIT, supported by the Ministry of Science and Higher Education in Poland.

References

1. Barbedo, J.G.A., Tzanetakis, G.: Musical Instrument Classification Using Individual Partials. IEEE Trans. Audio, Speech, Lang. Process. 19(1), 111–122 (2011)
2. Bosch, J.J., Janer, J., Fuhrmann, F., Herrera, P.: A Comparison of Sound Segregation Techniques for Predominant Instrument Recognition in Musical Audio Signals. In: 13th Int. Society for Music Information Retrieval Conf., pp. 559–564 (2012)
3. Breiman, L.: Random Forests. Machine Learning 45, 5–32 (2001)
4. Cont, A., Dubnov, S., Wessel, D.: Realtime multiple-pitch and multiple-instrument recognition for music signals using sparse non-negativity constraints. In: 10th Int. Conf. Digital Audio Effects, pp. 85–92 (2007)
5. Eggink, J., Brown, G.J.: Application of missing feature theory to the recognition of musical instruments in polyphonic audio. In: 4th International Conference on Music Information Retrieval (2003)
6. Essid, S., Richard, G., David, B.: Instrument recognition in polyphonic music based on automatic taxonomies. IEEE Trans. Audio, Speech, Lang. Process 14(1), 68–80 (2006)
7. Fuhrmann, F.: Automatic musical instrument recognition from polyphonic music audio signals. PhD Thesis. Universitat Pompeu Fabra (2012)
8. Goto, M., Hashiguchi, H., Nishimura, T., Oka, R.: RWC Music Database: Popular, Classical, and Jazz Music Databases. In: Proceedings of the 3rd International Conference on Music Information Retrieval, pp. 287–288 (2002)

9. Goto, M., Hashiguchi, H., Nishimura, T., Oka, R.: RWC Music Database: Music Genre Database and Musical Instrument Sound Database. In: 4th International Conference on Music Information Retrieval, pp. 229–230 (2003)
10. Heittola, T., Klapuri, A., Virtanen, A.: Musical Instrument Recognition in Polyphonic Audio Using Source-Filter Model for Sound Separation. In: 10th Int. Society for Music Information Retrieval Conf. (2009)
11. Herrera-Boyer, P., Klapuri, A., Davy, M.: Automatic Classification of Pitched Musical Instrument Sounds. In: Klapuri, A., Davy, M. (eds.) Signal Processing Methods for Music Transcription. Springer Science+Business Media LLC (2006)
12. ISO: MPEG-7 Overview, http://www.chiariglione.org/mpeg/
13. Jiang, W., Wieczorkowska, A., Raś, Z.W.: Music Instrument Estimation in Polyphonic Sound Based on Short-Term Spectrum Match. In: Hassanien, A.-E., Abraham, A., Herrera, F. (eds.) Foundations of Comput. Intel. Vol. 2. SCI, vol. 202, pp. 259–273. Springer, Heidelberg (2009)
14. Kashino, K., Murase, H.: A sound source identification system for ensemble music based on template adaptation and music stream extraction. Speech Commun. 27, 337–349 (1999)
15. Kirchhoff, H., Dixon, S., Klapuri, A.: Multi-Template Shift-Variant Non-Negative Matrix Deconvolution for Semi-Automatic Music Transcription. In: 13th International Society for Music Information Retrieval Conference, pp. 415–420 (2012)
16. Kitahara, T., Goto, M., Komatani, K., Ogata, T., Okuno, H.G.: Instrument identification in polyphonic music: Feature weighting to minimize influence of sound overlaps. EURASIP J. Appl. Signal Process 2007, 1–15 (2007)
17. Kubera, E.z., Kursa, M.B., Rudnicki, W.R., Rudnicki, R., Wieczorkowska, A.A.: All That Jazz in the Random Forest. In: Kryszkiewicz, M., Rybinski, H., Skowron, A., Raś, Z.W. (eds.) ISMIS 2011. LNCS (LNAI), vol. 6804, pp. 543–553. Springer, Heidelberg (2011)
18. Kubera, E., Wieczorkowska, A.: Mining Audio Data for Multiple Instrument Recognition in Classical Music. In: New Frontiers in Mining Complex Patterns NFMCP 2013, International Workshop, held at ECML-PKDD (2013)
19. Little, D., Pardo, B.: Learning Musical Instruments from Mixtures of Audio with Weak Labels. In: 9th International Conference on Music Information Retrieval (2008)
20. Martins, L.G., Burred, J.J., Tzanetakis, G., Lagrange, M.: Polyphonic instrument recognition using spectral clustering. In: 8th International Conference on Music Information Retrieval (2007)
21. Max-Planck-Institut Informatik: Chroma Toolbox: Pitch, Chroma, CENS, CRP, http://www.mpi-inf.mpg.de/resources/MIR/chromatoolbox/
22. Müller, M.: Information Retrieval for Music and Motion. Springer, Heidelberg (2007)
23. Niewiadomy, D., Pelikant, A.: Implementation of MFCC vector generation in classification context. J. Applied Computer Science 16(2), 55–65 (2008)
24. Opolko, F., Wapnick, J.: MUMS — McGill University Master Samples. CD's (1987)
25. Subrahmanian, V.S.: Principles of Multimedia Database Systems. Morgan Kaufmann, San Francisco (1998)
26. The University of IOWA Electronic Music Studios: Musical Instrument Samples, http://theremin.music.uiowa.edu/MIS.html
27. Vincent, E., Rodet, X.: Music transcription with ISA and HMM. In: Puntonet, C.G., Prieto, A.G. (eds.) ICA 2004. LNCS, vol. 3195, pp. 1197–1204. Springer, Heidelberg (2004)

Multi-label Ferns for Efficient Recognition of Musical Instruments in Recordings

Miron B. Kursa[1] and Alicja A. Wieczorkowska[2]

[1] Interdisciplinary Centre for Mathematical and Computational Modelling (ICM),
University of Warsaw, Pawińskiego 5A, 02-106 Warsaw, Poland
[2] Polish-Japanese Institute of Information Technology, Koszykowa 86,
02-008 Warsaw, Poland
M.Kursa@icm.edu.pl, alicja@poljap.edu.pl

Abstract. In this paper we introduce multi-label ferns, and apply this technique for automatic classification of musical instruments in audio recordings. We compare the performance of our proposed method to a set of binary random ferns, using jazz recordings as input data. Our main result is obtaining much faster classification and higher F-score. We also achieve substantial reduction of the model size.

1 Introduction

Music Information Retrieval (MIR) is a hot research topic last years [23], [26], with quite a successful solving of such problems as automatic song identification through query-by-example, also using mobile devices [25], [28], and finding music works through query-by-humming [18]. Still, one of the unattainable goals of MIR research is automatic score extraction from audio recordings, which is especially difficult for polyphonic data [8], [12]. Multi-pitch tracking combined with assignment of the extracted notes to particular voices (instruments) is a way to approach score extraction. Therefore, identification of instruments can be used to assign each note in a polyphonic and polytimbral sound to the appropriate instrument. However, the recognition of all playing instruments from recordings in polyphonic environment is still a challenging and unsolved task, related to multi-label classification of audio data representing a mixture of sounds.

In our work, the target is to recognize all instruments playing in the analyzed audio segment. No initial segmentation nor providing external pitch is required. The instruments identification is performed on short sound frames, without multi-pitch tracking. In our previous works, we were using sets (which we called batteries) of binary classifiers to solve the multi-label problem [13], [30] of identification of instruments in polyphonic environment. Random forests [2] and ferns [21], [22] were applied as classification tools. Recently, we have shown that random ferns are a good replacement for random forests in music annotation tasks, as this technique offers similar accuracy while being much more computationally efficient [15]. In this paper we propose a generalized version of random ferns, which can natively perform multi-label classification. Using real

T. Andreasen et al. (Eds.): ISMIS 2014, LNAI 8502, pp. 214–223, 2014.
© Springer International Publishing Switzerland 2014

musical recording data, we will show that our approach outperforms a battery of binary random ferns classifiers in every respect: in terms of accuracy, model size and prediction speed.

1.1 Background

The difficulty level of automatic instrument recognition in audio data depends on the polyphony level, and on the preprocessing performed. The simplest polyphonic research case is instrument identification in duets (2 instruments) [4], [10], [29], and the most complex one for symphonies, with high polyphony level (i.e. high number of instrument sounds played together). Since the sound waves of instruments overlap, so harmonic spectral components (partials) do, to a certain — sometimes large — extent. For single isolated sounds the instrument identification can even reach 100% for a few classes, but it decreases to about 40% for 30 or more classes [8]). For polyphonic input even labeling of ground truth data is difficult, so mixes and single sounds are commonly applied to facilitate the research on polyphonic audio data. The identification of instruments in polyphony is often supported with external provision of pitch data, but automatic multi-pitch tracking problem is addressed too [7]. Another simplified approach aims at the identification of a predominant instrument [1]. Multi-target identification of multiple instruments is performed as well, although this research is done on various sets of data, so the results cannot be directly compared. This section presents a general view of methods and results obtained in the research addressing this subject.

Audio data are usually parameterized before further processing in the classification procedure, and pure data representing amplitude changes of a complex audio wave are rarely used. Preprocessing usually consists in calculation of parameters describing audio signal, or (more often) spectral features. Still, direct spectrum/template matching can be also applied to instrument identification, without feature extraction [10], [11]. This approach can result in good accuracy; in [11], 88% was obtained for the polyphony of 3 instruments: flute, violin and piano, supported with integrating musical context into the system.

The higher the polyphony level and number of instruments considered in the recognition procedure, the lower usually accuracy of instrument identification is. In [12], 84.1% was obtained for duets, 77.6% for trios, and 72.3% for quartets, using LDA (Linear Discriminant Analysis) based approach. In [31], LDA yielded 60% average precision for instrument pairs (300 pairs, 25 instruments), and much a higher recall of 86–100%. Other techniques used in multiple instrument identification include SVM (Support Vector Machine), decision trees, and k-NN (k-Nearest Neighbor) classifiers [5], [16]. For the polyphony of up to four jazz instruments, the average accuracy of 53% was obtained in [5], whereas [17] obtained 46% recall and 56% precision for the polyphony of up to 4 notes for 6 instruments, based on spectral clustering, and PCA (Principal Component Analysis). The problematic overlapping partials are sometimes omitting in the instrument identification process [4], resulting in about 60% accuracy using GMM (Gaussian Mixture Models) for duets from 5-instrument set. Another interesting approach to multiple-instrument recognition is presented in [3]; their approach was inspired by non-negative matrix factorization, with an explicit sparsity control.

The research on instrument identification is often incorporated in studies addressing automatic score extraction. The experiments described in [17] aimed at sound separation, which is usually performed as an intermediate step in automatic music transcription, and then each separated sound can be independently labeled. Semi-automatic music transcription is addressed in [32] through shift-variant non-negative matrix deconvolution (svNMD) and k-means clustering; the accuracy dropped below 40% for 5 instruments, analyzed in form of mixes. However, we should be aware that music transcription is a very difficult problem, and such results are not surprising.

2   Data

The data we used originate from various recordings, all recorded at 44.1kHz/16-bit, or converted to this format. Testing was performed on recordings as well, not on mixes of single sounds, as often happens in similar research. This was possible because we used recordings especially prepared for research purposes, the original tracks for each instruments were available, and thus ground truth labeling was facilitated. Both training and testing data were used as mono input, although some of them were originally recorded in mono or stereo format. In the case of stereo data, mixes of the left and right channel (i.e. the average value of samples in both channels) were taken.

Sound parametrization was performed as a preprocessing in our research, for 40-ms frames. Spectrum was calculated first, using FFT (Fast Fourier Transform) with Hamming window, and various spectral features were extracted. No pitch tracking was performed nor required as preprocessing. Both training and testing data were labeled with instruments playing in a given segment. In the testing phase, the identification of instruments was performed on frame by frame basis, for consequent 40-ms frames, with 75% overlap (10 ms hop size).

2.1   Feature Set

The feature vector consists of parameters describing properties of a 40-ms audio frame, and differences of the same parameters but calculated between for a 30 ms sub-frame starting from the beginning of the frame and a 30 ms sub-frame with 10 ms offset. The features we used are mainly MPEG-7 low-level audio descriptors, are often used in audio research [9], and other features applied in instrument recognition research. The following 91 parameters constitute our feature set [13], [30]:

- *Audio Spectrum Centroid* — the power weighted average of the frequency bins in the power spectrum, with coefficients scaled to an octave scale anchored at 1 kHz [9];

- *Audio Spectrum Flatness*, $flat_1, \ldots, flat_{25}$ — features parameter describing the flatness property of the power spectrum within a frequency bin for selected bins; we used 25 out of 32 frequency bands;
- *Audio Spectrum Spread* — RMS (root mean square) of the deviation of the log frequency power spectrum wrt. *Audio Spectrum Centroid* [9];
- *Energy* — energy of the spectrum, in log scale;
- *MFCC* — 13 mel frequency cepstral coefficients. The cepstrum was calculated as the logarithm of the magnitude of the spectral coefficients, and then transformed to the mel scale, reflecting properties of the human perception of frequency. 24 mel filters were applied, and the results were transformed to 12 coefficients, and the logarithm of the energy was taken as 13th coefficient (0-order coefficient of MFCC) [19];
- *NonMPEG7 - Audio Spectrum Centroid* — a linear scale version of *Audio Spectrum Centroid*;
- *NonMPEG7 - Audio Spectrum Spread* — a linear scale version of *Audio Spectrum Spread*;
- *Roll Off* — the frequency below which an experimentally chosen percentage (85%) of the accumulated magnitudes of the spectrum is concentrated; parameter originating from speech recognition, applied to distinguish between voiced and unvoiced speech;
- *Zero Crossing Rate*, where zero-crossing is a point where the sign of the sound wave in time domain changes;
- changes (differences) of the above features for a 30 ms sub-frame of the given 40 ms frame (starting from the beginning of this frame) and the next 30 ms sub-frame (starting with 10 ms offset);
- *Flux* — the sum of squared differences between the magnitudes of the DFT points calculated for the starting and ending 30 ms sub-frames within the main 40 ms frame; this feature works on spectrum directly, not on its parameters.

2.2   Audio Data

In our experiments we focused on wind instruments, typically used in jazz music. Training data for clarinet, trombone, and trumpet were taken from three repositories of single, isolated sounds of musical instruments: McGill University Master Samples (MUMS) [20], The University of Iowa Musical Instrument Samples (IOWA) [27], and RWC Musical Instrument Sound Database [6]. Since no sousaphone sounds were available in these sets, we additionally used sousaphone sounds recorded by R. Rudnicki [24]. Training data were in mono format in RWC data and for sousaphone, and in stereo for the rest of the data. Training was performed on single sounds and mixes. Our classifiers were trained to work on larger instrument sets, so additionally sounds of 5 other instruments were used in the training. These were instruments also typical for jazz recordings: double bass, piano, tuba, saxophone, and harmonica. RWC, IOWA and MUMS repositories were used to collect these sounds. The testing data were taken from the following jazz band stereo recordings by R. Rudnicki [13], [24]:

- *Mandeville* by Paul Motian,
- *Washington Post March* by John Philip Sousa, arranged by Matthew Postle,
- *Stars and Stripes Forever* by John Philip Sousa, semi-arranged by Matthew Postle — Movement no. 2 and Movement no. 3.

These recordings contain pieces played by clarinet, trombone, trumpet, and sousaphone, which are our target instruments.

3   Classification

In the previous works, we have been solving the multi-label problem of recognizing instruments with the standard binary relevance approach. Namely, we were building a battery of binary models, each capable of detecting the presence or absence of a single instrument; for prediction, we were applying all the models to the sample and combining their predictions.

Unfortunately, this approach is not computationally effective, ignores the information about instrument-instrument interactions and requires sub-sampling of the training data to make balanced training sets for each battery member. Thus, we attempted to modify the random ferns classifier used in our methodology to natively support multi-label classification.

3.1   Multi-label Random Ferns

Random ferns classifier is an ensemble of K ferns, simple base classifiers equivalent to a constrained decision tree. Namely, the depth of a fern (D) is fixed and the splitting criteria on a given tree level are identical. This way, a fern has 2^D leaves and directs object x into a leaf number $F(x) = 1 + \sum_{i=1}^{D} 2^{i-1}\sigma_i(x) \in 1..2^D$, where $\sigma_i(x)$ is an indicator variable for a result of the i-th splitting criterion. We use the rFerns implementation of random ferns [14] which generates splitting criteria entirely at random, i.e. randomly selects both a feature on which the split will be done and the threshold value. Also, rFerns builds a bagging ensemble of ferns, i.e. each fern, say k-th, is not directly build on a whole set of objects but on a *bag* B_k, a multiset of training objects created by random sampling with replacement the same number of objects as in the original training set.

The leaves of ferns are populated with *scores* $S_k(x, y)$, indicating the confidence of a fern k that an object x falling into a certain leaf $F_k(x)$ belongs to the class y. The scores are generated based on a training dataset $X^t = \{x_1^t, x_2^t, \ldots\}$, and are defined as

$$S_k(x, y) = \log \frac{1 + |L_k(x) \cap Y_k(y)|}{C + |L_k(x)|} - \log \frac{1 + |Y_k(y)|}{C + |B_k|}, \qquad (1)$$

where $L_k(x) = \{x^t \in B_k : F_k(x) = F_k(x^t)\}$ is a multiset of training objects from a bag in the same leaf as a given object and $Y_k = \{x^t \in B_k : y \in Y(x^t)\}$ is a multiset of training objects from a bag that belong to a class y. $Y(x)$ denotes a set of true classes of an object x, and is assumed to always contain a single

element in a many-classes case; C is the number of all classes. The prediction of
the whole ensemble for an object x is $Y^p(x) = \arg\max_y \sum_{k=1}^{K} S_k(x, y)$.

Our proposed generalization of random ferns for multi-label classification is
based on the observation that while the fern structures are not optimized to a
given problem, the same set of F_k functions can serve all classes rather than
being re-created for each one of them. In the battery classification, we create
virtual *not-class* classes to get a baseline score value used to decide whether a
class of a certain score value should be reported as present or absent. With multi-
class random ferns, however, we can incorporate this idea as a normalization of
scores so that the sign of their value will become meaningful indicator of a class
presence. We call such normalized scores *score quotients* $Q_k(x, y)$, and define
them as

$$Q_k(x, y) = \log \frac{1 + |L_k(x) \cap Y_k(y)|}{1 + |L_k(x) \setminus Y_k(y)|} - \log \frac{1 + |Y_k(y)|}{1 + |B_k \setminus Y_k(y)|}. \tag{2}$$

The prediction of the whole ensemble for an object x naturally becomes $Y^p(x) = \{y : Q_k(x, y) > 0\}$.

4 Experiments

When preparing training data, we start with single isolated sounds of each target
instrument. After removing starting and ending silence [13], each file representing
the whole single sound is normalized so that the RMS value equals one. Then, we
create the training set of sounds by mixing random 40 ms frames extracted from
the recordings of 1 to 4 randomly chosen instruments; the mixing is done with
random weights and the result is normalized again to get the RMS value equal to
one. Finally, we convert the sound into a vector of features by applying previously
described sound descriptors. The multi-label decision for such an object is a set
of instruments which sounds were used to create the mix. We have repeated this
procedure 100 000 times to prepare our training set.

This set is used directly to generate the model with the multi-label random
ferns approach. When creating the battery of random ferns, we are splitting this
data into a set of binary problems. Each one is devoted to one instrument and
contains 3000 positive examples where this instrument contributed to the mix
and 3000 negative when it was absent.

In both cases, we used $K = 1000$ ferns and scanned depths $D = 5, 7, 10, 11, 12$.
As the random ferns is a stochastic algorithm, we have replicated training and
testing procedure 10 times.

Both models are tested on real jazz recordings described in Section 2.2 and
their predictions assessed with respect to the annotation performed by an expert.
The accuracy was assessed via precision and recall scores; these measures were
weighted by the RMS of a given frame, in order to diminish the impact of softer
frames which cannot be reasonably identified as their loudness approaches the
noise level. Our true positive score T_p for an instrument i is a sum of RMS of
frames which are both annotated and classified as i. Precision is calculated by

dividing T_p by the sum of RMS of frames which are classified as i; respectively, recall is calculated by dividing T_p by the sum of RMS of frames which are annotated as i.

As a general accuracy measure we have used F-score, defined as a harmonic mean of such generalised precision and recall.

5   Results

The results of accuracy analysis are presented in Figure 1. One can see that for fern depth greater than 7 the multi-label ferns achieved both significantly better precision and recall that the battery classifier; obviously this also corresponds to a higher F-score. The precision of both methods seems to stabilize for greater depth, while the recall and so F-score of multi-class ferns raise steadily and may be likely further improved. The variation of the results is also substantially smaller for multi-class ferns, showing that the output of this approach is more stable and thus more predictable.

Table 1 collects the sizes of created models and the speed with which they managed to predict the investigated jazz pieces. One can see that the utilization of multi-label ferns results in substantially greater prediction speed, on average

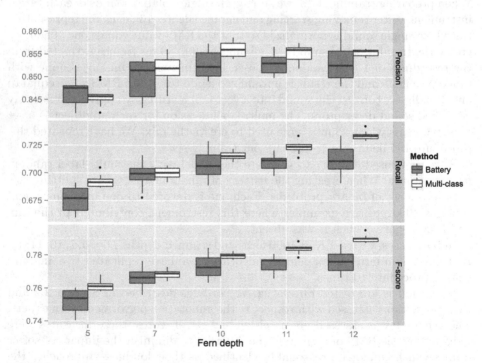

Fig. 1. Overall precision, recall and F-score for all the investigated jazz recordings and all the instruments for a battery of binary random ferns and for multi-label ferns

Table 1. Comparison of model size and prediction speed for a random ferns battery and multi-label random ferns. The speed is expressed as the total playing time of all investigated jazz recordings divided by the CPU time required to classify them.

Fern depth	Model size		Prediction speed	
	Battery	Multi-label	Battery	Multi-label
5	5MB	2MB	54×	359×
7	19MB	9MB	42×	301×
10	149MB	74MB	33×	238×
11	297MB	148MB	30×	216×
12	592MB	295MB	26×	204×

7 times better than the speed achieved by the battery of binary ferns. Theoretically, this factor should be equal to the number of classes because each object is predicted by a single classifier instead of a battery of them, so should be equal to 9 in our case. The difference is caused by a more subtle effects connected to a higher sophistication of multi-label code and should diminish with an increasing number of classes.

The difference between model sizes is less pronounced, with multi-label models being on average two times smaller than battery models. This is because the multi-label ferns model mainly consists of $2^D CK$ scores quotients, while the ferns battery $2^{D+1}CK$ score quotients (the models are binary but there is C of them).

There is a negative correlation between the achieved F-score and both prediction speed and model size, though, with the fern depth controlling the speed-quality trade-off. However, this way a user may utilize this parameter to flexibly adjust the model to the constraints of the intended implementation.

6   Summary and Conclusions

In this paper we introduce multi-label random ferns as a tool for automatic identification of musical instruments in polyphonic recordings of a jazz band. The comparison of performance of multi-label random ferns and sets of binary ferns shows that the proposed multi-label ferns outperform the sets of binary ferns in every respect. Multi-label ferns are much faster, achieve higher F-score, and the model size increase with increasing complexity also compares favorably with the set of binary random ferns. Therefore, we conclude that multi-label random ferns can be recommended as a classification tools in many applications, not only for instrument identification, and this technique can also be applied on resource-sensitive devices, e.g. mobile devices.

Acknowledgments. This project was partially supported by the Research Center of PJIIT, supported by the Ministry of Science and Higher Education in Poland, and the Polish National Science Centre, grant 2011/01/N/ST6/07035. Computations were performed at the ICM UW, grant G48-6.

References

1. Bosch, J.J., Janer, J., Fuhrmann, F., Herrera, P.: A Comparison of Sound Segregation Techniques for Predominant Instrument Recognition in Musical Audio Signals. In: 13th International Society for Music Information Retrieval Conference (ISMIR), pp. 559–564 (2012)
2. Breiman, L.: Random Forests. Machine Learning 45, 5–32 (2001)
3. Cont, A., Dubnov, S., Wessel, D.: Realtime multiple-pitch and multiple-instrument recognition for music signals using sparse non-negativity constraints. In: Proc. 10th Int. Conf. Digital Audio Effects (DAFx-2007), pp. 85–92 (2007)
4. Eggink, J., Brown, G.J.: Application of missing feature theory to the recognition of musical instruments in polyphonic audio. In: 4th International Society for Music Information Retrieval Conference, ISMIR (2003)
5. Essid, S., Richard, G., David, B.: Instrument recognition in polyphonic music based on automatic taxonomies. IEEE Trans. Audio, Speech, Lang. Process. 14(1), 68–80 (2006)
6. Goto, M., Hashiguchi, H., Nishimura, T., Oka, R.: RWC Music Database: Music Genre Database and Musical Instrument Sound Database. In: 4th International Society for Music Information Retrieval Conference (ISMIR), pp. 229–230 (2003)
7. Heittola, T., Klapuri, A., Virtanen, A.: Musical Instrument Recognition in Polyphonic Audio Using Source-Filter Model for Sound Separation. In: 10th International Society for Music Information Retrieval Conference, ISMIR (2009)
8. Herrera-Boyer, P., Klapuri, A., Davy, M.: Automatic Classification of Pitched Musical Instrument Sounds. In: Klapuri, A., Davy, M. (eds.) Signal Processing Methods for Music Transcription. Springer (2006)
9. ISO: MPEG-7 Overview, http://www.chiariglione.org/mpeg/
10. Jiang, W., Wieczorkowska, A., Raś, Z.W.: Music Instrument Estimation in Polyphonic Sound Based on Short-Term Spectrum Match. In: Hassanien, A.-E., Abraham, A., Herrera, F. (eds.) Foundations of Computational Intelligence Volume 2. SCI, vol. 202, pp. 259–273. Springer, Heidelberg (2009)
11. Kashino, K., Murase, H.: A sound source identification system for ensemble music based on template adaptation and music stream extraction. Speech Commun. 27, 337–349 (1999)
12. Kitahara, T., Goto, M., Komatani, K., Ogata, T., Okuno, H.G.: Instrument identification in polyphonic music: Feature weighting to minimize influence of sound overlaps. EURASIP J. Appl. Signal Process. 2007, 1–15 (2007)
13. Kubera, E.z., Kursa, M.B., Rudnicki, W.R., Rudnicki, R., Wieczorkowska, A.A.: All That Jazz in the Random Forest. In: Kryszkiewicz, M., Rybinski, H., Skowron, A., Raś, Z.W. (eds.) ISMIS 2011. LNCS, vol. 6804, pp. 543–553. Springer, Heidelberg (2011)
14. Kursa, M.B.: Random ferns method implementation for the general-purpose machine learning (2012), http://arxiv.org/abs/1202.1121v1 (submitted)
15. Kursa, M.B.: Robustness of Random Forest-based gene selection methods. BMC Bioinformatics 15(8(1)), 1–8 (2014)
16. Little, D., Pardo, B.: Learning Musical Instruments from Mixtures of Audio with Weak Labels. In: 9th International Society for Music Information Retrieval Conference, ISMIR (2008)
17. Martins, L.G., Burred, J.J., Tzanetakis, G., Lagrange, M.: Polyphonic instrument recognition using spectral clustering. In: 8th International Society for Music Information Retrieval Conference, ISMIR (2007)

18. MIDOMI: Search for Music Using Your Voice by Singing or Humming, http://www.midomi.com/
19. Niewiadomy, D., Pelikant, A.: Implementation of MFCC vector generation in classification context. J. Applied Computer Science 16(2), 55–65 (2008)
20. Opolko, F., Wapnick, J.: MUMS — McGill University Master Samples. CD's (1987)
21. Özuysal, M., Fua, P., Lepetit, V.: Fast Keypoint Recognition in Ten Lines of Code. In: 2007 IEEE Conference on Computer Vision and Pattern Recognition. IEEE (2007)
22. Özuysal, M., Calonder, M., Lepetit, V., Fua, P.: Fast Keypoint Recognition using Random Ferns. Image Processing (2008)
23. Ras, Z.W., Wieczorkowska, A.A. (eds.): Advances in Music Information Retrieval. SCI, vol. 274. Springer, Heidelberg (2010)
24. Rudnicki, R.: Jazz band. Recording and mixing. Arrangements by M. Postle. Clarinet — J. Murgatroyd, trumpet — M. Postle, harmonica, trombone — N. Noutch, sousaphone – J. M. Lancaster (2010)
25. Shazam Entertainment Ltd, http://www.shazam.com/
26. Shen, J., Shepherd, J., Cui, B., Liu, L. (eds.): Intelligent Music Information Systems: Tools and Methodologies. Information Science Reference, Hershey (2008)
27. The University of IOWA Electronic Music Studios: Musical Instrument Samples, http://theremin.music.uiowa.edu/MIS.html
28. TrackID, https://play.google.com/store/apps/details?id=com.sonyericsson.trackid
29. Vincent, E., Rodet, X.: Music transcription with ISA and HMM. In: 5th International Conference on Independent Component Analysis and Blind Signal Separation (ICA), pp. 1197–1204 (2004)
30. Wieczorkowska, A.A., Kursa, M.B.: A Comparison of Random Forests and Ferns on Recognition of Instruments in Jazz Recordings. In: Chen, L., Felfernig, A., Liu, J., Raś, Z.W. (eds.) ISMIS 2012. LNCS (LNAI), vol. 7661, pp. 208–217. Springer, Heidelberg (2012)
31. Barbedo, J.G.A., Tzanetakis, G.: Musical Instrument Classification Using Individual Partials. IEEE Transactions on Audio, Speech & Language Processing 19(1), 111–122 (2011)
32. Kirchhoff, H., Dixon, S., Klapuri, A.: Multi-Template Shift-Variant Non-Negative Matrix Deconvolution for Semi-Automatic Music Transcription. In: 13th International Society for Music Information Retrieval Conference (ISMIR), pp. 415–420 (2012)

Computer-Supported Polysensory Integration Technology for Educationally Handicapped Pupils

Michal Lech[1], Andrzej Czyzewski[1], Waldemar Kucharski[2], and Bozena Kostek[3]

[1] Multimedia Systems Department, Gdansk University of Technology, 80-233 Gdansk, Poland
[2] Learnetic S.A., Gdansk, Poland
[3] Audio Acoustics Lab., Gdansk University of Technology, 80-233 Gdansk, Poland
bokostek@audioacoustics.org

Abstract. In this paper, a multimedia system providing technology for hearing and visual attention stimulation is shortly presented. The system aims to support the development of educationally handicapped pupils. The system has been presented in the context of its configuration, architecture, and therapeutic exercise implementation issues. Results of pupils' improvements after 8 weeks of training with the system are also provided. Training with the system led to the development of spatial orientation and understanding cause-and-effect relationships.

Keywords: Polysensory Stimulation, Computer-based therapeutic exercises, Educationally handicapped pupils.

1 Introduction

The most important issue of education today is to increase its effectiveness. Meanwhile, we are witnessing a crisis of education based on the assumption that we all learn in the same way, equally fast whereas being stimulated in a similar way. The main reasons for the diversity of cognitive abilities of children of a similar age, are congenital or acquired cognitive preferences, and a variety of cognitive dysfunction (e.g. dyslexia). The number of students diagnosed with special educational needs is estimated at 10-25% of the population. This large spread of estimation is associated with individual qualification methods used in different countries. Recent data concerning the UK indicate the result of various learning disabilities at the rate of 21% . The data even more drawing attention to the apparent problem come from the 2011th WHO report. Screening tests show up to 35% percent of children in many countries are at risk of dysfunction impeding school activity. Assuming these facts, teaching methods in education is likely to be developed in two dimensions: individualized teaching and training of cognitive capacity.

During the past two decades, much attention has been devoted to developing didactic computer exercises; collaborative learning (Lingnau et al., 2003) mixed-reality technology such as motion capture or up-to-date robots, and games that develop both mental and manual skills in schoolchildren and students (Gider et al., 2012)(Sugimoto, 2011)(Sung et al., 2011). There are also many examples in game technology that are

T. Andreasen et al. (Eds.): ISMIS 2014, LNAI 8502, pp. 224–233, 2014.
© Springer International Publishing Switzerland 2014

created to assist students other than school activities, such as for example learning to dance. Dance training systems are however designed for users with sufficient skills to perform the dance movements. Although not very difficult for intellectually agile students, such solutions may be undeniably too sophisticated for mentally retarded pupils.

Since the common cause of many cognitive deficits, often diagnosed at school age, the disrupted communication between the hemispheres of the brain, which occurs often in conjunction with laterality disorders, in such cases, polysensory integration training may be a method supporting early learning. When planning work with pupils, one needs to define aims and choose educational materials associated both with polysensory acquisition of knowledge of the surrounding world and with translating these knowledge and skills into everyday life in later stages of development. The planning should be preceded by diagnosing the pupils' level of functioning and determining the nearest development zone (Petrushin, 2004). Education of mentally retarded pupils involves four stages: stimulating senses, sensory-kinesthetic integration, developing somatognosia and developing readiness to learn. In many studies one may find a statement that perceptual motor skills, both fine and gross, require the integration of sensory input (visual, auditory and kinesthetic) (Ayres, 1972)(Dore, 2006). It was Ayres who introduced the concept of sensory integration, i.e. most of incoming messages to the brain are received by the senses (Ayres, 1972). It has been applied in the therapy of various disabilities, including intellectual ones (Arendt et al., 1988).

In this paper, a multimedia system with several computer exercises/lessons addressing pupils with severe mental retardation is presented. The system described is aimed at stimulating the development of kinesthetic-perceptual functions. Section 2 presents a short review of some educational aspects of children with moderate or severe mental retardation. Section 3 provides details of the system architecture and software of the Multimedia System of Polysensory Integration (MSPI). Section 4 shows the results of the MSPI system after several weeks of use in a school for educationally handicapped and educable mentally retarded pupils. The final section provides our conclusions and outlines an idea for system improvement.

2   System Overview

2.1   System Components and Interaction

The system consists of a personal computer with an application installed, two monitors, a therapeutic mat, two USB cameras, 4 surround sound system speakers, and a stand for the speakers and one of the cameras. Such a hardware platform has also been utilized during the development of the MSPI for providing exercises for intellectually well-developed pupils (Czyżewski and Kostek, 2012) (Lech and Kostek, 2009) (Lech and Kostek, 2012). The monitors are placed back to back, with one of them displaying exercise screens to the pupil and the other displaying modified exercise screens and controls to the therapist. The therapeutic area is designated by a stand construction with a square therapeutic mat positioned in its centre on the floor.

The therapeutic mat has 9 square areas separated by straight lines. One of the cameras is placed on the floor in such a location that the pupil, when walking on the therapeutic mat, is always visible in the frame. The second camera is placed over the middle position of the therapeutic mat at such a height that the same requirement as for the floor camera is met. The speakers are hung in the corners of the stand construction.

During the application initialization the pupil does not occupy the field of view of any of the cameras. Thus, initial video streams used for image processing during exercising can be retrieved. After initialization, before running the chosen exercise, the pupil is asked to occupy the middle area of the therapeutic mat. The interaction with the system lies in walking on the mat and thus choosing one of the square areas at a time and bouncing to confirm the choice. Occupying a particular area causes an associated image to be displayed or a sound generated, according to the exercises described in Section 2.4. Bouncing produces an image display and also sound generation.

2.2   System Architecture and Implementation

The system developed by the authors has previously been described very thoroughly (Lech and Kostek, 2009) (Lech and Kostek, 2012), thus only a short review of its architecture (Fig. 1), technical basis and main functionalities are recalled here. The system has been implemented using JAVA SE language and C++ with OpenCV library (Bradski and Kaehler, 2008). The C++ / OpenCV part of the system is responsible for retrieving and processing the video streams from cameras and detecting the pupil's feet and position on the therapeutic mat. The remaining system functionalities have been implemented in JAVA with Swing packages used for GUI (Graphical User Interface). These functionalities consist in the graphical controls, enabling the exercises to be configured and run. Generation of sounds has also been contained in the JAVA part of the system. The functionality provided by the C++ / OpenCV part has been compiled as a dynamic link library and connected with JAVA code using JNI (Java Native Interface).

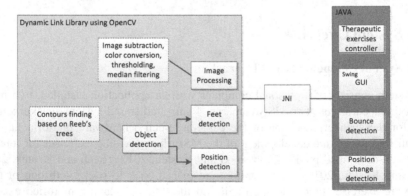

Fig. 1. System architecture

2.3   Technical Basis

The system is based on processed video streams, resulting from subtracting the initial video streams that do not contain the pupil from the video streams retrieved during training. This principle regards both the floor and overhead cameras. The subtraction method is absolute difference. After subtraction, the video streams are converted from BGR color space to grey scale and binary thresholded with a default threshold value equal to 50. The threshold values can be changed using application control panels, separately for each camera. Thus a teacher/therapist can adjust the processing, for example to pupils clothing, raising the threshold when the contrast between the pupil and the therapeutic mat or walls is high. This results in an increase in pupil position detection efficacy. After applying the binarization, the video streams are median filtered with a mask of a size equal to 9.

Detection of pupil silhouettes is performed using a contour detection algorithm implemented in OpenCV library (Bradski and Kaehler, 2008). The algorithm is based on the method proposed by Reeb (Reeb, 1946) and further developed by Bajaj et al. (1997) and Carr et al. (2004). Each object in the video streams is represented by a sequence of points connected by segments creating its shape. The Teh-Chin chain approximation algorithm in L1 metric flavor (Chan and Chin, 1992) has been used for contour approximation. The sequences are found using the method proposed by Suzuki and Abe (Suzuki and Abe, 1985), and as a consequence the extracted objects are represented in the form of trees. In each tree the root is a sequence representing the outer contour and the children are sequences designating inner contours of successive depth levels.

The algorithm utilized, developed by Suzuki and Abe (Suzuki and Abe, 1985), is based on two principle identifiers denoted NBD and LNBD. The NBD identifier stands for the sequential number of the newly-found border and LNBD denotes a sequential number of the border recently encountered during the searching process. Thus, the border identified by LNBD is either the parent border of the newly-found border or a border which shares the common parent with the newly-found border. The searching process is performed in a scanning manner. The input image $F = \{f_{ij}\}$ is scanned with a given raster starting with NBD set to 1. LNBD is reset to 1 every time a new row of the image is to be scanned. The following steps of the algorithm are performed for each pixel such that $f_{ij} \neq 0$.

(1) Select one of the following:

(a) If $f_{ij} = 1$ and $f_{i,\ j-1} = 0$, then decide that the pixel (i, j) is the border following the starting point of an outer border, increment NBD, and $(i_2, j_2) \leftarrow (i, j - 1)$.

(b) Else if $f_{ij} \geq 1$ and $f_{i,j+1} = 0$, then decide that the pixel (i, j) is the border following the starting point of a hole border, increment NBD, $(i_2, j_2) \leftarrow (i, j + 1)$, and LNBD $\leftarrow f_{ij}$ in the case $f_{ij} > 1$.

(c) Otherwise, go to (4).

(2) Depending on the type of the newly-found border and the border with the sequential number LNBD (i.e. the last border met on the current row), decide the parent of the current border as shown in Table 1.

(3) From the starting point (i, j), follow the detected border: this is done by the following substeps (3.1) through (3.5).

(3.1) Starting from (i_2, j_2), look around clockwise at the pixels in the neighborhood of (i, j) and find a nonzero pixel. Let (i_1, j_1) be the first nonzero pixel found. If no nonzero pixel is found, assign -NBD to f_{ij} and go to (4).

(3.2) $(i_2, j_2) \leftarrow (i_1, j_1)$ and $(i_3, j_3) \leftarrow (i, j)$.

(3.3) Starting from the next element of the pixel (i_2, j_2) in counterclockwise order, examine counterclockwise the pixels in the neighborhood of the current pixel (i_3, j_3) to find a nonzero pixel and let the first one be (i_4, j_4).

(3.4) Change the value $f_{i3, j3}$ of the pixel (i_3, j_3) as follows:

(a) If the pixel $(i_3, j_3 + 1)$ is a 0-pixel examined in substep (3.3), then $f_{i3, j3} \leftarrow$ -NBD.

(b) If the pixel $(i_3, j_3 + 1)$ is not a 0-pixel examined in substep (3.3) and

$f_{i3, j3} = 1$, then $f_{i3, j3} \leftarrow$ -NBD.

(c) Otherwise, do not change $f_{i3, j3}$.

(3.5) If $(i_4, j_4) = (i, j)$ and $(i_3, j_3) = (i_1, j_1)$ (coming back to the starting point), then go to (4); otherwise, $(i_2, j_2) \leftarrow (i_3, j_3), (i_3, j_3) \leftarrow (i_4, j_4)$, and go back to (3.3).

(4) If $f_{ij} \neq 1$, then LNBD $\leftarrow |f_{ij}|$ and resume the raster scan from pixel $(i, j + 1)$. The algorithm terminates when the scan reaches the lower right corner of the picture.

Amongst all the objects detected in the video streams retrieved from the floor and overhead cameras, objects that are marked as silhouettes must have a width, height and number of pixels above the set thresholds. Thus, possible noise which was not eliminated in the processing phase can be reduced. For images of size 320 x 240 pixels, the default thresholds for the number of pixels constituting the silhouette, width and height, are 50, 10 and 10, respectively.

Table 1. Decision rules for the parent border of newly-found border B (Bradski and Kaehler, 2008)

Type of B \ \ Type of border B'	Outer border	Hole border
Outer border	The parent border of border B'	Border B'
Hole border	Border B'	The parent border of border B'

For a silhouette in the image captured by the floor camera, the initial y position of the feet is determined. This process is performed before running the chosen exercise when the middle area is occupied by the pupil. The position is identified by the bottom edge of the shoes. Relative to this value bouncing is detected. The bounce threshold, i.e. the distance between the feet y initial position and the position reflecting the highest point of the bounce, can be adjusted by the therapist. Thus for pupils bouncing high, the bounce detection efficacy can be increased. The bounce is recognized after the system detects that the feet have left the ground, exceeded the bounce threshold, and returned to their initial position. The position of a pupil on the therapeutic mat is determined by the centre of gravity of the silhouette detected in the image captured by the overhead camera. The boundaries of each square of the therapeutic mat are determined manually by the therapist soon after the application is initialized. This operation is performed using the therapeutic mat calibrator (Fig. 2).

3   Therapeutic Exercises

The system contains 11 exercises/lessons, diversified in terms of the level of difficulty, which are simple for intellectually developed pupils but challenging for those that are mentally retarded. The exercises involve various combinations of the task of searching for an image associated with a particular therapeutic mat area (Fig. 2). By changing the position on the mat the pupil changes the displayed image and sound generated. A description of the exercises is given below.

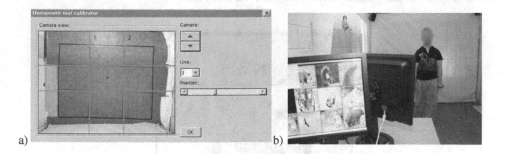

Fig. 2. View from the ceiling camera in therapeutic mat calibrator (a) and general view of the system during therapy (b)

3.1   Exercises – Searching for the Same Picture

Exercise 1 – Searching for the Same Picture
A pupil occupies the middle area of the therapeutic mat and looks at the screen. In the bottom area of the screen an image is displayed. This image is associated with one of the therapeutic mat areas and thus constitutes the pattern to be found by the pupil.

Walking on the mat causes the image associated with the area occupied at the moment to be displayed. The task of the pupil is to watch the screen and react with a bounce when both images are identical. In such a case a sound associated with the pattern is additionally played to stimulate the hearing attention. When the task is performed appropriately the pupil is awarded with a smiley and applause generated from the speakers. The exercise presented develops perception and visual memory (recognizing individuals, animals, objects) as well as forms an understanding of the term "the same". A therapist additionally stimulates the pupil's hearing attention and develops orientation in space and body schema by using verbal hints, such as "one step forward".

Exercise 2 – Associating the Image with Sound

The difference between this exercise and the previous one lies in the lack of image displayed above the pattern. Instead, a sound associated with each therapeutic mat square is played. The task of the pupil is to bounce when the sound reflects the pattern displayed. The exercise mainly develops audio perception and auditory memory (recognizing sounds made by animals, objects, etc.) A pupil is motivated to concentrate on auditory stimuli, isolate them, associate in memory the sound with its source and make a decision whether this is the proper sound.

Exercise 3 – Finding a Path to the Image

In the bottom area of the screen the image pattern is displayed. Above it there are 9 images associated with the therapeutic mat's squares (Fig. 3). The pupil's task is to find among these nine images the image that is the same as the pattern. Next, the pupil plans a path to the image. In certain aspects the exercise is similar to the exercises already described. The exercise develops visual and audio perception as well as space orientation and designating directions from the body axis.

Fig. 3. A view of the screen for a pupil conducting exercise 3

4   Pupils' Performance Results

The research on the effectiveness of the system has been carried out for 8 weeks in Primary Special School No. 26 in the Polish city of Torun. Eight pupils, aged 8–17, took part in the therapy. The pupils were chosen based on their psychomotor skills, the degree of understanding tasks, readiness to perform the exercises and willingness to work with a therapist. Depending on the developed scale for the assessment of progress in various skills, preliminary diagnosis has been performed. Selected skills, for which the results have been presented in the paper are described in Table 2. Therapeutic sessions were conducted regularly, once per week for each pupil. A single session lasted 45 minutes. After 8 weeks the progress was measured and registered on the development assessment scale mentioned.

Table 2. Description of selected skills

ID	Description of skills exercised by pupils
2C	Moves in a controlled manner
3C	Moves forward between therapeutic mat squares and does not step on lines
3D	Looks at the screen and at the mat squares and decides where to go
3E	Moves onto therapeutic mat square indicated by a therapist
4C	Moves right between therapeutic mat squares and does not step on lines
4D	Moves between therapeutic mat squares according to therapist's indications
4E	Moves between therapeutic mat squares according to the therapist's indications (naming direction, showing changes on the screen, confirming choice)
5B	Understands single words associated with the system and exercises
5D	Moves between mat squares according the to direction indicated by therapist
6D	Indicates directions: "left", "right", "forward", "back"
8C	Bounces with both legs and falls on the same therapeutic mat square
8D	Indicates direction in which he / she should move
10E	Assesses the correctness of choosing the mat square and bounces to confirm

In Fig. 4 the progress achieved by the pupils, for the best developed skills, listed in Table 2, has been presented.

Overall, pupils showed great enthusiasm for the proposed system and performing the exercises; they were amazingly cooperative. The high degree of motivation and interest in the equipment resulted in relatively fast progress. This progress was especially noticeable in two spheres: spatial orientation / directions understanding and sense of causation, i.e. concentration of attention and understanding of cause-and-effect relationships while interacting with the system. It is unlikely that the progress was due to the random factor as the therapists had never observed such improvements in skills when using a standard therapy. In none of the cases skill deterioration had occurred.

Also, the transfer of skills developed during therapy to everyday situations was noticed. For example, spontaneous utterances associated with spatial orientation and directions naming, like "Ann is sitting at the back of bus", were formulated.

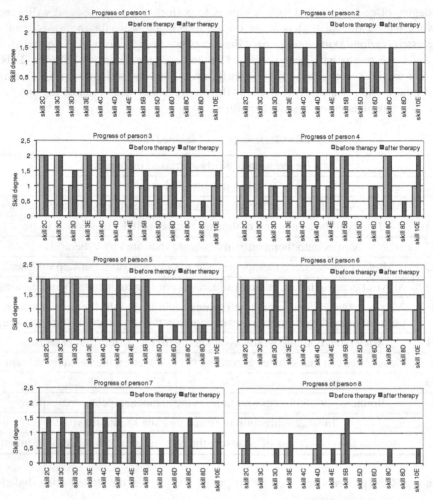

Fig. 4. Progress achieved by each pupil participating in the therapy exercises

5   Conclusions

In this paper, a multimedia system providing exercises that stimulate the hearing and visual attention of pupils with moderate or severe mental retardation has been presented. The system can facilitate individual educational-therapeutic programs designed for each pupil. The scheme of the exercises provided by the system is fixed. A therapist/teacher can, however, adjust the images and sounds to suit the educational needs, age or interests of a particular pupil.

One of the system's drawbacks is its partial susceptibility to contrast between a pupil's clothes and the therapeutic mat. It was observed that temporary errors in the position detection occurred when the color of the pupil's clothes were close to that of the particular jumping mat area, resulting in the colors blending. This resulted in a flash of a picture associated with an area of the mat other than that occupied. The proposed solution to this problem is to apply a Kalman filter to smooth the motion trajectory and thus eliminate erroneous position detection.

Acknowledgments. Research partially funded within the project No. POIG.01.03.01-22-017/08, entitled "Elaboration of a series of multimodal interfaces and their implementation to educational, medical, security and industrial applications". The project is subsidized by the EU regional development fund and by the Polish State budget.

References

Arendt, R.-E., MacLean, W.-E., Baumeister, A.-A.: Sensory integration theory and practice: An uncertain connection. American Journal on Mental Retardation 95, 427–429 (1988)

Ayres, A.-J.: Sensory integration and learning disorders. Western Psychological Services, Los Angeles (1972)

Bajaj, C.-L., Pascucci, V., Schikore, D.-R.: The contour spectrum. IEEE Visualization 1997, 167–173 (1997)

Bradski, G., Kaehler, A.: Learning OpenCV: Computer Vision with the OpenCV Library. O'Reilly, Sebastopol (2008)

Carr, H., Snoeyink, J., van de Panne, M.: Progressive topological simplification using contour trees and local spatial measures. In: 15th Western Computer Graphics Symposium, British Columbia (2004)

Chan, W.-S., Chin, F.: Approximation of Polygonal Curves with Minimum Number of Line Segments. In: Proc. 3rd Annu. Internat. Sympos. Algorithms and Computation, pp. 378–387 (1992)

Czyżewski, A., Kostek, B.: Intelligent Video and Audio Applications for Learning Enhancement. Journal of Intelligent Information Systems 38(3), 555–574 (2012)

Dore, W.: Dyslexia: The Miracle Cure. John Blake Publishing, London (2006)

Gider, F., Likar, B., Kern, T., Miklavcic, D.: Implementation of a Multidisciplinary Professional Skills Course at an Electrical Engineering School. IEEE Transactions on Education 55(3), 332–340 (2012)

Michal, L., Bozena, K.: Human-Computer Interaction Approach Applied to the Multimedia System of Polysensory Integration. In: Damiani, E., Jeong, J., Howlett, R.J., Jain, L.C. (eds.) New Directions in Intelligent Interactive Multimedia Systems and Services - 2. SCI, vol. 226, pp. 265–274. Springer, Heidelberg (2009)

Lech, M., Kostek, B.: Hand Gesture Recognition Supported by Fuzzy Rules and Kalman Filters. International Journal of Intelligent Information and Database Systems 6(5), 407–420 (2012)

Lingnau, A., Hoppe, H.U., Mannhaupt, G.: Computer supported collaborative writing in an early learning classroom. Journal of Computer Assisted Learning 19, 186–194 (2003)

Petrushin, V.-A.: Knowledge-Based Approach for Testing and Diagnosis. In: Kommers, P.-A.-M. (ed.) Cognitive Support for Learning: Imaging the Unknown, pp. 173–188 (2004)

Reeb, G.: Sur les points singuliers d'une forme de Pfaff completement integrable ou d'une fonction numerique. Comptes Rendus de l'Academie des Sciences 222, 847–849 (1946)

Sugimoto, M.: A Mobile Mixed-Reality Environment for Children's Storytelling Using a Handheld Projector and a Robot. IEEE Transactions on Learning Technologies 4(3), 249–260 (2011)

Sung, K., Hillyard, C., Angotti, R.-L., Panitz, M.-W., Goldstein, D.-S., Nordlinger, J.: Game-Themed Programming Assignment Modules: A Pathway for Gradual Integration of Gaming Context Into Existing Introductory Programming Courses. IEEE Transactions on Education 54(3), 416–427 (2011)

Suzuki, S., Abe, K.: Topological structural analysis of digital binary images by border following. Computer Vision, Graphics and Image Processing 30, 32–46 (1985)

Integrating Cluster Analysis to the ARIMA Model for Forecasting Geosensor Data

Sonja Pravilovic[1,2], Annalisa Appice[1], and Donato Malerba[1]

[1] Dipartimento di Informatica, Università degli Studi di Bari Aldo Moro
via Orabona, 4 - 70126 Bari - Italy
[2] Faculty of Information Technology, Mediterranean University
Vaka Djurovica b.b. 81000 Podgorica - Montenegro
{sonja.pravilovic,annalisa.appice,donato.malerba}@uniba.it

Abstract. Clustering geosensor data is a problem that has recently attracted a large amount of research. In this paper, we focus on clustering geophysical time series data measured by a geo-sensor network. Clusters are built by accounting for both spatial and temporal information of data. We use clusters to produce globally meaningful information from time series obtained by individual sensors. The cluster information is integrated to the ARIMA model, in order to yield accurate forecasting results. Experiments investigate the trade-off between accuracy and efficiency of the proposed algorithm.

1 Introduction

The pervasive ubiquity of geosensor networks, which measure several physical variables (e.g. atmospheric temperature, pressure, humidity or energy production), enable us to monitor and study dynamic physical phenomena at granularity details that were never possible before. A major challenge posed by a geosensor network is to combine the sensor nodes in computational infrastructures, that are able to produce globally meaningful information from time series data obtained by individual sensor nodes. These infrastructures should use appropriate primitives to account for both the temporal dimension of data, which determines the ground time of a reading, and the spatial dimension of data, which determines the ground location of a sensor. These primitives are investigated for several data mining tasks. In this paper, we focus on the task of forecasting the future values of time series of geosensor data.

The temporal dimension of data encourages the consideration of the temporal correlation of the data. This is a measure of how many future observations can be predicted from past behavior. The spatial location of a sensor inspires inferences on the spatial correlation of the data. This is a measure of how many data, taken at a relatively close location, behave similarly to each other. Similar behavior, which is very common in the geophysical field, is well described by the first law of Geography of Tobler [14], according to which "everything is related to everything else, but near things are more related than distant things".

T. Andreasen et al. (Eds.): ISMIS 2014, LNAI 8502, pp. 234–243, 2014.
© Springer International Publishing Switzerland 2014

All statistical time series forecasting methods explicitly account for temporal correlation of data. They include exponential smoothing methods, regression methods and auto-regressive integrated moving average (ARIMA) methods. In particular, ARIMA methods are known for the accuracy of forecasts. They give a forecast as a linear function of past observations (or the differences of past observations) and error values of the time series itself. Nevertheless, they overlook, in general, the spatial correlation of time series measured at nearby locations. In theory, ARIMA methods are applied to each sensor node independently. This leads to wasting power computing similar models on separate sensors and obfuscate important insights into the forecasting models.

In this paper, we propose an algorithm, called CArima (Cluster based ARIMA), that leverages the power of cluster analysis, in order to model prominent spatio-temporal dynamics in time series of geosensor data. It uses cluster patterns, in order to speed-up the learning process of traditional ARIMA methods and yield accurate forecasts. We begin observing that spatially related sensors are expected to measure similar values over the time. When this happens, information generated by sensors can be summarized by spatio-temporal clusters. Based upon this idea, we propose to determine the ARIMA model of a cluster. Parameters of this model can be estimated globally over the time series of the cluster. Global parameter are, then, used to determine local coefficients for the linear combination according to the forecast of a clustered time series can be produced.

The paper is organized as follows. Section 3 introduces the ARIMA model. In Section 2, related works on spatio-temporal clustering are reported. Section 4 describes the algorithm to integrate the cluster analysis to the ARIMA model. An experimental study is presented in Section 5 and conclusions are drawn.

2   ARIMA

ARIMA [3] is one of the most popular and powerful forecasting technique in time series analysis. Let Z be a time series, the ARIMA model describes the (weakly) stationary $(1 - L)^d Z$ in terms of two polynomials, one for the auto-regression and the second for the moving average. $(1 - L)^d$ is the differencing operator to transform a non-stationary time series into a stationary one.

$$(1 - L)^d z(t) = c + \epsilon(t) + \sum_{i=1}^{p} \phi(i) L^i ((1 - L)^d z(t)) + \sum_{i=1}^{q} \sigma(i) L^i \epsilon(t), \quad (1)$$

where L^i is the time lag operator, or backward shift. It translates the time series backwards, in order to observe the series from i positions shifted on the left (i.e. $L^i((1 - L)^d z(t)) = (1 - L)^d z(t - i)$). The parameter p is the auto-regression order, the parameter q is the moving average order, $\phi(i)$ and $\sigma(i)$ are the model coefficients, c is a constant, and t $\epsilon(\cdot)$ is the white noise. If the time series is stationary then $d = 0$ so that $(1 - L)^0 z(t) = z(t)$.

The selection of the triple of parameters (p, d, q) is not trivial [13] and a good practice is to search for the smallest p, d and q. Brockwell and Davis [4] recommend using AICc (AIC with correction c) to select p and q. Following this idea,

Hyndman and Khandakar [5] propose a stepwise algorithm, called Auto.ARIMA, that conducts a search over all possible models beginning with selection of parameter d using unit-root (KPSS) test, and then p and q by minimizing the AICc. However, they take into account a single time series.

Although ARIMA is broadly used in time series analysis, still there are few studies referring to spatial extensions of ARIMA. Kamarianakis ct al. [6] define Space-Time ARIMA that accounts for the property of spatial correlation by expressing each data point at the time t and the location (x, y) as a linearly weighted combination of data lagged both in space and time. More recently, Pravilovic et al. [9] define sARIMA that accounts for spatial lags of nearby time series when determining the parameters (p, d, q) for a specific time series. Both these approaches take into account spatial correlated time series, but they assume that, for each time series, spatial correlation is stationary over a circular neighborhood with fixed radius. This static model suffers of serious limitations as it ignores that spatial correlation can be manifested with different underlying latent structure of the network that varies among its portions in terms of density of measures and geometry of the neighborhood.

3   Spatio-temporal Clustering

The goal of clustering is to identify a structure in an unlabeled data set by organizing data into homogeneous groups where the within-group-object similarity is minimized and the between-group-object dissimilarity is maximized. Just like general clustering, time series clustering requires a clustering algorithm to form clusters given a set of unlabeled time series. Various algorithms have been developed to cluster different types of time series data. Some of them also account for spatial location of time series. Putting their differences aside, clustering time series of geosensor data requires to compare the way their time series evolve and to relate that to their spatial position. A classical approach consists in detecting the correlations (and therefore forming clusters) among different time series trying to filter out the effects of spatial autocorrelation, i.e., the mutual interference between objects due to their spatial proximity [16]. Andrienko et al. [1] apply density based clustering and interactive visual displays to objects having spatial and temporal positions. Distances are computed in a specific way for each type of objects. In [11], the authors propose progressive clustering approach that applies different distance functions that work with spatial, temporal, spatio-temporal data to gain understanding of the underlying data in a stepwise manner. Qin et al. [10] formulate a dissimilarity measure that includes both time-series dissimilarity and spatial dissimilarity, and then they incorporate the proposed dissimilarity into fuzzy C-means clustering. Birant and Kut [2] extend density based clustering by plugging-in both a spatial distance and a temporal distance. Spatial filters and temporal filters are used separately in the density estimate with both a spatial and a temporal threshold. In [15], the author considers spatial and temporal information as multi-variate property of an object and uses a multi-variate distance measure to compare them.

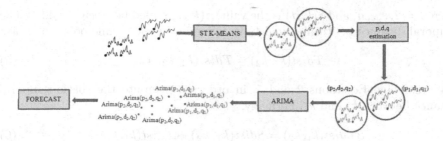

Fig. 1. Block diagram of CARIMA

4 Cluster-Based ARIMA

CArima is a two stepped time series forecasting algorithm (see Figure 1). The first step computes clusters by grouping time series at nearby locations, which measure similar data over time. A distance measure is defined to compare time series by accounting for both temporal and spatial information. A traditional clustering algorithm, k-means, is used to cluster time series according to this spatio-temporal distance. The second step determines a global triple (p, d, q) for the ARIMA model of each cluster. This is done by looking for the triple that optimizes the values of average AICc of the ARIMA models of the times series in the cluster. Finally, for each cluster, the model coefficients are defined locally to each time series. Details of both steps are described in the next subsections.

4.1 Spatio-Temporal Clustering

Let (Z, K, T) be the spatio-temporal system with Z a geophysical numeric variable, K a set of sensors and T a time line discretized in n equal-spaced time points. Each sensor $k_i \in K$ is geo-referenced with a 2D point (x, y) of an Euclidean space. Each measure of Z is taken by a sensor $k_i \in K$ at a time point of T and feeds a specific time series $Z(k_i)$. Let k_1 and k_2 be two sensors $(k_1, k_2 \in K)$. $Z(k_1)$ and $Z(k_2)$ be the two time series of Z produced by k_1 and k_2 over T.

The spatial distance between k_1 and k_2 can be computed by comparing the geographic coordinates of k_1 and k_2, that is,

$$Sdist(k_1, k_2) = (x_2 - x_1)^2 + (y_2 - y_1)^2. \tag{2}$$

For this computation, geographic coordinates are scaled between -1 and 1.

The temporal distance between k_1 and k_2 can be computed by comparing the time series $Z(k_1)$ and $Z(k_2)$. We use an exponential smoothing mechanism, in order to increase the similarity of time series which exhibit similar values closely to the present. Let us consider the recursive formula:

$$Tdist(k_1, k_2, 1) = (z(k_1, 1) - z(k_2, 2))^2 \tag{3}$$

$$Tdist(k_1, k_2, t) = Tdist(k_1, k_2, t - 1)\alpha + (z(k_1, t) - z(k_2, t))^2(1 - \alpha) \tag{4}$$

where $t = 2, \ldots, n$, and $z(k_i, t)$ is the value $z(k_i, t)$ scaled between -1 and 1. The temporal distance between k_1 and k_2 is defined over the n time points of T as:

$$Tdist(k_1, k_2) = Tdist(k_1, k_2, n). \tag{5}$$

We combine Equations 2 and 5, in order to compute the spatio-temporal distance between k_1 and k_2. Formally,

$$STdist(k_1, k_2) = Sdist(k_1, k_2) + Tdist(k_1, k_2). \tag{6}$$

We use the k-means algorithm to determine clusters of (Z, K, T). Sensor nodes are grouped in clusters according to the spatio-temporal distance measure formulated in Equation 6. The clustering algorithm inputs $numberOfClusters$, that is, the number of clusters to be discovered. To automate the choice of the number of cluster, we use the Silhouette index [12]. This index is here adapted, in order to measure of how well each time series lies within its cluster. Let k_i be a sensor, the Silhouette index of k_i is computed as follows:

$$Silhouette(k_i) = \frac{b(k_i) - a(k_i)}{\max\{b(k_i), a(k_i)\}} \tag{7}$$

where $a(k_i)$ is the average spatio-temporal distance of k_i from all other time series in the same cluster, while $b(k_i)$ is the lowest average spatio-temporal distance of k_i from a cluster which k_i is not member. $Silhouette(k_i)$ assumes values in the range $[-1, 1]$. A value of $Silhouette(k_i)$ closer to one means that the time series is appropriately clustered. The average index over all time series of a cluster is a measure of how tightly are grouped all the time series in the cluster.

$$Silhouette(K) = \frac{1}{\sharp K} \sum_{k_i \in K} Silhouette(k_i) \tag{8}$$

This average index is used as follows. We start with $numberOfClusters = MIN$, apply k-means with the specified $numberOfClusters$ and use the discovered clusters to compute $Silhouette(K)$. We iterate the k-means computation by increasing $numberOfClusters$ until $Silhouette(K)$ achieves a local maximum.

4.2   ARIMA Modeling

Let C be a cluster of sensors. CArima determines global parameters (p, d, q) for C and local model coefficients $\phi(k_i)$ and $\sigma(k_i)$ for each sensor $k_i \in C$. The algorithm to estimate (p, d, q) is two stepped.

The first step determines d. This step is performed by using successive KPSS unit-root test [7]. This test looks for stationary data or seasonally differential data in a time series. The test is computed for each time series in the cluster. The null hypothesis is that the time series is stationary around a deterministic trend. The alternate hypothesis is that the time series is difference-stationary. Every time series is expressed as the sum of deterministic trend, random walk, and stationary error, and the test is the LM test of the hypothesis that random walk has zero variance. Initially, $d = 0$. If the test hypothesis is rejected for at least one time series in the cluster, we increment d and test the difference data of each time series for a new unit root; and so on. We stop the procedure when do not reject the test hypothesis for any time series in the cluster and output d.

The second step determines p and q. This step is performed by computing AICs for all time series in the cluster and using a stepwise search to traverse the model space of (p, q). For each time series in the cluster, we compute four initial ARIMA models by using $(0, d, 0)$, $(1, d, 0)$, $(0, d, 1)$ and $(2, d, 2)$ as suggested in [5]. For every possible combination of parameters, we determine the best initial parameter model via the AICc information criterion. We consider the lower average AICc for the time series in the cluster, the better the parameter model. The average AICc per cluster is computed as follows:

$$AICc(\theta, \mathcal{C}) = \frac{1}{\sharp\mathcal{C}} \sum_{k \in \mathcal{C}} L^*(Z(k), \theta) + \frac{2h(h+1)}{n-h-1}, \qquad (9)$$

where θ is the parameter pair (p, q), h is the number of parameters in θ; n is the length of the time series $Z(k)$ and $L^*(\cdot)$ is the maximum likelihood estimate of θ on the initial states $Z(k)$. The initial parameter model, that minimizes the average AICc among all the starting models is selected. Variations on the current parameter model are considered. This is done by varying p and/or q from the current model by ± 1. The best parameter model (with the lower average AICc) considered so far (either the current model or one of its variation) becomes the new current parameter model. This step is repeated until no better model (with lower average AICc) can be found.

Once (p, d, q) are globally determined for the cluster $\mathcal{C}$, the model coefficients are locally determined per each time series $Z(k_i)$ with $k_i \in \mathcal{C}$. We use p, d and q to fit a least square regression model to $Z(k_i)$ according to Equation 1. The local model coefficients $\phi(k_i)$ and $\sigma(k_i)$ are outputted and they allow us to produce forecasting for k_i as many steps ahead can be required.

5   Empirical Evaluation

We have implemented CArima as a function in R and evaluated it by considering several geosensor data collections. Our goal is to analyze clusters discovered by the algorithm, efficiency of learning process when leveraging the power of clustering to select the parameters of the ARIMA model and accuracy of forecasts.

5.1   Data Description

We considered six sets of geosensor data collected via four sensor networks. **Eco-Texas** network is used to collect measurements of Wind speed, Temperature and Ozone through 26 sensors installed in Texas (http://www.tceq.state.tx.us/). Time series data are measured hourly in the period May 5-19, 2009. Wind

speed data range in $[0.3, 29.5]$. Temperature data range in $[0, 89]$. Ozone data
range $[47.9, 104.5]$. We used the period May 5-18, 2009 for the training phase,
and May, 19 2009 for the testing phase. **MESA** network is used to collect measurements of air pollution through 20 sensors installed in California (`http://`
`depts.washington.edu/mesaair/`). Time series data are measured every two
weeks in the period January 13, 1999 - September 23, 2009. Air pollution data
range in $[0.0004, 34.0042]$. We used the period January 13, 1999 - April 8, 2009
for the training phase, and April 22, 2009 - September 23, 2009 for the testing phase. **EasternWind** network is used to collect measurements of the wind
speed through 1326 sensors installed in the eastern region of the United States
(`http://www.nrel.gov/`). Time series data are measured every 30 minutes, at
80 meters above sea level, in the period January 1-8,2004. Wind Speed data range
$[0.119, 30.414]$. We used the period January 1-7, 2004 for the training phase, and
January 8-9, 2004 for the testing phase. **AirClimateSouthAmerica** (SAC) network is used to collect measurements of the air temperature through 900 sensors
installed in South America (`http://climate.geog.udel.edu/~climate/html_`
`pages/archive.html`). Each time series collected monthly-measures in the period January 1999 until December 2010. Air temperature data range $[-7.6, 32.9]$.
We used the period January-1999 until December 2009 for the training phase,
and January 2010 until December 2010 for the testing phase.

5.2   Experimental Methodology

For each data collection, time series are split in training set and testing set. We
compare CArima to auto.ARIMA, as well as to its spatial competitor sARIMA.
We run the three algorithms on the training time series and use the learned
models to forecast the testing values. We use the non-parametric Wilcoxon two-
sample paired signed rank test [8], in order to compare predictive capabilities of
the learned models. We perform this test for the pair of test predictions generated
for each sensor by setting the significance level at 0.05.

5.3   Discussions of the Results

Table 1 collects errors of forecasts produced by CArima and auto.ARIMA. We
remember that auto.ARIMA completely neglects information of spatially close
time series in the learning phase. For each dataset, we report the number of
clusters (column 2, Table 1) and average Silhouette index (column 3, Table 1) of
clusters detected by CArima. The maps of the clusters detected for both EasternWind and SAC are depicted in Figure 2. Both maps show that the use of
a spatio-temporal distance measure allows us to group similar time series that
are measured at nearby locations. In this way, we are able to depict spatial
correlations of time series throughout clusters. In addition, we can observe that
knowledge enclosed in clusters can really optimize the choice of ARIMA parameters. This is confirmed by results on accuracy (columns 4-5, Table 1). In general,
CArima yields more accurate predictions in average than auto.ARIMA when
both spatial information and temporal knowledge is processed. The analysis of

Table 1. CArima vs auto.ARIMA: number of clusters (column 2), average Silhouette index (column 3), average rmse (columns 4-5), and results of pairwise signed rank test (columns 6-10). + (++) is the number of sensors where accuracy of CArima (statistically) outperforms accuracy of auto.ARIMA, - (−) is the number of sensors where auto.ARIMA (statistically) outperforms CArima, (=) is the number of sensors where the two functions perform statistically equally.

	Clusters	Silhouette	avgRMSE		Wilcoxon				
			CArima	auto.ARIMA	+	++	-	−	=
EcoTexas Wind Speed	5	0.19	**2.57**	2.70	5	8	5	6	2
EcoTexas Temperature	5	0.28	**7.23**	9.08	1	14	1	6	4
EcoTexas Ozone	4	0.27	**21.48**	22.85	1	8	0	10	7
MESA	6	0.24	**0.84**	0.99	3	6	5	5	1
EasternWind	10	0.23	**2.77**	2.91	173	546	190	417	0
SAC	30	0.16	**1.61**	4.99	45	612	57	174	12

Table 2. CArima vs sARIMA: average rmse (columns 2-3), and results of pairwise signed rank test (columns 4-8). + (++) is the number of sensors where accuracy of CARIMA (statistically) outperforms accuracy of sARIMA, - (−) is the number of sensors where sARIMA (statistically) outperforms CArima, (=) is the number of sensors where the two functions perform statistically equally.

	avgRMSE		Wilcoxon				
	CArima	sARIMA	+	++	-	−	=
EcoTexas Wind Speed	**2.57**	2.61	4	7	4	8	3
EcoTexas Temperature	**7.23**	7.72	0	25	0	1	0
EcoTexas Ozone	**21.48**	22.96	1	9	0	9	7
MESA	0.84	**0.76**	2	1	0	1	16
EasternWind	**2.77**	2.85	193	511	196	426	0
SAC	1.61	**1.59**	123	59	187	122	409

pairwise signed rank Wilcoxon tests provides a deeper insights in the accuracy of forecasts. We performed a test for each sensor, by comparing forecasts produced by the compared algorithms for the testing period. In all data sets, the number of sensors where error of CArima is (statistically) lower than error of Auto.ARIMA is greater than the number of sensors where error of Auto.ARIMA is (statistically) lower than error of CArima. This statistical analysis allows us to show the viability of our idea of leveraging the power of spatial correlation though cluster analysis, in order to enhance accuracy of ARIMA modeling.

Table 2 collects errors of forecasts produced by CArima and sARIMA. We remember that sARIMA can account for spatially close time series in the learning phase. However, this information is looked for in a sphere of fixed radius, without accounting for the similarity between time series. Although, sARIMA (column 3, Table 2) can yield more accurate forecasts than auto.ARIMA (column 5, Table 1), it is, in general, outperformed by CArima (columns 2-3, Table 2). The only exceptions are MESA and SAC. However, the statistical comparison reveals that the two algorithms (columns rows 4-8, Table 2) have comparable errors on

Table 3. CArima vs auto.ARIMA vs sARIMA: computation time (in seconds)

	Auto.ARIMA	sARIMA	CArima		
			STdist matrix	k-means	ARIMA
EcoTexas Wind Speed	8	11	2	1	6
EcoTexas Temperature	6	13	2	1	5
EcoTexas Ozone	7	15	1	1	5
MESA	10	12	2	1	7
EasternWind	1332	1837	1490	35	365
SAC	1104	1536	1337	65	236

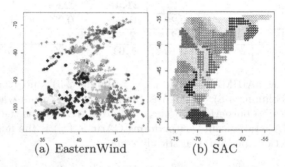

(a) EasternWind   (b) SAC

Fig. 2. Spatio-temporal clusters discovered by CARIMA

MESA, while sARIMA really outperforms CArima only on SAC. We ascribe this result to the structure of SAC. It contains interpolated data collected on a regular grid of sensors. So, we can expect that, the spherical regular neighborhood that sARIMA computes around each time series, well captures, in this dataset, the spatial autocorrelation information. In any case, CArima provides free insights in the cluster information, which are unavailable with its competitors.

Final considerations concern the analysis of computation times collected in Table 3. The most of the time spent for clustering is spent for the computation of the distance (column 4, Table 3). In any case, when clusters are detected, they speed up computation of ARIMA models (columns 2 and 6, Table 3).

6   Conclusions and Future Work

This paper describes a two-stepped algorithm that accounts for the spatio-temporal correlation of time series of geosensor data in forecasting analysis. The first step computes spatio-temporal clusters. The second step computes the ARIMA model whose parameters (p, d, q) are global to the clusters, while the models coefficients ϕ and σ are local to the sensor nodes. The empirical evaluation is performed, in order to investigate the effectiveness of proposed algorithm in several real world forecasting applications. We compare CArima forecasts against results of the baseline auto.ARIMA, as well as the spatial-aware sARIMA. For future work, we plan to investigate cluster analysis in multivariate time series and to extend our analysis to hybrid forecasting methods.

Acknowledgments. We would like to acknowledge the support of the European Commission through the project MAESTRA - Learning from Massive, Incompletely annotated, and Structured Data (Grant number ICT-2013-612944).

References

1. Andrienko, G., Andrienko, N.: Interactive cluster analysis of diverse types of spatiotemporal data. SIGKDD Explor. Newsl. 11(2), 19–28 (2010)
2. Birant, D., Kut, A.: St-dbscan: An algorithm for clustering spatial temporal data. Data and Knowledge Engineering 60(1), 208–221 (2007)
3. Box, G.E.P., Jenkins, G.M.: Time Series Analysis: Forecasting and Control, 3rd edn. Prentice-Hall (1994)
4. Brockwell, P., Davis, R.: Time Series: Theory and Methods, 2nd edn. Springer (2009)
5. Hyndman, R., Khandakar, Y.: Automatic time series forecasting: The forecast package for r. Journal of Statistical Software 26(3) (2008)
6. Kamarianakis, Y., Prastacos, P.: Space-time modeling of traffic flow. Comput. Geosci. 31(2), 119–133 (2005)
7. Kwiatkowski, D., Phillips, P.C., Schmidt, P., Shin, Y.: Testing the null hypothesis of stationarity against the alternative of a unit root. Journal of Econometrics (54), 159–178 (1992)
8. Orkin, R.D.M.: Vital Statistics. McGraw-Hill, New York (1990)
9. Pravilovic, S., Appice, A., Malerba, D.: An intelligent technique for forecasting spatially correlated time series. In: Baldoni, M., Baroglio, C., Boella, G., Micalizio, R. (eds.) AI*IA 2013. LNCS, vol. 8249, pp. 457–468. Springer, Heidelberg (2013)
10. Qin, K., Chen, Y., Zhan, Y., Cheng, F.: Spatial clustering considering spatio-temporal correlation. In: 2011 19th International Conference on Geoinformatics, pp. 1–4 (2011)
11. Rinzivillo, S., Pedreschi, D., Nanni, M., Giannotti, F., Andrienko, N., Andrienko, G.: Visually driven analysis of movement data by progressive clustering. Information Visualization 7(3), 225–239 (2008)
12. Rousseeuw, P.J.: Silhouettes: A graphical aid to the interpretation and validation of cluster analysis. Computational and Applied Mathematics 20, 53–65 (1987)
13. Sershenfeld, N.A., Weigend, A.S.G.: The future of time series. In: Gershenfeld, A.N., Weigen, A.S. (eds.) Time Series Prediction: Forecasting the Future and Understanding the Past, pp. 1–70 (1993)
14. Tobler, W.: A computer movie simulating urban growth in the Detroit region. Economic Geography 46(2), 234–240 (1970)
15. Trasarti, R.: Mastering the Spatio-Temporal Knowledge Discovery Process. PhD thesis. University of Pisa Department of Computer Science, Italy (2010)
16. Zhang, P., Huang, Y., Shekhar, S., Kumar, V.: Correlation analysis of spatial time series datasets: A filter-and-refine approach. In: Whang, K.-Y., Jeon, J., Shim, K., Srivastava, J. (eds.) PAKDD 2003. LNCS (LNAI), vol. 2637, pp. 532–544. Springer, Heidelberg (2003)

Unsupervised and Hybrid Approaches for On-line RFID Localization with Mixed Context Knowledge

Christoph Scholz, Martin Atzmueller, and Gerd Stumme

Knowledge & Data Engineering Group, University of Kassel, Germany
{scholz,atzmueller,stumme}@cs.uni-kassel.de

Abstract. Indoor localization of humans is still a complex problem, especially in resource-constrained environments, e. g., if there is only a small number of data available over time. We address this problem using active RFID technology and focus on room-level localization. We propose several unsupervised localization approaches and compare their accuracy to state-of-the art unsupervised and supervised localization methods. In addition, we combine unsupervised and supervised methods into a hybrid approach using different types of mixed context knowledge. We show, that the new unsupervised approaches significantly outperform state-of-the-art supervised methods, and that the hybrid approach performs best in our application setting. We analyze real world data collected at a two days evaluation of our working group management system MyGroup.

1 Introduction

For indoor positioning usually special hardware installations and extensive system configurations are required, since typical methods for outdoor positioning, e. g., the Global Positioning System (GPS), do not work. A typical problem is that signals are blocked by most construction materials. In this paper, we analyze indoor localization approaches (at room level basis) for humans. We utilize resource-aware and cost-effective active RFID technology consisting of a set of active RFID tags that send out signals which are received by a set of RFID readers who are located in different rooms. Using this technology is challenging, because only limited information is available for the localization task. In most scenarios where a positioning is required, unsupervised approaches seem appropriate: They do not require a special training period, are accordingly not dependent on a large number of training data, and can be instantly applied. Furthermore, new rooms can be simply added without any further learning phase. A straight forward unsupervised approach, for example, localizes a person to that room whose RFID readers receive the strongest signal strength from the person's RFID tag.

In this paper, we propose several unsupervised approaches that use background knowledge about the positions of the individual RFID readers. We analyze their quality with respect to the localization task, and compare these to supervised methods. Both, supervised and unsupervised methods make use of mixed context knowledge relating to the positioning of the utilized RFID readers as well as proximity (contact) data of the applied RFID tags. In particular, we study whether the knowledge about the RFID reader's positions used by the unsupervised approaches outweighs the training data information used by the supervised approaches. Furthermore, we propose a hybrid method that combines

T. Andreasen et al. (Eds.): ISMIS 2014, LNAI 8502, pp. 244–253, 2014.
© Springer International Publishing Switzerland 2014

unsupervised and supervised techniques. Our application context is given by our working group management system MyGroup [1]. In the MyGroup setting, localization at room-level is sufficient, for supporting group organization and everyday work processes. In our experiments, we apply real-world data collected using our working group management system. Our results indicate, that the new proposed unsupervised and hybrid methods significantly outperform state-of-the-art baseline approaches.

Our contribution can be summarised as follows:

1. We propose several unsupervised localization approaches based on a technique presented in [9] utilizing the positioning context of the individual readers. To the best of the authors' knowledge, this is the first such evaluation using real-world data.
2. In addition, we compare the accuracy of the unsupervised localization approaches to supervised localization methods, and show that the unsupervised techniques significantly outperform the supervised ones.
3. Furthermore, we introduce a novel unsupervised indoor localization method that uses proximity information as context knowledge to further improve the localization accuracy. We show that this unsupervised method outperforms the given state-of-the-art unsupervised baseline methods.
4. We finally present a new hybrid indoor localization method that combines unsupervised and supervised localization approaches making use of mixed context data. Overall, this novel method scores best in our experiments.

The rest of the paper is structured as follows: The next section discusses related work. After that, we describe the RFID setup and the collected RFID dataset used for the evaluation of our approach. We then present the novel localization approaches as well as the baseline approaches, before we perform an in-depth evaluation of these approaches. Finally, the paper concludes with a short summary.

2 Related Work

In the field of indoor localization several algorithms have been proposed, usually based on angle of arrival (AoA) [12], time of arrival (ToA) [8] or time difference of arrival (TDoA) [13] methodologies. On the one hand, these methods are highly accurate in user positioning, while on the other hand, they are rather costly and consume a lot of energy. Another group of localization algorithms uses the received signal strength to estimate the position of a person or object [7]. After calculating the distance of the object to at least three reference points the object's position can be estimated using triangulation. Another prominent approach that exists in the literature is known as fingerprint localization, e. g., [2, 4, 11].

In this paper, we also used fingerprints for the supervised indoor localization approach. Usually, this technique works in two phases, the off-line training phase and the online localization phase. In the off-line phase, data about the received signal strengths (RSS) for each point in the localization area is stored in a database to save the localization points (fingerprints). In the online phase the model then selects the location with the highest probability.

The RFID technology used in our experiments does not provide information (like ToA, TDoA, AoA, RSSI, etc.) that is normally used for positioning in traditional localization algorithms. Instead we use the number of packages the installed RFID readers receive from an RFID tag. This technique is based on an idea presented in [9]. Here the authors introduced an approach that uses the number of packages each RFID reader received from an RFID tag (in a specific time interval) to determine the person's position. In [9] the person is then allocated to the room whose RFID readers receive the most packages with the weakest signal strength. However, this approach has never been evaluated in a real world setting. In this work, we refine the localization approach discussed in [9] and propose several variants of this technique. Furthermore, we develop a novel unsupervised localization technique that uses proximity information as context knowledge to achieve a better localization accuracy. The idea to use proximity information of other proximity tags to improve the localization accuracy was first presented in [14] using a supervised approach. In contrast to the work of [14] in this paper we focus on the development and analysis of unsupervised localization methods that use proximity information to further improve the localization accuracy. The approaches presented in this paper aim at providing instant online localization services, such that no preceding training phase is required.

3   Positioning Data

In this section, we first describe the framework used for collecting the positioning data. We then describe the collected RFID dataset.

3.1   RFID Setup

In our application setting, we asked each participant to wear an active RFID tag. One decisive factor of these tags is the possibility to detect other active RFID tags within a range of up to 1.5 meters. The RFID tags have been developed by the SocioPatterns project (http://www.sociopatterns.org). In the following, we will refer to these active RFID tags as *proximity tags*. Each proximity tag sends out two types of RFID-signals, proximity signals and tracking signals. A proximity signal is used for contact sensing which is achieved by using signals with very low radio power levels [3]. The proximity tag sends out (every two seconds) one tracking signal in four different signals strengths (-18dbm, -12dbm, -6dbm, 0dbm) to RFID readers placed at fixed positions in the localization area. The tracking signals are used to transmit proximity information to a central server and to determine the position of each participant [9,14]. Depending on the signal strength, the range of a tracking signal inside a building is up to 25 meters. Each signal contains the signal strength and ID of the reporting tag and the IDs of all RFID tags in proximity. We here note that the used RFID readers are not able to measure the signal strength of one RFID signal. Therefore, we consider the sent signal strength (encoded in each signal) and not the received signal strength. For more information about the proximity tags we refer to Barrat et al [3] and the OpenBeacon website (http://www.openbeacon.org).

3.2   RFID Dataset

For our localization experiment we consider two kinds of tags: *user tags* and *object tags*: A user tag is a proximity tag worn by a participant during the evaluation study.

With an object tag we denote a proximity tag fixed to an unmovable object. In total, we fixed 21 object tags (in 12 rooms) on the monitor of each participant.

Ground truth: In summary, 28 people took part in our localization experiment. We collected their tag data over a duration of two days. To evaluate the accuracy of our algorithms we needed to determine the positions (rooms) of the participants, for which we applied the object tags. Since the tags detect other proximity tags only within a range of up to 1.5 meters, whenever a contact between a participant's tag and an object tag was recorded, we could infer that this participant was in the same room as the object tag (ground truth). In the experiments, we predicted the rooms for those locations where the precise location could be verified with the ground truth data.

4 Algorithms

In this section we first describe the unsupervised and supervised state-of-the art algorithms used for the localization of participants in indoor environments. We then present our new unsupervised and our new hybrid localization approach. We start by defining the *package count vector* for one proximity tag.

4.1 Package Count Vector

As in [14] we define a package count vector as follows: Given R RFID readers and P proximity tags, each transmitting on S different signal strengths, and l denotes the length of a time window and t a point in time. Then, let $V_r^l(p,t) \in \mathbb{N}^s$ ($1 \leq r \leq R$, $1 \leq p \leq P$) be an S-dimensional vector where the s-th entry is the number of packages that RFID reader r received from proximity tag p with signal strength s in the time interval $[t - l, t]$. The vector $V^l(p,t) = \left(V_1^l(p,t), V_2^l(p,t), \cdots, V_R^l(p,t)\right)$, i.e., the concatenation of the vectors $V_r^l(p,t)$ over all readers r, is called the *package count vector* of the proximity tag p at time t. The dimension of vector $V^l(p,t)$ is $S \cdot R$. The parameter l controls the influence of older signals, i.e., for longer intervals the probability increases that packages of a previous location influence the vector at the current point of time t.

4.2 Unsupervised Room-Level Localization

For our experiment we placed RFID readers in each room of the localization area. In [9] the authors presented an unsupervised method where the person is allocated to the room whose RFID readers receive the most packages with the weakest signal strength from the person's RFID tag. In our experiments, we analyze the following unsupervised localization strategies.

1. *MaxP:* In this localization strategy the person is allocated to the room, in which one of the RFID readers receives the most packages (independent of their individual signal strengths). If the resulting room-IDs are not unique, then we choose one room at random from the obtained set of IDs.
2. *MaxP.Sum:* In this localization strategy the person is allocated to the room, all of whose RFID readers receive the most packages (independent of their individual

signal strengths). This means that (in contrast to the *MaxP*-method) we add up all signals from all RFID-readers belonging to one room. If the resulting room-IDs are not unique, then we choose one room-ID at random from this resulting set.

3. *MaxPMinSS:* In this localization strategy the person is allocated to the room, in which one of the RFID readers receives the most packages with the weakest signal strength. If the resulting room-IDs (say R) are not unique, then after considering the i-th signal strength we continue with the algorithm restricted to the subset R (of all rooms-IDs) with signal strength $i + 1$.

4. *MaxPMinSS.Sum:* In this localization strategy the person is allocated to the room, all of whose RFID readers receive the most packages with the weakest signal strength. This means that (in contrast to the *MaxPMinSS*-method) we add up (for each signal strength separately) all signals from all RFID-readers belonging to one room. If the resulting room-IDs (say R) are not unique after considering the i-th signal strength, then we continue with the algorithm restricted to the subset R (of all rooms-IDs) with signal strength $i + 1$.

4.3 Supervised Room-Level Localization

In Section 5.2, we will analyze whether a supervised localization approach can help to further improve the localization accuracy. For this purpose we create a set of *package count vectors* (training data) for each room, and learn a classification model based on these vectors. In the online classification (localization phase) we determine the position of a participant from his current *package count vector*, using the classification model. As machine learning method we used the Random Forest classifier (RF) [5], Decision Trees (DT) [6] and the K-Nearest Neighbor (KNN) [10] classifier.

4.4 Advanced Unsupervised Room-Level Localization Using Proximity Data

Below, we describe a new approach which uses proximity information to improve the accuracy of the unsupervised localization algorithms. Let $C(p, t)$ denote the set of users (proximity tags) that were in contact with user p at time t. Assume, that we want to predict the position (room) of user p at time t. As input for the unsupervised algorithms we use the following vector:

$$V_+^l(p, t) = V^l(p, t) + \sum_{q \in C(p,t)} V^l(q, t). \tag{1}$$

Thus, the new vector $V_+^l(p, t)$ of user p is the sum over all package count vectors of the contacts of user p and of user p himself.

4.5 Hybrid Room-Level Localization Approach

We also study whether a combination of supervised and unsupervised localization approaches can further improve the localization quality. Hence, we present a novel approach that combines supervised and unsupervised localization approaches, called

Hybrid Indoor Localization Approach (HILA): Assume that we want to predict the location for person p. In this localization strategy we first determine the position for person p and all current contacts of person p using an unsupervised localization approach. If the majority vote (say room m) of all predictions is unique, then we allocate person p to room m. If the majority vote is not unique, then we use a supervised machine learning approach and place the person in the room where the prediction of the supervised approach has the highest confidence.

5 Evaluation

In our experiments, we evaluate the accuracy of unsupervised and supervised localization strategies, and compare the new unsupervised localization methods with some baseline methods. We start with an analysis of the different signal strengths.

5.1 Unsupervised Room-Level Localization

In this section we evaluate the defined unsupervised localization methods, cf. Section 4.2. In our experiments each participant wore a proximity tag. This tag sends out tracking signals in four different signal strengths. With the parameter l (see definition of a package count vector) we can control the influence of older signals. Since shorter values of l are more suitable to recognise more frequent changes of location. In our experiments we choose the parameter $l = 4$ (seconds). A proximity tag sends out one package in four different signals strengths every two seconds. This means that every RFID reader receives at most 8 signals (at most two signals per signal strength) from one RFID tag within the last 4 seconds. For the experiments we placed two RFID readers in each room. In the following, we first study the localization accuracy for each individual signal strength. Here we allocate a person to the room whose RFID reader receives the most packages from the concerning signal strength (like *MaxP*, but for only one signal strength). In Figure 1(a) we see that the stronger the signal strength the weaker the localization accuracy. However, this could be expected because stronger signals also reach more distant RFID readers with less package loss. This leads to a higher complexity in positioning. In Figure 1(b) we analyze the probability that, for a given signal strength s, one rooms's RFID reader receives a signal with signal strength s from an RFID tag located in the same room as the RFID reader. As expected, we observe that the probability is increased that RFID readers receive stronger signals.

In Figure 2 (see Table 5.1 for the rooms' individual localization results) we analyze the different unsupervised localization strategies. First, we note that a second reader significantly increases the localization accuracy. Furthermore we observe that the summation of all package count vectors received by RFID readers belonging to one room leads to a significantly higher positioning accuracy. However, it seems that there exists no significant difference between the *MaxP*- and the *MaxPMinSS*-approach. This means that the focus on weaker signals does not significantly improve the localization accuracy. However, when we add up all package count vectors from all readers belonging to one room we observe that the *MaxPMinSS.Sum*-approach works significantly better than the *MaxP.Sum*-approach. We also analyze the performance of an unsupervised approach that determines the room of a person based on the majority vote of all

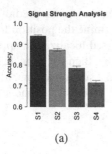
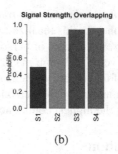

Fig. 1. (a) Accuracy (of positioning) using just one signal strength for the determination of current position. S1 here means signal strength 1, S2 signal strength 2, ... (b) The y-axis shows the probability that one room's RFID reader receives a signal (with signal s) from a person's RFID tag, where the person is in the same room as the RFID reader.

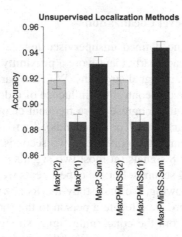

Fig. 2. Accuracy results for the unsupervised localization methods. MaxP(N), $N \in \{1, 2\}$, here means that we use exactly N RFID reader(s) in each room. In Table 5.1 we also give an overview of the rooms' individual localization results.

unsupervised localization approaches. We observe that this approach (93.81%) works marginally worse than the *MaxPMinSS.Sum*-approach (94.4%).

5.2 Comparing Unsupervised and Supervised Room-Level Localization

In this section we compare the accuracy of unsupervised and supervised localization methods. For this, we divide our dataset in training data and test data. We use the first three hours of the two days localization experiment as training data and the remaining 8.5 hours as test data. For the evaluation process we start by analysing different combinations of the two input parameters $mtry$ (denoting the number of predictors sampled for splitting at each node) and $ntree$ (the number of trees) for Random Forest (RF). Furthermore, we determine the best input parameter k (number of neighbors considered for each prediction) for k-nearest neighbor (KNN). For our experiments we

Table 1. Localization accuracy for each individual room

	MaxP(1)	MaxP(2)	MaxP.Sum	MaxPMinSS(1)	MaxPMinSS(2)	MaxPMinSS.Sum	RF	KNN	DT
0400	99.25	99.71	99.71	99.24	99.71	99.71	**99.82**	99.73	97.71
1170	65.09	51.34	79.17	65.00	51.26	64.23	95.60	**95.81**	54.16
1180	87.84	90.16	96.70	87.92	90.16	**96.88**	88.91	94.18	76.70
2150	69.60	86.89	59.44	69.47	**87.03**	83.79	71.92	60.58	53.85
2180	81.14	93.60	95.59	81.31	93.86	**95.70**	73.01	49.82	36.60
2190	89.01	93.35	95.06	89.20	95.44	**96.03**	93.68	91.14	93.62
2270	99.19	**99.58**	99.54	99.01	99.49	99.41	98.99	98.45	95.79
2290	100.00	100.00	100.00	100.00	100.00	100.00	99.97	99.90	86.37
3160	76.92	86.98	93.57	76.98	86.90	**98.24**	83.54	95.99	60.45
3190	99.08	99.97	100.00	99.07	99.97	100.00	99.90	98.86	78.40
3250	99.09	100.00	100.00	99.08	100.00	100.00	97.60	97.10	77.01
3320	96.85	98.49	98.50	96.86	**98.51**	**98.51**	98.30	97.51	71.44
∅	88.58	91.67	93.19	88.60	91.86	**94.38**	91.77	89.92	73.51

use the R-implementation of Random Forest and KNN. In order to estimate the possible potential of the supervised approaches, we aim to optimize the model parameters for the best possible case. Thus, we applied the test data to find optimal parameter combinations for the supervised approaches. As we will see below, also using these parameter combinations does not perform better than the unsupervised approaches. We found that Random Forest works best with $mtry = 6$ and $ntree = 400$, the KNN classifier works best with $k = 13$. Furthermore, we tested the dependence between training size and localization accuracy. In our experiments, a training size of 600 samples (sampled out of the 3 hours training data) performs very well. We use these parameter combinations to compare the performance of supervised and unsupervised approaches. We learn a classification model (on the training data) based on the package count vectors observed in the first three hours of the experiment. We then use the test data to analyze and compare the performance of supervised and unsupervised localization approaches. Figure 3(a) compares the accuracy of unsupervised and supervised approaches. For the supervised case, Random Forrest performs best. In addition, the unsupervised *MaxPMinSS.Sum*-method outperforms the supervised methods. In general we observe here that the knowledge about the RFID readers' location put into the unsupervised approaches leads to an outperformance of the supervised approaches. In addition the results indicate the weakness of supervised approaches, because due to external influences (number of persons in room, other placement of furnitures, etc.) the characteristic vectors (used to train the model) could change over time. In Table 5.1, it is interesting to see that in room 2290 the unsupervised methods perform accurately, whereas the supervised methods have (minimal) problems. Due to fact that room 2290 was separated from the other rooms we conclude that the unsupervised methods benefit from this separation.

5.3   Advanced Unsupervised Room-Level Localization Using Proximity Data

In order to show that proximity data helps to further improve the localization accuracy in an unsupervised setting, we compare our new unsupervised approaches to the unsupervised approaches (baseline approaches) presented before. For the analysis, we collected all package count vectors where we could detect a face-to-face proximity contact between participants of the experiment. We note here that the proximity contacts between a person and an object tag is just used for the ground truth. In Figure 3(b), we observe that our approach significantly outperforms the respective baseline approaches. The highest increase in accuracy could be observed when we used just one reader per room. Surprisingly also the MaxPMinSS.Sum approach works much better when we use our new approach. We note here that we just used rooms where an adequate number of proximity contacts (between persons) could be detected. Unfortunately, not in all rooms enough proximity contacts could be observed. Therefore, the baseline accuracy results of Figure 2 and Figure 3(b) are different.

5.4   Hybrid Room-Level Localization

For evaluating our new hybrid room-level localization approach (HILA), we used the MaxPMinSS.Sum-method as unsupervised approach and the RF-classifier as supervised approach. We choose these two approaches, because the MaxPMinSS.Sum-method performed best in the unsupervised case and the Random Forrest in the supervised case. We note here that we use the default parameters for the Random Forest classifier (as set in the R-library). Figure 3(b), we observe that this approach scores best in our experiments. Using the RF-classifier as decision support here improves the accuracy from 94.57% (MaxPMinSS.Sum approach) to 95.57% (HILA approach). In addition, using the optimized parameter combinations for Random Forest (cf., Section 5.2) further improves the accuracy to 96.31%.

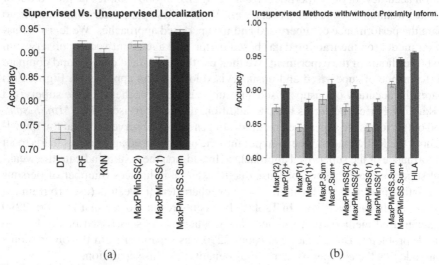

(a)   (b)

Fig. 3. a) Accuracy results for unsupervised and supervised localization methods. b) Comparison between the unsupervised approaches with and without proximity data. The + indicates that we included proximity data to the corresponding baseline approach.

6   Conclusions

In this paper, we proposed and analyzed several unsupervised localization approaches. We showed, that a second RFID reader in each room significantly increases the prediction accuracy. The localization approaches, where the package count vectors of all

RFID readers belonging to one room are added up, perform best. Furthermore, we studied unsupervised localization strategies that allocate a person to the room, whose RFID readers receive the most packages with the weakest signal strength. This technique performs significantly better when the package count vectors of all rooms RFID readers are added up. In addition, we compared the localization accuracy between unsupervised and supervised localization methods. We observed that unsupervised methods perform better than supervised localization strategies. We argue, that when we put the knowledge about the RFID readers' location into the unsupervised approaches they outperform the supervised approaches. Furthermore we presented two new approaches that use proximity information to further improve the localization accuracy. We showed that our new unsupervised localization methods that use proximity information significantly outperform the baseline methods. Furthermore, the novel HILA approach scored best.

References

1. Atzmueller, M., Becker, M., Kibanov, M., Scholz, C., Doerfel, S., Hotho, A., Macek, B.E., Mitzlaff, F., Mueller, J., Stumme, G.: Ubicon and its Applications for Ubiquitous Social Computing. New Review of Hypermedia and Multimedia 20(1), 53–77 (2014)
2. Bahl, P., Padmanabhan, V.N.: RADAR: An In-Building RF-Based User Location and Tracking System. In: Proceedings IEEE INFOCOM 2000, pp. 775–784 (2000)
3. Barrat, A., Cattuto, C., Colizza, V., Pinton, J.F., den Broeck, W.V., Vespignani, A.: High Resolution Dynamical Mapping of Social Interactions with Active RFID. CoRR (2008)
4. Bekkali, A., Sanson, H., Matsumoto, M.: RFID Indoor Positioning Based on Probabilistic RFID Map and Kalman Filtering. In: Third IEEE International Conference on Wireless and Mobile Computing, Networking and Communications, WiMob 2007, p. 21 (2007)
5. Breiman, L.: Random forests. Machine Learning 45(1), 5–32 (2001)
6. Breiman, L., Friedman, J.H., Olshen, R.A., Stone, C.J.: Classification and Regression Trees. Chapman & Hall, New York (1984)
7. Hightower, J., Vakili, C., Borriello, G., Want, R.: Design and Calibration of the SpotON Ad-Hoc Location Sensing System. Tech. rep. (2001)
8. Li, X., Pahlavan, K.: Super-Resolution TOA Estimation with Diversity for Indoor Geolocation. IEEE Transactions on Wireless Communications 3(1), 224–234 (2004)
9. Meriac, M., Fiedler, A., Hohendorf, A., Reinhardt, J., Starostik, M., Mohnke, J.: Localization Techniques for a Mobile Museum Information System. In: Proceedings of WCI (2007)
10. Mitchell, T.: Machine Learning, 1st edn. McGraw-Hill Education, ISE Editions (October 1997)
11. Ni, L.M., Liu, Y., Lau, Y.C., Patil, A.P.: LANDMARC: Indoor Location Sensing Using Active RFID. Wireless Networks 10(6), 701–710 (2004)
12. Niculescu, D., Badrinath, B.R.: Ad Hoc Positioning System (APS) Using AOA. In: Proceedings IEEE INFOCOM (2003)
13. Priyantha, N.B., Chakraborty, A., Balakrishnan, H.: The Cricket Location-Support System. In: Proceedings of the Sixth Annual International Conference on Mobile Computing and Networking (MOBICOM 2000), pp. 32–43 (2000)
14. Scholz, C., Doerfel, S., Atzmueller, M., Hotho, A., Stumme, G.: Resource-Aware On-Line RFID Localization Using Proximity Data. In: Gunopulos, D., Hofmann, T., Malerba, D., Vazirgiannis, M. (eds.) ECML PKDD 2011, Part III. LNCS (LNAI), vol. 6913, pp. 129–144. Springer, Heidelberg (2011)

Mining Surgical Meta-actions Effects with Variable Diagnoses' Number

Hakim Touati[1], Zbigniew W. Raś[1,2],
James Studnicki[3], and Alicja A. Wieczorkowska[4]

[1] Univ. of North Carolina, College of Comp. and Informatics, Charlotte, NC 28223
[2] Warsaw Univ. of Technology, Inst. of Computer Science, 00-665 Warsaw, Poland
[3] Univ. of North Carolina, College of Health and Human Serv., Charlotte, NC 28223
[4] Polish-Japanese Institute of Information Technology, 02-008 Warsaw, Poland
{htouati,ras,jstudnic,awieczor}@uncc.edu

Abstract. Commonly, information systems are organized by the use of tables that are composed of a fixed number of columns representing the information system's attributes. However, in a typical hospital scenario, patients may have a variable number of diagnoses and this data is recorded in the patients' medical records in a random order. Treatments are prescribed based on these diagnoses, which makes it harder to mine meta-actions from healthcare datasets. In such scenario, the patients are not necessarily followed for a specific disease, but are treated for what they are diagnosed for. This makes it even more complex to prescribe personalized treatments since patients react differently to treatments based on their state (diagnoses). In this work, we present a method to extract personalized meta-actions from surgical datasets with variable number of diagnoses. We used the Florida State Inpatient Databases (SID), which is a part of the Healthcare Cost and Utilization Project (HCUP) [1] to demonstrate how to extract meta-actions and evaluate them.

Keywords: Meta-actions, Actionable rules, Surgical treatments.

1 Introduction

Meta-actions are a higher level concept used to model a generalization of action rules [2]. They are actions taken by deciders to trigger transitions in some flexible attributes' values. Those transitions will eventually cascade to a change in values of some decision feature to model an action rule. In other words, action rules are specialized meta-actions that are associated with a decision attribute value transition. Meta-actions are commonly used to acquire knowledge about possible transitions in the information system and their causes. This knowledge is used to trigger action rules.

Meta-actions are commonly used in the healthcare, business, and social media domains. In the healthcare arena, meta-actions represent treatments prescribed by doctors to their patients. In this paper, we are mining meta-actions' effects to discover surgical treatment effects on patients.

T. Andreasen et al. (Eds.): ISMIS 2014, LNAI 8502, pp. 254–263, 2014.
© Springer International Publishing Switzerland 2014

Meta-actions were first introduced in [3] as role models to mine actionable patterns, then formally defined by Raś et. al and used to discover action rules based on tree classifiers in [4]. They were also used to personalize action rules based on patients side effects in [5]. In these papers, the authors assumed that meta-action effects and their side effects are known. In addition, meta-actions were mined in [6] for action rules personalization and reduction based on a utility function; however, they were mined from traditional information systems with a fixed number of attributes. There are multiple techniques to mine action rules [7,8,9] and actionable patterns [10]; however, mining meta-actions directly from information systems with variable number of attributes was not studied in earlier work. In this paper, we present a meta-actions mining technique for datasets with variable number of attributes, and we apply this work for non-traditionally structured (varied number of attributes for objects in the system) healthcare dataset models.

2 Preliminaries

Data is commonly represented statically in an information system. In this section, we will present the static data model and define the action model.

2.1 Static Representation

In this section, we give a brief description of how static data is represented and stored in information systems. We also describe the state of an object in the context of information systems.

Definition 1 (Information System). *By information system [11] we mean a triple of the form $S = (X, A, V)$ where:*

1. *X is a nonempty, finite set of objects.*
2. *A is a nonempty, finite set of attributes of the form $a : X \to 2^{V_a}$, which is a function for any $a \in A$, where V_a is called the domain of a.*
3. *V is a finite set of attribute values such as: $V = \bigcup \{V_a : a \in A\}$.*

If $a(x)$ is a singleton set, then $a(x)$ is written without parentheses (for instance, $\{v\}$ will be replaced by v). Table 1 represents an information system S with a set of objects $X = \{x_1, x_2, x_3, x_4, x_5\}$, a set of attributes $A = \{a, b, c, d\}$, and a set of attribute values $V = \{a_1, a_2, b_1, b_2, b_3, c_1, c_2, c_3, d_1, d_2\}$.

In practice, data is not commonly well organized, and information systems may not only have multivalued attributes but also missing data and/or variable number of attributes. To simplify the concept of objects with variable number of attributes and attributes with several values, we introduce the notion of object state and we define it as follows:

Definition 2 (Object State). *An object state of $x \in X$ is defined by the set of attributes A_x that the object x is characterized by, and their respective values $A(x) = \bigcup \{a(x) : a \in A_x\}$*

Table 1. Information System Example

	a	b	c	d
x_1	a_1	b_2	c_2	d_1
x_2	a_2	b_2	c_2	d_2
x_3	a_2	b_1	c_3	d_1
x_4	a_1	b_3	c_1	d_2
x_5	a_2	b_1	c_1	d_1

2.2 Actions Representation

In this section, we will define a few concepts that will help us represent actionable data with regards to an information system. Those concepts will be used in the following section to define meta-action effects and their extraction.

Definition 3 (Stable Attributes). *Stable attributes are object properties that we do not have control over in the context of an information system. In other words, actions recommending changes of these attributes will fail. For example, a birth date is a stable attribute.*

This type of attribute is not used to model actions since their values do not change. They are commonly used to cluster the dataset.

Definition 4 (Flexible Attributes). *Flexible attributes are object properties that can transition from one value to another triggering a change in the object state. For instance, salary and benefits are flexible attributes since they can change values.*

Flexible attributes are the only possible attributes that can inform us about the possible changes an object may go through. However, to model possible actions, values transition of attribute, we need another concept which is defined as:

Definition 5 (Atomic Action Terms). *Atomic action term, also called elementary action term in S, is an expression that defines a change of state for a distinct attribute in S.*

For example, $(a, v_1 \rightarrow v_2)$ is an atomic action term which defines a change of value for the attribute a in A from v_1 to v_2, where $v_1, v_2 \in V_a$. In the case when there is no change, we omit the right arrow sign, so for example, (a, v_1) means that the value of attribute a in A remains v_1, where $v_1 \in V_a$. We use atomic action terms to model a single attribute value transition; however, to model transitions for several attributes, we use Action Terms defined as:

Definition 6 (Action Terms). *Action terms are defined as the smallest collection of expressions for an information system S such that:*

- *If t is an atomic action term in S, then t is an action term in S.*
- *If t_1, t_2 are action terms in S and $\wedge$ is a 2-argument functor called composition, then $t_1 \wedge t_2$ is a candidate action term in S.*
- *If t is a candidate action term in S and for any two atomic action terms $(a, v_1 \rightarrow v_2), (b, w_1 \rightarrow w_2)$ contained in t we have $a \neq b$, then t is an action term in S.*

3   Meta-actions

In order to move objects from their current population state to a more desirable population state, deciders need to acquire knowledge on how to perform the necessary changes in objects' state. For instance, moving a patient from the sick population state to the healthy population state requires the practitioner to use a treatment such as a surgery. This actionable knowledge is represented by meta-actions that are defined as follows:

Definition 7 (Meta-actions). *Meta-actions associated with an information system S are defined as higher level concepts used to model certain generalizations of actions rules [3]. Meta-actions, when executed, trigger changes in values of some flexible attributes in S.*

Let us define $\mathbf{M}(S)$ as a set of meta-actions associated with an information system S. Let $a \in A$, $x \in X$, and $M \subset \mathbf{M}(S)$, then, applying the meta-actions in the set M on an object x will result in $M(a(x)) = a(y)$, where object x is converted to object y by applying all meta-actions in M to x. Similarly, $M(A(x)) = A(y)$, where $A(y) = \{a(y) : a \in A\}$ for $y \in X$, and object x is converted to object y by applying all meta-actions in M to x for all $a \in A$.

The changes in flexible attributes, triggered by meta-actions, are commonly represented by action terms for the respective attributes, and reported by an influence matrix presented in [3]. However, when an information system contains multivalued attributes where each attribute takes a set of values at any given object state and transitions to another set of values in a different object state, it is better to represent the transitions between the attribute initial set of values and another set of values by action sets that are defined as:

Definition 8 (Action Set). *An action set in an information system S is an expression that defines a change of state for a distinct attribute that takes several values (multivalued attribute) at any object state.*

For example, $\{a_1, a_2, a_3\} \rightarrow \{a_1, a_4\}$ is an action set that defines a change of values for attribute $a \in A$ from the set $\{a_1, a_2, a_3\}$ to the set $\{a_1, a_4\}$ where $\{a_1, a_2, a_3, a_4\} \subseteq V_a$. Action sets are used to model meta-action effects for information systems with multivalued attributes. In addition, action sets' usefulness is best captured by the set intersection that models neutral action sets and set minus that models positive action sets between the two states involved. In the previous example, neutral and positive action sets are respectively computed as follows: $\{a_1, a_2, a_3\} \rightarrow [\{a_1, a_2, a_3\} \cap \{a_1, a_4\}]$ and $\{a_1, a_2, a_3\} \rightarrow [\{a_1, a_2, a_3\} \setminus \{a_1, a_4\}]$.

In this paper, we are studying surgical meta-action effects that trigger a change in the patients' state. The patients are in an initial state where the meta-actions are applied and move to a new posterior state. We use the set minus (positive action set) between two patients' states to observe the diagnoses that disappeared as a positive effect of applying meta-actions in the initial state.

Furthermore, we use set intersection (neutral action set) to observe the diagnoses that remain the same; in other words, meta-actions applied had a neutral effect on these diagnoses. This type of information concerning meta-actions is represented by an ontology (personalized) [12]. For instance, the example shown in Figure 1 models a meta-action composed of positive action sets which are labeled *positive* and neutral action sets which are labeled *neutral*. In addition, *positive* and *neutral* are composed of action sets respectively labeled As_n and $\overline{As_n}$, which in turn are composed of diagnoses labeled Dx_n.

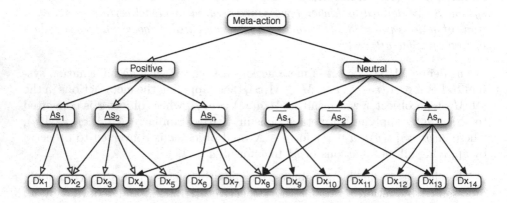

Fig. 1. Ontology Representation of a Meta-action

4   Meta-action Extraction

Meta-actions effects in the context of healthcare represent the patient's state transition from an initial state to a different state. Those effects are mined from large datasets for each patient separately then merged together based on their common subsets to form state transition patterns. In other words, each patient's state is extracted from a uniquely identified transaction and patients' visit transactions are clustered by the patients' identifier. In this paper, each state transaction for a patient represents a doctor consultation (patient visit to the doctor).

For each patient cluster, each transactions should be ordered based on temporal sequential order. Every two consecutive patient's transactions will be paired for every meta-action based on a temporal precedence relationship. The resulting pairwise partition will model the effects of the meta-actions taken.

Given real life data representation in an information system, we defined two methods to extract meta-actions effects. The first methods was defined in [6] and is used to extract meta-actions effects from traditional informations systems. In this method, since each attribute has a different meaning and a single value at any given object state in the information system, meta-actions effects are represented with action terms and saved in an influence matrix to be used by practitioners. The second method is defined in this section and used to extract meta-actions effects from information systems with variable number of attributes and multivalued attributes. This method is best suited for the surgical meta-actions mining problem since patients are diagnosed with several diagnoses at any consultation and have a different number of diagnoses.

4.1   Extracting Meta-actions with Variable Number of Attributes

Let us assume that $\mathbf{M}(S)$, where $S = (X, A, V)$, is a set of meta-actions associated with an information system S. In addition, we define the set $T = \{v_{i,j} : j \in J_i, x_i \in X\}$ of ordered transactions, patient visits, such that $v_{i,j} = [(x_i, A(x_i)_j)]$. The set $A(x_i)_j$ is defined as the set of attribute values $\{a(x_i) : a \in A\}$ of the object x_i for the visit uniquely represented by the visit identifier j. Each visit represents the current state of the object (patient) when recorded with respect to a temporal order based on j for all $v_{i,j} \in T$. For any particular visit, the patient state is characterized by a set of diagnoses. Each diagnosis is seen as an attribute, and each visit may have a different number of diagnoses.

For each patient's two consecutive visits $(v_{i,j}, v_{i,j+1})$, where meta-actions were applied at visit j, we can extract an action set. Let us define the set $P(S)$ of patient's two consecutive visits as $P(S) = \{(v_{i,j}, v_{i,j+1}) : x_i \in X, j \in J_i\}$. The corresponding action sets are: $\{(A(x_i)_j \rightarrow A(x_i)_{j+1}) : x_i \in X, j \in J_i\}$. We also define neutral action sets noted as $\overline{AS}$, and positive action sets noted as $\underline{AS}$. These action sets are: $\{(A(x_i)_j \rightarrow (A(x_i)_j \cap A(x_i)_{j+1})) : x_i \in X, j \in J_i\}$ and $\{(A(x_i)_j \rightarrow (A(x_i)_j \setminus A(x_i)_{j+1})) : x_i \in X, j \in J_i\}$ correspondingly, where $A(x_i)_j$ represents the set of diagnoses for a patient x_i at visit j.

The action sets resulting from the application of meta-actions represent the actionable knowledge needed by practitioners. However, patients do not have the same preconditions and do not react similarly to the same meta-actions. In other words, some patients might be partially affected by the meta-actions and might have other side effects not intended by the practitioners. For this reason, we need to extract the historical patterns in action sets. Let us assume that $\overline{as}_{i,j} = A(x_i)_j \cap A(x_i)_{j+1}$ and $\underline{as}_{i,j} = A(x_i)_j \setminus A(x_i)_{j+1}$, for any $x_i \in X$ and $j \in J_i$. Now, we define some properties for both the neutral and positive action sets extracted as follows:

1. $(\forall W)[W \subset \overline{as} \Rightarrow \overline{W} \in \overline{AS}]$
2. $(\forall W)[W \subset \underline{as} \Rightarrow \underline{W} \in \underline{AS}]$
3. $(\forall x_i \in X)(\forall j \in J_i)[\overline{as}_{i,j} \cup \underline{as}_{i,j} \subseteq A(x_i)_j]$
4. $(\forall x_i \in X)(\forall j \in J_i)[\overline{as}_{i,j} \cap \underline{as}_{i,j} = \emptyset]$

From the property number 1 and 2, given that any subset of an action set is an action set of the same meta-action, we can extract all action sets present in any pair of patient's visits using power sets. Let us define $\overline{P_{i,j}}$ as the power set of neutral action set $\overline{as}_{i,j}$ such that $\overline{P_{i,j}} \in \overline{AS}$. Similarly, we can define the set $\underline{P_{i,j}}$ as power set of positive action set $\underline{as}_{i,j}$ such that $\underline{P_{i,j}} \in \underline{AS}$. Hence, we can have all possible action sets composing a meta-action using power sets.

4.2   Meta-actions and Action Set Evaluation

To evaluate these actions set patterns, we need to compute their frequency of occurrence for all patients. A good measure of frequency is the support and it is seen here as the likelihood of the occurrence for a specific action set (set of diagnoses disappearing or remaining). The likelihood $Like(\overline{as})$ of a neutral action set $\overline{as}$ is defined as follows:

$$Like(\overline{as}) = card(\{(v_{i,j}, v_{i,j+1}) \in P(S) : \overline{as} \in \overline{P_{i,j}}\}) \qquad (1)$$

The likelihood $Like(\underline{as})$ of a positive action set $\underline{as}$ is defined as follows:

$$Like(\underline{as}) = card(\{(v_{i,j}, v_{i,j+1}) \in P(S) : \underline{as} \in \underline{P_{i,j}}\}) \qquad (2)$$

The likelihood support of action sets measures the likelihood of attributes being affected by the meta-actions applied, but it does not give a sense of how confident is the action set. A more sophisticated way to evaluate action sets is by computing their likelihood confidence. The intuition behind the action set confidence lies in the normalization of the action set with regards to the patient's precondition. The likelihood confidence of a neutral action set $\overline{as}$ is computed as follows:

$$ActionConf(\overline{as}) = \frac{Like(\overline{as})}{card(\{v_{i,j} : \overline{as} \subseteq A(x_i)_j, \ \forall x_i \in X\})} \qquad (3)$$

The likelihood confidence of a positive action set $\underline{as}$ is computed as follows:

$$ActionConf(\underline{as}) = \frac{Like(\underline{as})}{card(\{v_{i,j} : \underline{as} \subseteq A(x_i)_j, \ \forall x_i \in X\})} \qquad (4)$$

Depending on the objects' states, some of the action sets in AS may not be triggered by meta-actions. To be more precise, for a given meta-action m, only objects $x_l \in X$ that satisfy the following condition will be affected:

$$(\exists(v_{i,j}, v_{i,j+1}) \in P(S))(\exists v_{l,k} \in T)[A(x_l)_k \cap A(x_i)_j \neq \emptyset]$$

Given the action sets composing a meta-action m, we can define the global confidence of m as the weighted sum of its action sets likelihood confidences where the weights represent action sets likelihood support. The intuition behind the meta-action confidence is in defining how efficient is the application of a meta-action for any patient's precondition. The meta-action confidence $MetaConf(m)$ is computed for both neutral and positive action sets as follows:

$$MetaConf(m) = \frac{\sum_{i=1}^{n} Like(as_i) \cdot ActionConf(as_i)}{\sum_{i=1}^{n} Like(as_i)} \qquad (5)$$

where n is the number of action sets in m.

5   Experiments

5.1   Dataset Description

In this paper, we used the Florida State Inpatient Databases (SID) that is part of the Healthcare Cost and Utilization Project (HCUP) [1]. The Florida SID dataset contains records from several hospitals in the Florida State. It contains over 2.5 million visit discharges from over 1.5 million patients. The dataset is composed of five tables, namely: AHAL, CHGH, GRPS, SEVERITY, and CORE. The main table used in this paper is the *Core* table. The *Core* table contains over 280 attributes; however, many of those attributes are repeated with different codification schemes. In the following experiments, we used the Clinical Classifications Software (CCS) that consists of over 260 diagnosis categories, and 231 procedure categories. In our experiments, we used fewer attributes that are described in this section. Each record in the Core table represents a visit discharge. A patient may have several visits in the table. One of the most important attributes of this table is the *VisitLink* attribute, which describes the patient's ID. Another important attribute is the *Key*, which is the primary key of the table that identifies unique visits for the patients and links to the other tables. As mentioned earlier, a *VisitLink* might map to multiple *Key* in the database. This table reports up to 31 diagnoses per discharge as it has 31 diagnosis columns. However, patients' diagnoses are stored in a random order in this table. For example, if a particular patient visits the hospital twice with heart failure, the first visit discharge may report a heart failure diagnosis at diagnosis column number 10, and the second visit discharge may report a heart failure diagnosis at diagnosis column number 22. It is worth mentioning that it is often the case where patients examination returns less than 31 diagnoses. The *Core* table also contains 31 columns describing up to 31 procedures that the patient went through. Even though a patient might go through several procedures in a given visit, the primary procedure that occurred at the visit discharge is assumed to be the first procedure column. The *Core* table also contains an attribute called *DaysToEvent*, which describes the number of days that passed between the admission to the hospital and the procedure day. This field is anonymized in order to hide the patients' identity. There are several demographic data that are reported in this table as well such as: race and gender. Table 2 maps the *Core* table features to concepts described in this paper.

Table 2. Mapping Between Attributes and Concepts

Attributes	Concepts
VisitLink	Patient Identifier
DaysToEvent	Temporal visit ordering
DXCCSn	n^{th} Diagnosis, flexible attributes
PRCCSn	n^{th} Procedure, meta-actions
Race, Age Range, Gender,..	Stable attributes

5.2   Evaluation

We used our technique to extract meta-actions effects on the Florida SID dataset for several meta-actions. In this paper, we reported the confidence *ActionConf(as)* and likelihood *Like(as)* of few action sets for four different meta-actions. You can note from Table 3 that the positive action sets *ActionConf* is very high, which means that patients' diagnoses disappear after applying meta-actions. In other words, surgeries applied are very successful in curing patients disease. On the other hand, the neutral action sets *ActionConf* is small, which confirms the assumption that patients react in a consistently different way to meta-actions with regards to attributes that remain unchanged. In addition, the likelihood of neutral action sets extracted is small, which means that very few diagnoses remain unchanged after the surgeries. Table 3 represents the meta-actions (procedures) and action sets elements (diagnoses) with their CCS codification [1].

Table 3. Meta-actions' Action Sets Confidence and Likelihood

Meta-action	Action set	Type	ActionConf	Likelihood
34	{127, 106}	Positive	88.88%	32
	{108}	Neutral	16.57%	29
43	{59, 55, 106]}	Positive	85.71%	18
	{106}	Neutral	11.76%	18
44	{62, 106, 55}	Positive	94%	16
	{257, 101}	Neutral	15.62%	10
45	{59, 55}	Positive	89%	33
	{58}	Neutral	14.97%	28

In addition, we report in Figure 2 the meta-action confidence for 15 different meta-actions. We show in Figure 2 that the meta-actions are consistently successful for all their action sets regardless of the patients preconditions for these meta-actions. Figure 2 shows meta-actions with their CCS codes [1].

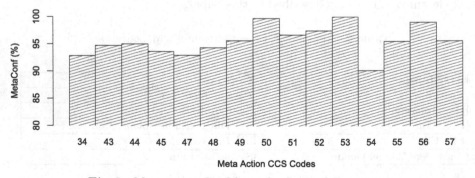

Fig. 2. Meta-action Confidence for Surgical Treatments

6   Conclusion

Mining surgical meta-actions is a hard task because patients may react differently to meta-actions applied, and surgery outcomes are different from one patient to another. In this paper, we presented a meta-action effects mining technique for surgical datasets with variable number of diagnoses (multivalued attributes). Furthermore, we presented the ontology representation of meta-action effects, and used the SID dataset that is part of HCUP to demonstrate the usefulness of our methodology in comparison with the action terms based techniques.

Acknowledgments. This project was partially supported by the Research Center of PJIIT, supported by the Ministry of Science and Higher Education in Poland.

References

1. Clinical classifications software (ccs) for icd-9-cm, http://www.hcup-us.ahrq.gov
2. Raś, Z.W., Wieczorkowska, A.A.: Action-rules: How to increase profit of a company. In: Zighed, D.A., Komorowski, J., Żytkow, J.M. (eds.) PKDD 2000. LNCS (LNAI), vol. 1910, pp. 587–592. Springer, Heidelberg (2000)
3. Wang, K., Jiang, Y., Tuzhilin, A.: Mining actionable patterns by role models. In: Proceedings of the 22nd International Conference on Data Engineering, ICDE 2006, pp. 16–26 (2006)
4. Raś, Z.W., Dardzińska, A.: Action rules discovery based on tree classifiers and meta-actions. In: Rauch, J., Raś, Z.W., Berka, P., Elomaa, T. (eds.) ISMIS 2009. LNCS, vol. 5722, pp. 66–75. Springer, Heidelberg (2009)
5. Touati, H., Kuang, J., Hajja, A., Raś, Z.W.: Personalized action rules for side effects object grouping. International Journal of Intelligence Science (IJIS) 3(1A), 24–33 (2013); Special Issue on "Knowledge Discovery", G. Wang (Ed.)
6. Touati, H., Ras, Z.W.: Mining meta-actions for action rules reduction. Fundamenta Informaticae 127(1-4), 225–240 (2013)
7. Raś, Z.W., Dardzinska, A., Tsay, L.S., Wasyluk, H.: Association action rules. In: IEEE International Conference on Data Mining Workshops, ICDMW 2008, pp. 283–290 (2008)
8. Qiao, Y., Zhong, K., Wang, H., Li, X.: Developing event-condition-action rules in real-time active database. In: Proceedings of the 2007 ACM Symposium on Applied Computing, SAC 2007, pp. 511–516. ACM, New York (2007)
9. Rauch, J., Šimůnek, M.: Action rules and the guha method: Preliminary considerations and results. In: Proceedings of the 18th International Symposium on Foundations of Intelligent Systems, ISMIS 2009, pp. 76–87. Springer (2009)
10. Yang, Q., Chen, H.: Mining case for action recommendation. In: Proceedings of ICDM, pp. 522–529 (2002)
11. Pawlak, Z.: Information systems - theoretical foundations. Information Systems Journal 6, 205–218 (1981)
12. Fensel, D.: Ontologies: A Silver Bullet for Knowledge Management and Electronic Commerce. Springer, Syracuse (2003)

A System for Computing Conceptual Pathways in Bio-medical Text Models

Troels Andreasen[1], Henrik Bulskov[1],
Jørgen Fischer Nilsson[2], and Per Anker Jensen[3]

[1] Computer Science, Roskilde University
[2] Mathematics and Computer Science, Technical University of Denmark
[3] International Business Communication, Copenhagen Business School
{troels,bulskov}@ruc.dk, jfni@dtu.dk, paj.ibc@cbs.dk

Abstract. This paper describes the key principles in a system for querying and conceptual path finding in a logic-based knowledge base. The knowledge base is extracted from textual descriptions in bio-, pharma- and medical areas. The knowledge base applies natural logic, that is, a variable-free term-algebraic form of predicate logic. Natural logics are distinguished by coming close to natural language so that propositions are readable by domain experts. The natural logic knowledge base is accompanied by an internal graph representation, where the nodes represent simple concept terms as well as compound concepts stemming from entire phrases. Path finding between concepts is facilitated by a labelled graph form that represents the knowledge base as well as the ontological information.

Keywords: Semantic text processing in bio-informatics and medicine, natural logic knowledge bases in bio-informatics, generative ontologies, path-finding in graph knowledge bases.

1 Introduction

In [2] we outlined a proposal for a software system for computational path finding in bio-scientific texts. Given two terms or phrases, perhaps being far from each other and seemingly unrelated, the system, called ONTOSCAPE, is to find conceptual connections between the terms by linking sentences in the text and assisted by ontological background information. Ideally such a system should be capable of tracing conceptual connections as humans do, taking into account the meanings of the sentences in the text rather than merely superficially spotting "stepping stone" words.

The principles introduced are general in nature. However, the bio-sciences in particular abound with textual descriptions of convoluted relationships between components and processes giving rise to bio-pathways.

In the present paper we detail the internal, general principles and methods for achieving the intended functionality. In the companion paper [1] we exemplify bio-models, and discuss the desired path-finding functionality in the bio-science domain.

T. Andreasen et al. (Eds.): ISMIS 2014, LNAI 8502, pp. 264–273, 2014.
© Springer International Publishing Switzerland 2014

The key principle of the system is a partial, one-by-one compilation of the sentences in the source text into a "semantic graph" of concepts and their conceptual relationships. This graph is subsequently used for path finding among some given terms. The graph comprises ontological information as well. Moreover, as a key component we introduce a form of so-called natural logic [10,7,11,4], which may be conceived of as a fragment of natural language with a well-defined logical semantics. The computational text analysis consists in translation of as much information as possible into the natural logic form. The natural logic propositions are then further decomposed and compiled into the semantic graph for path finding. The natural logic serves to provide a clear logical semantics for the semantic net representation, making the individual natural logic propositions readily recoverable from the graph form where the propositions are woven together.

The paper is structured as follows: In sections 2 - 6 we introduce the so-called reference natural logic language, which is the internal target language for the text processing systems part. In section 7 we outline the computational linguistic analysis. In section 8 we describe the applied graph representation and in section 9 we describe the pathfinding principles, followed by the concluding summary in section 10.

2 The Natural Logic Language

As mentioned, we identify and specify formally a fragment of natural language as an intermediate logical representation of the sentences in the text as a step towards the final graph representation used for path finding. This formal language is actually a form of natural logic with a well-defined logical semantics as to be explicated; hence the term NATURALOG. At the same time it can be read by humans almost like natural language, even though it may seem somewhat constrained and stereotypical.

Obviously, the formal language is only capable of covering part of the meaning content in the source texts. However, it is our working hypothesis that a substantial part of the meaning content pertaining to querying and path finding can be expressed in NATURALOG. Rather than viewing NATURALOG as a Procrustean bed, where limbs of the sentences are cut off arbitrarily, it should be seen as a means of abstracting and extracting the essential information from the sentences in a readable form. NATURALOG comes in a basic form described in this section and in an extended form taking into account a number of semantically equivalent linguistic expressions, cf. section 6.

The basic form of NATURALOG propositions is

Nounterm Verbterm Nounterm

as stipulated by the following grammar:

Prop   ::= [*Quant*] *Nounterm Verbterm* [*Quant*] *Nounterm*
Quant   ::= some | all
Nounterm ::= {*Adjective*}* (*Noun*|*CompNoun*) [*Relatorterm*]
Relatorterm ::= *RelClauseterm* {and *RelClauseterm* }* | *Prepterm**

RelClauseterm ::= {[that] *Verbterm* [*Nounterm*]}
CompNoun ::= { *Noun* }* *Noun Noun*
Verbterm ::= *Verb* { *Prepterm* }*
Prepterm ::= *Preposition Nounterm*

By default, the first quantifier, if absent, is taken to be all, and the second one to be some. In [1] we give a complementary natural logic account of the language.

By way of example, the sentence *cells that produce insulin are located in the pancreas* is formalized as all cell that produce some insulin located-in some pancreas, equivalent to cell that produce insulin located-in pancreas

The figure below illustrates the mediating role of NATURALOG.

Sentence from the text
↓
NATURALOG proposition
↕
Edges (triples) in the semantic graph

3   The Underlying "Naturalistic" Ontology

In NATURALOG the textual sentences are turned into propositions expressing a relationship between two relata concepts with a relator derived from the main verb. In the example above, the first concept is the compound concept cell-that-produce-insulin and the second is the atomic concept pancreas, with the relator located-in coming from the verb phrase *are located in*.

In the graph view, nodes are associated with concepts and directed edges (edges) with relators. Compound concepts like cell-that-produce-insulin are connected to their constituent concept nodes as to be explained.

This representation scheme conforms with a simplified, "naturalistic" view of the target domain comprising:

- Classes of physical objects, substances, states, processes and events and their properties
- Relationships between instances of the classes such as taxonomic, partonomic and causal relationships. Domain specific effect relationships such as "transports", "promotes", "reduces", and "prevents". Relationships appearing in a weakened form such a "may prevent"
- General logical reasoning principles for the relationships such as e.g. transitivity.

More subjective "higher level" categories such as purpose and intention fall outside this "mechanistic" ontological view. So do metaphorical explications such as the heart being a pump etc. The top ontology may stipulate admissible relationships e.g. endorsing only material parts of material wholes etc.

Compound concepts, coming from e.g. compound noun terms, relative clauses or prepositional phrases, are generated from classes and relationships exemplified by the compound cell-that-produce-insulin. Compound relationships may be formed by way of adverbial constructions giving rise to compound relationships like produces-in-liver. Conditional sentences using "if" or "when" may be represented by compounded, restricted relationships such as e.g. secretes-when-(infection-in-pancreas), which is a specialization of secretes-when-infection, which is in turn a specialization of the relation secretes. Relational terms such as "produce" often have a corresponding concept expressed as a nominalization, cf. "production". These semantic correspondences are recorded and used in the path-finding process.

4   Representing Ontological Information

Ontological information in the form of class inclusion relationships is expressible within the reference language simply in the form of copula sentences. For the copula sentence *insulin is a hormone* we have insulin isa hormone, where isa is the class inclusion relation used internally in NATURALOG. The system recognizes transitivity of inclusion as one of its reasoning principles: From insulin isa hormone together with hormone isa protein follows insulin isa protein.

We apply the so-called generative ontology introduced in [5], reflecting the well-known notion of generative grammar. This notion of ontology admits compound terms such as cell-that-produce-insulin, which obtains its proper meaning by being situated below cell via isa as well as connected to insulin via the relator produce, cf. figure 1. By appropriate deductive means it must be ensured that it is class-included also in cell-that-produce-hormone. The generative ontology accommodates all compound concept terms (whatever their complexity) encountered in the compiled part of the source text with their accompanying ontological relationships.

The content in the generative ontology consists of classifications and definitions, whereas the knowledge base part proper comprises the relationships stemming from assertions in the text sentences which are not definitional. However, these two parts are interwowen in the final semantic graph representation. Accordingly, the source text may contribute to and extend the generative ontology with propositions such as betacell isa cell. The ontology further comprises more *ad hoc* term definitions like

     pancreatic cell isa (cell that located-in pancreas) and conversely
     (cell that located-in pancreas) isa pancreatic cell

Such synonym terms achieved by mutual inclusion may use the designated relation syn as in

     pancreas syn pancreatic gland

By further development of our framework, such definitions may be generalized to (pancreatic-X) isa (X-located-in pancreas) and conversely (X located-in pancreas) isa (pancreatic X), where X is a variable ranging over concepts.

5   Logical Explication of our Natural Logic

Consider again the sample NATURALOG proposition

(cell that produce insulin) located-in pancreas

which in first order predicate logic is

$$\forall x (cell(x) \land \exists y (insulin(y) \land produce(x, y))) \rightarrow \exists z (pancreas(z) \land locatedin(x, z))$$

One may observe that this sentence falls outside the definite clause subset if rewritten into clause form. In description logic: $cell \sqcap \exists produce.insulin \sqsubseteq \exists locatedin.pancreas$. As a fundamental difference from description logic, in NATURALOG the main verb is kept as the main relator. In description logic, sentences are invariably turned into their copula form with the operator $\sqsubseteq$ corresponding to the inclusion relation isa, see further [6].

6 Extensions of the Reference Language

Though it might be considered semantically adequate for the purpose at hand, the NATURALOG language is syntactically severely restricted. In order to accommodate an enriched and less awkward syntax, we admit a broader range of phrases comprising i.a. the following extensions:

- Distributive phrases giving rise to parallel propositions
- Appositions and parenthetical relative clauses giving rise to separate propositions
- Inter-sentential anaphora, which is dealt with by caching "hooks" in the form of candidate antecedent concepts from the previous sentence during the analysis of a sentence
- Passive voice sentences, which are interpreted by reversing the corresponding active relator, cf. [4]
- An extended syntax admitting topicalization.

All of these extensions are conservative in the sense that they do not extend the semantic range of NATURALOG. Possible other topics for semantic extensions are use of various determiners including generalized quantifiers such as "most" and "few" and certain forms of intra-sentential anaphora, like the so-called Donkey-sentences.

7 Outline of the Linguistic Analysis

The translation of the text sentences to NATURALOG propositions is carried out sentence by sentence in two phases, a scanning phase and a parsing phase.

The output from the scanning is a token list, where a token is a concept (basic form or stem of the word) adorned with 0, 1 or more POS-tags retrieved from a vocabulary of terms. The case 0 is by default assumed to be an unknown noun. The token list also includes semantically relevant punctuation signs, notably commas for distinguishing between restrictive and parenthetical relative clauses and for recognizing distribution lists.

The parser conducts a DCG-based parsing of the token list. Simplistically, the sample sentence *cells that produce insulin are located in the pancreatic gland* is translated into cell (that produce insulin) located-in pancreas recognizing the relative clause and appealing to the previously introduced synonym pancreas for pancreatic gland.

Our linguistic sentence analysis naively assumes a simple correspondence between the targeted sentence and the above NATURALOG reference grammar. A fundamental complication is, however, the various sources of ambiguity, e.g. POS-ambiguity, and lexical and structural ambiguities inherent in the reference grammar. These ambiguities have to be resolved in combination. To this end we appeal to a relational compositionality principle, cf. [12]. The semantically fundamental functional compositionality principle in our approach is generalized to relational composition. This means that the meaning and outcome of a phrase is not composed and provided as a unique function of its constituents but rather as a relation, thereby postponing ambiguity resolution. The argument structure of verbs is to be used in the disambiguation process.

In a first version we simply adopt as a default principle alignment of sub-phrases rather than nesting, e.g. for two adjacent adnominal prepositional phrases. Alternatively one may choose to represent all structural alternatives. More sophisticated attempts at structural disambiguation may draw on general ontological knowledge, cf. [13,14] such as material objects and substances being partonomically incompatible with processes.

The grammar admits skipping of tokens (words) but tries to avoid this to the extent possible in order to capture as much of the original sentence as possible. Robustness is sought achieved by fall-back principles: If the verbal structure cannot be handled properly, in the worst case, appeal is made to a "is-related-to" relator at the top of the relator ontology.

8   The Semantic Graph Representation

The applied semantic graph representation takes the form of a finite, directed, labeled graph in which the nodes represent simple or compound concepts. Concepts, whether simple or compound, are uniquely represented by nodes in the graph. We assume here that the propositions have the determiner/quantifier form all C_1 R some C_2. There are two forms of labelled directed edges: definitional and assertional . These edges are labelled by relator names and connect two concepts and express primitive facts. The relator names are not uniquely represented. In general, a given relationship (e.g. derived from a verb) may appear on multiple edges whereas concepts appear only once. In particular, the isa relationship is ubiquitous in those parts of the graph which constitute the ontology.

The definitional edges are relationships connecting one concept node c_0 to two or more definientia concept nodes $c_1, ..., c_k$ for $k \geq 2$, each concept node contributing to the definition of the definiendum c_0 via a relator r_i. Logically, this means that

$$c_0 \text{ iff } (r_1 \ c_1) \text{ and } (r_2 \ c_2) \ ... \text{ and } (r_k \ c_k)$$

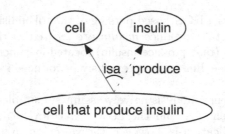

Fig. 1. A definition cell-that-produces-insulin *iff* (isa cell) *and* (produces insulin) rendering *iff* by the "⌢" arc

(where r_1 is isa) exemplified by cell-that-produces-insulin *iff* (isa cell) *and* (produces insulin). Here cell is the genus and (produces insulin) is the distinguishing feature (differentia). In the graph rendition, edges acquire their definitional (contrary to assertional) status by the "⌢" arc shown adjacent to the definiens node as illustrated in figure 1. Omitting the arc would give merely cell-that-produce-insulin isa cell) and cell-that-produce-insulin produces insulin, but not the converse "if" part. The "iff" form is crucial for the semantics of concepts stemming from noun phrases in the logical reasoning during the path finding process. An assertional edge (usually) stems from the main verb in an assertive sentence in the text. In particular, the copula sentence form gives rise to an inclusion relationship which is not an "iff" relationship. An "iff" definition is achievable, however, with a pair of inclusion relationships in opposite directions forming synonyms, e.g.:

        cell-that-produces-insulin isa betacell

        betacell isa cell-that-produces-insulin

A simple example of an assertional contribution rather than a definition is:

        cell-that-produces-insulin located-in islet-of-langerhans

The edge labels, i.e. the relators, may themselves be simple or compound. Adverbs and adverbial prepositional phrases give rise to compound relators as exemplified by

        secretes-from-pancreas isa secretes

Observe that the isa relationship is overloaded in that here it is placed between relationships (*in casu* stemming from the verb "secrete") rather than between concepts. These relationships between relations are situated in an ontology represented in a separate graph.

The NATURALOG compound terms are decomposed into triples in the graph as elucidated above. In the resulting ontology, additional isa inclusion edges are computed prior to path-finding computations in order to ensure that all sub-concepts are properly included in superior concepts. This is exemplified by considering the concept cell-that-produce-insulin together with the concept cell-that-produce-hormone and the relationship insulin isa hormone. Here the compilation process has to add the inclusion relationship cell-that-produce-insulin isa cell-that-produce-hormone.

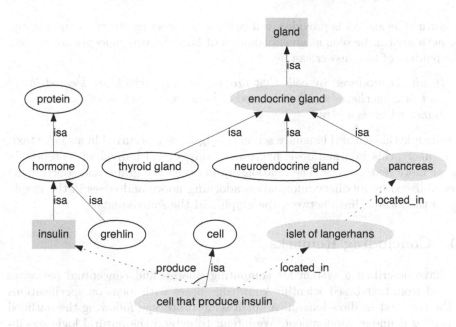

Fig. 2. A miniature ontology corresponding to a subset of a bio-model. A pathway connecting "insulin" and "gland" is shown. This pathway provides a candidate answer to a query specifying the two concepts (apparently the only one in this case).

9   Reasoning and Pathfinding in the Semantic Graph

The computation of pathways is supported by logical inference rules, and inferred propositions may constitute shortcuts, as it were, in the graph view. The pathfinding is assisted by the rule of transitivity for class inclusion and the so-called monotonicity rules stating that the subject part of the relationship may be specialized, and the object part may be generalized[1]. For instance, the transitivity of inclusion, isa, conceptually shortens the distance from a concept to a superior concept in the ontology via intermediate concepts. Similarly for partonomic, causative and effect relations. In [2], the path finding is explained more abstractly as application of appropriate logical comprehension principles supporting the relation composition.

A miniature ontology, corresponding to a bio-model covering among others the concept cell that produce insulin is visualised in the graph in figure 2. In addition, a candidate answer to the pathway query

insulin, gland?

---

[1] This is for the default determiners all c_1 r some c_2. Interestingly, some verbs exemplified by "prevent" invite the form all c_1 prevents all c_2.

is shown. The answer is provided as a pathway connecting the two concepts and the pathway can be seen as an explanation of how the two concepts are related. The reading of the answer can be

> *Insulin is produced by cells that produce insulin, which are located in islet of langerhans, which is located in pancreas, which is an endocrine gland, which is a gland.*

In principle this natural language sentence might have occurred in a source text. More likely, this information is harvested from several individual sentences in one or more text sources. The relationship between the graph and source texts is established by auxiliary information adorning nodes and edges in the graph, e.g. in the form of links between the graph and the source texts.

10   Concluding Remarks

We have described a system for computing queries and conceptual pathways derived from text-based scientific knowledge bases with focus on specifications in the bio- and medico-domain. A small scale prototype following the outlined principles is under development. We intend to extend the natural logic specification language so that it covers more natural language formulations. We also intend to provide enriched semantic features e.g. for dealing with various forms of property ascriptions such measurements, scales and intervals.

References

1. Andreasen, T., Bulskov, H., Fischer Nilsson, J., Jensen, P.A.: Computing Pathways in Bio-Models Derived from Bio-Science Text Sources. In: Proceedings of the IWBBIO International Work-Conference on Bioinformatics and Biomedical Engineering, Granada, April 7-9, pp. 217–226 (2014) ISBN 84-15814-84-9
2. Andreasen, T., Bulskov, H., Fischer Nilsson, J., Anker Jensen, P., Lassen, T.: Conceptual Pathway Querying of Natural Logic Knowledge Bases from Text Bases. In: Larsen, H.L., Martin-Bautista, M.J., Vila, M.A., Andreasen, T., Christiansen, H. (eds.) FQAS 2013. LNCS, vol. 8132, pp. 1–12. Springer, Heidelberg (2013)
3. Andreasen, T., Bulskov, H., Jensen, P.A., Lassen, T.: Extracting Conceptual Feature Structures from Text. In: Kryszkiewicz, M., Rybinski, H., Skowron, A., Raś, Z.W. (eds.) ISMIS 2011. LNCS, vol. 6804, pp. 396–406. Springer, Heidelberg (2011)
4. Fischer Nilsson, J.: Diagrammatic Reasoning with Classes and Relationships. In: Moktefi, A., Shin, S.-J. (eds.) Visual Reasoning with Diagrams. Studies in Universal Logic. Springer, Birkhäuser (2013)
5. Andreasen, T., Fischer Nilsson, J.: Grammatical Specification of Domain Ontologies. Journal: Data & Knowledge Engineering 48(2), 221–230 (2004)
6. Fischer Nilsson, J.: On Reducing Relationships to Property Ascriptions. In: Proc. of the 18th European-Japanese Conference Information Modelling and Knowledge Bases, pp. 249–256 (2008)
7. van Benthem, J.: Essays in Logical Semantics. Studies in Linguistics and Philosophy, vol. 29. D. Reidel Publishing Company (1986)

8. Smith, B., Rosse, C.: The Role of Foundational Relations in the Aligment of Biomedical Ontologies. In: Fieschi, M., et al. (eds.) MEDINFO 2004 (2004)
9. Andreasen, T., Bulskov, H., Jensen, P.A., Lassen, T.: Extracting Conceptual Feature Structures from Text. In: Kryszkiewicz, M., Rybinski, H., Skowron, A., Raś, Z.W. (eds.) ISMIS 2011. LNCS, vol. 6804, pp. 396–406. Springer, Heidelberg (2011)
10. Muskens, R.: Towards Logics that Model Natural Reasoning, Program Description Research program in Natural Logic (2011),
    http://lyrawww.uvt.nl/~rmuskens/natural/
11. Sanchez Valencia, V.: The Algebra of Logic. In: Gabbay, D.M., Woods, J. (eds.) Handbook of the History of Logic. The Rise of Modern Logic: From Leibniz to Frege, vol. 3. Elsevier (2004)
12. Fischer Nilsson, J.: Reconciling Compositional and Holistic Views on Semantics through Relational Compositionality - exemplified with the case of complex noun-noun compounds Approaches to the Lexicon,
    https://conference.cbs.dk/index.php/lexicon/lexicon/paper/view/886/488
13. Fischer Nilsson, J., Szymczak, B.A., Jensen, P.A.: ONTOGRABBING: Extracting Information from Texts Using Generative Ontologies. In: Andreasen, T., Yager, R.R., Bulskov, H., Christiansen, H., Larsen, H.L. (eds.) FQAS 2009. LNCS, vol. 5822, pp. 275–286. Springer, Heidelberg (2009)
14. Antoni Szymczak, B.: Computing an Ontological Semantics for a Natural Language Fragment, PhD-thesis, Technical University of Denmark (DTU), 295 p. (IMM-PHD-2010; No. 242) (2010)

Putting Instance Matching to the Test: Is Instance Matching Ready for Reliable Data Linking?

Silviu Homoceanu[1], Jan-Christoph Kalo[2], and Wolf-Tilo Balke[1]

IFIS TU Braunschweig, Mühlenpfordstraße 23, 38106 Braunschweig, Germany
{silviu,balke}@ifis.cs.tu-bs.de,
{j-c.kalo}@tu-bs.de

Abstract. To extend the scope of retrieval and reasoning spanning several linked data stores, it is necessary to find out whether information in different collections actually points to the same real world object. Thus, data stores are interlinked through owl:sameAs relations. Unfortunately, this cross-linkage is not as extensive as one would hope. To remedy this problem, instance matching systems automatically discovering owl:sameAs links, have been proposed recently. According to results on existing benchmarks, such systems seem to have reached a convincing level of maturity. But the evaluations miss out on some important characteristics encountered in real-world data. To establish if instance matching systems are really ready for real-world data interlinking, we analyzed the main challenges of instance matching. We built a representative data set that emphasizes these challenges and evaluated the global quality of instance matching systems on the example of a top performer from last year's Instance Matching track organized by the Ontology Alignment Evaluation Initiative (OAEI).

Keywords: Instance matching, owl:sameAs, link discovery, linked data.

1   Introduction

Fostered by the W3C Semantic Web group's initiative to build the "Web of Data", a massive amount of information is currently being published in structured form on the Web. Currently, more than 53 billion triples in over 300 data stores are available in the largest Virtuoso-based Semantic Web database (SWDB)[1]. The key point of this initiative and an important design principle of Linked Open Data (LOD) is that data from different sources is extensively inter-linked. This way queries can join information available in disjoint data stores with high precision. For example, for a query on biographic data and work of the film producer Martin Scorsese, biographic data could come from DBpedia while data about his work could come from LinkedMDB. All this is possible provided that DBpedia and LinkedMDB are inter-linked at least with respect to the entity of Martin Scorsese. The typical way for such cross-linkage

---

[1] The Virtuoso SWDB is accessible at http://lod.openlinksw.com/ through a SPARQL endpoint.

T. Andreasen et al. (Eds.): ISMIS 2014, LNAI 8502, pp. 274–284, 2014.
© Springer International Publishing Switzerland 2014

between LOD sources is through owl:sameAs links. Unfortunately, today cross-linkage is not nearly as extensive as one would hope: The number of unique owl:sameAs links we counted on the aforementioned SWDB, is about 570 million. Many links are missing and from the ones available, a large part are trivial links between DBpedia, Freebase, and YAGO ([1]).

Under the name of *entity reconciliation* or *instance matching* (usually mixed up with instance-based ontology matching because instance matching is often required for ontology matching), the problem of finding identity links (owl:sameAs) between identifiers of the same entity in various data stores has been heavily researched (see [2–7]). These systems make use of techniques like probabilistic matching, logic-based matching, contextual matching, or heuristic matching based on natural language processing (NLP). Each approach shows strengths and weaknesses. But these particularities are hard to assess, since each system was evaluated on different data samples. The choice of data for the evaluation has a big influence on the results. For instance, there is a large number of class equivalence links between DBpedia and YAGO. If these two data sources build a significant part of the evaluation data then approaches like the one presented in [5] are favored. The verbose nature of the URIs also helps shallow NLP techniques favoring for instance the system presented in [7]. The situation is different for other selections like LinkedMDB and YAGO since the URIs provided in LinkedMDB are more cryptic and links to and from YAGO are rare.

Of course, instance-matching approaches have to be able to work with all kinds of entities from multiple data-stores. Again, this may boost the performance of some systems, since different aspects of an entity can be learned iteratively from various stores. On the other hand it can be detrimental to the overall data quality, since the more entities and entity types are available, the more probable it becomes for systems to generate incorrect identity links. Take for instance LINDA [5] which heavily exploits transitive links to support the inter-linking process. When it was evaluated on the Billion Triple Challenge corpus comprising entities from various stores the respective precision was about 0.8. For relaxed similarity constraints the precision even drops to 0.66. But with every third identity link being incorrect, this level of quality does not seem satisfactory for performing join queries or reasoning. In contrast, SLINT+ [7] reports an average precision of 0.96 on DBpedia and Freebase data.

But does this really mean that SLINT+ performs better? The respective precision was achieved on a biased set, representing a highly inter-linked extract from DBpedia and Freebase! It is therefore impossible to directly compare the performance of the two systems. To make systems comparable to one another, the Ontology Alignment Evaluation Initiative (OAEI) organizes a yearly evaluation event including an Instance Matching track. For the last year's evaluation[2] there were evaluation tests involving data value differences, structural heterogeneity and language heterogeneity. With small data value and structure alterations and involving a small extract (1744 triples and 430 URIs) from a single high quality data source (DBpedia), we will show that the tests do not accurately reflect the problems encountered in real-world data.

---

[2] http://www.instancematching.org/oaei/imei2013/results.html

Actually, judging by the 2013's OAEI evaluation results (sustained precision of over 0.9), instance matching systems seem to have reached a level of maturity. But considering the modest precision achieved by systems like LINDA on real-world data, this raises the question: Is instance matching ready for reliable data inter-linking? To answer this question, we perform extensive real-world experiments on instance matching using a system which has proven very successful in OAEI tests. To the best of our knowledge, this is the first study that provides an in depth analysis over how effective instance matching systems are on real-world data.

2   The Instance Matching Problem

Instance matching is about finding and reconciling instances of the same entity in heterogeneous data. It is of special interest to LOD because the same entity may be identified with different URIs in different data stores and the owl:samesAs property useful for interlinking URIs of the same entity is not as wide-spread as needed.

In the context of LOD, given multiple sets of URIs D_1, D_2, ..., D_n, with each set comprising all unique URIs of a data store, *matching* two instances of an entity can formally be defined as a function *match:URI×URI*→{false, true} with:

$$match(URI_i, URI_j) := \begin{cases} true, if \; sim(URI_i, URI_j) > \theta \\ false, otherwise \end{cases} \quad with \; URI_i \in D_i, URI_j \in D_j$$

where $1 \leq i, j \leq n$, and *sim()* is a system dependent, complex similarity metric involving structural, value-based, contextual and other similarity criteria, and θ is a parameter regulating the necessary quality level for a match.

Based on this function, instance matching systems build an *equivalence class* for each entity. An equivalence class comprises all URIs used by any source to refer to some corresponding unique entity. For instance, considering only DBpedia, Freebase, YAGO and LinkedMDB, the equivalence class for the entity "Martin Scorsese" is:

> {http://dbpedia.org/resource/Martin_Scorsese,
> http://yago-knowledge.org/resource/Martin_Scorsese,
> http://rdf.freebase.com/ns/m.04sry,
> http://data.linkedmdb.org/resource/producer/9726,
> http://data.linkedmdb.org/resource/actor/29575,
> http://data.linkedmdb.org/resource/editor/2321}.

It's worth noticing that in contrast to general purpose knowledge bases like Free-base or DBpedia, specialized data stores like LinkedMDB have finer granularity, differentiating between Martin Scorsese as actor, editor, or producer. According to the owl:sameAs property definition in the OWL standard, all URIs referring to the same real world object should be connected through owl:sameAs. In consequence, all six URIs from the previous example should be linked by owl:sameAs relations. Of course one could argue that finer, context-based identity is required and that "Martin Scor-sese, the producer" may not be the same as "Martin Scorsese, the actor". For further discussions regarding context-based similarity and identity see [8]. In this paper we adopt the definition as provided by the OWL standard for the owl:sameAs property.

Instance matching is an iterative process. Once some of the instances are matched either manually or by some system and owl:sameAs links have been established, more identity links can be found by exploiting the transitivity inherent in identities: Given that URI_A and URI_B represent the same real world object, the same applying for URI_B and URI_C implies that also URI_A and URI_C represent the same real world entity. Consequently, an owl:sameAs link between URI_A and URI_C can be created. However, the actual process of discovering sameAs links *is based on some similarity function and not on identity*. Similarity functions, however, are usually not transitive!

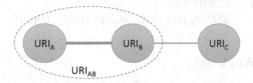

Fig. 1. Three URIs matching in a chain (URI_A and URI_C don't match). The similarity between URI_A and URI_B is stronger than the similarity between URI_B and URI_C.

Let us give a simplified example where the matching function relies on the Levenshtein distance on the rdfs:label property as similarity metric. Consider that a URI with rdfs:label "Scorsese, Martin" referring to the well-known movie producer, is matched with a URI with rdfs:label "Scorsese, Cartin" (which could be a typo). This last URI matches a URI with rdfs:label "Scorsese, Chartin" and the match process goes on up to a URI with rdfs:label "Scorsese, Charles". Charles Scorsese is an actor known for his role in Goodfellas and actually Martin's father. This problem is well known in the area of single link clustering: similarity clustering may lead to chains of URIs where neighboring URIs in the chain are similar, but for long enough chains the ends of the chain have almost nothing in common. Linking the URIs of Martin and Charles Scorsese with owl:sameAs would obviously be incorrect. Of course this example is constructed, but the danger of transitively matching unrelated instances in the context of large amounts of data is real. In consequence, evaluation data involving triples from multiple stores is necessary for exposing such weaknesses.

From the instance matching systems we found that only LINDA specifically addresses the problem of transitivity and selects only those matches consistent with transitivity as follows: On the example in Figure 1, considering that $sim(URI_A, URI_B) > sim(URI_B, URI_C)$, the equivalence class of URI_A comprises only URI_B and vice versa, i.e. both URIs refer the same entity and all properties valid for URI_A are also valid for URI_B and all properties valid for URI_B are also valid for URI_A. To express this we can denote the entity referred by URI_A and URI_B through URI_{AB}. Even though URI_A and URI_C don't show a large enough similarity, they are considered to refer the same entity if $match(URI_{AB}, URI_C)$ is true. Then, URI_C will also be added to the equivalence class. The process of finding identity links continues iteratively up to convergence.

Borrowing from hierarchical clustering, also the complete-linkage criteria could for instance be easily adopted to enforce transitivity. Assume after pairwise comparing all URIs we find three URIs matching in a chain like presented in Figure 1. Any set of n linked URIs satisfies the complete-linkage criteria, iff all n URIs match in a pairwise. Obviously this is not the case for chains. In consequence, chains are broken up by removing the weaker links. In the case of links of equal strength one of them is broken at random. Consider $sim(URI_A, URI_B) > sim(URI_B, URI_C)$. Since $match(URI_A, URI_C)$ is false, the link between URI_B and URI_C has to be removed. As a rule, the list of URIs being *weakly linked* to an URI_x is:

$$WL_{URI_x} = \{URI_y | \exists z: match(URI_x, URI_z) = true \land match(URI_y, URI_z)$$
$$= false \land sim(URI_x, URI_z) \geq sim(URI_x, URI_y)\}.$$

After all weak links are broken for all URIs, the *equivalence class* of an URI is given by a function $E:URI \rightarrow \{URIs\}$ where:

$$E(URI_k): = \{URI_l | match(URI_k, URI_l)\}$$

3   Related Work

Instance matching is crucial for several applications like data integration, identity recognition and more important, for ontology alignment. Recognizing the lack of evaluation data, OAEI provided a reference benchmark for ontology alignment since 2004. Probably fostered by advances in Linked Data, four years later, [9] is one of the first publications to address this problem for instance matching. The authors discuss the particularities of instance matching and name main challenges. Based on these challenges, they design a benchmark with movie data from IMDb that emphasizes on data value differences, structure and logical heterogeneity. Finally, they compare the results for two instance matching algorithms to show the applicability of the data set.

In 2009, OAEI introduced an instance matching track and provided first generated benchmarks[3]: One comprising three datasets with instances from the domain of scientific publications built on Digital Bibliography & Library Project (DBLP), one with three datasets covering several topics, structured according to different ontologies from DBpedia and one generated benchmark obtained by modifying a dataset according to the data value, structure and logical heterogeneity criteria introduced in [9]. Evaluation data has gradually improved and last year's benchmark comprised five test cases: One for value transformation, where the value of five properties was changed by randomly deleting or adding characters; one for structure transformation, where the length of property paths between resources and values has been changed; a languages test where comments and labels were provided in French instead of English; one set combining value and structure transformation using French text and one where besides the value, structure and language challenges, some entities have none or multiple counterparts (a cardinality test). The data for the tests was extracted from DBpedia: it comprised 1744 triples, 430 URIs and only 11 predicates. It involves only one type of entity: Personalities from the field of computer science like Alan Turing, Donald E. Knuth, or Grace Hopper and is limited to triples having such personalities as a subject. Four instance matching systems have been evaluated on this benchmark. Out of the four, SLINT+ [7] and RiMOM [4, 10, 11] achieved outstanding results with an average precision and recall over all test of more than 0.9.

---

[3] http://oaei.ontologymatching.org/2009/instances/

While these results are quite promising, similar systems have proven weaker performance on real-world larger in size and involving multiple data stores. To assess the performance of such systems with real-world data, we built an evaluation set comprising 90,000 entities, from four domains, extracted from five data stores. In contrast to the OAEI test cases, all domains were included in all tests rendering cross-domain false positive matches (e.g. person being matched to movie) possible. The data stores were all-purpose knowledge bases like DBpedia and Freebase as well as domain focused stores like LinkedMBD and DrugBase. Some sources have cryptic URI naming conventions while some are more explicit. Also the granularity of properties varies between sources. We believe this is a more appropriate way of measuring the success of instance matching algorithms.

Table 1. Number of entities and properties per data store and entity type

Types	Freebase	DBpedia	LMDB	NYT	DrugBase
		#entities / properties			
Person	10,000 / 1,006	10,000 / 2,537	10,000 / 10	4,979 / 11	0
Film	10,000 / 465	10,000 / 565	10,000 / 48	0	0
Drug	5,000 / 435	5,000 / 247	0	0	6,712 / 36
Org.	5,000 / 641	0	0	3,044 / 11	0
#entities	30,000	25,000	20,000	8,023	6,712
#triples	1,749,433	2,461,263	264,902	90,850	314,108

4   Evaluation Data

For evaluating instance matching systems we rely on real-world data comprising entities of types Person, Film, Drug and Organization. The data was extracted from five stores: Freebase, DBpedia, LinkedMDB, DrugBase and NewYork Times. A detailed description of the data set is presented in Table 1. Instance matching systems are quite resource demanding ([5, 7]). For this reason, the evaluation data has a manageable size of about 90 thousand entities. This translates to about 4.9 million triple representing all relations having one of the selected entities as a subject. Such volume can be matched in a matter of minutes on commodity hardware. A similar number of entities was selected from each data store. The size difference between entity types was considered, too: Overall, the data set comprises about 35 thousand entities of type person, 30 thousand entities of type film, about 15 thousand drug entities, and about 8 thousand organizations. To emphasize data value problems, entities were selected after alphabetically ordering them on their labels. This way, almost all entities have labels starting with the letter 'A'. Due to the small number of entities, DrugBase and NewYork Times have been selected in full. The number of properties per entity type is, with a maximum of 2,537 unique properties for persons, significantly higher than in the OAEI tests. This stresses out structure heterogeneity of real-world data. The ontology differences between data sources, different aggregation levels introduced by LinkedMDB, or the fact that persons are being matched with actors add to the challenges this data set poses. Furthermore, in contrast to OAEI tests, having data form multiple stores increases the risk of building wrong transitive links. At the same time, the fact that multiple domains are compared, the possibility of creating bad links between entities of different types also exists. Finally, the selected data is not heavily interlinked. There are 5,855 owl:sameAs links between entities in our data set. 5,264 of them are between DBpedia and Freebase entities, 548 between DBpedia and LinkedMDB entities and 43 between entities from DBpedia and the NewYork Times.

To encourage further research on this topic, we made this data set, and data generated by our experiments, available at: http://www.ifis.cs.tu-bs.de/node/2906.

Table 2. The number of owl:sameAs links, the number of owl:sameAs links between entities of different types and precision obtained by SLINT+ and by performing the transitive closure on inks created by SLINT+ respectively

θ	SLINT+			cl_{TR}		
	#sameAs	Inter-domain	Prec.	#sameAs	Inter-domain	Prec.
0.95	8,020	33	0.91	2,055	89	0.20
0.75	16,739	119	0.71	5,498	216	0.15
0.50	17,436	230	0.76	7,038	396	0.09
0.25	25,113	1,734	0.67	14,879	2,408	0.02

5   Instance Matching - Experiments

To assess the quality of instance matching systems, we performed instance matching on the data presented in the previous chapter and measured *sampled precision*. We computed the transitive closure of the resulting owl:sameAs links and measured the quality of the newly created links. We paid special attention to the resulting equivalence classes as well as to entities of different types that have been matched. All tests were performed for high to low similarity thresholds. Since one of the characteristics of the data set was that it is not highly interlinked, there were not enough owl:sameAs links available to also measure recall.

The instance matching system is a black box from our perspective. Any domain independent system can be used. SLINT+ is one of the systems to achieve exceptional results in instance matching tasks. It is training-free and domain-independent. It builds on thorough predicate alignment and selection, shallow NLP and correlation based instance matching. It has already been successfully tested on selections from DBpedia and Freebase and it is available online for download[4].

For a similarity threshold of 0.95, SLINT+ creates 8,020 owl:sameAs links (see Table 2). 33 of them link drugs or movies to persons. They are obviously wrong. Overall, we observed a sampled precision of 0.91 for this threshold. The lower the similarity threshold, the more links are found. For a similarity threshold of 0.25, 25,113 links are found. Even for such a low similarity threshold the precision is with a

---

[4] http://ri-www.nii.ac.jp/SLINT/index.html

value of 0.67 quite impressive. According to the OWL standard, owl:sameAs links are transitive. Like most instance matching systems, SLINT+ ignores this aspect, probably because few bad links may lead to an explosion of bad links through transitivity. On the other hand completely ignoring transitive links is dangerous since any query engine using the links created by SLINT+ may transitively link sources to solve join queries. Computing the transitive closure of the owl:sameAs relations discovered by SLINT+ for a threshold of 0.95 we obtained an additional 2,055 links. However, the precision measured for these transitive links is only 0.20.

Table 3. Number of equivalence classes per number of URIs in the equivalence class, for various similarity thresholds

#URIs per class	# equivalence classes			
	θ=0.95	θ=0.75	θ=0.5	θ=0.25
2	4,168	5,054	7,008	8,180
3	529	1,160	2,023	2,781
4	54	222	315	648
5	15	110	136	303
6	7	49	67	167
7	1	24	38	89
8	4	22	22	52
9	5	12	17	43
10	2	11	12	27
11	2	4	8	13
12	2	8	9	9
13	0	1	3	12
14	0	3	1	7
15	1	6	4	6
16	1	1	3	5
17	0	1	3	7
18	1	1	2	4
19	1	1	2	1
20	1	2	2	4
21	1	1	2	2
22	0	1	2	3
23	1	1	1	2
24	0	0	2	1
27	0	1	0	1
29	0	1	1	1
31	0	0	0	1
38	0	0	1	1

But how is this possible? As discussed in Section 2, due to the non-transitive nature of the similarity function, long chains of entities belonging to the same equivalence class may be created. The longer the chain, the higher the probability that URIs that are far apart in the chain refer different entities. Even for high precision oriented similarity thresholds like 0.95, SLINT+ produces 11 equivalence classes with more than 10 URIs each. Actually, the largest equivalence class has 23 URIs, while for lower similarity thresholds there are equivalence classes with 38 URIs (see Table 3). One false owl:sameAs link connecting two smaller equivalence classes in such a large class creates a huge explosion of false links. Assuming two equivalence classes each having 10 URIs, one false link created by SLINT+ connecting the two classes may generate up to 100 incorrect links (all pairwise combinations developing between the two classes: $C_2^{20} - 2 \cdot C_2^{10}$). Considering the high precision for 8,020 links but the low precision for all transitive links, the real, overall precision achieved by SLINT+ for a threshold of 0.95 is $\frac{8{,}020*0.91+2{,}055*0.20}{8{,}020+2{,}055} = 0.77$ and thus quite comparable to LINDA.

Not knowing all owl:sameAs links for all entities from our data set it is impossible to accurately measure *recall*. However, if we take into consideration that 25,113 entities were found with a precision of 0.67 and that an additional 14,879 were found with a precision of 0.02, we can assume that the data set should have, when correctly interlinked, at least 17,123 links (25,113 * 0.67 + 14,879 * 0.02). Assuming that 8,020 * 0.91 + 2,055 * 0.20 = 7,709 correct links have been discovered for a threshold of 0.95, this translates into a recall of at best 0.45. This is significantly lower than the results observed on the OAEI benchmark.

To sum up, results for today's instance matching systems seem quite impressive. But if the problem of transitivity is not properly considered, even for very high similarity thresholds the precision on links obtained through transitivity is catastrophic.

6   Conclusions and Future Work

The most important benefit of linked open data is that it creates a unified view of entities by tapping into information from different data stores. The standard mechanism for connecting instances of the same entity is to transitively exploit owl:sameAs properties. But to do this, first all individual instances of real-world entities have to be linked. Since manually creating all sameAs links is hardly feasible, instance matching systems mostly rely on similarities to automatically create sameAs links for subsequent traversal. The slight problem is that similarity functions are not transitive.

In this paper, we have shown that, even for high similarity thresholds ($\theta=0.95$), ignoring the missing transitivity may have catastrophic effects over the quality of the discovered links. In our experiments for a top-rated system it translated into an overall precision of less than 0.8 for a recall lower than 0.45. In conclusion, unfortunately, today, instance matching is not yet ready for reliable automatic data interlinking.

While our results on one hand call for ways of enforcing transitivity in instance matching systems, they also call for better evaluation within the OAEI instance matching track. A starting point is the data set constructed in this paper. But transitivity problems are by no means the only problems that have to be reflected in the evaluation benchmark. Similar challenges for instance matching, first introduced in [9], are:

- **Data Value Differences:** The same data may be represented differently in different sources. For instance a company's name may be "IBM" in one source and "International Business Machines Corporation" in another.
- **Structural Heterogeneity:** A data type property in one source may be defined as an object property in another source. Multiple properties from one source (first name and last name) may be composed into a single property in other sources (name). One source may have three values for a property while in another source the same property has just one value.
- **Logical Heterogeneity:** Instances of the same real-world object may belong to different concepts. These concepts may be subclasses of the same superclass. Two instances having the same property values may belong to disjoint classes.

Considering all this, a proper data set for evaluating instance matching systems should have triples from multiple stores for transitivity reasons, with a certain level of overlap between domains and different levels of data quality to address data value differences, and it should include sources having properties with different levels of cardinality and granularity to address structural and logical heterogeneity.

In the near future, after building a benchmark for proper matching evaluation we plan to analyze the transitive closures and their respective precision/recall also for RiMOM2013 and LogMap, two other state of the art systems. Moreover, we will thoroughly analyze ways of enforcing transitivity by design in instance matching.

References

1. Ding, L., Shinavier, J., Shangguan, Z., McGuinness, D.L.: SameAs Networks and Beyond: Analyzing Deployment Status and Implications of owl:sameAs in Linked Data. In: Patel-Schneider, P.F., Pan, Y., Hitzler, P., Mika, P., Zhang, L., Pan, J.Z., Horrocks, I., Glimm, B. (eds.) ISWC 2010, Part I. LNCS, vol. 6496, pp. 145–160. Springer, Heidelberg (2010)
2. Volz, J., Bizer, C., Gaedke, M., Kobilarov, G.: Discovering and Maintaining Links on the Web of Data. In: Bernstein, A., Karger, D.R., Heath, T., Feigenbaum, L., Maynard, D., Motta, E., Thirunarayan, K. (eds.) ISWC 2009. LNCS, vol. 5823, pp. 650–665. Springer, Heidelberg (2009)
3. Isele, R., Jentzsch, A., Bizer, C.: Silk Server - Adding missing Links while consuming Linked Data. In: Int. Workshop on Consuming Linked Data, COLD (2010)
4. Tang, J., Li, J., Liang, B., Huang, X., Li, Y., Wang, K.: Using Bayesian decision for ontology mapping. Journal Web Semant. Sci. Serv. Agents World Wide Web. 4, 243–262 (2006)
5. Böhm, C., de Melo, G.: LINDA: Distributed web-of-data-scale entity matching. In: Conf. Information and Knowledge Management, CIKM (2012)
6. Jiménez-Ruiz, E., Cuenca Grau, B.: LogMap: Logic-Based and Scalable Ontology Matching. In: Aroyo, L., Welty, C., Alani, H., Taylor, J., Bernstein, A., Kagal, L., Noy, N., Blomqvist, E. (eds.) ISWC 2011, Part I. LNCS, vol. 7031, pp. 273–288. Springer, Heidelberg (2011)

7. Nguyen, K., Ichise, R., Le, B.: Interlinking Linked Data Sources Using a Domain-Independent System. In: Takeda, H., Qu, Y., Mizoguchi, R., Kitamura, Y. (eds.) JIST 2012. LNCS, vol. 7774, pp. 113–128. Springer, Heidelberg (2013)
8. Halpin, H., Hayes, P.J.: When owl:sameas isn't the same: An analysis of identity links on the semantic web. In: Linked Data on the Web, LDOW (2010)
9. Alfio Ferrara, D.L.: Towards a Benchmark for Instance Matching. In: Int. Workshop on Ontology Matching, OM (2008)
10. Tang, J., Liang, B.-Y., Li, J., Wang, K.-H.: Risk Minimization based Ontology Mapping. In: Chi, C.-H., Lam, K.-Y. (eds.) AWCC 2004. LNCS, vol. 3309, pp. 469–480. Springer, Heidelberg (2004)
11. Zheng, Q., Shao, C., Li, J., Wang, Z., Hu, L.: RiMOM2013 Results for OAEI 2013. In: Int. Workshop on Ontology Matching, OM (2013)

Improving Personalization and Contextualization of Queries to Knowledge Bases Using Spreading Activation and Users' Feedback

Ana Belen Pelegrina[1], Maria J. Martin-Bautista[2], and Pamela Faber[1]

[1] Department of Translation and Interpreting
University of Granada
{abpelegrina,pfaber}@ugr.es
[2] Department of Computer Science and Artificial Intelligence
University of Granada
mbautis@decsai.ugr.es

Abstract. Facilitating knowledge acquisition when users are consulting knowledge bases (KB) is often a challenge, given the large amount of data contained. Providing users with appropriate contextualization and personalization of the content of KBs is a way to try to achieve this goal. This paper presents a mechanism intended to provide contextualization and personalization of queries to KBs based on collected data regarding users' preferences, both implicitly (users' profiles) and explicitly (users' feedback). This mechanism combines user data with a spreading activation (SA) algorithm to generate the contextualization. The initial positive results of the evaluation of the contextualization are presented in this paper.

Key words: Knowledge base, spreading activation, user feedback, contextualization, and personalization.

1 Introduction

When a knowledge base (KB) is well structured and contextualized, it can become an access point to domain knowledge and greatly facilitate knowledge acquisition. Such contextualization is crucial because the users of the resource should be able to easily access the information that is most relevant to their needs.

One way to supply users with relevant information is to filter it to provide information in consonance with their domain profile and preferences. This is a complex task and requires the use of methods capable of capturing the interests of potential users and incorporating these preferences in the software. Effective Computer Science methods for this purpose include clustering [1,2], fuzzy logic [3,4], data mining [5,6], and SA [7,8].

This paper describes a mechanism based on SA for personalized knowledge extraction, derived from user activities and feedback. In Computer Science, SA is a search method for associative and semantic networks as a way to simulate

T. Andreasen et al. (Eds.): ISMIS 2014, LNAI 8502, pp. 285–294, 2014.
© Springer International Publishing Switzerland 2014

a *'human-like'* search for information or documents (see Sections 2 and 3 for a more exhaustive explanation of SA and its applications). In the framework of the proposed mechanism, information concerning user interactions with the KB is combined with user feedback and integrated into a SA algorithm to personalize and contextualize the results of queries to the KB. The first implementation of this mechanism was presented in [8], although this first version did not make provision for explicit user feedback. The mechanism is applied and tested in EcoLexicon, a KB of environmental knowledge [9].

This paper is organized as follows, Section 2 provides the background of SA. Section 3 discusses related work and gives an outline of our proposal. Section 4 describes our approach and its implementation. The preliminary results of the application of the technique for a test implementation are shown in Section 5. Finally, Section 6 presents conclusions and future work.

2   Spreading Activation

SA theory was first proposed in the fields of psychology, semantic processing, and linguistics [10,11]. Its objective is to explain the way humans represent and retrieve the knowledge in the human brain. This theory states that knowledge is represented in the mind in the form of a graph or a semantic network composed of concepts (or nodes) and relations between concepts (links) in the form of a semantic network. According to this theory, humans recall interconnected concepts starting with the activation of an initial concept, which spreads to the adjacent nodes in a sequence of pulses in a decreasing gradient [11].

In Computer Science, the SA process was adopted as the basis for a search algorithm in semantic and associative networks. This algorithm assumes that each node in the network has an activation value and each edge has a weight. These values are usually initialized to certain values that represent certain properties of the information in the network, such as the strength of the relationship between the connected nodes or the relevance of certain nodes.

The activation process begins with a set of source nodes, for example, the search terms. First, the activation value of these nodes is updated. Later, this activation spreads through the neighboring nodes in a series of iterations. The activation value of these nodes is calculated by using (1) and (2) [12]. More specifically, the input value (I_j) is computed by accumulating the output value (O_i) of the nodes directly connected to it, adjusted by the weight of the relations between the nodes ($w_{i,j}$). The output value (O_j) is calculated by applying the activation function (f) to the input value I_j. The spreading process ends when at least one of the stop conditions (e.g., number of visited nodes) is achieved. The algorithm can also be limited by a set of restrictions that constrain the spreading activation process in order to adjust or improve the results of the algorithm. These restrictions thus compensate for some of the drawbacks of a pure spreading activation approach (e.g. saturation of the network) [12].

$$I_j = \sum_i O_i w_{i,j} \tag{1}$$

$$O_j = f(I_j) \qquad (2)$$

3 Related Work

In Computer Science, SA has been used to solve a wide range of problems in different areas. These include information and document retrieval [12,13], Web search [14,15,16], recommendation systems [17,18], data analysis [19], and data visualization [20,21]. This broad usage stems from benefits of this technique such as the following: (i) widespread availability of the underlying data structure required; (ii) its simplicity and customizability; (iii) the 'human-like' nature of the generated results.

In this paper, we integrate user feedback in the mechanism proposed in [8] as a way to improve the personalization of the results. The fact that such feedback is employed in a wide variety of computer systems from machine learning[22] to recommender systems [23], is an indication of its potential usefulness.

The combination of user feedback and the SA technique has been explored in a number of research papers [12,13,24,25,26]. For instance, the proposals in [13,24] concern the use of feedback and SA in information retrieval systems. However, none of them includes the contextualization and personalization of the knowledge proposed in this paper.

A similar approach to the work presented here can be found in [25]. In this paper, the author combines feedback from users with an SA algorithm to provide adaptation and harness knowledge from users. However the framework in [25] is designed for only one context (i.e., organizations), whereas our proposal can be applied to many different domains and contexts. In addition, it only provides global adaptation and is thus unable to adapt the results to the preferences of different users.

Other work relies on implicit user feedback in combination with SA, such as [26]. However, the lack of an option to provide explicit feedback might hinder the need for users to give negative feedback about certain results. For example, if a user makes two consecutive searches for two entities the system has no way of differentiating between (i) a user that is satisfied with the results of the query and proceeds with an unrelated query, and (ii) a user that is dissatisfied and makes another query with the hope of obtaining better results.

4 Personalization and Contextualization of Queries to KBs

The goal of this research is to show how feedback can be used in knowledge based systems to improve contextualization and personalization for users. In more depth, we propose a modification of the mechanism presented in [8], a four-step mechanism for personalization and knowledge extraction derived from the information about users' activities by means of SA. This proposal is applied to networked/graph-based knowledge bases (knowledge structured as a set of entities that are connected by different relations). When a user performs a query about a certain entity, the system shows him/her a network with this entity at the center of the network and linked to clusters of the most relevant entities.

The mechanism is modified by adding a new step to the process that provides with users an opportunity to offer feedback about the quality or adequacy of the results, based on their needs. The updated process is shown in Fig. 1. A brief explanation of each step is provided below.

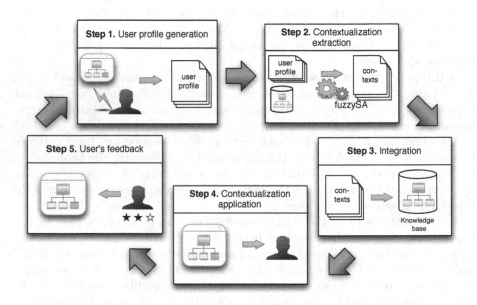

Fig. 1. The steps involved in the proposed mechanism

1. **User profiles generation.** The construction of this profile is based on the queries and browsing behavior of the user.
2. **Contextual information extraction.** By combining the user profile information and his/her feedback, a SA algorithm is used to generate the contextualizations.
3. **Integration of the contextualization in the knowledge base.** The resulting contextualization then becomes part of the knowledge base.
4. **Application of the contextualization.** When users are searching or browsing, the contextualization obtained in steps 2 and 3 is employed to personalize their activities.
5. **User's feedback.** The system presents the users with a mechanism to evaluate the results obtained for their queries. This feedback will be subsequently used to improve the results provided in step 3. This paper is focused on this step.

Section 4.1 provides a detailed explanation of how the new feedback step is incorporated in the mechanism. The modified SA algorithm is presented in Section 4.2.

4.1 Users' Feedback

When the system gives users the results of their queries, it also provides them with a set of three possible values for the feedback, listed below. We have chosen to only allow users a small number of options in order to facilitate their choice.

– 'Bad' (-1). The user judges the results provided by the system as unsatisfactory or incorrect. This will penalize the entities included in the result next time the user repeats the query.
– 'None/neutral' (0). The user judges the results provided by the system as neutral or does not offer any feedback for these results. This is the default value and will not influence the results in the next query.
– 'Good' (+1). The user judges the results provided by the system as acceptable. This will reward the entities included in the result the next time that the user repeats the query.

Once the user makes his/her evaluation, it is stored in the knowledge base. For each entity presented to the user, the evaluation is stored in relation to the central entity. For example, if the center of the network is the entity *'WATER'*, the entities *'RAIN'* and *'ICE'* are included in the network, and if the user provides a 'good' feedback, the system will store the following tuples in the knowledge base: *('WATER', 'RAIN', 'good')* and *('WATER', 'ICE', 'good')*. The next time the user performs a search for this entity the system will take into account this feedback to adapt the results to the user preferences.

4.2 New Spreading Activation Algorithm

As a means of incorporating the feedback, the SA algorithm in [8] was modified. The first modification affects the inputs of the algorithm, which now include the entities added to the user profile and the feedback provided by the user. The algorithm now also includes a call to a new procedure (see Alg. 1) that processes the feedback and profile data, and which computes the initial activation value for each node based in this data.

5   Experimental Example

EcoLexicon is a terminological knowledge base for the domain of the Environment [9]. It is composed of more than 4,000 concepts, 10,000 terms in different languages, multimedia resources, and linguistic data, as well as the relations between the aforementioned concepts, terms and resources. Users can browse and search the content of EcoLexicon through a web application, available at http://ecolexicon.ugr.es/.

Algorithm 1 Processing of the feedback and profiles

---

procedure PROCESSFEEDBACKANDPROFILE($NODES$, $ORIGIN$, $PROFILE$, $FEEDBACK$)

    for all $node \in NODES$ **do**

        $A_{node} \leftarrow 0$

        if $node \in PROFILE$ **then**

            $A_{node} \leftarrow 1$

        end if

        if $node \in FEEDBACK_{ORIGIN}$ **then**

            $A_{node} \leftarrow FEEBACK_{ORIGIN,node}$

        end if

    end for

    return A

end procedure

Algorithm 2 Spreading Activation Algorithm.

---

procedure SPREADINGACTIVATION($NODES$, $VERTEX$, $ORIGIN$, $PROFILE$, $FEEDBACK$, F, D)

    $A \leftarrow ProcessFeedbackAndProfile(NODES,ORIGIN,PROFILE,FEEDBACK)$

    $A_{ORIGIN} \leftarrow 1$

    $toFire \leftarrow ORIGIN$

    while $toFire \neq \varnothing$ **do**

        for all $i \in toFire$ **do**

            toFire.remove(i)

            for all $j \in adjacents(i)$ **do**

                $A_j \leftarrow i(A_{j-1})$

                $O_j \leftarrow f(A_j)$

                if $O_j > F$ **then**

                    toFire.add(j)

                end if

            end for

        end for

    end while

    return A, O

end procedure

EcoLexicon classifies the different concepts and relations between them in a set of domains (e.g., *geology, environmental sciences, soil sciences*, etc.) [27]. Nevertheless, these domains are manually defined and thus require the work of a number of experts in various areas to create them. The goal of this section is to present an alternative to these domains using the extended mechanism proposed and to test the validity of the hypothesis in this paper (i.e. the inclusion of user feedback to improve the results). With this purpose in mind, the visual tool was extended to include modules to build the user profiles and gather the user feedback. In the case of EcoLexicon, the user profiles contain every entity (concept or term) that the user has searched or interacted with in the web application.

Section 5.1 describes the experiment design. The results of the experiment are presented in Section 5.2.

5.1   Experimental Setup

In order to evaluate our proposed improvement to the mechanism, we chose the profile of a user with some expertise in the following domains: *'hydrology'*, *'physics'*, *'geology'*, *'coastal engineering'* and *'meteorology'*. We also selected a set of fifteen random concepts included in the aforementioned domains: *'breeze'*, *'earthquake'*, *'wind'*, *'dune'*, *'fog'*, *'relief'*, *'ocean waves'*, *'hurricane'*, *'fan'*, *'steel'*, *'gravity'*, *'water'*, *'cliff'*, *'seashore'* and *'osmosis'*. We then evaluated each possible pair of concepts and domains (DM, C) using the following steps:

1. We ran the algorithm presented in [8] using the concept C as the $START$.
2. We ran the algorithm introduced in Section 4.2 using the concept C as the $ORIGIN$, using the same parameters as in step 1 (decay factor, activation function, etc.).
3. We compared the results obtained in steps 1 and 2 with the domain DM, which had been manually defined by experts. The result was a percentage of correctly classified concepts for each algorithm.

The parameters for both SA algorithms were the following:

- Activation threshold (F): 0.65.
- Decay factor (D): 0.9.
- Activation function (see (3) and (4)): the activation value (A_j) for the entity j is computed using its previous activation value (A_{j-1}), the output value of node i (O_i), the weight of the relation between nodes i and j ($w_{i,j}$), and the decay factor (D). Subsequently, the output value (O_j), to be used in future pulses, is calculated with the activation threshold (F).

$$A_j = i(A_{j-1}) = A_{j-1} + O_i D w_{i,j} \tag{3}$$

$$O_j = f(A_j) = \begin{cases} 0 & \text{if } A_j < F \\ A_j & \text{otherwise} \end{cases} \tag{4}$$

5.2   Results

Table 1 shows the average difference for each concept obtained with the mechanism with and without feedback. As an example, Table 2 shows the detailed results for the concept 'breeze'. As can be seen in Table 1, the version of the algorithm with feedback improved the results of the algorithm without feedback by an average of 2.13%. However, for some concepts (e.g., 'earthquake') there is no improvement in the results. This can be explained by some incorrect feedback provided by the user. To address this possibility, it is necessary to provide users with a way to undo incorrect feedback.

Table 1. Results of the experiment. The table contains the average for the algorithm with no feedback, with feedback, and the average difference in the results for each concept. The average was calculated with the results of the comparison between the results and the experts domains.

Concept	Avg. no feedback	Avg. feedback	Difference
Breeze	93.85%	95.00%	1.22%
Earthquake	97.58%	96.73%	-0.85%
Wind	95.54%	96.53%	0.99%
Dune	91.65%	91.83%	0.18%
Fog	96.20%	96.75%	0.55%
Relief	90.80%	93.54%	2.74%
Ocean waves	89.20%	93.05%	3.85%
Hurricane	90.32%	95.33%	3.70%
Fan	95.36%	95.89%	0.53%
Steel	85.16%	89.36%	4.20%
Gravity	89.93%	93.21%	3.28%
Water	88.73%	92.89%	4.15%
Cliff	79.41%	81,53%	2,13%
Seashore	90.09%	92.99%	2.90%
Osmosis	86.83%	88.64%	1.81%
Average	90.40%	92.53%	2.13%

Table 2. Results for the concept 'Breeze' in each domain

Concept	Domain	No feedback	Feedback	Difference
	Hydrology	96.08%	96.03%	-0.05%
	Physics	91.50%	93.39%	1.89%
Breeze	Geology	98.04%	97.19%	-0.85%
	Coastal engineering	92.16%	94.21%	2.06%
	Meteorology	91.50%	94.55%	3.04%

6   Conclusions and Future Work

As a way to provide contextualized and personalized queries in KB, we have proposed a five-step mechanism based on the following: (i) user profiles; (ii) a SA algorithm; and (iii) user feedback. This mechanism is an extension of the one proposed in [8], which includes a new step for user feedback. By taking advantage of the feedback provided by the users of the KB, the results offered are better contextualized and personalized. In order to evaluate the improvement provided by the extension of the mechanism, we performed an experiment to test the improvement in the contextualization. The results of the experiment (an average improvement of 1.71%) indicate that using feedback is an effective way of improving the contextualization and personalization of queries in KBs.

Regarding future work, we plan to include more detailed user feedback (e.g. rate only one entity), better management of uncertainty, new parameters and constraints for the SA algorithm, and an evaluation of the contextualization and personalization by users and experts.

Acknowledgements. This research has been carried out within the framework of the project FFI2011-22397, funded by the Spanish Government; the project P11-TIC7460, funded by Junta de Andalucía; FPI research grant BES-2012-052718; and the project FP7-SEC-2012-312651, funded from the European Union in the Seventh Framework Programme [FP7/2007-2013] under grant agreement No 312651.

References

1. Ferragina, P., Gulli, A.: A personalized search engine based on web-snippet hierarchical clustering. Software: Practice and Experience 38(2), 189–225 (2008)
2. Han, L., Chen, G.: A fuzzy clustering method of construction of ontology-based user profiles. Advances in Engineering Software 40(7), 535–540 (2009)
3. Widyantoro, D., Yen, J.: Using fuzzy ontology for query refinement in a personalized abstract search engine. In: International Joint Conference on 9th IFSA World Congress and 20th NAFIPS 2001, vol. 1, pp. 610–615 (July 2001)
4. Kim, K.J., Cho, S.B.: Personalized mining of web documents using link structures and fuzzy concept networks. Applied Soft Computing 7(1), 398–410 (2007)
5. Mobasher, B., Cooley, R., Srivastava, J.: Automatic personalization based on web usage mining. Communications of the ACM 43(8), 142–151 (2000)
6. Mulvenna, M.D., Anand, S.S., Büchner, A.G.: Personalization on the net using web mining: Introduction. Communications of the ACM 43(8), 122–125 (2000)
7. Katifori, A., Vassilakis, C., Dix, A.: Ontologies and the brain: Using spreading activation through ontologies to support personal interaction. Cognitive Systems Research 11(1), 25–41 (2010)
8. Pelegrina, A.B., Martin-Bautista, M.J., Faber, P.: Contextualization and personalization of queries to knowledge bases using spreading activation. In: Larsen, H.L., Martin-Bautista, M.J., Vila, M.A., Andreasen, T., Christiansen, H. (eds.) FQAS 2013. LNCS, vol. 8132, pp. 671–682. Springer, Heidelberg (2013)
9. Reimerink, A., Faber, P.: Ecolexicon: A frame-based knowledge base for the environment. In: Proceedings of the International Conference "Towards eEnvironment", pp. 25–27 (2009)
10. Anderson, J.R.: A spreading activation theory of memory. Journal of Verbal Learning and Verbal Behavior 22(3), 261–295 (1983)
11. Collins, A.M., Loftus, E.F.: A spreading-activation theory of semantic processing. Psychological Review 82(6), 407 (1975)
12. Crestani, F.: Application of spreading activation techniques in information retrieval. Artificial Intelligence Review (1997)
13. Preece, S.E.: Spreading activation network model for information retrieval. Dissertation Abstracts International Part B: Science and Engineering 42(9) (1982)

14. Crestani, F., Lee, P.L.: Searching the web by constrained spreading activation. Information Processing & Management 36(4), 585–605 (2000)
15. Jiang, X., Tan, A.H.: Learning and inferencing in user ontology for personalized Semantic Web search. Information Sciences 179(16), 2794–2808 (2009)
16. Sieg, A., Mobasher, B., Burke, R.: Ontological User Profiles for Representing Context in Web Search. In: 2007 IEEE/WIC/ACM International Conferences on Web Intelligence and Intelligent Agent Technology - Workshops, pp. 91–94. IEEE (November 2007)
17. Sieg, A., Mobasher, B., Burke, R.: Improving the effectiveness of collaborative recommendation with ontology-based user profiles. In: Proceedings of the 1st International Workshop on Information Heterogeneity and Fusion in Recommender Systems - HetRec 2010, pp. 39–46. ACM Press, New York (2010)
18. Sieg, A., Mobasher, B., Burke, R.: Ontology-Based Collaborative Recommendation. In: ITWP (2010)
19. Teufl, P., Payer, U., Parycek, P.: Automated analysis of e-participation data by utilizing associative networks, spreading activation and unsupervised learning. In: Macintosh, A., Tambouris, E. (eds.) ePart 2009. LNCS, vol. 5694, pp. 139–150. Springer, Heidelberg (2009)
20. Collins, C., Carpendale, S.: Vislink: Revealing relationships amongst visualizations. IEEE Transactions on Visualization and Computer Graphics 13(6), 1192–1199 (2007)
21. Kuß, A., Prohaska, S., Meyer, B., Rybak, J., Hege, H.C.: Ontology-based visualization of hierarchical neuroanatomical structures. Proc. Vis. Comp. Biomed, 177–184 (2008)
22. Stumpf, S., Rajaram, V., Li, L., Burnett, M., Dietterich, T., Sullivan, E., Drummond, R., Herlocker, J.: Toward harnessing user feedback for machine learning. In: Proceedings of the 12th International Conference on Intelligent User Interfaces, IUI 2007, pp. 82–91. ACM, New York (2007)
23. Shapira, B.: Recommender systems handbook. Springer (2011)
24. Shoval, P.: Expert/consultation system for a retrieval data-base with semantic network of concepts. In: ACM SIGIR Forum, vol. 16, pp. 145–149. ACM (1981)
25. Hasan, M.M.: A spreading activation framework for ontology-enhanced adaptive information access within organisations. In: van Elst, L., Dignum, V., Abecker, A. (eds.) AMKM 2003. LNCS (LNAI), vol. 2926, pp. 288–296. Springer, Heidelberg (2004)
26. Hussein, T., Ziegler, J.: Adapting web sites by spreading activation in ontologies. In: Proceedings of International Workshop on Recommendation and Collaboration, New York, USA (2008)
27. León Araúz, P., Magaña Redondo, P.: Ecolexicon: Contextualizing an environmental ontology. In: Proceedings of the Terminology and Knowledge Engineering (TKE) Conference 2010, pp. 341–355 (2010)

Plethoric Answers to Fuzzy Queries:
A Reduction Method Based on Query Mining

Olivier Pivert and Grégory Smits

University of Rennes 1, Irisa
Lannion France
{olivier.pivert,gregory.smits}@univ-rennes1.fr

Abstract. Querying large-scale databases may often lead to plethoric
answers, even when fuzzy queries are used. To overcome this problem,
we propose to strengthen the initial query with additional predicates,
selected among predefined ones according mainly to their degree of semantic relationship with the initial query. In the approach we propose,
related predicates are identified by mining a repository of previously executed queries.

Keywords: Databases, fuzzy queries, plethoric answers, cooperative answering, query augmentation, query mining.

1 Introduction

The practical need for endowing intelligent information systems with the ability
to exhibit a cooperative behavior has been recognized since the early nineties. As
pointed out in [9], the main intent of cooperative systems is to provide correct,
non-misleading and useful answers, rather than literal answers to user queries.

Two dual problems are addressed in this field. The first one is known as the
empty answer set (EAS) problem, that is, the problem of providing the user
with alternative data when there is no item fitting his/her query. The second
one is the plethoric answer set (PAS) problem which occurs when the amount
of returned data is too large to be manageable. This paper focuses on this latter
issue in the context of fuzzy queries.

The PAS problem has been intensively addressed by the information retrieval
community and two main approaches have been proposed for Boolean queries.
The first one, that may be called data-oriented, aims at ranking the answers in
order to return the best k ones to the user, by using a complementary preference
clause [11,6,10,8]. These approaches assume that the preference clause makes
it possible in general to obviate the PAS problem, but this is of course not
always true. Here, we consider the case where a (fuzzy) preference query returns
a plethoric answer set, which corresponds to the case where there are too many
elements that belong to the class of the most preferred answers. In this data-
oriented approach, one may also include the works that aim at summarizing the
answer set to a query [20].

T. Andreasen et al. (Eds.): ISMIS 2014, LNAI 8502, pp. 295–304, 2014.
© Springer International Publishing Switzerland 2014

The second type of approach may be called query-oriented as it performs a modification of the initial query in order to make it more selective. For instance, a possible strategy is to strengthen (some of) the predicates involved in the query (for example, a predicate $A \in [a_1, a_2]$ may be transformed into $A \in [a_1+\gamma, a_2-\gamma]$) [2]. However, for some predicates, this type of strengthening may lead to a deep modification of the meaning of the initial predicate. Consider for instance a query looking for fast-food restaurants located in a certain district delimited by geographical coordinates. A strengthening of the condition related to the location could lead to the selection of restaurants in a very small area, and the final answers would not necessarily fit the user's need. Another category of query-oriented approaches aims at automatically completing the initial query with additional predicates to make it more demanding [12,5,4]. Our work belongs to this last family of approaches but its specificity concerns the way additional predicates are selected.

Indeed, we consider that the predicates added to the query must respect two properties: i) they must reduce the size of the initial answer set, ii) they must modify the scope of the initial query as little as possible. Based on a pre-defined vocabulary materialized by fuzzy partitions that linguistically describe the attribute domains, we propose to identify the predicates that are the most correlated to the initial query. Such correlation properties are inferred from a workload of previously submitted queries.

The remainder of the paper is structured as follows. Section 2 introduces the basic necessary notions and Section 3 describes the query augmentation approach we propose, based on the retrieval of the predicates the most related to a query. Section 4 discusses related work. Section 5 recalls the main contributions and outlines some perspectives for future research.

2   Preliminaries

2.1   The Plethoric Answer Problem in the Fuzzy Query Case

We consider a database fuzzy querying framework such as the SQLf language presented in [3,14] and whose implementation is described in [17]. Besides the fuzzy predicates used to express the user requirements, we also assume that he/she specifies a quantitative threshold defining the approximate number of expected results, denoted by k.

A typical example of a fuzzy query is: "retrieve the recent and low-mileage cars", where *recent* and *low-mileage* are gradual predicates represented by fuzzy sets as illustrated in Figure 1. Let Q be a fuzzy query. We denote by Σ_Q the answer set to Q when addressed to a regular relational database D. Σ_Q contains the items of the database that *somewhat* satisfy the fuzzy requirements involved in D. Formally,

$$\Sigma_Q = \{t \in D \mid \mu_Q(t) > 0\},$$

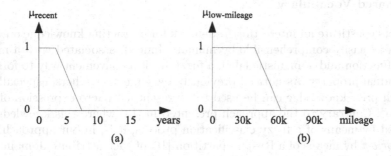

Fig. 1. Fuzzy predicates (a) *recent* and (b) *low-mileage* (where 30K means 30.000 km)

where t stands for a database tuple. Let h, $h \in (0, 1]$ be the height of Σ_Q, *i.e.* the highest membership degree assigned to an item of Σ_Q. Let now Σ_Q^* ($\subseteq \Sigma_Q$) denote the set of answers that satisfy Q to degree h.

$$\Sigma_Q^* = \{t \in D \mid \mu_Q(t) = h\}$$

Definition 1. *Let Q be a fuzzy query, we say that Q leads to a PAS problem if the set Σ_Q^* is too large, i.e., is significantly greater than k.*

To reduce Σ_Q^*, we propose an approach that integrates additional predicates as new conjuncts to Q. By doing so, we obtain a more restrictive query Q' which leads to a reduced set of answers $\Sigma_{Q'}^* \subseteq \Sigma_Q^*$. This strengthening strategy based on predicate correlation is mainly dedicated to what we call underspecified queries, i.e., queries which typically involve few predicates (between one and three) to describe an information need that could be more precisely described by using additional properties. For example, consider a user looking for *almost new* cars through a query like "find second hand cars which are very recent". The answer set to this query can be reduced through the integration of additional properties such as *low mileage, high security* and *high comfort level, i.e.*, properties that are usually possessed by very recent cars.

This strengthening approach aims at identifying correlation links between additional properties and an initial query. These additional correlated properties are suggested to the user as candidates for the strengthening of his/her initial query. This interactive process is iterated until the result is of an acceptable size for the user and corresponds to what he/she was really looking for.

As to the integration of this technique in a fuzzy querying system, several scenarios are possible. First, the query augmentation algorithm may be auto- matically executed when a query answer set satisfies the condition specified in Definition 1. Second, this approach may also be used as the basis of a faceted- search system where the user is guided step-by-step in the specification of his/her query (this aspect will not be detailed here due to lack of space).

2.2   Shared Vocabulary

Fuzzy sets constitute an interesting framework for extracting knowledge on data that can be easily comprehensible by humans. Indeed, associated with a membership function and a linguistic label, a fuzzy set is a convenient way to formalize a gradual property. As noticed previously by some researchers, especially in [13], such prior knowledge can be used to represent a "macro expression of the database". Contrary to the approach presented in [13] where this knowledge is computed by means of a fuzzy classification process, it is, in our approach, defined *a priori* by means of a Ruspini partition [16] of each attribute domain. Let us recall that a Ruspini partition is composed of fuzzy sets, where a set, say P_i, can only overlap with its predecessor P_{i-1} or/and its successor P_{i+1} (when they exist) and for each tuple t, $\sum_{i=1}^{n_i} \mu_{P_i}(t) = 1$, where n_i is the number of partition elements of the concerned attribute. These partitions are assumed to be specified by an expert during the database design step and represent "common sense partitions" of the domains instead of the result of an automatic process which may be difficult to interpret. Indeed, we consider that the predefined fuzzy sets involved in a partition constitute a shared vocabulary and that these fuzzy sets are used by the users when formulating their queries. Moreover, in our approach, the predicates that are added to the initial query also belong to this vocabulary.

Let us consider a relation R containing w tuples $\{t_1, t_2, \ldots, t_w\}$ defined on a set Z of q categorical or numerical attributes $\{Z_1, Z_2, \ldots, Z_q\}$. A shared predefined vocabulary on R is defined by means of partitions of the q domains. A partition $\mathscr{P}_i$ associated with the domain of attribute Z_i is composed of m_i fuzzy predicates $\{P_{i,1}, P_{i,2}, \ldots, P_{i,m_i}\}$. A predefined predicate denoted by $P_{i,j}$ corresponds to the jth element of the partition defined on attribute Z_i. Each $\mathscr{P}_i$ is associated with a set of linguistic labels $\{L_{i,1}, L_{i,2}, \ldots, L_{i,m_i}\}$, each of them corresponding to an adjective that gives the meaning of the fuzzy predicate, see Fig. 2.

A query Q to this relation R is composed of fuzzy predicates chosen among the predefined ones which form the partitions. If Q leads to a plethoric answer set, we propose to strengthen Q in order to obtain a more restrictive query Q' such that $\Sigma^*_{Q'} \subseteq \Sigma^*_Q$. Query Q' is obtained through the integration of additional predefined fuzzy predicates chosen from the shared vocabulary defined on R.

3   An Approach Based on Query Mining

3.1   Query Augmentation Based on Predicate Co-occurrence

In the approach we propose, the new conjuncts to be added to the initial query are chosen among a set of possible predicates pertaining to the attributes of the schema of the database queried (see Section 2.2). This choice is mainly made according to their (statistical) association with the initial query. A user query Q is composed of $n, n \geq 1$, specified fuzzy predicates, denoted by $P_{k_1,l_1}, P_{k_2,l_2}, \ldots, P_{k_n,l_n}$, which come from the shared vocabulary associated with the database. The

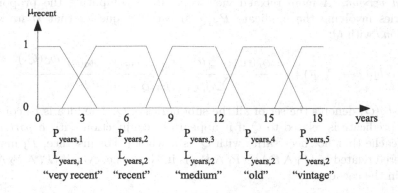

Fig. 2. A partition of the domain of attribute *years*

first step of the strengthening approach is to identify the predefined predicates the most related to the initial query Q.

The notion of relatedness is used to assess the extent to which two fuzzy conditions (one associated with a predefined predicate $P_{i,j}$, the other associated with the initial query Q) are somewhat "semantically" linked. Roughly speaking, we consider that a predicate $P_{i,j}$ is somewhat related to a query Q if user queries involving some conditions from Q often include condition $P_{i,j}$ also. For instance, one may assume that a fuzzy predicate "highly powerful engine" is more related to a query aimed at retrieving "fast cars" than a predicate such as "low consumption". Adding predicates that are related to the user-specified ones makes it possible to preserve the scope of the query (i.e., the user's intent) while making it more demanding.

3.2 Finding the Most Related Predicates

Let $Q = P_{k_1,l_1} \wedge \ldots \wedge P_{k_n,l_n}$ be the user query yielding a plethoric answer set. Let $\mathcal{C}$ be a query cache (i.e., a workload of past queries). The idea is to look in $\mathcal{C}$ for queries of the form $Q' \wedge P_{k'_1,l'_1} \wedge \ldots \wedge P_{k'_q,l'_q}$, where $Q' \sqsubseteq Q$ (seeing a query as a set of predicates linked conjunctively). For each such query, one has to assess the correlation between Q and each $P_{k'_i,l'_i}$, $i \in \{1..q\}$. Let us denote by $|\mathcal{C}|$ the number of queries in the cache. Let us denote by $nb_\mathcal{C}(Q_i)$ the number of times a query Q_i has been submitted to the system and $N = \sum_{i=1}^{|\mathcal{C}|} nb_\mathcal{C}(Q_i)$.

Drastic version. A first way of defining relatedness corresponds to the following formula:

$$rel_1(Q \rightarrow P_{k'_i, l'_i}) = \frac{\sum_{Q_i \in \mathcal{C} \mid Q_i \sqsupseteq (Q \wedge P_{k'_i, l'_i})} nb_\mathcal{C}(Q_i)}{\sum_{Q_i \in \mathcal{C} \mid Q_i \sqsupseteq Q} nb_\mathcal{C}(Q_i)}. \tag{1}$$

Here, the idea is to compute the proportion of queries involving the condition $P_{k'_i, l'_i}$ among the queries *more specific than* Q that have been run in the past.

Relaxed version. A more relxaed view consists in computing the proportion of queries involving the predicate $P_{k'_i, l'_i}$ among the queries that *share some conditions* with Q:

$$rel_2(Q \to P_{k'_i, l'_i}) = \frac{\sum_{Q_i \in \mathcal{C} \mid Q_i \sqsupseteq (Q' \wedge P_{k'_i, l'_i}) \text{ and } Q' \in sbq(Q)} nb_{\mathcal{C}}(Q_i)}{\sum_{Q_i \in \mathcal{C} \mid Q_i \sqcap Q \neq \emptyset} nb_{\mathcal{C}}(Q_i)} \tag{2}$$

where $sbq(Q)$ denotes the set of all the subqueries of Q. The idea is to consider that a predicate is related to Q if it appears in conjunction with *a part of Q* in the cache (but not necessarily with Q as a whole). For instance, P_3 may be considered related to $(P_1 \wedge P_2)$ if $(P_1 \wedge P_3)$ is in the cache, even if $(P_1 \wedge P_2 \wedge P_3)$ is *not* in the cache.

Weighted version. A possible refinement of the previous formula is:

$$rel_3(Q \to P_{k'_i, l'_i}) =$$

$$\frac{\sum_{Q_i \in \mathcal{C} \mid Q_i = (Q' \wedge P_{k'_i, l'_i}) \text{ and } Q' \in sbq(Q)} (nb_{\mathcal{C}}(Q_i) \times w(Q_i - \{P_{k'_i, l'_i}\}, Q))}{\sum_{Q_i \in \mathcal{C} \mid Q_i \sqcap Q \neq \emptyset} (nb_{\mathcal{C}}(Q_i) \times w(Q_i, Q))} \tag{3}$$

where $w(Q_i, Q)$ is a weight in $(0, 1]$ which depends on the number of atomic predicates in $Q_i \sqcap Q$. It may be defined for instance as:

$$w(Q_i, Q) = \frac{|Q \sqcap Q_i|}{|Q|}.$$

Then, if the considered query Q writes $(P_1 \wedge P_2 \wedge P_3)$ and the cache contains the queries $Q_1 = (P_1 \wedge P_2 \wedge P_4)$ and $Q_2 = (P_1 \wedge P_5)$, predicate P_4 will be considered a better candidate for augmenting Q than P_5 (assuming that $nb_{\mathcal{C}}(Q_1) = nb_{\mathcal{C}}(Q_2)$).

Remark 1. For a query $Q = P_{k_1, l_1} \wedge \ldots \wedge P_{k_n, l_n}$, the only candidate predicates $P_{k'_q, l'_q}$ that make sense are such that $k'_q \neq k_i, \forall i \in [1, n]$. Indeed, it would not be relevant to add a condition on an attribute that is already involved in a condition from Q. For example, if Q is "mileage is low and price is average", one must not add a condition about the mileage or the price since it would lead to a somewhat contradictory requirement (the maximal possible satisfaction degree would be 0.5 or 0 due to the definition of a Ruspini partition).

Using Formula (1), (2), or (3), one can identify the predefined predicates the most related to an under-specified query Q (i.e., those that appear the most frequently in conjunction with Q — or a part of Q — in the cache). Notice that these formulas can be simplified by ignoring the denominator, since it is a constant for a given query Q and the objective is only to rank the candidate predicates.

In practice, Formula (3) seems to be the most suitable since (i) it is not as restrictive as Formula (1), (ii) it allows for a better discrimination than Formula (2), and (iii) it guarantees that the queries that would have been selected by Formula (1) are at the top of the list.

3.3   Interactive Query Strengthening Process

The following algorithm is based on Formula (3), but it can be easily adapted to implement Formula (1) or (2) instead. In this algorithm $Q.predset$ denotes the set of predicates that compose query Q.

> **Input**: a query $Q = P_{k_1,l_1} \wedge \cdots \wedge P_{k_n,l_n}$ yielding a plethoric answer set; a cache C of previously submitted queries
>
> **Output**: a ranked list CP of predicates that may be used for augmenting Q

```
1.1  begin
1.2  │  CP ← ∅;
1.3  │  foreach predicate P_{k_i,l_i} from Q do
1.4  │  │  foreach query Q' from C containing P_{k_i,l_i} do
1.5  │  │  │  rest ← Q' − {P_{k_i,j} | ∃P_{k_i,q} ∈ Q};
1.6  │  │  │  foreach predicate P' of rest do
1.7  │  │  │  │  if P' ∉ CP then
1.8  │  │  │  │  │  nb(P') ← nb(Q') × w(Q', Q);
1.9  │  │  │  │  │  insert ⟨P', nb(P')⟩ into CP;
1.10 │  │  │  │  end
1.11 │  │  │  │  else
1.12 │  │  │  │  │  nb(P') ← nb(P') + nb(Q') × w(Q', Q);
1.13 │  │  │  │  │  rerank(CP);
1.14 │  │  │  │  end
1.15 │  │  │  end
1.16 │  │  end
1.17 │  end
1.18 end
```

Algorithm 1. Determining the candidate predicates

Line 1.5 stems from Remark 1, i.e., its purpose is to keep the sole candidate predicates that do not concern an attribute that is already filtered in Q.

In order for Algorithm 1 to have acceptable performances, it is of prime importance to have an efficient access to the queries in the cache (line 1.4). A simple solution is to maintain an index on the cache so as to have a direct access to the queries that contain a given linguistic term (fuzzy predicate) from the vocabulary.

Remark 2. Several augmentation steps may be needed if the augmented query has a result whose cardinality is still much greater than k.

4   Related Work

In their probabilistic ranking model, Chaudhuri *et al.* [7] also propose to use a correlation property and to take it into account when computing ranking scores. However, correlation links are identified between *attributes* and not *predicates*. As in our approach, the identification of these correlations relies on a workload of past submitted queries.

The approach advocated by Ozawa *et al.* [12,13] is based on the analysis of the database itself, and aims at providing the user with information about the data distributions and the most efficient constraints to add to the initial query in order to reduce the initial set of answers. The main limitation of the approach advocated in [12] is that the attribute chosen is the one which maximizes the dispersion of the initial set of answers, whereas most of the time, it does not have any semantic link with the predicates that the user specified in his/her initial query. To illustrate this, let us consider again a relation describing second hand cars. Let Q be the fuzzy query that aims to select *estate cars* that are *recent*, and let us assume that it yields a plethoric answer set. In such a situation, Ozawa *et al.* [12] first apply a fuzzy c-means algorithm [1] to classify the data, and each fuzzy cluster is associated with a predefined linguistic label. After having assigned a weight to each cluster according to its representativity of the initial set of answers, a global dispersion degree is computed for each attribute. The user is then asked to add new predicates on the attribute for which the dispersion of the initial set of answers is maximal. In this example, this approach may suggest that the user should add a condition on the attributes *mileage* or *brand*, on which the *recent estate cars* are probably the most dispersed. However, such conditions are not semantically linked to the initial query.

The problem of plethoric answers to fuzzy queries has been addressed in [2,4,5]. In [2], a query strengthening mechanism is proposed, that is based on a parameterized proximity relation. The idea is to apply an *erosion* operation so as to modify the membership function of one or several fuzzy predicates, thus making the query more drastic. However, as mentioned in Section 1, such an erosion-based approach may lead to a deep modification of the meaning of the user query. Moreover, it is not easy to know which predicate(s) of Q should be strengthened. The approaches proposed in [4,5] are closer to that advocated in the present paper, since they are based on a *query augmentation* mechanism. The main difference is that the extra predicates that are used to augment the initial query are determined on the basis of a different type of "semantic correlation". Basically, in [4,5], a predicate P is considered to be all the more correlated to a query Q as there are many items in the database that satisfy both P and Q. In other terms, this view is based on the *data* and not on the *queries*. The approaches described in [4,5] imply maintaining a summary of the database, whereas we avoid the cost related to this aspect in the present approach.

The approach that we propose fits in the category of "history-based" cooperative answering approaches in the classification established by Stefanidis et al. [19], i.e., approaches that exploit available information about the previous interactions of the users with the database, as in recommender systems (contrary to the techniques presented in [4,5] that fit in the category of "current-state approaches", i.e., that exploit the content and schema of the current query result and database instance). To the best of our knowledge, no such approach has been proposed yet in the literature for dealing with the plethoric answer set problem in a fuzzy querying framework.

5   Conclusion

The approach presented in this paper deals with the plethoric answer set problem in the context of fuzzy queries expressed using an expert-defined linguistic vocabulary. The idea is to determine the predicates that are the most suitable for strengthening the initial query. These predicates are selected among a set of predefined fuzzy terms, according to their degree of semantic correlation with the initial query. This latter degree, that evaluates the extent to which a predicate P is *related* to an initial query Q, is computed on the basis of the co-occurrence of P with parts of Q in a cache of previously executed queries.

This work opens many perspectives for future research. While preserving the general principle of this approach, it would be interesting, for instance, to let the user specify his/her own fuzzy predicates when querying a database instead of forcing him/her to use a predefined vocabulary. However, this would imply a serious modification of the algorithm inasmuch as one would have to assess the (fuzzy) resemblance between a predicate from the user query and any predicate involved in a query from the cache (in a somewhat similar spirit, measures for assessing the proximity between fuzzy predicates are used in [15] to repair failing fuzzy queries). This would likely make the approach both less effective (because of the approximate nature of the matching between predicates) and less efficient (because of the cost attached to the computation of the matching degrees).

Another perspective would be to exploit additional information in order to help the user choose the augmented query to run next. An idea would be to store in the cache, along with each query previously executed, the fuzzy cardinality of its result [18]. Then, one could try to evaluate the cardinality of the result (or at least a lower bound of it) of every augmented query produced by the algorithm.

Another important aspect concerns the qualitative assessment of the approach, which is not an easy issue in the absence of suitable benchmarks and "gold standards". To this end, we intend to define an evaluation protocol over the second hand cars database used as an illustration in the paper, in order to collect qualitative evaluations (about the relevance of the augmented queries obtained) from users, and compare the approach with those described in [4,5] in terms of user-perceived relevance. Defining a hybrid approach combining both query-based relatedness (as in the present paper) and data-based correlation (as in [4,5]) could also be an interesting topic to study.

References

1. Bezdek, J.: Pattern recognition with fuzzy objective function algorithm. Plenum Press, New York (1981)
2. Bosc, P., Hadjali, A., Pivert, O.: Empty versus overabundant answers to flexible relational queries. Fuzzy Sets and Systems 159(12), 1450–1467 (2008)
3. Bosc, P., Pivert, O.: SQLf: A relational database language for fuzzy querying. IEEE Transactions on Fuzzy Systems 3(1), 1–17 (1995)
4. Bosc, P., Hadjali, A., Pivert, O., Smits, G.: On the use of fuzzy cardinalities for reducing plethoric answers to fuzzy queries. In: Deshpande, A., Hunter, A. (eds.) SUM 2010. LNCS, vol. 6379, pp. 98–111. Springer, Heidelberg (2010)

5. Bosc, P., Hadjali, A., Pivert, O., Smits, G.: Trimming plethoric answers to fuzzy queries: An approach based on predicate correlation. In: Hüllermeier, E., Kruse, R., Hoffmann, F. (eds.) IPMU 2010. LNCS, vol. 6178, pp. 595–604. Springer, Heidelberg (2010)
6. Bruno, N., Chaudhuri, S., Gravano, L.: Top-k selection queries over relational databases: Mapping strategies and performance evaluation. ACM Transactions on Database Systems 27, 153–187 (2002)
7. Chaudhuri, S., Das, G., Hristidis, V., Weikum, G.: Probabilistic ranking of database query results. In: Nascimento, M.A., Özsu, M.T., Kossmann, D., Miller, R.J., Blakeley, J.A., Schiefer, K.B. (eds.) VLDB, pp. 888–899. Morgan Kaufmann (2004)
8. Chomicki, J.: Querying with intrinsic preferences. In: Jensen, C.S., Jeffery, K., Pokorný, J., Šaltenis, S., Bertino, E., Böhm, K., Jarke, M. (eds.) EDBT 2002. LNCS, vol. 2287, pp. 34–51. Springer, Heidelberg (2002)
9. Corella, F., Lewison, K.: A brief overview of cooperative answering. In: Technical report (2009),
   http://www.pomcor.com/whitepapers/cooperative_responses.pdf
10. Kießling, W.: Foundations of preferences in database systems. In: VLDB, pp. 311–322. Morgan Kaufmann (2002)
11. Lacroix, M., Lavency, P.: Preferences: Putting more knowledge into queries. In: Proc. of the 13rd VLDB Conference, pp. 217–225 (1987)
12. Ozawa, J., Yamada, K.: Cooperative answering with macro expression of a database. In: Proc. of IPMU 1994, pp. 17–22 (1994)
13. Ozawa, J., Yamada, K.: Discovery of global knowledge in database for cooperative answering. In: Proc. of Fuzz-IEEE 1995, pp. 849–852 (1995)
14. Pivert, O., Bosc, P.: Fuzzy Preference Queries to Relational Databases. Imperial College Press, London (2012)
15. Pivert, O., Jaudoin, H., Brando, C., Hadjali, A.: A method based on query caching and predicate substitution for the treatment of failing database queries. In: Bichindaritz, I., Montani, S. (eds.) ICCBR 2010. LNCS, vol. 6176, pp. 436–450. Springer, Heidelberg (2010)
16. Ruspini, E.: A new approach to clustering. Information and Control 15(1), 22–32 (1969)
17. Smits, G., Pivert, O., Girault, T.: Reqflex: Fuzzy queries for everyone. PVLDB 6(12), 1206–1209 (2013)
18. Smits, G., Pivert, O., Hadjali, A.: Fuzzy cardinalities as a basis to cooperative answering. In: Pivert, O., Zadrozny, S. (eds.) Flexible Approaches in Data, Information and Knowledge Management. SCI, vol. 497, pp. 261–289. Springer, Heidelberg (2014)
19. Stefanidis, K., Drosou, M., Pitoura, E.: "You may also like" results in relational databases. In: Proc. of PersDB 2009 (2009)
20. Ughetto, L., Voglozin, W.A., Mouaddib, N.: Database querying with personalized vocabulary using data summaries. Fuzzy Sets and Systems 159(15), 2030–2046 (2008)

Generating Description Logic $\mathcal{ALC}$ from Text in Natural Language

Ryan Ribeiro de Azevedo[1,2], Fred Freitas[2], Rodrigo Rocha[1,2],
José Antônio Alves de Menezes[1], and Luis F. Alves Pereira[1,2]

[1] Computer Science, UAG/Federal Rural University of Pernambuco, Garanhuns, Brazil
{ryan,rodrigo,jaam,lfap}@uag.ufrpe.br
[2] Center of Informatics, Federal University of Pernambuco, Recife, Brazil
{rra2,fred,rgcr,lfap}@cin.ufpe.br

Abstract. In this paper, we present a natural language translator for expressive ontologies and ensure that it is a viable solution to the automated acquisition of ontologies and complete axioms, constituting an effective solution for automating the expressive ontology building Process. The translator is based on syntactic and semantic text analysis. The viability of our approach is demonstrated through the generation of descriptions of complex axioms from concepts defined by users and glossaries found at Wikipedia. We evaluated our approach in an initial experiment with entry sentences enriched with hierarchy axioms, disjunction, conjunction, negation, as well as existential and universal quantification to impose restriction of properties.

Keywords: Description Logic (DL), Ontology, Ontology Learning, PLN.

1 Introduction

One of the subfields of Ontology that has been standing out along the last decade is Ontology Engineering. Its purpose is to create, represent and model knowledge domains, most of which are not trivial, such as Bioinformatics and e-business, among others. However, as pointed out by [9], the task of Ontology Engineering still consumes a big amount of resources even when principles, processes and methodologies are applied. Besides financially expensive, the ontology design is also an arduous and onerous task [5]. Thus, new technologies, methods and tools for overtaking these technical and economic challenges are necessary. This way, the need for highly specialized personnel and the manual efforts required can be minimized.

For this purpose, a research line that is gaining importance through the past two decades is the extraction of domain models from text written in natural language, using Natural Language Processing (NLP) techniques. The process of modeling a knowledge domain from text and the automated design of ontologies, through an analysis of a set of texts using NLP techniques, for example, is known as Ontology Learning and was first proposed by [7]. Even so, as affirmed by [11], despite the increasing interest and efforts taken towards the improvement of Ontology Learning methods based in NLP techniques [10] [1] [2][3], the notable potential of the

T. Andreasen et al. (Eds.): ISMIS 2014, LNAI 8502, pp. 305–314, 2014.
© Springer International Publishing Switzerland 2014

techniques and representations available to the learning process of expressive ontologies and complex axioms has not yet been completely exploited. In fact, there are still gaps and unanswered questions that need viable and effective solutions. Among them, these stand out [8][10]:

- The necessity of combining the knowledge of specialists of domain with the competencies and experience of ontology engineers in a single effort: there are scarce specialized resources and demand for professionals. This obstacle reduces the use of semantic ontologies by users and specialists in general.
- There is a considerable amount of tools and *frameworks* of *Ontology Learning* that have been developed aiming at the automatic or semi-automatic construction of ontologies based on structured, semi-structured or unstructured data. Nonetheless, although useful, the majority of these tools used in Ontology Learning are only capable of creating informal or unexpressive ontologies.
- Evaluating the consistency of ontologies automatically: it is necessary that the automatically created ontologies be assessed by the time of their development, minimizing the amount of errors committed by the ones involved in the development phase and verify whether or not the ontology is contradictory and free of inconsistencies.

All the questions and issues abovementioned justify the approach hereby proposed. It is based in a translator, which consists in the utilization of a hybrid method that combines syntactic and semantic text analysis both in superficial and in-depth approaches of NLP. Demonstrating that a translator for creating ontologies that formalizes and codifies knowledge in OWL DL $\mathcal{ALC}$ [6] from sentences provided by users is a viable and effective solution to the process of automatic construction of expressive ontologies and complete axioms.

2   The Approach and Example

One of the goals of this work consists in demonstrating that a translator, through the processing of sentences in natural language provided by users, is capable of creating – automatically and according to the discourse interpreted $\mathcal{ALC}$ [6] ontologies with minimal expressivity. An overview of the translator's architecture and function flow diagram are depicted in Figure 1 and described as follows.

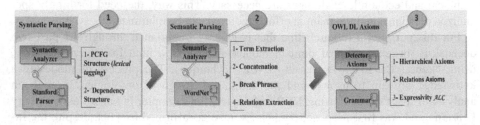

Fig. 1. Translator's architecture and function flow diagram

The architecture of our approach is composed by 3 modules: the **Syntactic Parsing Module (1)**, **Semantic Parsing Module (2)** and the **OWL DL Axioms Module (3)**. The activities executed in the respective modules and their functions are presented in the following sections.

2.1   Syntactic Parsing Module

The syntactic analysis of the sentences inserted by users takes place in the **Syntactic Parsing Module (1)**, this module uses Probabilistic Context-Free Grammars (PCFGs). For this purpose, we used the Stanford Parser 2.0.5[1]. A PCFG consists of:
1. A context-free grammar $G = (N, \Sigma, S, R, q)$ as follows:
    - a. N is the set of all non-terminals seen in the trees $t_1 \dots t_m$.
    - b. Σ is the set of all words seen in the trees $t_1 \dots t_m$.
    - c.   The start symbol S is taken to be S.
    - d.   The set of rules R is taken to be the set of all rules $\alpha \rightarrow \beta$ seen in the trees $t_1 \dots t_m$.
    - e.   The maximum-likelihood parameter estimates are

        $$qML(\alpha \rightarrow \beta) = \frac{Count(\alpha \rightarrow \beta)}{Count(\alpha)}$$

        where $Count(\alpha \rightarrow \beta)$ is the number of times that the rule $\alpha \rightarrow \beta$ is seen in the trees $t1 \dots tm$, and $Count(\alpha)$ is the number of times the non-terminal α is seen in the trees $t_1 \dots t_m$.
2. A parameter $q(\alpha \rightarrow \beta)$. For each rule $\alpha \rightarrow \beta \in R$. The parameter $q(\alpha \rightarrow \beta)$ can be interpreted as the conditional probabilty of choosing rule $\alpha \rightarrow \beta$ in a left-most derivation, given that the non-terminal being expanded is α. For any $X \in N$, we have the constraint:

$$\sum_{\alpha \rightarrow \beta \in R : \alpha = X} q(\alpha \rightarrow \beta) = 1$$

In addition we have $q(\alpha \rightarrow \beta) \geq 0$ for any $\alpha \rightarrow \beta \in R$. Given a parse-tree $t \in \mathcal{T}_G$ containing rules $\alpha 1 \rightarrow \beta 1, \alpha 2 \rightarrow \beta 2, \dots, \alpha n \rightarrow \beta n$, the probability of t under the PCFG is:

$$p(t) = \prod_{i=1}^{n} q(\alpha i \rightarrow \beta i)$$

Two activities are executed by this module, the lexical tagging and the dependence analysis (Using PCFG). The results obtained by this module are shown in Fig. 2 and 3. We used the sentence **(S1)**: "A self-propelled vehicle is a motor vehicle or road vehicle that does not operate on rails" to illustrate the results obtained by the translator's modules.

---

[1] http://nlp.stanford.edu/software/lex-parser.shtml

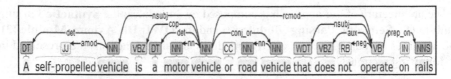

Fig. 2. Lexical tagging and dependence structure

Each word of the sentence (S1) above (Fig. 2) is grammatically classified according to their lexical categories and the dependence between them is attributed.

> (**NP** (DT A) (JJ self-propelled) (NN vehicle)) | (**VP** (VBZ is) (NP (DT a) (NN motor) (NN vehicle) | (CC or) (NN road) (NN vehicle)) | (**SBAR** (**WHNP** (WDT that)) |
> (**VP** (VBZ does) (RB not) (**VP** (**VB** operate) (**PP** (IN on) (NP (NNS rails))

Fig. 3. Classification in syntagmatic or sentential categories

Syntagmatic categories are in red and, in black, the lexicon to which each category pertains (See Fig. 3).

2.2   Semantic Parsing Module

The results of the activities carried out by the systems of the **Syntactic Parsing Module (1)** are used by the systems of the **Semantic Parsing Module (2)**, which carries out the activities shown in Figure 4 and are detailed as follows.

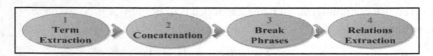

Fig. 4. Activities carried out in the Semantic Parsing Module

This module initiates its activities by assessing the entry sentence and the referred result of the syntactic analysis obtained in the previous module and then starts the extraction of terms (**Term Extraction**) that are fit to be concepts of the ontology (Activity (1)). In this phase, terms classified as prepositions (IN), conjunctions (CC), numbers (CD), articles (IN, CC, RB, DT+PDT+WDT) and verbs (EX+MD+VB+VBD+VBG+VBN+VBP+VBZ) are discarded, and the terms classified as nouns (NN+NNS+NNP+NNPS) and adjectives (JJ+JJR+JJS) are indicated as possible concepts of the ontology, therefore, the terms extracted, who are fit to be concepts were: motor/NN, vehicle/NN, road/NN, vehicle/NN, self-propelled/JJ, vehicle/NN e rails/NNS as presented in Figure 5.

Fig. 5. Result of the extraction of terms

After the term extraction activity is done, the Activity (2), called **Concatenation** is enabled. This activity uses the results of the dependences between the terms (See Figures 2 e 3) and makes the junction of NPs composed by two or more nouns and/ or adjectives inside the analyzed sentence and which, in fact, are related. In the example sentence (S1), the concatenation results in the junction of the terms (self-propelled/JJ ↔ vehicle/NN), (motor/NN ↔ vehicle/NN) and (road/NN ↔ vehicle/NN) into an only term, because they are dependent of one another, resulting in just 3 terms: *self-propelled-vehicle*, *motor-vehicle* e *road-vehicle*, and no longer 6 terms, as in the initial phase of the **Term Extraction** activity.

In Activity (3), **Break Phrases,** every time terms or punctuation marks like comma (,), period (.), *and, or, that, who* or *which* (what we call sentence breakers) are found, the sentences are divided into subsentences and analyzed separately, the result for (S1) was:

A self-propelled-vehicle is a motor-vehicle I or road-vehicle I that does not operate on rails

The last activity to take place in the **Semantic Parsing Module** is Activity (4), **Relations Extraction**. The relations between the terms are verified and validated through verbs found in the sentences and patterns observed in the translator's inner grammar. The verbs are separated and the terms dependent on verbs are extracted, resulting in:

self-propelled-vehicle **is a** motor-vehicle I self-propelled-vehicle **is a** road-vehicle I
self-propelled-vehicle **operate** on rails

Table 1. Patterns/Rules for transforming the hierarchical axioms of terms

Construction Patterns of hierarchical axioms of terms

(1) SN_1 (*is a* I *is an* I *is*) SN
(2) SN_1 (*are a* I *are an* I *are*) SN
(3) SN_1 *and* SN_2 *and* SN_3 *and* SNn (*are a* I *are an* I *are* I *is a* I *is an* I *is*) SN
(4) SN_1, SN_2 , SN_3, SN_n (*are a* I *are an* I *are* I *is a* I *is an* I *is*) SN
(5) SN_1, SN_2 *and* SN_3 (*are a* I *are an* I *are* I *is a* I *is an* I *is*) SN
(6) SN_1, SN_2 *and* SN_3, SN_4 (*are a* I *are an* I *are* I *is a* I *is an* I *is*) SN
(7) SN_1 (*is a* I *is an* I *is*) SN and/that/who/which (**is a** I **is an** I **is**) SN and (**is a** I **is an** I **is**) SN...
(8) SN_1 SNn (*are a* I *are an* I *are* I *is a* I *is an* I *is*) SN and SNn (*are a* I *are an* I *are* I *is a* I *is an* I *is*) SN_1 and SN_n...
(9) SN_1 *and* SN_2 *and* SN_3 *and* SNn (*are a* I *are an* I *are* I *is a* I *is an* I *is*) SN and (*are a* I *are an* I *are* I *is a* I *is an* I *is*) SN_1 and SN_n...
(10) SN_1, SN_2 *and* SN_3, SN_n (*are a* I *are an* I *are* I *is a* I *is an* I *is*) SN
(11) SN_1 (*are a* I *are an* I *are* I *is a* I *is an* I *is*) SN_2 *or/and* SN_3

This module detects the terms and the relations between them, both hierarchical and nonhierarchical. However, this module neither extracts disjunctions, conjunctions nor generates OWL code corresponding to the result obtained. The activity of this module is exclusively for detecting terms, their relations and validity. Some patterns/rules are used during the discovery and the learning of the **Semantic Parsing Module (2).** The patterns/rules used for learning hierarchical axioms are showed in Table 1.

The generated result for the above patters are the same, *i.e.*, {SN_1, SN_2, SN_3, SN_n} ⊑ SN, notice that the **and** conector for the mentioned patterns play the role of conector of concepts, establishing among them a dependency relation. The patterns above also are recognized as axioms OR, AND, and NOT and may be represented in the following way (See Table 2):

Table 2. Patterns/Rules for transforming the hierarchical axioms of relations using Inserction (and), Conjuction (or) and Nagation (not)

Construction Patterns using verbs, intersection (and), conjuction (or) and negation (not)

(1) SN_1 (*are a* | *are an* | *are* | *is a* | *is an* | *is*) SN *or/and* SN (*That/Who/Which*) (*has* **not**) SN
(2) SN_1 (*are a* | *are an* | *are* | *is a* | *is an* | *is*) SN *or/and* SN (*That/Who/Which*) (*has*) SN
(3) SN_1 (*are a* | *are an* | *are* | *is a* | *is an* | *is*) SN *or/and* SN (*That/Who/Which*) (*Verb*) SN
(4) SN_1 (Verb) SN (*or/and*) SN
(5) SN_1 (Verb) SN (*or/and*) (Verb) SN
(6) SN_1 *has* SN_n *and* SN_n
(7) SN_1 (*has* **not**) SN_n
(8) SN_1 (*does not* | *doesn't* | *is not* | *isn't*) SN_n
(9) SN_1 (*does not* | *doesn't* | *is not* | *isn't*) (Verb) SN_n
(10) SN_1 *has* SN_n *and has not* SN_n
(11) SN_1 (*are a* | *are an* | *are* | *is a* | *is an* | *is*) SN (*That/Who/Which*) (Verbo) (*or*) (Verb) SN
(12) SN_1 (*are a* | *are an* | *are* | *is a* | *is an* | *is*) SN *or/and* SN (*That/Who/Which*) (Verb) (*or/and*) (Verb) SN
(13) SN_1 (Verb) SN (*or/and*) (Verb) SN (*or/and*) (Verb) SN…

One should notice that new construction patterns of hierarchical axioms and relations may be inserted to the internal grammar of the described approach by human intervention. All the patterns presented in Table 2 composes disjunction (⊔), conjunction (⊓) and negation (¬) rules beyond the axioms with $\forall r.C$: universal restriction and $\exists r.C$: existencial restriction. In the next section, the operation of the **OWL DL Axioms Module** is described and the obtained results using its processing is showed.

2.3   OWL DL Axioms Module

The function of the **OWL DL Axioms Module** is to symbolically find/learn axioms that prevent ambiguous interpretations and limit the possible interpretations of the discourse, enabling systems to verify and disregard inconsistent data. The process of discovering the axioms is the hardest part of the process of creating ontologies. Here, the axioms discovered correspond to DL *ALC* expressivity. The module recognizes coordinating conjunctions (*OR* and *AND*), labeled CC, indicating the union (disjunction) and intersection (conjunction) respectively for concepts and/or properties, recognizes linking verbs followed by negations, like *does not* (or *doesn't*), *has not* (or *hasn't*), and *is not* (or *isn't*) for negation axioms (¬), besides generating universal quantifiers (∀) and existential quantifiers (∃). In this module, we used Protégé-OWL API 3.5[2] and OWL API[3].

It also recognizes *is* and *are* as taxonomic relations (⊑ - hierarchical). The transformations occur in four steps and make use of the results obtained by the previous modules:

Step (1): construction of taxonomic/hierarchical relations. The basic pattern used here is **\<NPs\> \<VP\> \<NPs\> where \<VP\> in this case is a (*is a/an*, *is* or *are*).** Other patterns are possible (See Table 1). For all the transformations, the patterns are automatically chosen by the translator. For our example, the results of <u>Step (1)</u> were:

self-propelled-vehicle **is a** motor-vehicle ➔ *self-propelled-vehicle ⊑ motor-vehicle*
self-propelled-vehicle **is a** road-vehicle ➔ *self-propelled-vehicle ⊑ road-vehicle*

By subsumption reasoning the implicit hierarchical relations (motor-vehicle ⊑ vehicle) and (road-vehicle ⊑ vehicle) are discovered and created automatically.

Step (2): construction of nonhierarchical relations. The pattern used here is **\<NPs\> \<VP\> \<NPs\> where \<VP\>** in this case is a verb other than **(*is a/an*, *is* or *are*)**. Other patterns are possible (See Table 2).

*self-propelled-vehicle **operate** on rails ➔*
self-propelled-vehicle ≡ ∃operate.rails

Step (3): verification of conjunctions and disjunctions. The conjunctions OR and AND are verified and analyzed. They can be associated with concepts and/or properties. The pattern **\<NPs\> *is a/an* or *are* \<NPs\> \<CC\> \<NPs\>**, where **\<CC\>** is the conjunction *Or* or *And* that links two or more **\<NPs\>** is one of the patterns associated with union and intersections of concepts, and is chosen by the translator resulting in:

*A self-propelled-vehicle is a motor-vehicle **or** road-vehicle ➔*
self-propelled-vehicle ≡ (motor-vehicle ⊔ road-vehicle)

---

[2] http://protege.stanford.edu/plugins/owl/api/
[3] http://owlapi.sourceforge.net/

Step (4): detection of negations. The fourth analysis detects the negations, its dependences and classifies the sentence to apply the patterns. Two negations are possible: negations and disjunctions of concepts and negations of properties. Two patterns or a junction of these patterns are taken into consideration in the process of extraction of negation axioms for hierarchies: **<NPs> is not <NPs>** and the pattern **<NP>does not<VP><NP>** for negation of properties (See Table 2). For (S1), the following result was obtained:

$$\textit{self-propelled-vehicle that does not operate on rails} \rightarrow$$
$$\underline{\textit{self-propelled-vehicle} \sqcap \neg\exists \textit{operate.rails}}$$

The final result, after the integration of the partial results obtained by the three modules, for (S1) in OWL 2 code, was:

(S1): "*A self-propelled vehicle is a motor vehicle or road vehicle that does not operate on rails*"
$$\underline{\textit{self-propelled-vehicle} \equiv (\textit{motor-vehicle} \sqcup \textit{road-vehicle}) \sqcap \neg\forall \textit{operate.rails}}$$
$$\underline{(\textit{motor-vehicle} \sqcap \textit{road-vehicle}) \sqsubseteq \textit{vehicle}}$$

Our approach generated 4 axioms, 2 of which being hierarchical axioms, 1 being the union between concepts and 1 other of negation of properties (Universally Restricting Property Values). The approach proposed by us is effective in patterns like this and makes correct or approximately correct interpretations of what the user desires. This statement can be verified by observing the early results in the following section.

3   Results and Discussion

In order to validate the translator, sentences from various knowledge domains were used. The set of data utilized in the experiments contains a total of 120 sentences and in all of them there were negation axioms and in all of them there were negation, conjunction or disjunction axioms and/or two and/or three types of axioms in the same sentence, as well as axioms with definition of terms hierarchy. We opted for sentences found in *Wikipedia* glossaries because they offered in principle a controlled language without syntactic and semantic errors, besides providing a great opportunity for automatic learning.

For discussion and comparison purposes, some of the sentences analyzed were extracted from the work of [10], which were also obtained from *Wikipedia*. Some examples of sentences having negation, union and conjunction axioms used in the experiments and the respective results generated by the translator, along with a discussion on these results are shown as follows.

Processed Sentence (1): Juvenile is an young fish or animal that has not reached sexual maturity.

Result: → *juvenile* ≡ *(young_fish* ⊔ *young_animal)* ⊓ ¬ ∀*hasReached.Sexual-maturity* | → *young_fish* ⊑ *fish* | → *young_animal* ⊑ *animal*

Discussion: the result of the analysis of the sentence is different from the results of the processing performed by the LExO system [10]: Juvenile ≡ (young ⊓ (Fish ⊔ Animal) ⊓ ¬∃reached.(Sexual ⊓ Maturity). The compared system (LExO) classifies *young*, *fish* and *animal* as distinct terms, however, by the interpretation in natural language of the sentence in analysis, the word *young* is an adjective of the *fish* concept, thus, our approach classifies and represents '*young fish*' as a composite noun, that is, composing a single concept (*young_fish*). The same occurs for s*exual maturity*, being interpreted by LExO as distinct concepts when they are not, whereas in our approach these two terms are classified as a single concept in the same way as the classification of the previous concept (*young_fish*). We can also observe the creation of two axioms, one of union of concepts and one of negation of property. By subsumption reasoning (in the reasoning activities, we used the inference machine Pellet 2.2.2[4] [4]) the hierarchical axioms: young_fish ⊑ fish and young_animal ⊑ animal were discovered and automatic created by our approach.

Processed Sentence (2): Vector is an organism which carries or transmits a pathogen.

Result: → *vector* ≡ *organism* ⊓ *(∃carries.Pathogen* ⊔ *∃transmits.Pathogen)*

Discussion: the result obtained in (2) was also compared with resulted generated by LExO [10]: Vector ≡ (Organism ⊓ (*carries* ⊔ ∃*transmit.Pathogen*)). The verb *to carry* was not correctly classified as an existential restriction when analyzed by LExO, whereas in our approach, the sentence was coherently classified, the existential quantifier was created and the disjunction of the relations created was performed, where *carriesPathogen* and *transmitsPathogen* are disjoint (⊔), which evidences the accurate interpretation of the sentence in natural language.

Processed Sentence (3): whale is an aquatic animal and mammal.

Result: → whale ≡ (aquatic_animal ⊓ aquatic_mammal) | → *aquatic_animal* ⊑ *animal* / → aquatic_mammal ⊑ mammal | → whale ⊑ *animal* / → whale ⊑ mammal

Discussion: in this sentence, the concept *whale* forms a hierarchy with the other two concepts (*aquatic animal* and *aquatic mammal*). Besides, it generates an intersection of both terms, meaning that individuals pertaining to the concept *whale* pertain to the set of individuals of both concepts at the same time. By subsumption reasoning the hierarchical axioms: aquatic_animal ⊑ animal and aquatic_mammal ⊑ mammal were discovered and created automatically by our approach. Futhermore, our approach discovered, by deduction, the following axioms: whale ⊑ animal and whale ⊑ mammal, in this case, distinctly of sentence in (2), and as expected, the OWL 2 code associated to this deduction is not generated, the user is only informed that the translator was able to deduce according to the sentence processing and it is showed that it performs a reasoning although there are limitations.

---

[4] http://clarkparsia.com/pellet/

4   Final Remarks and Future Works

In this paper, we describe an approach to automatic development of expressive ontologies from definitions provided by users. The results obtained through the experiments evidence the need of automatic creation of expressive axioms, sufficient to creating ontologies with $\mathcal{ALC}$ expressivity, besides the success in the identification of rules and axioms pertaining to $\mathcal{ALC}$ expressivity. We also conclude that the translator can aid both experienced ontology engineers and developers and inexperienced users just starting to create ontologies. As future works, we include the integration of our approach with other existing approaches in the literature, the creation of a module for automatic inclusion of unprecedented patterns in the translator and one module for automatic insertion of individuals for terms of ontologies created by the translator.

References

1. Buitelaar, P., Cimiano, P.: Ontology learning and population: Bridging the gap between text and knowledge. In: Frontiers in Artificial Intelligence and Applications Series, vol. 167. IOS Press (2008)
2. Buitelaar, P., Cimiano, P., Magnini, B. (eds.): Ontology learning from text: Methods, applications and evaluation, pp. 3–12. IOS Press (2005)
3. Cimiano, P., Völker, J.: Text2Onto. NLDB, 227–238 (2005)
4. Sirin, E., Parsia, B., Grau, B.C., Kalyanpur, A., Katz, Y.: Pellet: A practical OWL-DL Reasoner. J. Web Sem. 5(2), 51–53 (2007)
5. Gómez-Pérez, A., Fernández-López, M., Corcho, O.: Ontological Engineering with examples from the areas of Knowledge Management, e-Commerce and the Semantic Web. In: Advanced Information and Knowledge Processing. Springer, London (2004)
6. Horrocks, I., et al.: OWL: A Description-Logic-Based Ontology Language for the Semantic Web. In: The Description Logic Handbook: Theory, Implementation and Applications, 2nd edn., pp. 458–486. Cambridge University Press (2007)
7. Madche, A., Staab, S.: Ontology learning for the semantic web. IEEE Intelligent Systems 16(2), 72–79 (2001)
8. Pease, A.: Ontology: A Practical Guide. Published by. Articulate Software Press. USA (2011)
9. Simperl, E., VTempich, C.: Exploring the Economical Aspects of Ontology Engineering synthetic. In: Handbook on Ontologies, International Handbooks on Information Systems, pp. 337–358. Springer, Berlin (2009)
10. Völker, J.: Learning Expressive Ontologies. No. 002. In: Studies on the Semantic Web. AKA Verlag / IOS Press (2009) ISBN: 978-3-89838-621-0
11. Zouaq, A.: An Overview of Shallow and Deep Natural Language Processing for Ontology Learning. In: Ontology Learning and Knowledge Discovery Using the Web: Challenges and Recent Advances, vol. 2, pp. 16–37. IGI Global. EUA (2011) ISBN 978-1-60960-625-1 (hardcover)

DBaaS-Expert: A Recommender
for the Selection of the Right Cloud Database

Soror Sahri[1], Rim Moussa[2], Darrell D.E. Long[3], and Salima Benbernou[1]

[1] Université Paris Sorbonnes Cité, Université Paris Descartes, France
{soror.sahri,salima.benbernou}@parisdescartes.fr
[2] LaTICE Laboratory, University of Tunis, Tunisia
rim.moussa@esti.rnu.tn
[3] Storage Systems Research Center, University of California, Santa Cruz, USA
darrell@cs.ucsc.edu

Abstract. The most important benefit of Cloud Computing is that organizations no longer need to expend capital up-front for hardware and software purchases. Indeed, all services are provided on a pay-per-use basis. The cloud services market is forecast to grow, and numerous providers offer database as a service (DBaaS). Nevertheless, as the number of DBaaS' offerings increases, it becomes difficult to compare various offerings through checking of a documentation ads-oriented. In this paper, we propose and describe *DBaaS-Expert* – a framework which helps a user to choose the right DBaaS Cloud Provider among DBaaS' offerings. The core components of *DBaaS-Expert* is first an ontology which captures cloud data management systems services concepts, and second a ranking core which scores each DBaaS offer in terms of criteria.

1 Introduction

Cloud computing has emerged as a new paradigm, which allows *enabling ubiquitous, convenient, on-demand network access to a shared pool of configurable computing resources that can be rapidly provisioned and released with minimal management effort or service provider interaction* [1]. Cloud providers typically publish their service description, pricing policies and Service-Level-Agreement (SLA) rules on their websites in various formats. A data management system is one of the applications that are deployed in the cloud, the service is denoted as database-as-a-service (DBaaS). Nevertheless, as the number of DBaaS offerings increases, with different cost plans and different services, it becomes necessary to be able to automate the ranking of DBaaS offerings along a company needs. Therefore, a company should have detailed knowledge of the offerings of cloud providers that can meet its operational needs. This is not obvious, since expertise in cloud computing is required but is lacking.

We propose *DBaaS-Expert*-a framework addressing the selection of the most suitable cloud-based data management system. In order to develop this framework, first we conducted a thorough DBaaS offerings review. Second, we propose a list of dimensions, which describe DBaaS offerings and an ontology for

T. Andreasen et al. (Eds.): ISMIS 2014, LNAI 8502, pp. 315–324, 2014.
© Springer International Publishing Switzerland 2014

DBaaS. Third, we compute the ranking values of database service candidates based on user requirements. In our work, we perform the ranking following a known method of multi-criteria decision-making (MCDM): *Analytic Hierarchy Process (AHP)*. To the best of our knowledge, there is no existing work which addressed the problem of selection of the right DBaaS offer. The remainder of the paper is organized as follows. A review of related work is provided in Section 2. In Section 3, *DBaaS framework* is presented. Section 4 presents *DBaaS ontology*. Section 5 presents the ranking core based on *AHP*. In Section 6, a use case is devised for validating our framework. Finally, we conclude the paper.

2   Related Work

The existence of multiple options and features of cloud services, makes the selection of the appropriate cloud provider very difficult and challenging. Several works have proposed to automate cloud service selection for IaaS and PaaS models. Next, we overview related work and highlight our contribution.

Some reviewed papers are based on benchmarking [2,3]. Most of them propose revolving well known TPC benchmarks into benchmarks for data management systems' assessment in the cloud. Curino et al. [4] propose *OLTP-Bench*, an open-source framework for benchmarking on-line transaction processing (OLTP) and web workloads. Other reviewed papers [5–7] adopted a different approach based on the proposal of a meta-data model for the description of Cloud Service Providers (CSP) offerings. Among ontology-based papers one cite Zhang et al. [7]. They implemented *CloudRecommender*-a system for infrastructure services (IaaS) selection. They propose a Cloud Computing Ontology to facilitate the discovery of IaaS services categorized into functional services and QoS data. Variability modeling is used to understand and define commonalities and variabilities in software product lines and to support product derivation. For instance, Wittern et al. [6] adopt feature modeling to capture aspects and configurations of cloud services. Quinton et al. [5] address services selection from multiple CSPs using Hybrid Modeling. Indeed, within a multi-cloud configuration, a *CSP A* is selected for hosting the database, while a *CSP B* is selected for hosting the application. They use feature models to describe cloud systems configurations. For specific description format, Arkaitz et al. [8], have defined an XML schema that guides the description of the different capabilities of cloud storage systems such Amazon, Azure.

Multi-Criteria Decision Making methods take into consideration multiple conflicting criteria (p.e., cost vs. quality) that need to be evaluated in making decisions. Menzel et al. [9] propose *CloudGenius framework* that provides a multi-criteria approach in decision support, namely AHP technique, to automate the selection process focusing on IaaS models. Garg et al. [10] provide a *SMICLOUD framework* measuring the quality of CSPs based on QoS attributes proposed by Cloud Service Measurement Index Consortium (CSMIC) [11] and ranking the cloud services according to these attributes. *SMICloud* considers only quantitative attributes (such that response time, cost) in the context of IaaS Clouds.

Previous overviewed research work addressed only IaaS and PaaS cloud models. In this paper, we propose *DBaaS-Expert* a framework for scoring and ranking DBaaS offers in terms of criteria. For this purpose, we first propose an ontology allowing a full description of any available DBaaS offer made by a Cloud Service Provider. Second, we propose a ranking core based on a well admitted mathematical method for multi-criteria decision-making, which is AHP.

3   DBaaS-Expert Framework

In this section, we first formulate DBaaSs' offerings ranking as a Multi-Criteria Decision Making (MCDM) problem. Then, we propose a framework and a system architecture for solutioning the problem.

3.1   DBaaSs' Offerings Ranking Problem Statement

The typical MCDM problem deals with the evaluation of a *set of alternatives* in terms of a *set of decision criteria*; where *alternatives* represent the different choices available to the decision maker and *criteria* (or attributes) represent the different dimensions from which the *alternatives* can be viewed. *Criteria* are rarely of equal importance, therefore *criteria* will be weighted in terms of their importance to the decision maker. When a suitable process is applied to the problem, a rating of the alternatives can be formed into a rank. The *DBaaSs' Offerings Ranking Problem* is an MCDM problem. Indeed, *first* the set of M DBaaSs' offerings, denoted as $DBaaS_1, DBaaS_2, \ldots DBaaS_M$ (e.g., Amazon RDS, Google BigQuery) map to alternatives. *Second*, DBaaSs' offerings are characterized by a set of N decision criteria $C_1, C_2, \ldots, C_N$ (e.g., Performance, high-availability capacity, elasticity, security, and so on). The objective of any solution addressing the problem is how to evaluate the set of offerings in terms of the set of criteria with two objectives (O_1) *maximize Quality and Capacity of Service* and (O_2) *minimize cost* under fully or partially satisfying a *set of user-requirements*.

3.2   Proposed Framework

DBaaS-Expert is a framework for cloud-based data management system selection. It helps to find relevant database criteria that meet users' requirements. The main component of *DBaaS-Expert* is the *DBaaS ontology*. The latter is detailed in Section 4. The main functionalities of *DBaaS-Expert* are ensured by the following modules: *application* module, *mapping* module and *ranking* module. These modules are part of the logical architecture of *DBaaS-Expert framework* depicted in Figure 1:

- *Application Module.* Users communicate with *DBaaS-Expert* via the application module. This module allows users to enter their business requirements. In order to enable the matching between user requirements and the ontology, the application module presents the DBaaS dimensions (as defined in the ontology). Consequently, for each dimension, users can choose among its possible concepts or individuals with respect to the ontology.

- *Selection Module.* This module first maps the entered user requirements to concepts present in the DBaaS ontology, and then according to the mapping, it selects DBaaS offers which satisfy the user business requirements.
- *Ranking Module.* The obtained DBaaSs offers from the selection module are ranked. We perform the ranking following a well known mathematical method of multi-criteria decision-making: *Analytic Hierarchy Process (AHP)*.

Notice that the input of the *ranking module* is already filtered in the *selection module*. That is the application of AHP method is adapted to the selection from DBaaS ontology. Consequently, the ranking is done dynamically according to the user query. At the best of our knowledge, our work is the first to combine the use of ontology and an MCDM method for selection purpose.

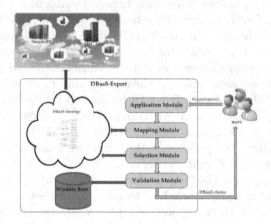

Fig. 1. Logical Architecture of *DBaaS-Expert* Framework

4   DBaaS Ontology

We propose an ontology derived from our review of most known DBaaS providers and their offerings. To the best of our knowledge, this work is the first to report a *DBaaS ontology*. Due to the lack of space, we only describe the relevant concepts of *DBaaS ontology*. Concepts are divided into four categories which relate to (i) basic concepts, (ii) quality of service concepts, (iii) capacity of service concepts and (iv) cost of service concepts. More details about the ontology concepts and its description using *Protégé ontology editor* are in [12].

4.1 Basic Concepts

- *DBaaS Offer*: Examples of DBaaS offer individuals are: Herokupostgres [13], Oracle DBaaS [14], Amazon SimpleDB [15], Google BigQuery [16].
- *Cloud Service Provider*: Each *DBaaS offer* admits a unique CSP. Examples of CSPs individuals are: Amazon, Google, Micorsoft.

4.2 General Concepts

General concepts describe each DBaaS offer.

- *Workload Type*: Workload type is *On-Line Transaction Processing* (OLTP), or *On-Line Analytical Processing* (OLAP).
- *Storage Model*: It is either traditional Hard Disk Drives (HDDs) systems, In-memory data systems or Solid State Devices (SSDs) systems.
- *Data Model*: Numerous systems exist such as Relational stores, Key-value stores, Document-oriented stores, Graph DBMS, etc.
- *Consistency Model*: Consistency guarantees affect latency and system response to concurrent read and write requests. A non-exhaustive list of consistency models are causal, eventual, strict and weak consistency models.
- *System Constraints*: Constraints are related to data volume handled, cluster size, ease of manageability as scripts running tools, etc.
- *Resource*: Resources are specific hardware configurations that the customer chooses for running its workload, such as virtual machine instance with CPU and RAM characteristics or a network with a bandwidth characteristic.
- *Trial Version*: some DBaaSs are available for free trial during a period of time.

4.3 DBaaS -Quality of Service Dimensions

- *Service Level Agreement*: SLAs capture the agreed upon guarantees between a service provider and its customer. They define the characteristics of the provided service including service level objectives, as maximum response times, minimum throughput rates and data consistency, and define penalties if these objectives are not met by the service provider.
- *Client Support*: Ideally, every customer receives 24 × 7 × 365 support from the Cloud Service Provider.

4.4 DBaaS -Capacity of Service Dimensions

- *High-Availability*: In order to overcome hardware failures, data storage systems implement redundancy through replication, erasure-resilient codes or both. Redundant data is refreshed either synchronously or asynchronously. Also, in order to overcome a whole data center outage, some cloud service providers afford data distribution across different geographical zones.
- *Security*: To ensure confidentiality of sensitive data in the cloud, it is important that the data be encrypted.

- *Elasticity*: Elasticity is the ability to scale-up (provision new nodes) and scale-down (release nodes) a data storage system when the underlying application demands it.
- *Scalability*: Scalability is the ability of a system, to increase total throughput under an increased load when hardware resources are added.
- *Interoperability and Portability*: For long viability, the company should be able to easily migrate to another CSP, and gets its data back in a standard format. Hence, cloud providers and customers must consider all legal issues.

4.5   DBaaS -Cost of Service Dimensions

- *Cost Model*: Even though, many services look similar from the outside, CSPs have different pricing models for storage, CPU, bandwidth, and services for DBaaSs. Indeed, (i) *service cost* is either usage-based or subscription-fee based; (ii) for *bandwidth cost*, most CSPs provide data transfer to their data centers at no cost, while data download is priced; (iii) for *CPU cost*, there are two types of providers charging. The first is *instance-based* for which the CSP charges the customer for the number of allocated instances and how long each instance is used. This is regardless of whether the instances are fully utilized or under utilized. The second is *CPU cycles-based*, for which the CSP charges the customer for the number of CPU cycles a customers application consumes; and finally (iv) the *storage cost* is either *block-rate pricing* or *bundling pricing*. CSPs adopt *storage block-rate pricing* where the range of consumption is subdivided into subranges and the unit price is held constant over each subrange. Other providers adopt instead a *bundling pricing* mode, also called quantity discount.

5   DBaaS-Expert Ranking Core

In order to solve the *DBaaSs' Offerings Ranking Problem*, our approach uses *Analytic Hierarchy Process* (AHP) [17] for DBaaS rating and scoring. In this Section, we describe the application of AHP for solutioning *DBaaSs' Offerings Ranking Problem*.

5.1   AHP-Based Solution for DBaaSs' Offerings Ranking Problem

The main stages of the ranking process are (i) devise the AHP tree, (ii) depict criteria weights, (iii) assess selected offers of DBaaSs satisfying user requirements, and finally (iv) compute the score of each offer.

AHP Tree for DBaaS Offerings' Scoring Problem. Figure 2 illustrates the AHP tree proposed for DBaaSs' scoring and ranking along a set of criteria. The set of proposed criteria derives directly from the *DBaaS Ontology* proposed in section 4. The apex of the DBaaS hierarchy consists of three main criteria:

best quality service, best capacity of service and most affordable service. Each criterion parents criteria from the second layer. For instance, in Figure 2, the *quality of service (QoS) criterion* parents *client support, SLA fullfilment* and *dispute resolution* sub-criteria.

Depiction of Relative Importance of Criteria for DBaaS Selection. First of all, we assume that criteria weights are specified by the user of *DBaaS-Expert*. The weights of importance of the criteria are determined using *pairwise comparisons*. Weights are chosen in a scale of 1 to 9, as recommended by Saaty [17]. For instance, *Criterion C_j* is between to be classified as *equally important* (corresponding value is 1) and *moderate more important* (corresponding value is 3) than *Criterion C_i*. Thus, the corresponding comparison assumes the value of 2. A similar interpretation is true for the rest of the entries of Table 1-(a). The size of the matrix is N^2 (9 for $N = 3$ where N is the number of criteria). The diagonal values are unity (i.e., $criteria\_weights[i,i] = 1$) and the matrix is such that cells in the upper diagonal and cells in the lower diagonal are in inverse relationship (i.e., $criteria\_weights[i,j] = \frac{1}{criteria\_weights[i,j]}$). Notice that, given N criteria, $\frac{N \times (N-1)}{2}$ comparisons are performed. The next step is to compute the *eigenvector* of the *squared criteria' weights matrix* in order to extract the relative importances implied by the previous comparisons. Given a judgement matrix with pairwise comparisons, the corresponding *eigenvector* is obtained by *first* raising the pairwise matrix to powers that are successively squared each time, *second* calculating sums over rows and normalizing values by dividing sum over rows by the sum over column (i.e., geometric mean of each row). This process is iterated until the *eigenvector* does not change from previous iteration (i.e., Δ is negligible). Since, criteria are organized in a hierarchical way, there are two types of weights, namely *local weights* and *global weights*. *Local weights* correspond to weights of criteria of same level, and are calculated as demonstrated in Table 1. The Global weight of a sub-criterion $sub - C_j$ is equal to the product of its local weight and the global weight of C_j.

Table 1 demonstrates pairwise-comparisons of criteria. First, the user enters *weight_of_C_i* compared to *weight_of_C_j* (i.e., half important), *weight_of_C_i* compared to *weight_of_C_k* (i.e., three times more important), and *weight_of_C_j* compared to *weight_of_C_k* (i.e., four times more important). In the example, two iterations are performed. At each iteration, the *squared matrix, eigen vector*, and Δ*eigenvectors* are calculated. The example shows that *Criterion C_i* is the second most important criterion ($W_i = 0.3196$), *Criterion C_j* is the most important criterion ($W_j = 0.5584$) and *Criterion C_k* is the least important criterion.

Assessment of DBaaS' Offerings. First of all, the assessment of DBaaSs' offerings in terms of criteria is performed by experts. Assessment of DBaaSs' offerings in terms of criteria may be done by two ways, (i) *pairwise comparisons* as for criteria: assessments are chosen in a scale of 1 to 9; or (ii) direct assignment through metering of each criterion. The latter is challenging. Indeed, criteria have different types of data (cost: \$, transaction throughput: Tps, feature: bool) and

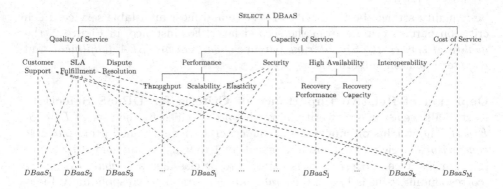

Fig. 2. AHP Tree for DBaaS Offerings' Ranking Problem

Table 1. Example of Transformation of the Matrix *Weights of Criteria* into a *Priority Vector*

(a) 1st Matrix.

	C_i	C_j	C_k
C_i	1	$\frac{1}{2}$	3
C_j	2	1	4
C_k	$\frac{1}{3}$	$\frac{1}{4}$	1

(b) 1st Squared Matrix.

	C_i	C_j	C_k
C_i	3.0000	1.7500	8.0000
C_j	5.3332	3.0000	14.0000
C_k	1.1666	0.6667	3.0000

(c) 1st eigenvector.

	C_i	C_j	C_k	$\sum$	eigenvector
C_i	3.0000	1.7500	8.0000	12.7500	0.3194
C_j	5.3332	3.0000	14.0000	22.3132	0.5595
C_k	1.1666	0.6667	3.0000	4.8333	0.1211
				$\sum$ 39.9155	1.0000

(d) 2nd Squared Matrix.

	C_i	C_j	C_k
C_i	27.6653	15.8330	72.4984
C_j	48.3311	27.6662	126.6642
C_k	10.5547	6.0414	27.6653

(e) 2nd eigenvector.

	C_i	C_j	C_k	$\sum$	eigenvector	Δ
C_i	27.6653	15.8330	72.4984	115.9967	0.3196	-0.0002
C_j	48.3311	27.6662	126.6642	202.6615	0.5584	0.0011
C_k	10.5547	6.0414	27.6653	44.2614	0.1220	-0.0009
				$\sum$ 362.9196	1.0000	

Table 2. Example of Assessment of 3 DBaaSs along 2 criteria in order to obtain columns *i* and *j* of the *Decision Matrix*

(a) DBaaSs' assessment along $Criterion_i$.

	$DBaaS_x$	$DBaaS_y$	$DBaaS_z$	eigenvector
$DBaaS_x$	1	6	6	0.75
$DBaaS_y$	$\frac{1}{6}$	1	1	0.13
$DBaaS_z$	$\frac{1}{6}$	$\frac{1}{1}$	1	0.13

(b) DBaaSs' assessment along $Criterion_j$.

	$DBaaS_x$	$DBaaS_y$	$DBaaS_z$	eigenvector
$DBaaS_x$	1	1	9	0.48
$DBaaS_y$	$\frac{1}{1}$	1	8	0.46
$DBaaS_z$	$\frac{1}{9}$	$\frac{1}{8}$	1	0.06

different considerations (cost: lower is better, throughput: greater is better, ...). For the sake of simplicity, DBaaS offers are assessed in terms of criteria along pairwise comparisons. Table 2 illustrates DBaaS offerings assessment along two criteria through pairwise comparisons.

Table 3. Decision Matrix

	Criteria					
	C_1	C_2	...	C_i	...	C_M
	W_1	W_2	...	W_i	...	W_M
$DBaaS_1$	$a_{1,1}$	$a_{1,2}$	...	$a_{1,i}$	...	$a_{1,M}$
$DBaaS_2$	$a_{2,1}$	$a_{2,2}$	...	$a_{2,i}$	...	$a_{2,M}$
Alternatives $DBaaS_3$	$a_{3,1}$	$a_{3,2}$	...	$a_{3,i}$	...	$a_{3,M}$
⋮	⋮	⋮	⋮	⋮	⋮	⋮
$DBaaS_N$	$a_{N,1}$	$a_{N,2}$	...	$a_{N,i}$	...	$a_{N,M}$

DBaaS Score Calculus. After the alternatives are compared with each other in terms of each one of the decision criteria and the individual *priority vectors* are derived; the priority vectors become the columns of the decision matrix as shown in Table 3, and the score of each $DBaaS_i$ is calculated as follows, $Score(DBaaS_i) = \sum_{j=1}^{M} a_{i,j} W_j$

6   Use Case

Hereafter, we describe the outline of typical use case of *DBaaS-Expert*.

- *(Step 1)* The user enters business requirements via the application module of *DBaaS-Expert*. For instance, he may choose *OLAP* as workload type and *Relational store* as a data model.
- *(Step 2)* *DBaaS-Expert* maps the user requirements to concepts from the DBaaS ontology and filters all relevant DBaaS offers. The outcome of this step is a list of offers of DBaaSs satisfying the user business requirements.
- *(Step 3)* corresponds to AHP run,
  - *(Step 3-a)* *DBaaS-Expert* builds a tree, with a subset of DBaaSs selected from *Step 2*. Notice that, all other alternatives are discarded from selection,
  - *(Step 3-b)* *DBaaS-Expert* allows the user to perform paiwise comparisons of the different criteria, in order to obtain a *priority vector* for criteria.
  - *(Step 3-c)* *DBaaS-Expert* uses its expert knowledge materialized in the wisdom base for the assessment of the selected DBaaS offerings among criteria, and builds the *Decision Matrix*,
  - *(Step 3-d)* *DBaaS Expert* rates and ranks the offers and returns the result to the user,

7   Conclusion

In this paper, we propose DBaaS-Expert framework, which allows a user to choose the most suitable DBaaS. Our contribution is two fold: First, we propose a *DBaaS ontology*. Second, ranking DBaaS offers using a well known AHP - mathematical method for multi-criteria decision-making. In the future, we plan to continue the current work research to include DBaaS assessment using user feedbacks and past experiences.

References

1. Mell, P., Grance, T.: The NIST definition of cloud computing (2011),
   http://csrc.nist.gov/publications/nistpubs/800-145/SP800-145.pdf
2. Binnig, C., Kossmann, D., Kraska, T., Loesing, S.: How is the weather tomorrow?:
   Towards a benchmark for the cloud. In: DBTest (2009)
3. Kossmann, D., Kraska, T., Loesing, S.: An evaluation of alternative architectures
   for transaction processing in the cloud. In: SIGMOD Conference, pp. 579–590
   (2010)
4. Carlo, C., Djellel, D., Andrew, P., Philippe, C.M.: Benchmarking oltp/web
   databases in the cloud: The oltp-bench framework. In: Proc. of 4th CloudDB,
   pp. 17–20 (2012)
5. Quinton, C., Haderer, N., Rouvoy, R., Duchien, L.: Towards multi-cloud configurations using feature models and ontologies. In: Proc. of the Intl. Workshop on
   Multi-Cloud Apps and Federated Clouds, pp. 21–26 (2013)
6. Wittern, E., Kuhlenkamp, J., Menzel, M.: Cloud service selection based on variability modeling. In: Liu, C., Ludwig, H., Toumani, F., Yu, Q. (eds.) Service Oriented
   Computing. LNCS, vol. 7636, pp. 127–141. Springer, Heidelberg (2012)
7. Zhang, M., Ranjan, R., Nepal, S., Menzel, M., Haller, A.: A declarative recommender system for cloud infrastructure services selection. In: Vanmechelen, K., Alt-
   mann, J., Rana, O.F. (eds.) GECON 2012. LNCS, vol. 7714, pp. 102–113. Springer,
   Heidelberg (2012)
8. Ruiz-Alvarez, A., Humphrey, M.: An automated approach to cloud storage service
   selection. In: Proc. of 2nd ScienceCloud Workshop, pp. 39–48 (2011)
9. Menzel, M., Ranjan, R.: Cloudgenius: Decision support for web server cloud mi-
   gration. In: Proc. of 21st WWW, pp. 979–988 (2012)
10. Garg, S.K., Versteeg, S., Buyya, R.: Smicloud: A framework for comparing and
    ranking cloud services. In: Proc. of 4th IEEE UCC, pp. 210–218 (2011)
11. CSMIC: Cloud service measurement index consortium,
    http://www.cloudcommons.com/fr/about-smi
12. DBaaS-expert: Technical report (2014),
    https://sites.google.com/site/rimmoussa/investig_dbaas
13. Heroku: Herokupostgres: Sql   database-as-a-service   (2013),
    https://postgres.heroku.com/
14. Oracle: Database as a service: Reference architecture: An overview (2011),
    http://www.oracle.com/technetwork/topics/entarch/
    oes-refarch-dbaas-508111.pdf
15. Amazon: SimpleDB (2013), http://aws.amazon.com/documentation/simpledb/
16. Sato, K.: An inside look at google bigquery (2013),
    https://cloud.google.com/files/BigQueryTechnicalWP.pdf
17. Saaty, T.: The Analytic Hierarchy Process: Planning, Priority Setting, Resource
    Allocation. McGraw-Hill (1980)

Context-Aware Decision Support in Dynamic Environments: Methodology and Case Study

Alexander Smirnov[1,2], Tatiana Levashova[1], Alexey Kashevnik[1], and Nikolay Shilov[1]

[1] SPIIRAS, 39, 14th line, St. Petersburg, 199178, Russia
[2] University ITMO, 49, Kronverkskiy pr., St.Petersburg, 197101, Russia
{smir,tatiana.levashova,alexey, nick}@iias.spb.su

Abstract. Dynamic environments assume on-the-fly decision support based on available information and current situation development. The paper addresses the methodology of context-aware decision support in dynamic environments. Context is modeled as a "problem situation." It specifies domain knowledge describing the situation and problems to be solved in this situation. The context is produced based on the knowledge extracted from an application ontology, which is formalized as an object-oriented constraint network. The paper proposes a set of technologies that can be used to implement the ideas behind the research. An application of these ideas is illustrated by an example of decision support for tourists travelling by car. In this example, the proposed system generates ad hoc travel plans and assists tourists in planning their attraction attending times depending on the context information about the current situation in the region and its foreseen development.

Keywords: Context-aware decision support, ontology, in-car information.

1   Introduction

Current developments of in-vehicle information systems (e.g., Ford's AppLink, Chrysler's UConnect, Honda's HomeLink, etc.) make it possible to benefit from integration of new decision support methodologies into cars to provide richer driving experience and seamless integration of information from various sources.

Context-aware decision support is required in situations happening in dynamic, rapidly changing, and often unpredictable distributed environments such as, for example, roads. Such situations can be characterized by highly decentralized, up-to-date data sets coming from various information sources. The goals of context-aware support to operational decision making are to timely provide the decisions maker with up-to-date information, to assess the relevance of information & knowledge to a decision, and to gain insight in seeking and evaluating possible decision alternatives.

The present research addresses methodological foundations of context-aware decision support. The theoretical fundamentals are built around ontologies. The ontologies are a widely accepted tool for modeling context information. They provide efficient facilities to represent application knowledge, and to make objects of the dynamic environments context-aware and interoperable.

T. Andreasen et al. (Eds.): ISMIS 2014, LNAI 8502, pp. 325–334, 2014.
© Springer International Publishing Switzerland 2014

The proposed fundamentals are supported by advanced intelligent technologies with their application to Web. The developed context-aware decision support system (DSS) has a service-oriented architecture. Such architecture facilitates the interactions of service components and the integration of new ones [1, 2, 3].

The rest of the paper is structured as follows. Section 2 proposes the methodology of context-aware decision support. Section 3 illustrates the application of the methodology in the area of tourism. Main research results are summarized in the conclusion.

2   Methodology

The developed DSS is intended to support decisions (plans of actions) according to the current situation. The system operates in a dynamic environment. Two kinds of resources are distinguished in the environment: *information* and *acting*. The *information resources* are various kinds of sensors, electronic devices, databases, etc. that provide data & information and perform computations (also referred to as information sources). Particularly, some information resources are responsible for problem solving. The *acting resources* are people and /or organizations that can perform certain actions affecting the environment.

The present research follows the knowledge-based methodology to building DSSs. The idea behind the research is to describe the behavior of DSS by means of two independent types of reusable components: "domain ontologies" and "task ontologies" (defining domain-independent algorithms that describe abstract methods for achieving solutions to problems occurring in the application domain). The both components constitute an application ontology. This ontology represents non-instantiated knowledge. The components are interrelated in a way that indicates what domain knowledge is used by a certain problem. Context knowledge accumulates up-to-date information from the environment. To take this information into account, the domain and task ontologies have to clearly define the context and related information sources.

Context management concerns organization of contextual information for the use in a given application. Identification of context relations enables context arrangements in the knowledge base. Context is defined as any information that can be used to characterize the situation of an entity. An entity is a person, place or object that is considered relevant to the interaction between a user and an application, including the user and application themselves [4].

A DSS's user (decision maker) indicates the type of the current situation and/or formulates the problem(s) to be solved in this situation. Five fundamental categories can be defined for the context [5]: individuality (personal user preferences), time (the moment of decision making), location (location related information affecting the decision, e.g., traffic intensity, objects nearby, etc.), activity (what the user is currently doing, which task is being solved), and relations (between the decision maker / other objects considered in the situation and other persons / objects).

Context is suggested being modeled at two levels: abstract and operational. These levels are represented by abstract and operational contexts, respectively (Fig. 1).

Abstract context is an ontology-based model integrating information and knowledge relevant to the current problem situation. Such knowledge is extracted from the application ontology. As the two components make up the application ontology, the context specifies domain knowledge describing the situation and problems to be solved in this situation. The abstract context reduces the amount of knowledge represented in the application ontology to the knowledge relevant to the current problem situation. In the application ontology this knowledge is related to the resources via the alignment of their descriptions and ontology elements, therefore the abstract context allows the set of resources to be reduced to the resources needed to instantiate knowledge specified in the abstract context. The reduced set of resources is referred to as contextual resources. The ontology alignment model developed by the authors is protected by USA patent US 2012/0078595 A1 [6].

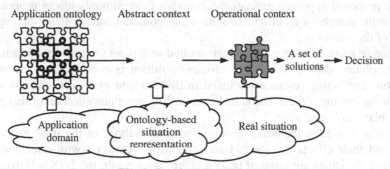

Fig. 1. Ontology-based context

Operational context is an instantiation of the domain constituent of the abstract context with data provided by the contextual resources. This context reflects any changes in environmental information, so it is a near real-time picture of the current situation. The context embeds the specifications of the problems to be solved. The input parameters of these problems, which correspond to properties of the classes of the domain constituent, are instantiated.

Constraint satisfaction techniques [7, 8] can be used to take into account dynamic environmental conditions and other possible constraints that have an impact on the problem. These techniques are naturally combined with ontology-based problem definition, and allow to set context parameters so that they would be taken it into account when the current situation constraints are applied.

Thus, the problems embedded in the operational context are processed as constraint satisfaction problems in its enumeration form (the result is a set of alternative feasible solutions). As a result, a set of feasible (alternative) solutions in the current situation is produced. Each solution is a plan of actions for acting resources in the current situation. Decision making is regarded as a choice between these alternatives.

Table 1 shows the correspondences between the steps of the described above approach and the phases of the Simon's model [9], which specifies decision making consisting of "intelligence", "design", and "choice" phases. Unlike other decision making models, Simon suggests to minimize the efforts of decision makers to evaluate consequences of the possible alternatives and his/her dependence on the multiple factors influencing the choice. Simon proposed a 'satisfactory' decision as a result of decision making. That is a decision that is not efficient or optimal, but the decision that satisfies all the stakeholders interested in it. This is especially important for complex and operational problem solving.

The constrained-based approach enables to express the multiple influencing factors (e.g., the preferences of the stakeholders interested in the decision, intervals of the resources' availabilities, the resources' costs, etc.) by means of constraints. The factors' constraints along with the constraints specified in the operational context are processed as a constraint satisfaction problem solving. A set of feasible (satisfactory) plans is the result of problem solving. At that, these plans do not depend on decision makers' attentions, information they have, or stress. Moreover, the decision makers are saved from information overload. The decision maker can choose any plan from the set or take advantage of some predefined efficiency criteria.

The proposed approach exceeds the bounds of the Simon's model proposing two more steps: search for an efficient solution and communications about the implementation of this solution.

If one or more efficiency criteria are applied to this set of feasible solutions, an efficient solution can be found. The efficient solution is considered as the workable decision. The acting resources included in the efficient plan communicate with the DSS in the person of the decision maker regarding acceptance/refusal of this plan, i.e. on the plan implementation.

In order to enable capturing, monitoring, and analysis of the implemented decisions and their effects the abstract and operational contexts with references to the respective decisions are retained in an archive. As a result, the DSS is provided with the reusable models of problem situations. Based on an analysis of the operational contexts together with the implemented decisions the user preferences are revealed.

Table 1. Three-phase model

Phase	Phase content	Steps of Simon's model	Methodology steps
Intelligence	Finding, identifying, and formulating the problem or situation that calls for a decision	1. Fixing goals	• Abstract context creation
		2. Setting goals	• Operational context producing
Design	Search for possible decisions	3. Designing possible alternatives	• Constraint-based generation of feasible (satisfactory) alternatives
Choice	Evaluation of alternatives and choosing one of them	4. Choice of a satisfactory decision	• Choice of a satisfactory decision

3   Case Study "Tourist Information Support"

Recently, the tourism business is getting more and more popular. People travel around the world and visit museums and other places of interests. They have a restricted amount of time and usually would like to see many museums. In this regard a system is needed, which would allow assisting visitors (using their mobile devices), in planning their museum attending time depending on the context information about the current situation in the museums (amount of visitors around exhibits, closed exhibits, reconstructions and other), traffic situation, and visitor's preferences.

Infomobility is a relatively new term assuming such operation of a decision support system and service provision whereby the use and distribution of dynamic and selected multi-modal information to the users, both pre-trip and, more importantly, on-trip, play a fundamental role in attaining higher traffic and transport efficiency as well as higher quality levels in travel experience by the users [10].

Mobile devices interact with each other through a smart space [11] (a virtual space enabling devices to share information independntly of their locations) implemented via an open source software platform (Smart-M3) [12] that aims at providing a Semantic Web information sharing infrastructure between software entities and devices in the form of blackboard. In this platform, the ontology is represented via RDF triples (more than 1000 triples). Every visitor installs a smart space client to his/her mobile device. This client shares needed information with other mobile devices in the smart space. As a result, each mobile device can acquire only shared information from other mobile devices. When the visitor registers in the environment, his/her mobile device creates the visitor's profile (which is stored in a cloud and contains long-term context information of the visitor such as his/her preferences). The information storage cloud (not computing, which is distributed among the services of the smart space) might belong to the system or be a public cloud. The only requirement is providing for the security of the stored personal data. The profile allows specifying visitor requirements in the smart space and personalizing the information and knowledge flow from the service to the visitor.

The ontology slice that describes a service at a certain point of time is its abstract context (Fig. 2). It is formed automatically (or reused) applying ontology slicing and merging techniques [13]. The purpose of the abstract context formation is to collect and integrate knowledge relevant to the current task (situation). The information sources defined in the abstract context provide the information that instantiates the context and forms the operational context. A concrete description of the current situation is formed, and the problem at hand is augmented with additional data. On the knowledge representation level, the operational context is a set of RDF triples to be added to the smart space by an appropriate service. Therefore, other services can discover these RDF triples and understand the current problem.

In Fig. 2, service i queries up-to-date information from the operational context through smart space in accordance with the task specified in the service's ontology. Services j and n are involved in solving a particular task. They form the operational context related to this task and based on the abstract contexts of the services. This

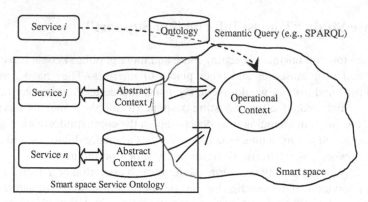

Fig. 2. Ontology and context models in smart space

operational context is described by the smart space service ontology, which also corresponds to the current task and integrates abstract contexts of the involved services.

A prototype of the context-aware decision support system has been implemented based on the proposed approach. Maemo 5 OS-based devices (Nokia N900) and Python language are used for the implementation.

The tourist downloads software for getting intelligent tourist support. Installation of this software takes a few minutes depending on operating system of mobile device. When the tourist runs the system for the first time, the profile has to be completed. This procedure takes not more than 10 minutes. The visitor can fill the profile or can use a default profile. In case of default profile, the system cannot propose preferred exhibitions to the visitor.

The in-vehicle information system is also connected to the smart space (SS in Fig. 3) for a higher interaction level of decision support. For example, let us consider Ford's AppLink system. The AppLink provides the current user location and other car information to the operational context of the smart space automatically. Based on the oprtational context complemented with information from other sources, the attraction information service together with the recommendation service propose an attraction visiting plan and put it into the AppLink's in-car navigation system (Fig. 3).

Other services contributing to the creation of the operational context and involved in the scenario include: attraction information system (AIS) providing textual and multimedia information about the attractions, attraction recommendation service (RS) analysing appropriateness of attractions and points of interests to user preferences, croud sourcing service (CroudS) analysing feedback of other users with similar interests related to corresponding attractions and points of interests, motorway related information services (MSA), and car dealer service (CarS) monitoring the car condition, reminding about required servicing, etc. Any other services can also be connected to the smart space to enrich the decision support possibilities. For example, the car diagnostic system can perform firmware update while the user is visiting a museum if the expected museum vistitng time is longer than time required for the update.

The overall scenario of the case study is shown in Fig. 4. Though the scenario is shown as linear, in fact, it can be interactive and iterative if the user wishes to adjust generated solution via adding additional constraints or preferences.

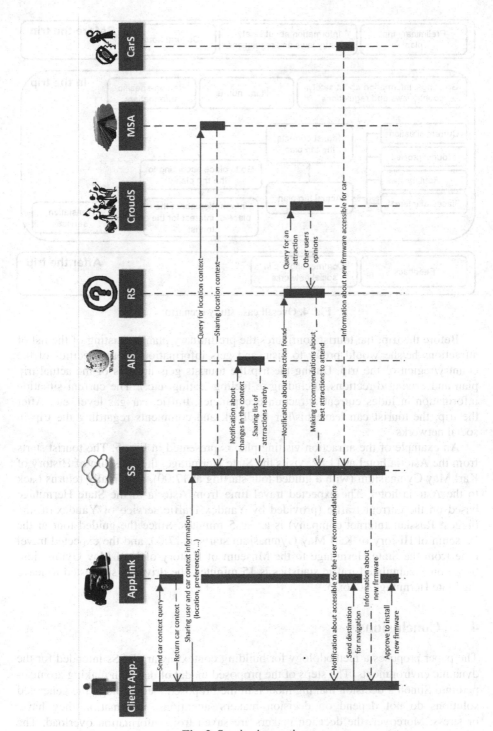

Fig. 3. Service interaction

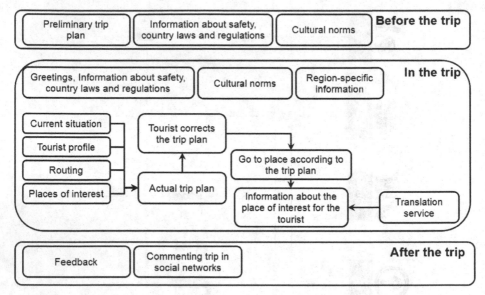

Fig. 4. Overall case study scenario

Before the trip, the tourist configures the preliminary plan consisting of the list of attractions he/she would prefer to visit, and gets information about specifics of the country/region of the trip. During the trip the tourists gets updates of the actual trip plan and driving directions (including re-fueling, eating, etc.). The current situation information includes current attraction occupancies, traffic, car gas level, etc. After the trip, the tourist can leave his/her feedback and comments regarding the trip in social networks.

An example of the attraction visiting plan is presented in Fig. 5. The tourist starts from the Astoria hotel at 11:52, visits the State Hermitage, the Museum of History of Karl May Gymnasium (with a guided tour starting at 17:00), and finally returns back to the Astoria hotel. The expected travel time from Astoria to the State Hermitage based on the current traffic (provided by Yandex.Traffic service of Yandex.ru, the biggest Russian Internet company) is about 5 minutes. Since the guided tour at the Museum of History of Karl May Gymnasium starts at 17:00, and the expected travel time from the State Hermitage to the Museum of History of Karl May Gymnasium based on accumulated traffic statistics is 15 minutes, the driver is suggested to leave the State Hermitage at 16:35.

4   Conclusions

The paper proposes a methodology for building context-aware DSSs intended for the dynamic environments. The steps of the proposed model of decision making are mapped into Simon's decision making model. In the proposed methodology the generated solutions do not depend on decision makers' attentions, information they have, or stress. Moreover, the decision makers are saved from information overload. The

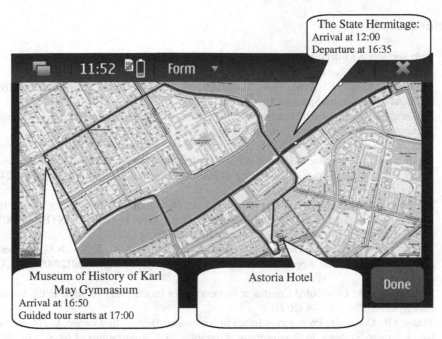

Fig. 5. A sample of attraction visiting plan at the center of St. Petersburg

decision maker can choose any solution from the set (make a decision) taking or not an advantage of the proposed efficiency criteria.

For illustrative purposes, the methodology is implemented in a decision support system for tourists visiting attractions by car. The system helps tourists to plan their attraction attending time depending on the context information about the current situation in the region, its foreseen development and tourists' preferences, using their mobile devices and in-car information systems.

At the moment, if the user wishes to adjust generated solutions via adding additional constraints or preferences the solutions will have to be generated from scratch (only the context is reused). Future work will address incremental solution adaptation via iterative interaction with the user based on planning models (e.g., opportunistic planning [14]).

Acknowledgements. The research was supported partly by projects funded by grants # 13-07-13159, # 13-07-12095, # 14-07-00345, # 12-07-00298, # 14-07-00427, and # 13-07-00336 of the Russian Foundation for Basic Research, project 213 (program 15) of the Presidium of the Russian Academy of Sciences, and project #2.2 of the basic research program "Intelligent information technologies, system analysis and automation" of the Nanotechnology and Information technology Department of the Russian Academy of Sciences. This work has also been partly financially supported by Government of Russian Federation, Grant 074-U01.

References

1. Web Services Architecture, W3C Working Group Note, Web (2004),
   `http://www.w3.org/TR/ws-arch/`
2. Alonso, G., Casati, F., Kuno, H.A., Machiraju, V.: Web Services – Concepts, Architectures and Applications. Springer, Heidelberg (2004)
3. Papazoglou, M.P., van den Heuvel, W.-J.: Service Oriented Architectures: Approaches, Technologies and Research Issues. VLDB Journal 16(3), 389–415 (2007)
4. Dey, A.K., Salber, D., Abowd, G.D.: A Conceptual Framework and a Toolkit for Supporting the Rapid Prototyping of Context-Aware Applications. In: Moran, et al. (eds.) Context-Aware Computing, A Special Triple Issue of Human-Computer Interaction 16, pp. 229–241 (2001)
5. Zimmermann, A., Lorenz, A., Oppermann, R.: An Operational Definition of Context. In: Kokinov, B., Richardson, D.C., Roth-Berghofer, T.R., Vieu, L. (eds.) CONTEXT 2007. LNCS (LNAI), vol. 4635, pp. 558–571. Springer, Heidelberg (2007)
6. Balandin, S., Boldyrev, S., Oliver, I.J., Turenko, T., Smirnov, A.V., Shilov, N.G., Kashevnik, A.M.: Method and Apparatus for Ontology Matching, US Patent 2012/0078595 A1 (2012)
7. Baumgaertel, H.: Distributed Constraint Processing for Production Logistics. IEEE Intelligent Systems 15(1), 40–48 (2000)
8. Tsang, J.P.: Constraint Propagation Issues in Automated Design. In: Gottlob, G., Nejdl, W. (eds.) Expert Systems in Engineering: Principles and Applications. LNCS, vol. 462, pp. 135–151. Springer, Heidelberg (1991)
9. Simon, H.A.: Making Management Decisions: The Role of Intuition and Emotion. Academy of Management Executive 1, 57–64 (1987)
10. Ambrosino, G., Boero, M., Nelson, J.D., Romanazzo, M. (eds.): Infomobility systems and sustainable transport services, ENEA Italian National Agency for New Technologies, Energy and Sustainable Economic Development, 336 (2012),
    `http://old.enea.it/produzione_scientifica/pdf_volumi/`
    `V2010_09-Infomobility.pdf`
11. Kashevnik, A., Teslya, N., Shilov, N.: Smart Space Logistic Service for Real-Time Ridesharing. In: Proceedings of 11th Conference of Open Innovations Association FRUCT, pp. 53–62 (2012)
12. Honkola, J., Laine, H., Brown, R., Tyrkko, O.: Smart-M3 information sharing platform. In: Proceedings of the 1st Int'l Workshop on Semantic Interoperability for Smart Spaces (SISS 2010), Electronic Proceedings (2010)
13. Smirnov, A., Pashkin, M., Chilov, N., Levashova, T.: Constraint-driven methodology for context-based decision support. Design, Building and Evaluation of Intelligent DMSS 14(3), 279–301 (2005)
14. Hayes-Roth, B.: Human Planning Processes. Scientific Report (1980),
    `http://www.rand.org/content/dam/rand/pubs/reports/`
    `2007/R2670.pdf`

Unsupervised Aggregation of Categories
for Document Labelling

Piotr Borkowski, Krzysztof Ciesielski, and Mieczysław A. Kłopotek

Institute of Computer Science, Polish Academy of Sciences,
ul. Ordona 21, 01-237 Warszawa, Poland
kciesiel,piotrb,klopotek@ipipan.waw.pl

Abstract. We present a novel algorithm of document categorization, assigning multiple labels out of a large set of hierarchically arranged (but not necessarily tree-like) set of possible categories. It extends our Wikipedia-based method presented in [1] via unsupervised aggregation (generalization) of document categories. We compare resulting categorization with the original (not aggregated) version and with the variant which transforms categories to a manually selected set of labels.

1 Introduction

Automated (text) document categorization consists in assignment of one or more labels stemming from a pre-defined hierarchically arranged set of categories. Its challenge lies in the number of labels, amounting up to hundreds of thousands [3,5,2,1] (e.g. in Wikipedia ($\mathfrak{W}$) or MeSH ontology ($\mathfrak{M}$)).

To surpass the bad balance between the training data and the number of categories, diverse techniques have been applied. In [3], removal of categories responsible for highest mis-classification risk is suggested. The paper [6] proposes pooling together training data from various sources to get a bigger training set and suggests a technique to handle contradictions in the training sets. Instead of a training set labeled by expert, [5,2] exploit $\mathfrak{W}$ resources by a combined random walk through the categorized document and the $\mathfrak{W}$. Work [5] suggests usage of a directed graph to get deliberately more general categories while [2], by using an undirected graph for random walk, seeks best fitting.

Our methodology differs from the above ones in that it seeks to find most specific generalizations based on directed category graph (like $\mathfrak{W}$ category graph) and with very limited usage of $\mathfrak{W}$ pages assigned to the categories (in fact only their titles are used). In this way, we avoid the noise generated by exploitation of $\mathfrak{W}$ articles content, and do not loose the information on meaningfulness of each category. Furthermore, we spread the authority of an initial categorization in a different way from [5,2]

In section 2 we briefly recall our algorithm [1] for initial category assignment. In section 3 we introduce our new categorization aggregation algorithm. In section 4 we summarize the experimental results on the effectiveness of the new algorithm.

T. Andreasen et al. (Eds.): ISMIS 2014, LNAI 8502, pp. 335–344, 2014.
© Springer International Publishing Switzerland 2014

2   Our Basic Taxonomy-Based Document Categorization Method

Though our agglomerative categorization methods would work with any basic taxonomy-based categorization method, let us briefly explain our basic approach described in [1].

We assume that there is a well-defined taxonomy available that is a hierarchy of categories (like $\mathfrak{W}$ category graph or MeSH ontology) to be assigned to documents. Further there are also available sets of concepts (like $\mathfrak{W}$ pages) which are assigned to one or more categories from the taxonomy. There may exist (unidirectional) links between the concepts of the taxonomy, expressing their relatedness. Both the category names and concept names are assumed to be strings of words (in the same natural language). For example the name of a $\mathfrak{W}$ page ($\mathfrak{W}$ concept) would be its title, as well as the text in the link (anchor text) to this $\mathfrak{W}$ page, and also texts related to so-called disambiguation pages that are used in $\mathfrak{W}$. The $\mathfrak{W}$ category name is just the string being the category page title. It will be used as a label assigned to a document by categorization algorithm.

Our method consists essentially of the following steps. We assign concepts to the terms in the document. In case of homonyms (that is homonym concepts assigned to terms), we disambiguate the concept assignments using the category graph in the context of the rest of the document. Thereafter we move from taxonomy concepts to categories and we compute the weight of each category assigned in this way and select a limited number of highest weight categories.

Let us outline these steps.

After each term in the document content is assigned with a set of concepts ($\mathfrak{W}$ pages) and disambiguated, concepts are mapped to $\mathfrak{W}$ categories. At this stage each term is assigned with a vector of categories. The vectors differ in length as the concepts may be assigned to the different number of categories. To avoid unnecessary impact of the vector length, the vector of categories is normalized for each term, so that the sum of weights assigned to the categories is equal to $tfidf$ value for the associated term

Thus, the impact of each term is proportional to its normalized $tfidf$ weight. We compute the category vector for the document by summing up the vectors for the individual terms, weighting the summands by the respective weight. Finally we obtain a weighted ranking of categories.

The above-mentioned steps rely on our measures of similarity between terms, concepts and categories.

Following guidelines of [4], we introduced two basic measures: IC (*Information Content*) and $MSCA$ (*Most Specific Common Abstraction*), with slight modification that take into account the number of concepts (pages) related to a given category in a taxonomy. So, $IC(k) = 1 - log\,(1 + s_k)\,/log\,(1 + N)$, where s_k is the number of taxonomy concepts in the category k and all its subcategories, and N is the total number of taxonomy concepts. The main category has the lowest value of $IC = 0$.

For the given two categories k_1 and k_2, the $IC_{MSCA}(k_1, k_2) = \max\{IC(k) : k \in CA\}$, where CA is the set of super-categories for both categories k_1 and k_2. The properties of $IC(k)$ measure ensure that the category chosen is most specific amongst the common super-categories (we will denote category $k^* \in CA(k_1, k_2)$ maximizing IC value as $MSCA(k_1, k_2)$).

In the literature dealing with Wordnet many measures based on IC and $MSCA$ have been proposed [4], including LIN similarity:

$$sim_{\text{CATEG}_{\text{Lin}}}(k_1, k_2) = \frac{2 \cdot IC_{MSCA}(k_1, k_2)}{IC(k_1) + IC(k_2)} \tag{1}$$

3 Unsupervised Adaptive Aggregation of Categories

As mentioned earlier, we identified the need to create an algorithm to cluster adaptively the categories assigned to documents. In the earlier version of the algorithm a plain projection onto a pre-defined subset was performed. Now the new algorithm groups together low-level categories assigned to documents into more general ones, by exploiting relationships between them.

3.1 Mapping to the Predefined Set of Labels

First aggregation algorithm to be presented is the one that generalizes the original ranking of categories $k_1, k_2, \ldots, k_R$ by transforming it to the set of **manually** selected target labels $l_1, l_2, \ldots, l_T$. The purpose of the algorithm is to assign a weight to each of the target labels so that the total weight of the original and target categories remains the same (i.e. original weights are redistributed).

Suppose that the category k_i with weight (in the original ranking of categories) w_i is a sub-category (not necessarily direct) of a subset of the target categories, $l_{k_i}^1, l_{k_i}^2, \ldots, l_{k_i}^S$. Then each of the target categories has its weight increased by w_i/S. In other words, the weight w_i is proportionally distributed between all super-categories.

The algorithm passes through each of the original categories k_i, $i = 1, \ldots, R$, repeating the above-described procedure. Weights propagated from each of the original categories are summed up, which results in the weighted ranking of the target categories.

3.2 Unsupervised Mapping

The unsupervised mapping algorithm is the extension of the mapping described in 3.1. Target set $l_1, l_2, \ldots, l_T$ is constructed in unsupervised manner relaying on the input categories that are to be aggregated. As previously, input categories are equipped with weights (k_i, w_i) $i = 1, \ldots, R$. Below we explain detailed construction of the target set, its size depends on the parameter T given by a user. For a given set of input categories $K_R = \{k_1, k_2, \ldots, k_R\}$ we construct a set of all its $MSCA$ categories and denote it by $M(K_R) = \{MSCA(k_i, k_j); 1 \leq i <$

$j \leq T\}$. We repeat the procedure by iterative aggregation of $MSCA$ categories, $M_2(K_R) = M(M(K_R))$, until we get a singleton set (M_S) consisting of the most general (root) category, i.e. $M_S = M(\cdots(M(K_R)))$. We add all the sets, obtaining candidate superset of target categories: $\mathcal{M} = M(K_R) \cup M_2(K_R) \cup \ldots \cup M_S$. For a given T we choose a subset $\{l_1, l_2, \ldots, l_T\} \subset \mathcal{M}$ that will be the target set of labels, in the following way:

1. For each category $l \in \mathcal{M}$ we compute $weight(l) = \sum_{i=1}^{R} w_i \cdot sim(k_i, l)$
2. We sort all the categories according to the descending value of $weight(l)$; as the target set of labels we choose the first T categories, obtaining subset $\mathcal{M'} \subset \mathcal{M}$

Above described method of constructing the target set of labels finds quite general categories topically related to the original categories. Still, to obtain a set of weights which are proportional to the original weights w_i (and further to the $tdidf$ weights of content terms) we have to perform additional step: we redistribute weights w_i by mapping original categories to the set $\mathcal{M'}$ using the algorithm described in section 3.1.

The complexity of all the presented algorithms is linear with respect to the number of the documents (and their length). The complexity of aggregation process is limited by the number of categories in the original ranking and by the length of the longest path in the Wikipedia category graph.

4   Experimental Results

4.1   Benchmark Dataset

In order to measure the efficiency of presented algorithms we used four benchmark sets. Two of them were made of html documents downloaded from the web. They were preprocessed by heuristically extracting their *main content*. In particular, html tags, menus and user comments were removed.

Benchmark documents were collected from the Polish Internet, the task was a part of the ongoing Nekst project. We used Polish Wikipedia from June 2013 with several manual fixes (especially aimed at removal of cycles and technical categories from the category graph). It should be noted here that any other Wikipedia version can be directly used, in order to categorize documents in other languages.

All documents were divided into separate groups based on their text length measured by the number of characters (C), in the following way: *short* $(1000 \leq C < 2000)$, *medium* $(2000 \leq C < 10,000)$, *long* $(10,000 \leq C)$. Files shorter then 1000 characters were not processed. Benchmark sets are presented below.

Our first test set is based on the DMOZ[1] taxonomy and consists of 2804 text files. Each of them includes complete text from a single Polish language web page just with html tags removed but without any additional text preprocessing. The files are divided into 34 groups with a single label having their representation

---

[1] Open Directory Project http://www.dmoz.org

in 𝔚 (and in the DMOZ taxonomy as well). Some of the categories are a subcategory of another, therefore we cannot use this benchmark set when we us categorization with an aggregation to the predefined set of real labels (3.1).

This was the motivation for using a part of the above collection as a separate benchmark, called *DMOZ 15 Categories Corpus* later on. This benchmark consists of 1121 text files and is divided into 15 labeled groups. After mapping these labels to 𝔚 categories, any category is not a subcategory of another.

The next benchmark was constructed using documents from the popular science web portal http://kopalniawiedzy.pl. This domain contains news from various fields of science divided into separate labeled groups. In August 2013 a collection of 13,099 html documents was downloaded.

Documents belonging to some very general categories (e.g. *Humanistyka (Humanities), Ciekawostki (Trivia)*) were omitted. The labels originally assigned by the web authors had to be mapped to the categories from 𝔚 taxonomy. However, it is not a trivial task because the structure of the target set can gravely affect the process of mapping to the predefined set of labels (see 3.1). In our first attempt two of the selected labels appeared to be too general and a great deal of predicted categories were mapped into them (it is another motivation for adaptive, unsupervised aggregation). The results for this benchmark collection (*kopalniawiedzy.pl – 2nd set of labels*) is presented in the result section.

We slightly modified a set of labels (*Natural science* and *Technology* were substituted by more specific *Astronomy* and *Engineering disciplines*) and finally a collection of pages tagged with one of six labels was acquired. This collection is denoted by *kopalniawiedzy.pl – 1st set of labels*. Below we present a total number of documents in the group (the first number) followed by the specific number of *short/medium/long* documents.

- *Bezpieczenstwo komputerowe (Computer security) (121: 85/36/0)*
- *Medycyna (Medicine) (2979: 1235/1744/0)*
- *Nauki biologiczne (Life Sciences) (3122: 1205/1917/0)*
- *Astronomia (Astronomy) (283: 135/148/0)*
- *Psychologia (Psychology) (1733: 700/1033/0)*
- *Nauki techniczne (Engineering disciplines) (4861: 3321/1540/0)*

Last benchmark set, denoted *6 selected domains*, was created of 8,503 web pages downloaded from various domains in March 2013. Labels in square brackets represent corresponding 𝔚 category. They were manually assigned and reflect the subject of domains. As above, numbers represent a total number of documents and a division by length.

- *atlas-zwierzat.pl [Zwierzęta (Animals)] (1070: 547/493/30)*
- *filmweb.pl [Filmy (Films)] (2484: 730/1565/189)*
- *otomoto.pl [Motoryzacja (Automotive)] (1212: 368/771/73)*
- *ptaki.info [Ptaki (Birds)] (1251: 647/532/72)*
- *www.agatameble.pl [Meble (Furniture)] (1428: 245/1179/4)*
- *www.gotujmy.pl [Sztuka kulinarna (Cuisine)] (1058: 313/741/4)*

Benchmark documents transformed into pure text format (after extraction of the *main content* of the html pages) can be found at http://www.ipipan.waw.pl /~kciesiel/ismis/benchmark_datasets.zip.

4.2 Categorization Efficiency Measures

Because of the taxonomical properties of the category graph, neither classical ("flat") classification efficiency measures, nor hierarchy-based classification measures are not suited to the document categorization evaluation, since categories and labels are not independent, like in classification case. Therefore a generalized precision measure has been proposed. We consider the set of categories predicted by our algorithm and the real ones (that is, labels assigned by a human editor).

For each predicted category k_i we compute $sim'(k_i) = \max\{sim(k_i, k_{real}) : k_{real}\}$, (similarity to the most similar real category). Then we average over all documents to obtain a *generalized precision* saying how close on average a predicted category matches its most similar document label.

Note that this measure is parametrized with a chosen *similarity measure*. Experimental results have been computed in three variants:

- the **LIN PRECISION** similarity measure, from Eq. (1) see Section 2.
- the **SHORTEST-PATH** similarity measure, equal to $1/2^{(SP/2)}$, where SP is the length of the shortest path in the $\mathfrak{W}$ category graph between the real and the predicted category
- the **STANDARD PRECISION** measure, which is equal to the binary similarity measure, equal to 1 if predicted category matches real one, and 0 otherwise.

4.3 Results and Discussion

In the presented experiments, we restricted categorization algorithm results to just one category with the highest weights and we compared it with manually assigned document labels. We compared the results for several domains of varying complexity and diversity of content and thematic labels.

We compared precision measures defined earlier for:

- **no aggregation**: categorization without aggregation generalizing assigned categories
- **adaptive aggregation**: unsupervised aggregation that does not use any additional information – in a few different variants that varies the number of target labels on which the original ranking is mapped (cf. constant T in section 3.2).
- **transformation to user-defined labels**: partially supervised aggregation that uses a user-defined set of target labels (in our case, corresponding to the user-defined labels of documents), see section 3.1 – these results can be seen as the upper limit of quality that can be achieved by introducing additional supervised information to the unsupervised aggregation process

Beside quality evaluation, exemplary categorization results (original and aggregated) are presented in Table 1 (the first three most common categories among the group of documents with a given label are shown).

Table 1. Categorization results for benchmark sets: *6 selected domains* and *kopalniawiedzy.pl – 1st set of labels*. For each group three most frequent predicted category names were presented.

Domain's category	Three most frequent category names
6 selected domains; no agregation	
cuisine	thai cuisine, cakes, spices
furniture	furniture, furniture company, japanese tanks of world war II
birds	birds of Europe, ornithology, gulls
animals	biology, felines, vesper bats
automotive	vehicle electrical system, vehicle safety, pricing
movies	film genre, american film actors, polish movies
6 selected domains; adaptive agregation $T = \infty$	
cuisine	diet and food, dishes by main ingredients , asian cuisine
furniture	furniture, company, domestic life
birds	birds, birds of the world, conservation
animals	polish biogeography, mammals, birds
automotive	auto parts, vehicles, taxes
movies	american movies, cinematography, polish movies
kopalniawiedzy.pl – 1st set of labels; no agregation	
computer security	anti-virus programs , russian companies, malware
life sciences	climate change, atmospheric pollutants, evolution
psychology	the brain, universities in canada, verification of stat. hypoth.
medicine	diabetology, immunology, diseases of the nervous system
engineering disciplines	intel processors, graphics, quantum computing
astronomy / physics	missions to mars, black holes, the chronology of the universe
kopalniawiedzy.pl – 1st set of labels; adaptive agregation $T = \infty$	
computer security	computer security, computer science, software
life sciences	biology, mammals, genetics
psychology	psychology, american films, medical specialties
medicine	medical specialties, diseases , immunology
engineering disciplines	computer equipment, physics, systems and electronics
astronomy / physics	physics, elementary particles, particle physics

In Table 2 we present average statistics for the three precision measures defined in section 4.2. As it was described in section 4.1, documents from every benchmark set were tagged with one label (however in general it is possible to evaluate multi-labelled benchmarks). We also took into account one category returned by the algorithm (the one with the highest weight in the ranking) and we compared it with the actual label of the document. The performed experimental evaluation leads us to the conclusions (Table 2):

- unsupervised adaptive aggregation improves original categorization results, with respect to any of the precision measures, for each benchmark sets, regardless of the document content length
- in case of the unsupervised aggregation, low values of the standard precision are noticeable. The standard measure does not take into account any similarity relationship between categories and there is almost 100 000 possible labels. Thus, a random chance of hitting the exact label is ca. 1000 times lower than achieved quality.
- as one could expect, the projection on a predefined set of labels, reflecting topics appearing in the document content, results in the highest precision of categorization for all benchmarks. However, in the context of categorization of a large collection of Internet pages, such result should be treated rather as a theoretical upper limit of achievable quality, since the target labels are not known a priori. Moreover, there is no single subset of labels which could be relevant for any given document.
- depending on the particular topics, a predefined set of candidate target labels can be significantly inadequate. We evaluated two variants of target labels for the same benchmark: *kopalniawiedzy*. Two label sets were similar, except for two labels. Nevertheless, the selection of a particular variant significantly affected precision values. In the *2nd set of labels*, we included *nauki przyrodnicze* (*natural science*) and *nauki techniczne* (*engineering disciplines*). These two categories covered topics of the majority of the benchmark documents. In the *1st set of labels* these two general categories were replaced with more precise alternatives: *technologia* (*technology*) and *astronomia* (*astronomy*). The average value of standard precision has increased from 0.208 to 0.712 for the short documents, and from 0.326 to 0.639 for the medium-length documents (there were no long documents in the benchmark set).
- it should be stressed here that categorization task differs from the unsupervised classification. Partially for this reason, the projection on a fixed set of user-defined document labels significantly improves standard precision. In such case, the resulting categorization space narrows to 6 or 15 (DMOZ) labels only. Indirectly, this shows that the original (not aggregated) categories found by unsupervised categorization are in fact subcategories of real document labels (thus, after casting up in the taxonomy we often hit the exact document label)
- unsupervised adaptive aggregation – despite the lack of any external information on the actual labels – also improves the standard precision values, even up to 10 times in comparison with the unaggregated variant.
- both unsupervised and supervised aggregation improve the value of the two other measures, i.e. generalized precision based on LIN similarity and shortest-paths precision.
- setting the limit on the number of candidate labels (cf. constant T in section 3.2) has positive impact on the unsupervised aggregation results. Preliminary results show that T should be close to the actual number of topics in a given collection of documents.

Table 2. Average values of various precision measures for the following benchmark sets: *DMOZ 34 Categories Corpus, DMOZ 15 Categories Corpus, kopalniawiedzy.pl* (with two versions of user-defined label sets) and *6 selected domains*

Experiment type	LIN PRECISION			SHORTESTPATH			PRECISION		
	short	medium	long	short	medium	long	short	medium	long
DMOZ 34 Categories Corpus									
no aggregation	0.287	0.266	0.289	0.261	0.26	0.259	0.006	0.006	0.005
adapt. aggreg. $T = 1$	0.388	0.372	0.437	0.344	0.332	0.366	0.055	0.061	0.042
adapt. aggreg. $T = 5$	0.343	0.327	0.394	0.331	0.332	0.358	0.045	0.056	0.045
adapt. aggreg. $T = \infty$	0.346	0.319	0.336	0.337	0.328	0.311	0.052	0.058	0.04
DMOZ 15 Categories Corpus									
no aggregation	0.164	0.19	0.229	0.257	0.27	0.293	0.004	0.008	0.008
adapt. aggreg. $T = 1$	0.235	0.301	0.378	0.275	0.303	0.348	0.023	0.046	0.063
adapt. aggreg. $T = 5$	0.242	0.3	0.383	0.313	0.328	0.386	0.066	0.065	0.109
adapt. aggreg. $T = \infty$	0.236	0.265	0.302	0.323	0.342	0.348	0.039	0.061	0.078
transf. to user-def. lab	0.423	0.46	0.529	0.556	0.593	0.641	0.395	0.428	0.5
kopalniawiedzy.pl – 1st set of labels									
no aggregation	0.205	0.245	–	0.225	0.23	–	0.001	0.001	–
adapt. aggreg. $T = 1$	0.318	0.375	–	0.355	0.375	–	0.059	0.067	–
adapt. aggreg. $T = 5$	0.302	0.352	–	0.353	0.374	–	0.054	0.064	–
adapt. aggreg. $T = \infty$	0.268	0.319	–	0.311	0.319	–	0.014	0.014	–
transf. to user-def. lab	0.725	0.65	–	0.803	0.748	–	0.712	0.639	–
kopalniawiedzy.pl – 2nd set of labels									
no aggregation	0.181	0.231	–	0.18	0.207	–	0.002	0.002	–
adapt. aggreg. $T = 1$	0.256	0.344		0.262	0.309		0.022	0.031	
adapt. aggreg. $T = 5$	0.233	0.311	–	0.278	0.332	–	0.023	0.037	–
adapt. aggreg. $T = \infty$	0.222	0.295	–	0.249	0.288	–	0.01	0.011	–
transf. to user-def. lab	0.23	0.353	–	0.459	0.571	–	0.208	0.326	–
6 selected domains									
no aggregation	0.481	0.515	0.473	0.321	0.336	0.309	0.036	0.081	0.000
adapt. aggreg. $T = 1$	0.66	0.661	0.761	0.441	0.419	0.499	0.126	0.052	0.06
adapt. aggreg. $T = 5$	0.673	0.637	0.706	0.482	0.455	0.53	0.133	0.096	0.16
adapt. aggreg. $T = \infty$	0.602	0.588	0.661	0.417	0.41	0.451	0.112	0.05	0.027
transf. to user-def. lab	0.868	0.889	0.938	0.86	0.88	0.928	0.846	0.862	0.919

We conclude with the performance and scalability remark. Presented experiments utilizes benchmark sets which are relatively small (up to 10,000 documents), since they required manually labeled documents. Nevertheless, the algorithms themselves are very efficient and their execution is easily parallelized (by processing individual documents in parallel). Categorization is a part of ongoing Nekst project, aimed at the creation of semantic search engine for the Polish Internet. There were nearly 150 million html pages categorized (source data took ca. 3 TB). The calculations were performed on a hadoop cluster of 68 computers (two 6-core processors & 64 GB RAM each). Categorization took less than 90 minutes and aggregation took less than 500 seconds.

5   Conclusions

The implemented solution has a number of advantages:

- unsupervised adaptive aggregation improves original categorization results, with respect to any of the precision measures, for each benchmark sets, regardless of the document content length
- the target label set is smaller (comparing to the original categorization) but in general case (unlabeled, multi-topical collection of documents) it fits the document categories better than mapping to the predefined set of labels
- the solution requires neither definition of the set of topics by the user nor additional parameters (the algorithm determines itself the aggregation process stopping condition, based on a dependence between labels),
- it is language-independent (as far as taxonomy of concepts is provided) and completely unsupervised
- it can be used to various categorization tasks. In a matter of fact, we conducted first experiments with medical documents categorization via *MeSH* ontology.

Acknowledgments. This research has been partly supported by the European Regional Development Fund with the grant no. POIG.01.01.02-14-013/09: *Adaptive system supporting problem solution based on analysis of textual contents of available electronic resources.*

References

1. Ciesielski, K., Borkowski, P., Kłopotek, M.A., Trojanowski, K., Wysocki, K.: Wikipedia-based document categorization. In: Bouvry, P., Kłopotek, M.A., Leprévost, F., Marciniak, M., Mykowiecka, A., Rybiński, H. (eds.) SIIS 2011. LNCS, vol. 7053, pp. 265–278. Springer, Heidelberg (2012)
2. Coursey, K., Mihalcea, R., Moen, W.: Using encyclopedic knowledge for automatic topic identification. In: CoNLL 2009, pp. 210–218. Association for Computational Linguistics (2009)
3. Dekel, O., Shamir, O.: Multiclass-multilabel classification with more classes than examples. In: Teh, Y.W., Titterington, D.M. (eds.) AISTATS. JMLR Proceedings, vol. 9, pp. 137–144. JMLR.org (2010)
4. Pirró, G., Seco, N.: Design, implementation and evaluation of a new semantic similarity metric combining features and intrinsic information content. In: Meersman, R., Tari, Z. (eds.) OTM 2008, Part II. LNCS, vol. 5332, pp. 1271–1288. Springer, Heidelberg (2008)
5. Syed, Z.S., Finin, T., Joshi, A.: Wikipedia as an ontology for describing documents. In: ICWSM 2008 (2008)
6. Titov, I., Klementiev, A., Small, K., Roth, D.: Unsupervised aggregation for classification problems with large numbers of categories. In: Teh, Y.W., Titterington, D.M. (eds.) AISTATS. JMLR Proceedings, vol. 9, pp. 836–843. JMLR.org (2010), http://dblp.uni-trier.de/db/journals/jmlr/jmlrp9.html#TitovKSR10

Classification of Small Datasets:
Why Using Class-Based Weighting Measures?

Flavien Bouillot[1,2], Pascal Poncelet[1], and Mathieu Roche[1,3]

[1] LIRMM, Univ. Montpellier 2, CNRS – France
[2] ITESOFT, Aimargues – France
[3] TETIS, Cirad, Irstea, AgroParisTech – France
{firstname.lastname}@lirmm.fr
http://www.lirmm.fr/

Abstract. In text classification, providing an efficient classifier even if the number of documents involved in the learning step is small remains an important issue. In this paper we evaluate the performance of traditional classification methods to better evaluate their limitation in the learning phase when dealing with small amount of documents. We thus propose a new way for weighting features which are used for classifying. These features have been integrated in two well known classifiers: Class-Feature-Centroid and Naïve Bayes, and evaluations have been performed on two real datasets. We have also investigated the influence on parameters such as number of classes, documents or words in the classification. Experiments have shown the efficiency of our proposal relatively to state of the art classification methods. Either with a very few amount of data or with a small number of features that can be extracted from poor content documents, we show that our approach performs well.

1 Introduction

Classification of documents is a topic addressed for a long time. Basically the problem can be summarized as follows: *How to efficiently assign a document to one or more classes according to its content?* Usually best results are obtained with an important number of examples (i.e. documents) and features (i.e. words) in order to build an efficient classification model.

However, more and more we need to provide a classifier even if the number of features is quite small [1]. For example, with the development of social networks, we need to use tools that classify tweets exchanged every days or every hours. Here we have to deal not only with the rapid rate but also with the poor content of exchanged texts (i.e. 140 characters). In this context the extraction of relevant and discriminative features represents a challenging issue. Another quite opportunity of applying classification on small number of documents is when the number of labeled documents is itself small. Basically labeling documents is a very time consuming process, requiring lots of efforts.

Recently, new approaches based on semi-supervised or active learning methods try to start a first classification over the small number of labeled documents and ask to the user to validate or not the model in a dynamic way (e.g. [2,3]). For instance, in *Semi-Supervised Learning and Active Learning* approaches, they require a few number of

T. Andreasen et al. (Eds.): ISMIS 2014, LNAI 8502, pp. 345–354, 2014.
© Springer International Publishing Switzerland 2014

labeled documents and a huge number of unlabeled documents in order to improve the model. They must deal with the problem of small amount of data in the first steps of the classification. In the same way, *Real Time learning*, i.e. in a data stream context, have only a few training examples to start building a classifier.

For text classification algorithms, the more the number of learning data is, the better classification results are. And obviously, the classification performance decreases with a reduced training data set. The main contribution of this paper consists in proposing a new way for weighting features which are used for classifying. These features have been integrated in two well known classifiers: Class-Feature-Centroid and Naïve Bayes.

The remainder of this paper is organized as follows. In Section 2 we present some weighting methods and discuss about their efficiency in our context. The new way for weighting features which are $TF\text{-}IDF$-based is presented in Section 3. We present how they have been integrated in two classification approaches in Section 4. Conducted experiments on two real datasets have been compared with traditional classification approaches and are described in Section 5. Finally, Section 6 concludes and presents future work.

2   Weighting Measures

The main goal of the classification is to assign a document to one or more classes according to the terms used in the document. Let $C = C_1, C_2, ..., C_n$ be a set of n classes and $D = d_1, d_2, ..., d_m$ a set of m documents. In the learning phase, each document is attached to one class and we note $D_{i,j}$ the ith document of the class j and $D_j = d_{1,j}, d_{2,j}, d_{3,j}$ the set of documents of the class j.

Texts are commonly represented in the *bag-of-words* model where all documents form a lexicon. $L = t_1, t_2, ..., t_{|L|}$ is a lexicon containing $|L|$ terms where t_i ($1 \le i \le |L|$) is a unique term in the lexicon. Each document is then represented by a weighted vector of terms without considering their position in the document. $\overrightarrow{D_j} = \{w_{1j}, w_{2j}, ..., w_{|L|j}\}$ is a vector representation of the document j where w_{ij} is a weighting factor (e.g. Frequency, Boolean, $TF\text{-}IDF$...) of the term t_i for the document j.

Traditionally, the $TF\text{-}IDF$ measure gives greater weight to specific terms of a document [4]. $TF\text{-}IDF$ is a weighting measure which has been proved to be well appropriate for text classification. It is obtained as follows. In a first step, the frequency of a term (*Term Frequency*) corresponding to its occurrences in the document is computed. It is called the *inner-document weight* of a term. Thus, for the document d_j and the term t_i, the frequency of t_i in d_j is given by the following equation:

$$TF_{i,j} = \frac{n_{i,j}}{\sum_k n_{k,j}}$$

where $n_{i,j}$ stands for the number of occurrences of the term t_i in d_j. The denominator is the number of occurrences of all terms in the document d_j.

The *Inverse Document Frequency* (IDF) measures the importance of the term in the corpus. It is the *inter-documents weight* of a term, obtained by computing the logarithm of the inverse of the proportion of documents in the corpus containing the term. It is defined as follows:

$$IDF_i = log_2 \frac{|D|}{|\{d_j : t_i \in d_j\}|}$$

where $|D|$ stands for the total number of documents in the corpus and $|\{d_j : t_i \in d_j\}|$ is the number of documents having the term t_i. Finally, the $TD\text{-}IDF$ is obtained by multiplying inner-document weight and inter-documents weight as follows:

$$TF - IDF_{i,j} = TF_{i,j} \times IDF_i$$

Traditionally the $TF\text{-}IDF$ measure is used to evaluate the weight of a term within a document and does not take into account the weight of a term by considering the class of the document rather than the document itself. The weight of a term i for the class j, called w_{ij}, depends both on both inner-class term weight and inter-classes term weight. The *inner-class weight* measures the importance of a term within a class (Is this term representative of the class j?) while the *inter-classes weight* measures is used to evaluate if a term is discriminative relatively to other classes. Such weighting measures are, for instance, presented in [5] and [6] but they suffer the following drawbacks:

- When two classes are semantically close, they usually consider a term as representative even if it is used only once. In other words, the number of occurrences of a term in a class is not considered.
- When the class is composed of unbalanced documents (i.e. long vs. short documents for instance), they tend to give a higher weight to terms occurring in the longest documents rather than the shortest ones.

3 New Measures for Inner-Class and Inter-Classes Weighting

In order to weight terms in a class, we propose new inner-class and new inter-classes weighting measures which are based on $TF\text{-}IDF$.

3.1 Inner-class Measures: $Inner\text{-}Weight^{Tf}$ and $Inner\text{-}Weight^{Df}$

First we propose a weighting measure based on the term-frequency as described in the previous section. This measure is called $inner\text{-}weight^{Tf}$ and, for a term t_i in class j, $inner\text{-}weight_{ij}^{Tf}$ is obtained as follows:

$$inner\text{-}weight_{ij}^{Tf} = \frac{TF_{ti}^j}{|n_j|} \tag{1}$$

where TF_{ti}^j stands for the number of terms t_i in C_j and $|n_j|$ is the number of terms in C_j.

$Inner\text{-}weight^{Tf}$ has the same limits than those previously described when considering unbalanced documents within the class. We thus propose another weighting measure that focuses on the document.

In the following, we assume that: the most frequent term in a class is not the most representative term for this class. Now, we consider Document Frequency instead of Term Frequency. The $inner\text{-}weight^{Df}$ for a term t_i in class j is thus defined as follows:

$$inner\text{-}weight_{ij}^{Df} = \frac{DF_{ti}^{j}}{|d_j|} \tag{2}$$

where DF_{ti}^{j} is the number of documents containing t_i in C_j and $|d_j|$ is the number of documents in C_j.

Example 1. Let C_0, C_1 and C_2 be three classes. Each class is composed by three documents called respectively $d_{j,1}$, $d_{j,2}$ and $d_{j,3}$ where j refer to the class (i.e. $d_{0,1}$ refers to the document d_1 in class C_0). Each document is composed by several terms called respectively Term A, Term B and *Others*.

Class	Document	Terms A	Terms B	Others
	$d_{0,1}$	4	1	10
C_0	$d_{0,2}$	2	2	10
	$d_{0,3}$	0	1	10
	$d_{1,1}$	1	3	10
C_1	$d_{1,2}$	3	0	10
	$d_{1,3}$	2	0	10
	$d_{2,1}$	0	0	10
C_2	$d_{2,2}$	0	3	10
	$d_{2,3}$	0	0	10

In the following, we focus on the weighting factor of the terms A and B only for the class C_0. So first we compute $inner\text{-}weight_{i0}^{Tf}$ and $inner\text{-}weight_{i0}^{Df}$ for terms A and B.

Class	Document	Terms A	Terms B	Others
	$d_{0,1}$	4	1	10
C_0	$d_{0,2}$	2	2	10
	$d_{0,3}$	0	1	10
inner-weights Computation				
$inner\text{-}weight_{i0}^{Tf}$	$\frac{TF_{ti}^{j}}{\lceil n_0 \rceil}$	$\frac{6}{40} = 0,15$	$\frac{4}{40} = 0,10$	
$inner\text{-}weight_{i0}^{Df}$	$\frac{DF_{ti}^{j}}{\lceil d_0 \rceil}$	$\frac{2}{3} = 0,66$	$\frac{3}{3} = 1$	

3.2   Inter-classes Measures: $Inter\text{-}Weight^{Class}$ and $Inter\text{-}Weight^{Doc}$

The $inter\text{-}weight^{class}$ measures consider the number of classes containing a term. It is different from traditional approaches that focus on the number of documents and is obtained by:

$$inter\text{-}weight_{ij}^{class} = log_2(1 + \frac{|C|}{C_{ti}}) \tag{3}$$

where $|C|$ is the number of classes and C_{ti} is the number of classes containing the term t_i.

Considering only the presence or the absence of a term in a class might be too restrictive when:

- there are very few classes. The less the number of classes, the more important the inter-classes influence in the global weighting is.
- there are semantically close classes. Semantically close classes result in a high number of common terms between classes.
- there are a huge number of terms in classes (due to very long documents or a large number of documents). The higher the number of terms by class, the higher the probability to have a term appearing at least once in a class is.

We thus propose, as in the inner-class measure, to consider documents instead of classes. However here documents are documents within other classes. Otherwise, by taking into account all the documents, terms which are very frequent and discriminative of one class are underestimated.

So we define $inter\text{-}weight^{doc}$ as follows:

$$inter\text{-}weight_{ij}^{doc} = log_2(\frac{|d| - |d \in C_j| + 1}{|d : t_i| - |d : t_i \in C_j| + 1})$$ (4)

where $|d|$ stands for the number of documents in all classes; $|d \in C_j|$ the number of documents in C_j; $|d : t_i|$: the number of documents in all classes containing the term t_i and $|d : t_i \in C_j|$ is the number of documents in C_j containing the term t_i. Adding the number of documents in all classes prevent the case where t_i is only used in C_j (when $|d : t_i| - |d : t_i \in C_j| = 0$).

Example 2. Let us compute $inter\text{-}weight_{i0}^{Class}$ and $inter\text{-}weight_{i0}^{Doc}$ for terms A and B.

Class	Document	Terms A	Terms B	Others
C_0	$d_{0,1}$	4	1	10
	$d_{0,2}$	2	2	10
	$d_{0,3}$	0	1	10
C_1	$d_{1,1}$	1	3	10
	$d_{1,2}$	3	0	10
	$d_{1,3}$	2	0	10
C_2	$d_{2,1}$	0	0	10
	$d_{2,2}$	0	3	10
	$d_{2,3}$	0	0	10
inter-weightsComputation				
$inter\text{-}weight_{i0}^{class}$	$log_2(\frac{|C|}{C_{ti}})$	$log_2(\frac{3}{2}) = 0,58$	$log_2(\frac{3}{3}) = 0$	
$inter\text{-}weight_{i0}^{doc}$	$log_2(\frac{|d| - |d \in C_0| + 1}{|d:t_i| - |d:t_i \in C_0| + 1})$	$log_2(\frac{7}{4}) = 0,81$	$log_2(\frac{7}{3}) = 1,22$	

4 Integration of the New Measures in Classification Algorithms

Usually SVM (Support Vector Machine) and Naïve Bayes are recognized as two of the most effective text classification methods. Nevertheless they are not well adapted to small learning datasets [7]. Sometimes, for instance, they require complicated adaptation such as the definition of a new kernel for SVM [8]. Our new weighting measures can be used in Naïve Bayes [9] and Class-Feature-Centroid [5] approaches which offer the following advantages: (i) they have been recognized as very efficient for text classification; (ii) based on weighted features, they can easily be modified; (iii) finally the obtained models are easy to interpret for users.

First, for each class of the learning set, we compute $C_j = \{w_{1j}, w_{2j}, ..., w_{|L|j}\}$ for different combinations of inner- and inter-classes weights:

- $w_{ij}^{Tf-Class}=inner\text{-}weight^{Tf} \times inter\text{-}weight^{class}$
- $w_{ij}^{Df-Class}=inner\text{-}weight^{Df} \times inter\text{-}weight^{class}$
- $w_{ij}^{Tf-Doc}=inner\text{-}weight^{Tf} \times inter\text{-}weight^{doc}$
- $w_{ij}^{Df-Doc}=inner\text{-}weight^{Df} \times inter\text{-}weight^{doc}$

Example 3. For example with our running example, for term A and class C_0, we have:
$w_{A0}^{Tf-Class}=inner\text{-}weight_{A0}^{Tf} \times inter\text{-}weight_{A0}^{class} = 0,15 \times 0,58 = 0,09$

The integration of our measures in Naïve Bayes is done as follows. After computing $C_j = \{w_{1j}, w_{2j}, ..., w_{|L|j}\}$ where $w_{i,j}$ is the weight of the ith term in the class C_j, we estimate the probability that an unlabeled document d belongs to a class C_j : $P(d \in C_j) = P(C_j) \prod_i (w_{i,j})$. Experiments combining $w_{ij}^{Tf-Class}$ weighting method and Naïve Bayes approach is called $Nb^{Tf-Class}$ in the following (resp $Nb^{Df-Class}$, Nb^{Tf-Doc}, and Nb^{Df-Doc}).

Class-Feature-Centroid is a recent learning model presented in [5]. In Class-Feature-Centroid, each class is considered as a Vector Space Model [4] which is based on the *bags of words* representation. Each class is represented as a term-vector and a class-vector is a centroid. $\overrightarrow{C_j} = \{w_{1j}, w_{2j}, ..., w_{|L|j}\}$ is the centroid representation of the class j where w_{ij} is the weight of the term t_i for the class j. For classifying an unlabeled document d, the document is also considered as a term-vector ($\overrightarrow{d} = \{w_{1j}, w_{2j}, ..., w_{|L|j}\}$) and the distance (e.g. cosine) between document-vector $\overrightarrow{d}$ and all centroids $\overrightarrow{C_j}$ is compared. Then, we compute $Class\text{-}Feature\text{-}Centroid$ approach with each w_{ij}. Experiments are called $Cfc^{Tf-Class}$, $Cfc^{Df-Class}$, Cfc^{Tf-Doc}, and Cfc^{Df-Doc}.

5   Experiments

5.1   Experimental Protocol and Datasets

In order to evaluate our proposal we selected two different datasets:

- The *Reuter* dataset is the Reuters-21578[1] frequently used for evaluating classification and information retrieval algorithms. It contains news from Reuters newswire and proposes different categories such as sugar, gold, soybean, etc. Documents have been manually classified.

---

[1] http://trec.nist.gov/data/reuters/reuters.html/

– The *Tweet* dataset is composed of French tweets that have been collected for the Polop project[2] during the French Presidential and Legislative elections in 2012. It is composed of more than 2 122 012 tweets from 213 005 users. We consider the set of tweets of a user as one document rather than one document per tweet.

In order to evaluate the efficiency of our proposal, experiments have been done by comparing the results of different supervised classification methods[3] with features weighted by the classical $TF\text{-}IDF$: (i) Two different versions of SVM: *SMO*, using a polynomial kernel [10] and *LibSVM*, using a linear kernel [11]; (ii) One original Naïve Bayes approach known as very efficient for text classification: *DMNB* [12] and one Decision Tree: LadTree [13]. Other comparisons have been performed including Naive-Bayes [14], NaiveBayes Multinomial [15], LibSVM with Radial Basis Function and Polynomial kernel [11], J48 and RepTree [16]. As they clearly performed very badly with our datasets, results are not presented in this paper[4]. For each dataset, we remove stop words and words having less than 3 characters. We choose to not apply lemmatization tools since they are not really well adapted to Tweets.

On first experiments for each series, we did a classical 3 fold cross validation. We consider the test dataset as represented in Figure 1. Three iterations have been chosen in order to have reliable results and we decide not to apply a classical cross validation in order to keep sufficient numbers of documents in the test datasets. Moreover, using the same test dataset over all experiments allows us to ensure that modifications on performance are only due to changes made on the learning dataset.

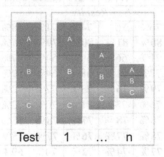

Fig. 1. Validation Process

In this paper, the quality of the classification is evaluated by using micro-averaged F-measure[5].

---

[2] http://www.lirmm.fr/%7Ebouillot/polop

[3] These methods are available on Weka. http://www.cs.waikato.ac.nz/ml/weka/

[4] Interested reader may found all the results at the following URL: www.lirmm.fr/%7Ebouillot/weipond

[5] Results with Macro and Micro-averaged Precision, Macro and Micro-averaged Recall and Macro-averaged F-measure are available in the web page.

5.2 Results

In order to study the behavior of Naïve Bayes and Class-Feature-Centroid approaches and other supervised algorithms according to the reduction of the dataset's volume, we realized three series of experiments on *"Reuter"* and *"Tweet"* dataset. The first series of experiments evaluate the impact on classification results when the number of classes decreases, the second when the number of documents decreases, and the third, when the number of words decreases. As expected, the impact on the decreasing number of classes is consistent. Detailed results of the experiments can be found on the results webpage. Ultimately, we can say that LadTree is a bit more impacted by a large number of classes, and Class-Feature-Centroid and Naïve Bayes approaches used with new weighting method slightly outperform classical algorithms.

How the Number of Documents Impacts Classification? Our second series of experiments focuses on the number of documents per class. We set the number of classes (i.e. 10) and we reduce the number of documents per class from 50 to 3. Nine experiments have been performed (see Table 1).

Table 1. Micro-Averaged F-Measure when Decreasing the Number of Documents per Class

Datasets								
Classes	10	10	10	10	10	10	10	10	10
Doc	500	450	390	330	270	210	150	90	30
Terms	62808	57336	47753	42219	33572	26040	17596	9641	3023
Results								
DMNB	76%	76%	76%	74%	71%	68%	67%	54%	38%
LadTree	**80%**	**80%**	**81%**	**80%**	**79%**	**76%**	**78%**	51%	16%
LibSVM	69%	71%	66%	59%	54%	47%	45%	30%	21%
SMO	73%	72%	71%	68%	64%	59%	57%	41%	22%
$Cfc^{Df-Class}$	78%	79%	76%	73%	72%	72%	69%	56%	36%
Cfc^{Df-Doc}	75%	75%	73%	71%	70%	72%	67%	55%	36%
$Cfc^{Tf-Class}$	77%	78%	78%	77%	78%	75%	72%	**64%**	**45%**
Cfc^{Tf-Doc}	77%	78%	77%	76%	77%	75%	70%	63%	**45%**
$Nb^{Df-Class}$	77%	77%	74%	72%	70%	69%	66%	53%	36%
Nb^{Df-Doc}	73%	72%	71%	69%	67%	68%	65%	51%	36%
$Nb^{Tf-Class}$	78%	78%	78%	77%	78%	75%	71%	**65%**	**49%**
Nb^{Tf-Doc}	77%	78%	77%	76%	77%	75%	71%	**64%**	**49%**

From these experiments, we can conclude that our new weighting measures with Class-Feature-Centroid and Naïve Bayes approaches (1) slightly outperform other algorithms (except LadTree), (2) are more resistant than most of other algorithms when the number of documents decreases dramatically.

How the Number of Words Impacts Classification? On the *"Tweet"* dataset, we decide to set the number of classes (5) and documents (1 186) and to randomly remove words in order to decrease the number of terms available per document. We assume that randomly removing terms may change the document nature. Seven experiments have been done (see Table 2).

Table 2. Micro-averaged F-Measure when Decreasing the Number of Words per Documents

Datasets							
Classes	5	5	5	5	5	5	5
Documents	1186	1186	1186	1186	1186	1186	1186
Terms	1579374	1322148	613777	264025	202166	157177	76851
Results							
DMNB	**93%**	**92%**	**87%**	**77%**	**78%**	70%	60%
LadTree	72%	72%	67%	56%	56%	53%	49%
LibSVM	67%	51%	50%	51%	51%	30%	22%
SMO	**91%**	**90%**	**82%**	**71%**	70%	61%	51%
$Cfc^{Df-Class}$	79%	79%	71%	57%	57%	57%	53%
Cfc^{Df-Doc}	38%	38%	37%	37%	37%	37%	38%
$Cfc^{Tf-Class}$	**86%**	**86%**	**82%**	**72%**	**72%**	**72%**	**67%**
Cfc^{Tf-Doc}	57%	56%	55%	52%	52%	52%	50%
$Nb^{Df-Class}$	80%	79%	71%	57%	55%	53%	47%
Nb^{Df-Doc}	37%	37%	37%	37%	37%	38%	38%
$Nb^{Tf-Class}$	**88%**	**87%**	**83%**	**74%**	**71%**	**68%**	**59%**
Nb^{Tf-Doc}	57%	56%	55%	52%	51%	50%	46%

Conclusions on these experiments are (1) Naïve Bayes and Class-Feature-Centroid with new weighting measures give results slightly better than SVM and DMNB when the number of terms is low (experiments 6 and 7), and slightly worse otherwise (experiments 1 and 2), (2) Naïve Bayes and Class-Feature-Centroid with new weighting measures outperform LadTree and LibSVM. It is interesting to underline that results are similar on an English corpus (i.e. *"Reuter"*) and on a French corpus (i.e. *"Tweet"*).

6 Conclusions and Future Work

Dealing with few data in text classification still remains an important issue which is little addressed in the literature. However, there are many applications where having a large amount of data is not possible or desirable. In this paper, we introduced new weighting measures that we apply in Class-Feature Centroid-based and Naïve Bayes approaches. Experiments show that these new measures are well adapted for dealing with small set of learning data. We compared its efficiency relatively to seven other supervised learning approaches and showed that it performed very well even for different languages. We also investigated the impact of different parameters by varying the number of classes, documents, and words. Conducted experiments have highlighted that best results are obtained when the number of features is reduced. Furthermore, generated models are easy to validate and interpret. This interpretation is important, for instance, when we have to face with news or tweets evolving at a rapid rate and then obtain the top-k relevant terms for a class over time. The generated models could be used to automatically extract trends over time as the ones that have been proposed during the US elections. In future work, we plan to adapt our approach with $Okapi_{BM25}$ measure. We also want to better investigate the *semantically closeness properties* of classes in order to better evaluate among inner-class and inter class weights which one is the most appropriate.

References

1. Forman, G., Cohen, I.: Learning from little: Comparison of classifiers given little training. In: Boulicaut, J.-F., Esposito, F., Giannotti, F., Pedreschi, D. (eds.) PKDD 2004. LNCS (LNAI), vol. 3202, pp. 161–172. Springer, Heidelberg (2004)
2. Zeng, H.J., Wang, X.H., Chen, Z., Lu, H., Ma, W.Y.: Cbc: Clustering based text classification requiring minimal labeled data. In: Proceedings of the Third IEEE International Conference on Data Mining, ICDM 2003, p. 443 (2003)
3. Lin, F., Cohen, W.W.: Semi-supervised classification of network data using very few labels. In: Proceedings of the 2010 International Conference on Advances in Social Networks Analysis and Mining, ASONAM 2010, pp. 192–199 (2010)
4. Salton, G., McGill, M.J.: Introduction to Modern Information Retrieval. McGraw-Hill, Inc., New York (1986)
5. Guan, H., Zhou, J., Guo, M.: A class-feature-centroid classifier for text categorization. In: Proceedings of the 18th International Conference on World Wide Web, WWW 2009, pp. 201–210. ACM, New York (2009)
6. Zhang, X., Wang, T., Liang, X., Ao, F., Li, Y.: A class-based feature weighting method for text classification. Journal of Computational Information Systems 8(3), 965–972 (2012)
7. Kim, S.B., Han, K.S., Rim, H.C., Myaeng, S.H.: Some effective techniques for naive bayes text classification. IEEE Transactions on Knowledge and Data Engineering 18(11), 1457–1466 (2006)
8. Joachims, T.: A statistical learning learning model of text classification for support vector machines. In: Proceedings of the 24th Annual International ACM SIGIR Conference on Research and Development in Information Retrieval, SIGIR 2001, pp. 128–136. ACM (2001)
9. Lewis, D.D.: Naive (bayes) at forty: The independence assumption in information retrieval. In: Nédellec, C., Rouveirol, C. (eds.) ECML 1998. LNCS, vol. 1398, pp. 4–15. Springer, Heidelberg (1998)
10. Platt, J.C.: Advances in kernel methods, pp. 185–208. MIT Press, Cambridge (1999)
11. Chang, C.C., Lin, C.J.: Libsvm: A library for support vector machines. ACM Trans. Intell. Syst. Technol. 2(3), 27:1–27:27 (2011)
12. Su, J., Zhang, H., Ling, C.X., Matwin, S.: Discriminative parameter learning for bayesian networks. In: Proceedings of the 25th International Conference on Machine Learning, pp. 1016–1023. ACM (2008)
13. Holmes, G., Pfahringer, B., Kirkby, R., Frank, E., Hall, M.: Multiclass alternating decision trees. In: Elomaa, T., Mannila, H., Toivonen, H. (eds.) ECML 2002. LNCS (LNAI), vol. 2430, pp. 161–172. Springer, Heidelberg (2002)
14. John, G.H., Langley, P.: Estimating continuous distributions in bayesian classifiers. In: Proceedings of the Eleventh Conference on Uncertainty in Artificial Intelligence, pp. 338–345. Morgan Kaufmann Publishers Inc. (1995)
15. McCallum, A., Nigam, K., et al.: A comparison of event models for naive bayes text classification. In: AAAI 1998 Workshop on Learning for Text Categorization, pp. 41–48 (1998)
16. Quinlan, J.R.: C4.5: Programs for machine learning. Morgan Kaufmann Publishers Inc., San Francisco (1993)

Improved Factorization of a Connectionist Language Model for Single-Pass Real-Time Speech Recognition

Łukasz Brocki, Danijel Koržinek, and Krzysztof Marasek

Polish-Japanese Institute of Information Technology, Warsaw, Poland

Abstract. Statistical Language Models are often difficult to derive because of the so-called "dimensionality curse". Connectionist Language Models defeat this problem by utilizing a distributed word representation which is modified simultaneously as the neural network synaptic weights. This work describes certain improvements in the utilization of Connectionist Language Models for single-pass real-time speech recognition. These include comparing the word probabilities independently between the words and a novel mechanism of factorization of the lexical tree. Experiments comparing the improved model to the standard Connectionist Language Model in a Large-Vocabulary Continuous Speech Recognition (LVCSR) task show the new method obtains about a 33-fold speed increase while achieving a minimally worse word-level speech recognition performance.

Keywords: Connectionist language model, real-time single-pass automatic speech recognition, lexical tree factorization.

1 Introduction

In Speech Recognition, language modeling is usually handled using Statistical Language Models [1]. These, conceptually simple models, attempt to derive the probability distribution of word sequences by estimating it statistically on a large collection of textual data, however, for any task with a significantly large vocabulary, this estimate becomes unreliable and various tricks need to be employed to make them usable [2–5]. Artificial Neural Network (ANN) based, also known as Connectionist Language Models, have been researched for a while [6, 7] but have not been used much in real-world situations, mostly because of the performance cost that these solutions entailed. Recently, there has been a renewed interest in Connectionist models for Speech Recognition thanks to some ground-breaking improvements in the field of ANNs [8]. Some solutions to the problem of complexity of ANN based Language models suggest using hierarchical decomposition based on existing databases, like Wordnet [9]. Our method still computes actual word probabilities but does so implicitly by modeling factorization on a lexical tree, instead.

The paper explains some of the basics behind the Connectionist Language models (section 2) and the problems related to their performance (sections 3

T. Andreasen et al. (Eds.): ISMIS 2014, LNAI 8502, pp. 355–364, 2014.
© Springer International Publishing Switzerland 2014

and 4). Following that, it presents our improvement to the performance (section 5) and finally shows some experimental results (section 6).

2   Neural Network as a Language Model

The general approach in using neural networks in language modeling is as follows: the neural network calculates the contextual word probability at time t using words that occur previously $(t - 1, t - 2, \ldots)$:

$$P(w_t | w_{t-1}, w_{t-2}, \ldots, w_{t-n}) \tag{1}$$

The neural network uses what is called a distributed representation for each word in the dictionary. The distributed representation is a vector of real numbers of a specified length used to describe each word in the dictionary. Usually, several dozen dimensions are used in this representation. The information about word features is thus distributed over several independent components. The neural network maps the real vectors representing the previous words in the sequence, given at input, into the contextual probability of the words present in the dictionary. During training, the values of the features of individual words in the dictionary are modified simultaneously with the synaptic weights of the neural network. The word features are not predefined in any way but changed only during the training phase. The features are not derived from clustering but using gradient descent learning to capture correspondences between words [6]. Because the Connectionist Language Model denotes each word as a vector of real values, it compresses a multidimensional local representation, sometimes reaching even 100 thousand dimensions, into a low-dimensional distributed representation (most commonly up to several dozen dimensions). The objective of the neural network is to minimize the language model error. After the training phase, the words that occur in similar contexts will also occur in a close Euclidean distance in the feature space.

3   Computational and Memory Complexity

The computation complexity of calculating a single probability of word occurrence in a given context using a single hidden layer feed-forward neural network will be analyzed below. The probability of the word represented by the neuron with index k when using the Softmax activation function is given by the following formula:

$$P(w_t = k | w_{t-n}, w_{t-n+1}, \ldots, w_{t-1}) = \frac{e^{net_k}}{\sum_{i=1}^{N} e^{net_i}} \tag{2}$$

where net is the weighted sum of signals reaching the neuron. To calculate the probability of any of the words, the weighted sum of all the neurons in the network needs to be calculated first. This is the first weakness of the Connectionist Language Model. The volume of computation with a dictionary of 100k words is so large that the implementation of such a neural network in a real time environment is practically impossible. Let us, however, perform a complete analysis of the algorithm complexity while computing individual words in context:

$$P(w_t|w_{t-n+1}, \ldots, w_{t-1}) \tag{3}$$

Unlike the output layer, the excitation of the input layer can be computed in constant time, independent of the dictionary size, because it is sufficient to copy the word feature values representing the context. The computation of the hidden and output layer excitations amounts to simple matrix multiplication and sum operations. This weighted sum is then modified by a non-linear activation function. The computational complexity of neural networks is often expressed in the amount of "floating point operations" (FLOP). In the case of a single word in context, this can be calculated using the following formula:

$$(n - 1) * P * H + H + (H * N) + N \tag{4}$$

where H is the number of neurons in the hidden layer, N is the number of neurons in the output layer (i.e. equals the number of words in the dictionary), P is the number of features used to describe a word and n is the range of the n-gram context. The most significant part of the formula is the $(H * N)$ portion, representing the computation of the denominator of the Softmax function. It should be also mentioned that in order to compute the Softmax function, a very costly exponential function needs to be executed N times. That is why the above formula underestimates the actual computational complexity. For a typical LVCSR task, the values of these parameters may be the following: $n = 5$, $P = 100$, $H = 200$, $N = 100000$. Using the formula above, the computational complexity using these parameters would amount to 20,180,200 FLOPs, which is an insurmountable obstacle, given that speech recognition systems test thousands of hypotheses (utilizing this whole computation) each second. As a side note, according to the formula and the parameters, the computation of the output layer takes 20,100,000 FLOPs, which is 99.6% of the complete calculation. This includes the computation of the weighted sum of all the signals reaching all the neurons in the network, as well as executing the exponential function N times.

With regards to the memory complexity, the number of synaptic weights and the total number of features used to represent words need to be counted:

$$P * N + (n - 1) * P * H + H * N \tag{5}$$

Using the example parameters from the previous paragraph, the number of synaptic weights and word features will be 120,080,000, which takes around 458 MB of memory (assuming 32 bits per value). Naturally, all the values have to be present in the working memory in order to achieve reasonable performance. When using BPTT (backpropagation through time) to train this model, the memory requirements can be several times larger because additional information about error and gradient for each parameter needs to be stored. Furthermore, the precision of computation may also present an issue when calculating a gradient using a large number of samples. The usage of double precision floating point numbers may therefore become necessary [10].

4 Lexical Trees and Factorization

To speed up the speech recognition process, all the data structures, especially lexical trees, state graphs and search algorithms need to be optimized as much as possible. The basic data structure used in speech recognition is the word lexicon. In most situations, this structure is stored in the form of a tree. Lexical trees share phoneme states that are common among different words, thus reducing the search space. Ney [11] shows that a lexical tree based on a vocabulary of 12306 words contained only 43k phonetic nodes. A linear lexicon built on the same vocabulary had as many as 100800 nodes. Further experiments were performed to establish the advantage of the tree over the linear lexicon with regards to the speed of computation during hypothesis search. The experiments show a seven fold increase in speed during search compared to the linear lexicon.

Unfortunately, the biggest advantage of the lexical tree – the compression of the prefixes – is its greatest flaw. A single node that is a part of more than one prefix automatically belongs to more than one word. Only the leaves belong to a single word every time. Therefore, a hypothesis that occupies a node within the tree cannot definitely determine which word it belongs to. Because of this, the word probability in such cases can be determined only at the end of the tree. The early application of the probability of the language model allows speeding up and improving the speech recognition accuracy because worse hypotheses can be rejected earlier in the search process. To avoid this problem, the factorization of the language model probabilities along the lexical tree can be employed.

Factorization is a process of decomposing a value into several components. The product of these components should be equal to the initial value. To improve the tree lexicon, the word probabilities need to be factorized with respect to the branching of the lexical tree. A lexical tree without factorization assigns a probability of 1 to each internal node in the tree and the actual word probability at the leaves. The factorization of a lexical tree works by spreading the individual word probabilities from the leaves onto the rest of the nodes in the tree. The hypothesis search algorithm then multiplies the individual factors along the path to obtain the actual word probability after reaching the end, leaf node.

In the case of a unigram language model, a single factorized lexical tree is completely sufficient to recognize speech. That is because the probabilities of a unigram language model are context free and do not require any information about the history of previous words to calculate the probability of the next. If n-grams of higher order are used, the state of the language model cannot be determined locally. That is why it is theoretically required to store a factorized lexical tree for each language model state. For a bigram model, that takes a context of one previous word, a copy of a factorized lexical tree is needed for each previous word. If the size of the search space for a unigram language model is assumed to be equal to P, in case of the bigram model, this size will be $V * P$ where V is the vocabulary size. For an n-gram model this would be $V^{n-1} * P$. The search space will grow exponentially with the increase of the context of the language model. This applies equally to the statistical language model and to the Connectionist Language Model that uses a feed-forward neural network with a specified context n. Unfortunately with a recurrent neural network, the search space grows theoretically into infinity, since the context used in the network is theoretically infinite. Another issue is its internal memory, which significantly complicates its implementation as a language model. Both statistical n-gram and Connectionist Language Models using a feed-forward neural network depend solely on the input and always provide the same value for the same input. A recurrent neural network does not have this guarantee and therefore an instance of internal network states is necessary for each of the currently active lexical trees (word histories).

Even though the search space grows exponentially with the length of the context, many language models allow skipping most of the unlikely hypotheses. Research shows [11] that the amount of computation that is required during speech recognition is only double for a bigram language model compared to the unigram. In other words, even if the static search space is large, the actual search space checked by the decoder is similar or sometimes even smaller when using a better and richer language model.

5   Modified Factorization Routine

To improve the performance of the Connectionist Language Model, two problems need to be addressed. Firstly, the computation of probabilities of word occurrences for large vocabularies is currently unfeasible using a Connectionist Language Model in a real-time speech recognition system. This results from the way the Softmax activation works, as was mentioned in section 3. The other equally problematic issue is that even if it was somehow possible to easily compute all the word occurrence probabilities (even as fast as a statistical n-gram model), the complete factorization across the lexical tree would still need to be performed and this was also shown as unfeasible in the previous section.

The modified factorization routine described below uses an artificial neural network to calculate the factorized word occurrence probabilities directly across the lexical tree. The model consists of a dictionary containing the word features and a neural network using many groups of Softmax activation functions used in the problem decomposition. Each group is responsible for computing the transition probabilities between the nodes of the tree. Each group carries out its decision independent of all the other groups in the network.

The nodes of the tree that have more than one child are also associated with the output neurons of the neural network. Each neuron is then trained to calculate the factorized probability of the language model corresponding to the given node. The described model is several times faster than the standard model. There are two reasons behind this. Firstly, the calculation of factorized probability directly from the outputs of the network means that the full computation of tree

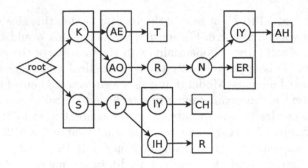

Fig. 1. A lexical tree containing the words: cat, corner, cornea, speech, spear (in ARPA-bet). The node groups marked in rectangles are part of the same Softmax function.

factorization can be omitted. Even if final word probabilities are still required, they can be calculated by simply multiplying all the factorized probabilities in the sequence. The second reason for its improvement in speed is that the model calculates the probabilities of individual word occurrences independently of each other. In standard Connectionist Language Models, the calculation of any word probability requires that all of them are calculated first, because of how the Softmax function computes its outputs. In the modified system, the Softmax function must be calculated only for the children of the branching nodes of the sequence belonging to the chosen word. For the example in figure 1, to calculate the probability of the word "speech", the Softmax function has to be calculated only 2 times (the group with phonemes K and S and the group with phonemes IY and IH) and only for 2 outputs each time (phonemes K and P). In the standard Connectionist Language Model, the Softmax function would need to be calculated for all the words in the vocabulary (5 classes).

For this simple example, the differences are not significant but given a vocabulary of 12k words, they are much more substantial. In the modified version of the Connectionist Language Model, for the given 12k vocabulary, the maximum number of units that need to be calculated in any given Softmax function is equal to the size of the first layer of the lexical tree. The latter layers have increasingly sparser branching, with fewer children. Most of the nodes in the deeper parts of the tree are not branching nodes and require no computation as their factorized probability is simply 1. In the standard model, for the same example, the Softmax function would have to be computed for all the 12k outputs every time. The larger the vocabulary, the greater the difference in speed between the two methods. The modified method benefits from the branching of the lexical tree in the same way as the tree structure iteslf benefits with respect to size when compared to a linear model. In other words, the computational complexity of the proposed solution increases logarithmically with vocabulary size compared to the standard method, where the complexity increases linearly with the vocabulary.

6 Speech Recognition Speed Evaluation

The proposed method in this paper was tested on a Polish LVCSR task. A language model estimating the factorized word occurrence probabilities was trained using the transcripts of the fourth term of the Senate found in the IPI PAN [12] language corpus. The training data was around 800k words and the test and validation around 100k words each. All the words occurring less than 9 times in the corpus were not added to the vocabulary, thus the vocabulary consisted of 10507 words. The lexical tree built using the above dictionary had 21 layers. The perplexity of the model was 161 which was worse than the connectionist language model using a recurrent neural network and estimating the word probabilities directly (which had a perplexity of 153). The difference may stem from the fact that the modified algorithm needs to multiply many factorized values in order to produce the word probabilities, which can introduce precision errors. Another reason is the fact that the training procedure always generates a different model, even when using the same data, due to random initialization and data shuffling.

Two experiments, comparing the standard and modified Connectionist Language Model used in speech recognition, were performed. Each word in the dictionary had a unique 50-element real number vector assigned to it representing the word features used as input to the neural network. The vectors were initialized by random values in the range of $< -0.01, 0.01 >$. Recurrent neural networks were used in the experiment. Both networks had 50 inputs. The standard network, predicting the regular word occurrence probabilities had 10508 outputs and the modified network, calculating the factorized probabilities had 58732 neurons in the output layer. Both layers were trained using gradient descent with momentum. The learning rate was set to 10^{-7} and the momentum to 0.99. The training of the language model estimating the straightforward word probabilities took 12 days. The modified model estimating the factorized probabilities took around 2 months to train.

Both language models were implemented in an in-house speech recognition system. This system uses a neural network based acoustic model trained on 12 MFCC and an energy feature with delta and acceleration of the 13 parameters, giving a feature set of 39 features. The acoustic model was trained on approximately 10 thousand several minute long dialogs from the public transport call center of the city of Warsaw [13]. All the data was manually transcribed by a group of 20 people [14]. The sampling frequency of these recordings was 8 kHz. The phrases used for testing were generated by the Loquendo speech synthesizer using the voice Zosia. The acoustic model was not adapted to the synthesized voice but previous experiments showed that the system performed well in recognizing this voice. The synthesized voice was chosen in order to rapidly generate substantial data in order to test the system and since the goal was to compare the quality of the language modeling, the quality of the audio was not of big concern. The texts used in the test data were from a part of the IPI PAN corpus that was not used in the training of the model. Both systems were tested using 100 utterances.

Table 1. Comparison of speed and accuracy of the speech recognition system using a standard and modified connectionist language model. The second to last row contains the overall speed of the system expressed in real-time. The last row shows the accuracy. Other rows show a break-down of the occupancy of the algorithm in performing various tasks.

	Standard	Modified
Factorization occupancy	76%	0%
LM processing occupancy	22%	52%
Other occupancy	2%	48%
Speed (xRT)	0.03	1.0
Word accuracy	78.31	76.19

The speech recognition system used a breadth-first search algorithm. The differences in the accuracy between the systems are minimal. Experiments show that the version of the system that uses the modified algorithm works many times faster than the traditional method that has to calculate the factorization. The faster version works exactly at real-time speed because that is how the heuristics of the system were tuned. It is possible to achieve even greater speeds but that would cause fewer hypotheses being searched and the system would make more errors. The real-time speed is entirely sufficient for any speech recognition application. Table 1 shows the results of the two experiments: the modified system clearly outperforms the traditional method by a factor of 33.

In addition to the overall speed, Table 1 also shows a break down of algorithm occupancy performing various tasks throughout the computation process. Given that the modified system computes the factorized probability directly from the network outputs, it does not need to spend any time calculating factorization manually. The row titled "Language model processing occupancy" contains the portion of time it took to compute all the outputs of the neural network, together with calculating the Softmax function. The "Other occupancy" row contains all the other tasks, like: signal feature extraction, acoustic model processing and hypothesis search. The values in the table were calculated using a standard profiling program.

The last row in Table 1 contains the word accuracy of the system. The reduction of over 2% may be explained by the difference in perplexity between the two systems but given the size of the test set and the performance of the language models, one can say that the performance of the two systems is comparable. If the speed is taken into account, however, the differences are much more evident: the traditional system would not be able to work in real-time even with the maximum pruning of the search space.

7   Conclusions and Discussion

This paper described a Connectionist Language Model capable of working in real-time. This was achieved by improving the factorization routines utilized in the factorized lexical tree used in the system. The modified language model performs implicit probability factorization, which speeds up the speech recognition process 33 times. The unique method of factorization in the lexical tree allows the creation of a LVCSR system that works in real-time on a standard modern home computer. Even though the experiment was performed using Polish data, the method is completely language independent and should work for any language and within any system.

It has also been suggested by others that GPU based techniques should be used to further optimize our work and although such libraries do exists for more coneventional ANN architectures (e.g. [15]), there are none easily attainable for the LSTM neural network. It is, however, something that is planned for the future.

The final system relies purely on ANN based models, both for acoustic and language modeling. ANN models have seen much improvement in the recent years [8]. Following the work of [16], the proposed method for language modeling could be a great candidate for creating a fully ANN based sequence transducer for use in connectionist LVCSR systems.

Acknowledgments. The work was sponsored by a research grant from the Polish Ministry of Science and Higher Education, no. N516 519 439. The research leading to these results has also received funding from the European Union Seventh Framework Programme (FP7/2007-2013) under grant agreement no 287658.

References

1. Jelinek, F.: Statistical methods for speech recognition. MIT press (1997)
2. Jelinek, F.: Interpolated estimation of markov source parameters from sparse data. Pattern Recognition in Practice, 381–397 (1980)
3. Katz, S.: Estimation of probabilities from sparse data for the language model component of a speech recognizer. IEEE Transactions on Acoustics, Speech and Signal Processing 35, 400–401 (1987)
4. Kneser, R., Ney, H.: Improved clustering techniques for class-based statistical language modelling. In: Third European Conference on Speech Communication and Technology (1993)
5. Kneser, R., Ney, H.: Improved backing-off for m-gram language modeling. In: IEEE International Conference on Acoustics, Speech, and Signal Processing, ICASSP 1995, vol. 1, pp. 181–184 (1995)
6. Bengio, Y., Ducharme, R., Vincent, P., Janvin, C.: A neural probabilistic language model. The Journal of Machine Learning Research 3, 1137–1155 (2003)
7. Blitzer, J., Weinberger, K., Saul, L.K., Pereira, F.C.: Hierarchical distributed representations for statistical language modeling. Advances in Neural Information Processing Systems 18 (2005)
8. Hinton, G., Deng, L., Yu, D., Dahl, G.E., Mohamed, A.R., Jaitly, N., Senior, A., Vanhoucke, V., Nguyen, P., Sainath, T.N., et al.: Deep neural networks for acoustic modeling in speech recognition: The shared views of four research groups. IEEE Signal Processing Magazine 29, 82–97 (2012)

9. Morin, F., Bengio, Y.: Hierarchical probabilistic neural network language model. In: Proceedings of the International Workshop on Artificial Intelligence and Statistics, pp. 246–252 (2005)
10. Bishop, C.M.: Neural networks for pattern recognition. Oxford university press (1995)
11. Ncy, H., Haeb-Umbach, R., Tran, B.H., Oerder, M.: Improvements in beam search for 10000-word continuous speech recognition. In: 1992 IEEE International Conference on Acoustics, Speech, and Signal Processing, ICASSP 1992, vol. 1, pp. 9–12. IEEE (1992)
12. Przepiórkowski, A.: Korpus ipi pan. Wersja wstepna. Instytut Podstaw Informatyki, Polska Akademia Nauk, Warszawa (2004)
13. Marasek, K., Brocki, Ł., Koržinek, D., Szklanny, K., Gubrynowicz, R.: User-centered design for a voice portal. In: Marciniak, M., Mykowiecka, A. (eds.) Bolc Festschrift. LNCS, vol. 5070, pp. 273–293. Springer, Heidelberg (2009)
14. Koržinek, D., Brocki, Ł., Gubrynowicz, R., Marasek, K.: Wizard of oz experiment for a telephony-based city transport dialog system. In: Proceedings of the IIS 2008 Workshop on Spoken Language Understanding and Dialogue Systems (2008)
15. Bergstra, J., Breuleux, O., Bastien, F., Lamblin, P., Pascanu, R., Desjardins, G., Turian, J., Warde-Farley, D., Bengio, Y.: Theano: A CPU and GPU math expression compiler. In: Proceedings of the Python for Scientific Computing Conference (SciPy) (2010) Oral Presentation
16. Graves, A.: Sequence transduction with recurrent neural networks. In: CoRR. Volume abs/1211.3711, Edinburgh, Scotland (2012)

Automatic Extraction of Logical Web Lists

Pasqua Fabiana Lanotte, Fabio Fumarola, Michelangelo Ceci,
Andrea Scarpino, Michele Damiano Torelli, and Donato Malerba

Dipartimento di Informatica, Universita degli Studi di Bari "Aldo Moro",
via Orabona, 4 - 70125 Bari - Italy
{pasqua.fabiana,ffumarola,ceci,malerba}@di.uniba.it,
{andrea,daniele}@datatoknowledge.it

Abstract. Recently, there has been increased interest in the extraction of structured data from the web (both "Surface" Web and "Hidden" Web). In particular, in this paper we focus on the automatic extraction of Web Lists. Although this task has been studied extensively, existing approaches are based on the assumption that lists are wholly contained in a Web page.They do not consider that many websites span their listing on several Web Pages and show for each of these only a partial *view*. Similar to databases, where a view can represent a subset of the data contained in a table, they split a *logical list* in multiple views (*view lists*). Automatic extraction of *logical lists* is an open problem. To tackle this issue we propose an unsupervised and domain-independent algorithm for *logical list extraction*. Experimental results on real-life and data-intensive Web sites confirm the effectiveness of our approach.

Keywords: Web List Mining, Structured Data Extraction, Logical List.

1 Introduction

A large amount of structured data on the Web exists in several forms, including HTML lists, tables, and back-end Deep Web databases (e.g. Amazon.com, Trulia.com). Caffarella et al. [3] estimated that there are more than one billion HTML tables on the Web containing relational data, and Elmeleegy et al. [5] suggested an equal number from HTML lists. Since the Web is a large and underutilized repository of structured data, extracting structured data from this source has recently received great attention [3,5]. Several solutions have been proposed to find, extract, and integrate structured data, which are used in few public available products like Google Sets and Google Fusion Tables. In addition, the analysis of large amount of structured data on the web has enabled features such as schema auto-complete, synonymy discovery [3], market intelligence [1], question answering [6], and mashup from multiple Web sources [15]. However, only few websites give access to their data through Application Programming Interfaces (e.g. Twitter, Facebook). The majority of them present structured data as HTML and/or backed through "Hidden" Web Database.

In this paper, we focus on the problem of automatic extraction of Web Lists. Several methods have been presented in the literature [13,16,10,9,14] as regards

T. Andreasen et al. (Eds.): ISMIS 2014, LNAI 8502, pp. 365–374, 2014.
© Springer International Publishing Switzerland 2014

Fig. 1. An example of Amazon Web page

to this task, but they fail to detect lists which span multiple Web pages. This is an open issue, because many web sites, especially data-intensive (e.g. Amazon, Trulia, AbeBooks,...), present their listings as *logical list*, that is, a list spanning multiple pages (*e.g.* computers, books, home listings)[1]. It is as if, each list represents a *view* of the same *logical list*. Similar to databases, where a *view* can represent a subset of the data contained in a table partitioned over a set of attributes, a *logical list* is split in multiple *views* (Web Pages) in order to avoid information overload and to facilitate users' navigation.

For example, Fig. 1 shows a Web page from Amazon.com that contains the results for the query "Computer". On this page, the boxes A, B, C, D, E, F are web lists. The list in the box A shows a *view* of the "Computers" products, that is the top six sorted by relevance, and F allows us to navigate to the other views of the products ordered by relevance. Thus navigating the links in F we can generate the *logical list* of the products for the query "Computer". Boxes B, C, D and E contain respectively the lists representing filters for "Department", "Featured Brands", "Computer Processor Type", and "Hard Disk Size", which are attributes of the *virtual* table "Computer". Moreover, the anchor-text links in boxes B, C, D and E stores valuable information which can be used to annotate data records, and thus to individuate new attributes. For example, the anchor-text links of web list C can be used to index data records based on "Computer brands". Traditionally, search engines use the proximity of terms on a page as a signal of relatedness; in this case the computer brand terms are highly related to some data records, even though they are distant.

Providing automated techniques for *logical list* extraction would be a significant advantage for data extraction and indexing services. Existing data record

---

[1] The motivations behind this approach are as well technical (reducing bandwidth and latency), and non technical as avoiding information overload or maximizing page views.

extraction methods [13,16,9,14] focus only in extracting *view* lists, while several commercial solutions[2] provide hand-coded rules to extract *logical lists*.

In this paper, we face this issue by proposing a novel unsupervised algorithm for automatic discovery and extraction of *logical lists* from the Web. Our method requires only one page containing a *view list*, and it is able to automatically extract the *logical list* containing the example *view list*. Moreover, during the process, it enriches the list's elements with the pair <*url, anchor-text*> used for the extraction task. We have validated our method on a several real websites, obtaining high effectiveness.

2 Definitions and Problem Formulation

In this section, we introduce a set of definitions we will use through the paper.

Definition 1. The Web Page Rendering *is the process of laying out a spatial position of all the text/images and other elements in a Web Page to be rendered.*

When an HTML document is rendered in a Web browser, the CSS2 visual formatting model [11] represents the elements of the document by rectangular boxes that are laid out one after the other or nested inside each other. By associating the document with a coordinate system whose origin is at the top-left corner, the spatial position of each text/image element on the Web page is fully determined by both the coordinates *(x,y)* of the top-left corner of its corresponding box, and the box's height and width.

Property 1. The spatial positions of all text/image elements in a Web Page define the *Web Page Layout.*

Property 2. As defined in [7], each *Web Page Layout* has a tree structure, called *Rendered Box Tree*, which reflects the hierarchical organization of HTML tags in the Web page.

As we can see in Fig. 2, on the left there is the Web Page Layout of the Web page in Fig. 1. On the right, there is its Rendered Box Tree. The technical details of building Rendered Box Trees and their properties can be found in [7]. Under the Web page layout model, and the Rendered Box Tree we can give the definition for Web lists.

Definition 2. *Web List: It is collection of two or more objects, under the same parent box and visually adjacent and aligned on a rendered web page. This alignment can occur via the x-axis (i.e. a vertical list), the y-axis (i.e. horizontal list), or in a tiled manner (i.e., aligned vertically and horizontally) [8].*

For example the list A in Fig. 1 is a tiled list, while B is a vertical list and F is a horizontal list.

---

[2] Lixto, Screen Scraper Studio, Mozenda Screen Scaper.

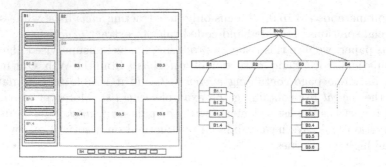

Fig. 2. An example of Rendered Box Tree

The list's elements can be called as *Data Records*. Similar to the concept of data records into database, data records into a web page are a set of similar and structured objects containing information. Typically, they are formatted using similar HTML tags (i.e. the same HTML structure).

Definition 3. *Logical List*: *It is a list whose Data Records are distributed on more then one Web Pages.*

An example is shown in Fig. 3, where the boxes A1 and A2 represent a part of a *logical list*.

Definition 4. *View List*: *It is a view of a logical list, whose Data Records are all contained in same Web page.*

For Example the list F in Fig. 1 is an example of a view list. In fact, it contains only some of data records belonging to its logical list (that is the *pagination list*).

Definition 5. *Dominant List*: *It is the view list of interest, containing data records from the logical list that we want to extract.*

For example the list A in Fig. 1 is the Dominant List for the given Web page.

3 Methodology

In this section we describe the methodology used for *logical list* extraction. The algorithm employs a three-step strategy. Let P a Web Page, it first extracts the set L^P of the lists contained in P; in the second step, it identifies the *dominant list* $l_{dom}^P \in L$; finally, it uses l_{dom}^P to discover the *logical list* LL which includes l_{dom}^P as sub-list. These steps are detailed in the following sub-sections.

3.1 List Extraction

Given a Web Page P as input, its lists $L = \{l_1, l_2, \ldots, l_n\}$ are extracted through running an improved version of HyLiEn [8]. With respect to HyLiEn, we made

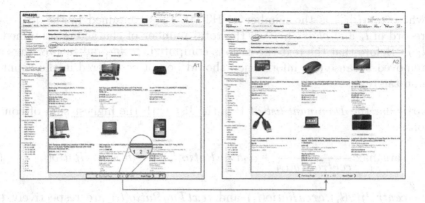

Fig. 3. An example of *logical list* for Amazon's products

several improvements. First, to render the Web pages we removed the dependency to the open source library *CSSBox*[3], because we found that this library was not able to correctly render several Web pages. We implemented a WebKit wrapper, called *WebPageTraverser*[4], which is released as open source project. Given as input the url of a Web page P, *WebPageTraverser* outputs a JSON[5] representation of the P using the *rendered box tree model*. Second, to compute the similarity of two sub-trees in *rendered box tree* (see Prop. 2) we adopted the HTML tag distance, presented in [2] instead of the string edit distance used by HyLiEn. Although, our current implementation uses the HyLiEn algorithm to obtain Web lists our solution is independent of any specific list-extraction algorithm. We used HyLiEn because it showed interesting result compared to the state of art algorithms for List Extraction [8].

3.2 Dominant List Identification

Given a Web Page P and the set of list $L = \{l_1, l_2, \ldots, l_n\}$ extracted in the first step, we use three measures to identify the *dominant list* of P:

- **Centrality.** Given a list $l_i \in L$, the *centrality* of l_i w.r.t P is obtained by computing the Euclidean distance between the center of the parent-box of l_i and the center of root-box of P.
- **Area Ratio.** Given a list $l_i \in L$, the *area ratio* of l_i w.r.t P is the size of the box containing l_i divided the size of root-box of P.
- **Text-Tag Ratio.** Given a list $l_i \in L$, and let m the length of l_i, the *text-tag ratio* of l_i is computed as:

$$\frac{1}{m} \sum_{j=0}^{m} \frac{chars(l_i[j])}{tag(l_i[j])} \tag{1}$$

---

[3] http://cssbox.sourceforge.net
[4] https://bitbucket.org/wheretolive/webpagetraverser
[5] http://www.json.org/

where $tag(l_i[j])$ is the number of HTML tags contained in the j-th data record of l_i and $chars(l_i[j])$ is the total number of characters contained in $l_i[j]$. Before that the text-tag ratio is computed, *script* and *remark* tags are removed because this information should be not considered in the count of non-tag text.

In particular the *Dominant list* of P is the list with the highest sum of contributions:

$$\arg\max_{l_i \in L} \frac{\alpha_1}{centrality(l_i)} + \alpha_2 areaRatio(l_i) + \alpha_3 textTagRatio(l_i) \qquad (2)$$

where $centrality(l_i), areaRatio(l_i)$ and $textTagRatio(l_i)$ are respectively the centrality measure, area ratio and text-tag ratio of a list l_i contained in L. $\alpha_1 = \alpha_2 = \alpha_3$ are set to 0.3 to give the same weight to each measure.

Algorithm 1. `LogicalListDiscovery`

    input : dominant list l_{dom}^P, set $L_- = \{L \setminus l_{dom}^P\}$
    output: logical list LL

1   $LL = \{l_{dom}^P\}$;
2   **forall the** $l \in L_-$ **do**
3   **forall the** $u \in l$ **do**
4           $L_u \leftarrow$ HyLiEn(u);
5   L_u.filterSimilarity(l_{dom}^P, α);
6   LL.add(L_u);

7   $LL \leftarrow LL$.flatMap();
8   $LL \leftarrow LL$.removeDuplicates();
9   **return** LL;

3.3 Logical List Discovery

Identified the dominant list l_{dom}^P of the Web Page P, the last step of the algorithm is to discover the logical list LL containing l_{dom}^P. This is done by taking advantage of the regularities of Web Sites. As described by Crescenzi et al. [4], Web page links reflect the regularity of the web page structure. In other words, links that are grouped in collections with a uniform layout and presentation usually lead to similar pages. Link-based approaches are used in the literature for tasks strictly related to the one solved by our method. For instance, Lin et al. [12] used Web links to discover new attributes for web tables by exploring hyperlinks inside web tables. Lerman et al. [10] uses out-links to "detail web pages" in order to segment Web tables. In this paper, we successfully use links grouped as lists to navigate Web pages and to discover *logical lists*.

The algorithm 1 describes the approach used. It takes as input the dominant list l^P_{dom}, the minimum similarity threshold α, and the set of the lists L extracted from P. It iterates over all the lists in the set $L_- = \{L \setminus l^P_{dom}\}$ (line 1), and, for each url u in l_i it alternates, (i) the extraction of the set list L_u contained in the Web Page U having u as url (line 4) to, (ii) the filtering of all the lists in L_u which have a similarity with l^P_{dom} lower than α (line 6). At each iteration, all the lists resulting from step (ii) are added to LL (line 7). Finally, LL is flattened and all the duplicate elements are merged (lines 8-9). Moreover, during the process all the anchor text of url u are used as attributes to annotate the discovered *view* lists and are reported in the final *logical list LL*.

4 Experiments

In this section we presents the empirical evaluation of the proposed algorithm. We manually generated and verified a test dataset. In particular, for the experiment, we select 40 websites in different application domains (music shops, web journals, movies information, home listings, computer accessories, etc.) with list elements presented in different ways. For the deep-web databases, we performed a query for each of them and collected the first page of the results list, and for others we manually select a Web page. Table 1 shows in the first column the ground truth, that is, the number of data records which belong to the *logical list* to be extracted. The dataset is composed of 66.061 list elements extracted from 4405 Web pages. We rendered each Web page and we manually identified (i) *dominant list* and, (ii) following the out-links of the other lists in the pages we annotated *logical lists*. This task required around 7 days of 4 people.

To the best of our knowledge the task of *Logical List Discovery* is novel, and there are not any other methods to compare with. So, we evaluated the effectiveness of our algorithm by using *precision, recall* and *f-measure* metrics, computed over the number of *logical list* elements to be extracted (manually verified) w.r.t to the algorithm results. In particular, the precision is the measure of, how many of the extracted *view* lists belong to a *logical list*. The recall allows us to measure how many of the discovered *view* lists are true positive element of a *logical list*. We also included the *F-Measure* which is the weighted harmonic means of *precision* and *recall*. These metrics are evaluated counting how many data records of the *logical list* are found in the *view* lists.

$$precision = \frac{TP}{TP + FP}, \; recall = \frac{TP}{TP + FN}, F - measure = \frac{2(precision \times recall)}{precision + recall}$$

$$(3)$$

4.1 Results

The execution of the algorithm requires two parameters which are empirically set to $\alpha = 0.6$ and $\beta = 50$. These parameters are need by the HyLiEn Algorithm. Our methods uses α during the *Logical List Discovery* step.

Table 1 presents the main results. The first column holds for each logical list the number of data records to extract. The second and the third columns contain the number of true positive and the number of false negatives data records. We do not plot the number of false positives, because our algorithm outputted always 0 false positives during the experiment evaluation. Finally, the fourth, fifth and sixth columns show the values for precision, recall and f-measure.

In general, the experimental results show that our algorithm is able to discover *logical lists* in a varying set of Web sites (that is, it is not domain dependent). Moreover, the quality of the results are not correlated to how the lists are rendered in Web pages (i.e. horizontal, vertical and tiled). In average, it achieves 100% for Precision, 95% for Recall and a F-Measure 97%. With respect to the ground truth, the algorithm does not extract any False Positive, and it outputs only 466 False Negatives. In general, it returns perfect results (100% precision and recall) for several kind of websites spanning different applications domain, but there are some of them which presents values for recall ranging from 81% and 91%. Considering "last.fm", which gave a recall equal to 81%, we found that the presentation of the data records is sometime quite different, because of the high variance in the number of the "similar to" tags (which are presented as HTML <a>) assigned to each listing. Analyzing other examples such as "Il-Sole24Ore.it" and "RealEstateSource.au" we found the same problem, that is, the presentation of the data records is quite variable across the Web pages, and so the HyLiEn algorithm sometimes misses some of the data records. Anyway we see that the proposed algorithms is effective is able to discover *logical lists* on different type of websites.

5   Conclusions and Future Works

In this paper, we have presented a new method for *Logical List Extraction*. Our method solves the open issue of discover and extract lists which spans multiple Web pages. These *logical lists* are quite common in many websites, especially data-intensive, where their listings are split on multiples pages in order to avoid information overload and to facilitate users' navigation. However, the data stored in such *logical list* need to be automatically extracted to enable building services for market intelligence, synonyms discovery, question answering and data mashup. Experimental results show that our new method is extremely accurate and it is able to extract *logical lists* in a wide range of domains and websites with high precision and recall. Part of this future work will involve tasks such as indexing the Web based on lists and tables, answering queries from lists, and entity discovery and disambiguation using lists.

Acknowledgements. This work fulfills the research objectives of the PON 02 00563 3470993 project "VINCENTE - A Virtual collective INtelligenCe ENvironment to develop sustainable Technology Entrepreneurship ecosystems" funded by the Italian Ministry of University and Research (MIUR).

Table 1. Discovered *logical list* elements for Web sites dataset

Website	Ground	TP	FN	Precision	Recall	F-measure
BariToday.it	904	904	0	100%	100%	100%
Subito.it	1000	1000	0	100%	100%	100%
GitHub.com	100	100	0	100%	100%	100%
TestoLegge.it	360	360	0	100%	100%	100%
Zoopla.co.uk	597	597	0	100%	100%	100%
FindAProperty.co.uk	60	60	0	100%	100%	100%
Savills.co.uk	232	232	0	100%	100%	100%
AutoTrader.co.uk	60	60	0	100%	100%	100%
EbayMotors.com	3925	3925	0	100%	100%	100%
Doogal.co.uk	38240	38240	0	100%	100%	100%
RealEstateSource.com	368	316	62	100%	85%	91%
AutoWeb.co.uk	180	180	0	100%	100%	100%
TechCrunch.com	434	422	12	100%	95%	98%
Landsend.com	1243	1243	0	100%	100%	100%
TMZ.com	300	300	0	100%	100%	100%
IlSole24Ore.it	510	445	65	100%	81%	86%
GoBari.it	350	340	10	100%	97%	98%
AGI.it	60	60	0	100%	100%	100%
BBCNews.co.uk	347	310	37	100%	89%	94%
milano.corriere.it	30	30	0	100%	100%	100%
torino.repubblica.it	70	68	2	100%	98%	99%
Ansa.it	1506	1479	27	100%	98%	99%
LeMonde.fr	445	418	27	100%	94%	97%
Time.com	377	377	0	100%	100%	100%
aur.ArchLinux.org	575	575	0	100%	100%	100%
Immobiliare.it	609	536	73	100%	86%	93%
bitbucket.org	130	130	0	100%	100%	100%
MyMovies.com	563	515	48	100%	92%	96%
Trulia.com	3300	3300	0	100%	100%	100%
YouTube.com	580	567	13	100%	98%	99%
FileStube.com	332	304	28	100%	91%	95%
Last.fm	60	41	19	100%	68%	81%
Bing.com	130	130	0	100%	100%	100%
addons.mozilla.org	984	939	45	100%	95%	97%
AutoScout24.com	840	840	0	100%	100%	100%
Facebook.com	2820	2820	0	100%	100%	100%
SlideShare.net	2037	2037	0	100%	100%	100%
Gazzetta.it	970	970	0	100%	100%	100%
ElPais.es	294	285	9	100%	98%	99%
StackOverflow	585	585	0	100%	100%	100%
Sums and Averages	66.527	66.061	466	100%	95%	97%

References

1. Baumgartner, R.: Datalog-related aspects in lixto visual developer. In: de Moor, O., Gottlob, G., Furche, T., Sellers, A. (eds.) Datalog 2010. LNCS, vol. 6702, pp. 145–160. Springer, Heidelberg (2011)
2. Bing, L., Lam, W., Gu, Y.: Towards a unified solution: Data record region detection and segmentation. In: Proceedings of the 20th ACM International Conference on Information and Knowledge Management, CIKM 2011, pp. 1265–1274. ACM, New York (2011)
3. Cafarella, M.J., Halevy, A., Madhavan, J.: Structured data on the web. Commun. ACM 54(2), 72–79 (2011)
4. Crescenzi, V., Merialdo, P., Missier, P.: Clustering web pages based on their structure. Data Knowl. Eng. 54(3), 279–299 (2005)
5. Elmeleegy, H., Madhavan, J., Halevy, A.: Harvesting relational tables from lists on the web. The VLDB Journal 20(2), 209–226 (2011)
6. Fader, A., Soderland, S., Etzioni, O.: Identifying relations for open information extraction. In: Proceedings of the Conference on Empirical Methods in Natural Language Processing, EMNLP 2011, pp. 1535–1545. Association for Computational Linguistics, Stroudsburg (2011)
7. Fumarola, F., Weninger, T., Barber, R., Malerba, D., Han, J.: Extracting general lists from web documents: A hybrid approach. In: Mehrotra, K.G., Mohan, C.K., Oh, J.C., Varshney, P.K., Ali, M. (eds.) IEA/AIE 2011, Part I. LNCS, vol. 6703, pp. 285–294. Springer, Heidelberg (2011)
8. Fumarola, F., Weninger, T., Barber, R., Malerba, D., Han, J.: Hylien: A hybrid approach to general list extraction on the web. In: Srinivasan, S., Ramamritham, K., Kumar, A., Ravindra, M.P., Bertino, E., Kumar, R. (eds.) WWW (Companion Volume), pp. 35–36. ACM (2011)
9. Gatterbauer, W., Bohunsky, P., Herzog, M., Krüpl, B., Pollak, B.: Towards domain-independent information extraction from web tables. In: Proceedings of the 16th International Conference on World Wide Web, WWW 2007, pp. 71–80. ACM, New York (2007)
10. Lerman, K., Getoor, L., Minton, S., Knoblock, C.: Using the structure of web sites for automatic segmentation of tables. In: Proceedings of the 2004 ACM SIGMOD International Conference on Management of Data, SIGMOD 2004, pp. 119–130. ACM, New York (2004)
11. Lie, H.W., Bos, B.: Cascading Style Sheets: Designing for the Web, 3rd edn., p. 5. Addison-Wesley Professional (2005)
12. Lin, C.X., Zhao, B., Weninger, T., Han, J., Liu, B.: Entity relation discovery from web tables and links. In: Proceedings of the 19th International Conference on World Wide Web, WWW 2010, pp. 1145–1146. ACM, New York (2010)
13. Liu, B., Grossman, R.L., Zhai, Y.: Mining web pages for data records. IEEE Intelligent Systems 19(6), 49–55 (2004)
14. Liu, W., Meng, X., Meng, W.: Vide: A vision-based approach for deep web data extraction. IEEE Transactions on Knowledge and Data Engineering 22(3), 447–460 (2010)
15. Maximilien, E.M., Ranabahu, A.: The programmableweb: Agile, social, and grassroot computing. In: Proceedings of the International Conference on Semantic Computing, ICSC 2007, pp. 477–481. IEEE Computer Society, Washington, DC (2007)
16. Miao, G., Tatemura, J., Hsiung, W.: Extracting data records from the web using tag path clustering. In: The World Wide Web Conference, pp. 981–990 (2009)

Combining Formal Logic and Machine Learning for Sentiment Analysis

Niklas Christoffer Petersen and Jørgen Villadsen*

Algorithms, Logic and Graphs Section
Department of Applied Mathematics and Computer Science
Technical University of Denmark
Richard Petersens Plads, Building 324, DK-2800 Kongens Lyngby, Denmark

Abstract. This paper presents a formal logical method for deep structural analysis of the syntactical properties of texts using machine learning techniques for efficient syntactical tagging. To evaluate the method it is used for *entity level* sentiment analysis as an alternative to pure machine learning methods for sentiment analysis, which often work on sentence or word level, and are argued to have difficulties in capturing long distance dependencies.

1 Introduction

There exist formal proofs that some natural language structures requires formal power beyond *context-free grammars* (CFG), i.e. [18] and [3]. Thus the search for grammars with more expressive power has long been a major study within the field of computational linguistics. The goal is a grammar that is as restrictive as possible, allowing efficient syntactic processing, but still capable of capturing these structures. The class of *mildly context-sensitive grammars* are conjectured to be powerful enough to model natural languages while remaining efficient with respect to syntactic analysis cf. [12]. This class includes *Tree Adjunct Grammar* (TAG) [11], *Head Grammar* (HG) [17] and *Combinatory Categorial Grammar* (CCG) [19]. It has been shown that these are all equal in expressive power [21].

The common problem for each of the above grammar formalisms is the syntatic tagging, i.e. for each token in the text to analyse to assign it the correct *type*. This is simply due to the fact that a lexical unit can entail different lexical categories (e.g. "service" is both a noun and a verb), and different semantic expressions (e.g. the noun "service" can both refer to assistance and tableware).

For instance in *CCGbank*, which is a CCG-version of The Penn Treebank [14], the expected number of lexical categories per token is 19.2 cf. [10]. This means that an exhaustive search of even a short sentence (seven tokens) is expected to consider over 960 million ($19.2^7 \approx 961\,852\,772$) possible taggings. This is clearly not a feasible approach, even if the parsing can explore all possible deductions in polynomial time of the number of possible taggings. The number of lexical categories assigned to each token needs to be reduced, but simple reductions as just

---

* Corresponding author: jovi@dtu.dk

T. Andreasen et al. (Eds.): ISMIS 2014, LNAI 8502, pp. 375–384, 2014.
© Springer International Publishing Switzerland 2014

assigning the most frequent category observed in some training set (for instance *CCGbank*) for each token is not a solution. This would fail to accept a large amount of valid sentences, simply because it is missing the correct categories.

2   Combinatory Categorial Grammar

Combinatory Categorial Grammar (CCG), pioneered largely by [20], adds a layer of combinatory logic onto pure Categorial Grammar, which allows an elegant and succinct formation of *higher-order* semantic expressions directly from the syntactic analysis.

A CCG lexicon, $\mathcal{L}_{\mathrm{CCG}}$, is mapping from a lexical unit, $w \in \Sigma^*$, to a set of 2-tuples, each containing a lexical category and semantic expression that the unit can entail cf. (1), where Γ denotes the set of lexical and phrasal categories, and Λ denotes the set of semantic expressions.

$$\mathcal{L}_{\mathrm{CCG}} : \Sigma^* \to \mathcal{P}(\Gamma \times \Lambda) \tag{1}$$

A category is either *primitive* or *compound*. The set of primitive categories, $\Gamma_{\mathrm{prim}} \subset \Gamma$, is language dependent and, for the English language, it consists of S (sentence), NP (noun phrase), N (noun) and PP (prepositional phrase). Compound categories are recursively defined by the infix operators $/$ (forward slash) and $\backslash$ (backward slash), i.e. if α and β are members of Γ, then so are α/β and $\alpha\backslash\beta$.

A *tagging* of a lexical unit $w \in \Sigma^*$ is simply the selection of one of the pairs yielded by $\mathcal{L}_{\mathrm{CCG}}(w)$. Thus for some given ordered set of lexical units, which constitutes the text $T \in \Sigma^*$ to analyse, the number of possible combinations might be very large.

2.1   Combinatory Rules

CCGs can be seen as a logical deductive proof system where the axioms are members of $\Gamma \times \Lambda$. A text $T \in \Sigma^*$ is accepted as a sentence in the language, if there exists a deductive proof for S, for some tagging of T. Rewrite is done using a language independant set of *combinatory rules*:

Here X, Y, Z and T are varibles ranging over categories (i.e. Γ), and f, a and g are variables over semantic expressions (i.e. Λ).

With only functional application, $(>)$ and $(<)$, the system is capable of capturing any context-free language cf. [20, p. 34]. Figure 1 shows the deduction of S from the simple declarative sentence "the hotel had an exceptional service" (semantics is omitted).

The functional composition, $(>_{\mathbf{B}})$ and $(<_{\mathbf{B}})$ is often used in connection with type-raising $(>_{\mathbf{T}})$ and $(<_{\mathbf{T}})$, for instance to allow relative clauses, coordination, while crossed functional composition, $(>_{\mathbf{B}\times})$ and $(<_{\mathbf{B}\times})$ are needed for more exotic linguistic phenomenons such as *heavy noun phrase shifting*.

$$X/Y : f \quad Y : a \quad \Rightarrow \quad X : f\,a \qquad\qquad (>)$$
$$Y : a \quad X\backslash Y : f \quad \Rightarrow \quad X : f\,a \qquad\qquad (<)$$
$$X/Y : f \quad Y/Z : g \quad \Rightarrow \quad X/Z : \lambda a.f(g\,a) \qquad (>_\mathbf{B})$$
$$Y\backslash Z : g \quad X\backslash Y : f \quad \Rightarrow \quad X\backslash Z : \lambda a.f(g\,a) \qquad (<_\mathbf{B})$$
$$X : a \quad \Rightarrow \quad T/(T\backslash X) : \lambda f.f a \qquad (>_\mathbf{T})$$
$$X : a \quad \Rightarrow \quad T\backslash(T/X) : \lambda f.f a \qquad (<_\mathbf{T})$$
$$X/Y : f \quad Y\backslash Z : g \quad \Rightarrow \quad X\backslash Z : \lambda a.f(g\,a) \qquad (>_{\mathbf{B}_\times})$$
$$Y/Z : g \quad X\backslash Y : f \quad \Rightarrow \quad X/Z : \lambda a.f(g\,a) \qquad (<_{\mathbf{B}_\times})$$

$$
\begin{array}{c}
\\
\\
\\
\text{the} \quad \text{hotel} \quad \text{had} \quad \dfrac{\text{an} \quad \dfrac{\text{exceptional}\ \ \text{service}}{N/N \qquad N}{>}}{}
\end{array}
$$

the hotel   had   an   exceptional service
$\dfrac{NP/N \quad N}{\dfrac{NP}{}{>}} \quad (S\backslash NP)/NP \quad \dfrac{\dfrac{NP/N \quad \dfrac{N/N \quad N}{N}{>}}{NP}{>}}{\dfrac{S\backslash NP}{}{>}}$

$$\dfrac{}{S}{<}$$

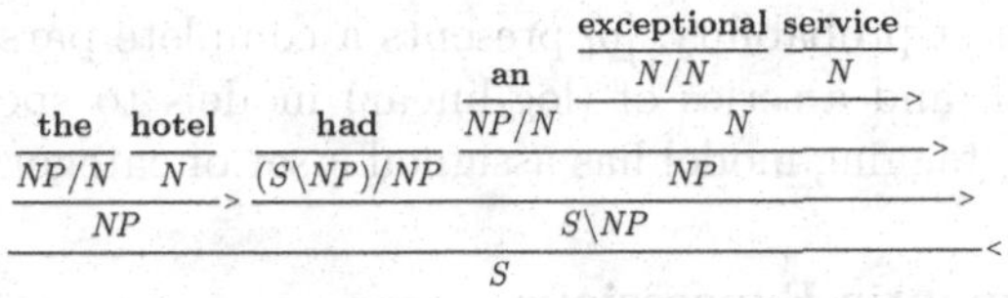

Fig. 1. Deduction of simple declarative sentence

2.2 Semantic Expressions

The syntactic categories constitute a type system for the semantic expressions, with a set of primitive types, $\mathcal{T}_{\text{prim}} = \{\tau_x \mid x \in \Gamma_{\text{prim}}\}$. Thus, if a lexicon entry has category $(N\backslash N)/(S/NP)$ then the associated semantic expression must honor this, and have type $(\tau_{\text{NP}} \to \tau_{\text{S}}) \to \tau_{\text{N}} \to \tau_{\text{N}}$. This is a result of the *Principle of Categorial Type Transparency* [15], and the set of all types are denoted $\mathcal{T}$. The set of semantic expressions, Λ, is a superset of the *simply-typed* λ-expressions, Λ', cf. Definition 1. In Section 4 this is extended to support the desired sentiment analysis.

Definition 1. *The set of simply typed λ-expressions, Λ', is defined recursively, where an expression, e, is either a variable x from an infinite set of typed variables $\mathcal{V} = \{v_1 : \tau_\alpha, v_2 : \tau_\beta, \ldots\}$, a functional abstraction, or a functional application. For futher details see for instance [1].*

$$x : \tau \in \mathcal{V} \quad \Rightarrow \quad x : \tau \in \Lambda' \qquad\qquad \text{(Variable)}$$
$$x : \tau_\alpha \in \mathcal{V},\ e : \tau_\beta \in \Lambda' \quad \Rightarrow \quad \lambda x.e : \tau_\alpha \to \tau_\beta \in \Lambda' \qquad \text{(Abstraction)}$$
$$e_1 : \tau_\alpha \to \tau_\beta \in \Lambda',\ e_2 : \tau_\alpha \in \Lambda' \quad \Rightarrow \quad (e_1 e_2) : \tau_\beta \in \Lambda' \qquad \text{(Application)}$$

3 Machine Learning Tagging

There exists some wide covering CCG lexicons, most notable *CCGbank*, compiled by [10] by techniques presented by [9]. It is essentially a translation of almost the entire Penn Treebank [14], which contains over 4.5 million tokens, and where each sentence structure has been analysed in full and annotated. The result is a highly covering lexicon, with some entries having assigned over 100 different lexical categories. Thus efficient syntactical CCG-tagging is required.

3.1   Maximum Entropy Tagging

The Supertagger [5] is a machine learning approach based on a *maximum entropy model* that estimate the probability that a token is to be assigned a particular category, given the *features* of the local context, e.g. the POS-tag of the current and adjacent tokens, etc. This is used to select a subset of possible categories for a token, by selecting categories with a probability within a factor of the category with highest probability. [6] presents a complete parser, which utilizes this tagging model, and a series of (log-linear) models to speed-up the actual deduction once the tagging model has assigned a set of categories to each token.

3.2   Adding Semantic Expressions

Clearly such lexicons only constitutes half of the previous defined $\mathcal{L}_{\text{CCG}}$ map, i.e. only the lexical categories, Γ. However due to the *Principle of Categorial Type Transparency* it is known exactly *what* the types of the semantic expressions should be.

The mechanism used to generate the semantic expression depends on the specific analysis requested. However one solution is to handle the cases that impact the result of the analysis, and then use a generic annotation algorithm for all other cases. Both the generic and the special case algorithms will be a transformation $(\mathcal{T}, \Sigma^\star) \to \Lambda$, where the first argument is the type, $\tau \in \mathcal{T}$, to construct, and the second argument is the lemma, $\ell \in \Sigma^\star$, of the lexicon entry to annotate.

Table 1 shows the result of applying such function on some lemmas and types. The result for a noun as "room" is simply the zero-argument functor of the same name. The transitive verb "provide" captures two noun phrases, and yields a functor with them as arguments. More interesting is the type for the determiner "every", when used for instance to modify a performance verb. It starts by capturing a noun, then a function over noun phrases, and lastly a noun phrase. The semantic expression generated for this type is a functor, with simply the noun as first argument, and where the second argument is the captured function applied on the noun phrase.

Table 1. Some input/output of a generic annotation algorithm

Lemma Type	Generic semantic expression
room τ_{N}	room
provide $\tau_{\text{NP}} \to \tau_{\text{NP}} \to \tau_{\text{S}}$	$\lambda x.\lambda y.\text{provide}(x, y)$
every $\tau_{\text{N}} \to (\tau_{\text{NP}} \to \tau_{\text{S}}) \to (\tau_{\text{NP}} \to \tau_{\text{S}})$	$\lambda x.\lambda y.\lambda z.\text{every}(x, y\ z)$

4 Case Study: Sentiment Analysis

The research in sentiment analysis has only recently enjoyed high activity cf. [13] and [16], which arguably is due to a combination of the progress in machine learning research, the availability of huge data sets through social networks on the Internet. The potential in such data are numerous, and has found applications in both commercial products and services, as well as the political and financial world cf. [7].

Most published research on sentiment analysis focusses on pure machine approches and builds on text classification principles. The methods often classify the entire text as either positive or negative, and thus no details are discovered about the entity of the opinions that are expressed by the text.

This case study presents a formal logical approach for *entity level* sentiment analysis which utilizes machine learning techniques for efficient syntactic tagging. The motivation for focusing on the formal approach is:

- Different domains can have very different ways of expressing sentiment. What is weighted significant as positive in one domain can be negative, or even pure nonsense in another cf. [2].
- Labeled training data are sparse, and since machine learning mostly assumes at least some portion of labeled target data are available this constitutes an issue with the pure machine learning approach.
- Pure machine learning methods will usually classify sentiment on document, sentence or simply on word level, but not on an entity level due to thier semantic weakness cf. [4]. This can have unintended results when trying to analyse sentences with coordination of sentiments for multiple entities.

The sentiment polarity model used in this case study is continuous, and can thus be seen as a weighted classification. Thereby is the polarity a value in some predefined interval, $[-\omega; \omega]$, as illustrated by Figure 2. An opinion with value close to $-\omega$ is considered highly negative, whereas a value close to ω is considered highly positive. Opinions with values close to zero are considered almost neutral.

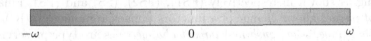

$-\omega$ 0 ω

Fig. 2. Continuous sentiment polarity model

4.1 Combinatory Categorial Grammar for Sentiment Analysis

In order to apply the CCG formalism to the area of sentiment analysis the expressive power of the semantics needs to be adapted to this task. The set of semantic expressions, Λ, is defined as a superset of simply typed λ-expressions cf. Definition 2.

Definition 2. *Besides variables, functional abstraction and functional application, which follows from simply typed λ-expressions cf. [1], the following structures are available:*

- A *n-ary* functor *(n ≥ 0) with name f from an infinite set of functor names, polarity* $j \in [-\omega; \omega]$, *and* impact argument *k* $(0 \le k \le n)$.
- A sequence *of n semantic expressions of the same* type.
- The change of impact argument.
- The change *of an expression's polarity.*
- The scale *of an expression's polarity. The magnitude of which an expression's polarity may scale is given by* $[-\psi; \psi]$.

Formally this can be stated:

$$e_1, \ldots, e_n \in \Lambda, 0 \le k \le n, \ j \in [-\omega; \omega] \quad \Rightarrow \quad f_j^k(e_1, \ldots, e_n) \in \Lambda \qquad \text{(Functor)}$$

$$e_1 : \tau, \ldots, e_n : \tau \in \Lambda \quad \Rightarrow \quad \langle e_1, \ldots, e_n \rangle : \tau \in \Lambda \quad \text{(Sequence)}$$

$$e : \tau \in \Lambda, 0 \le k' \quad \Rightarrow \quad e^{\rightsquigarrow k'} : \tau \qquad \text{(Impact change)}$$

$$e : \tau \in \Lambda, \ j \in [-\omega; \omega] \quad \Rightarrow \quad e_{\circ j} : \tau \in \Lambda \qquad \text{(Change)}$$

$$e : \tau \in \Lambda, \ j \in [-\psi; \psi] \quad \Rightarrow \quad e_{\bullet j} : \tau \in \Lambda \qquad \text{(Scale)}$$

The semantics includes normal α-conversion and β-, η-reduction as shown in the semantic rewrite rules for the semantic expressions given by Definition 3. More interesting are the rules that actually allow the binding of polarities to the phrase structures. The *change of a functor* itself is given by the rule (FC1), which applies to functors with, impact argument, $k = 0$. For any other value of k the functor acts like a non-capturing enclosure that passes on any change to its k'th argument as follows from (FC2). The *change of a sequence* of expressions is simply the change of each element in the sequence cf. (SC). Finally it is allowed to *push change* inside an abstraction as shown in (PC), simply to ensure the applicability of the β-reduction rule. Completely analogue rules are provided for the scaling as shown in respectively (FS1), (FS2), (SS) and (PS). Finally the *change of impact* allows change of a functors impact argument cf. (IC). Notice that these *change, scale, push* and *impact change* rules are type preserving, and for readability type annotation is omitted from these rules.

Definition 3. *The rewrite rules of the semantic expressions are given by the following, where* $e_1[x \mapsto e_2]$ *denotes the* safe *substitution of x with e_2 in e_1, and FV(e) denotes the set of free variables in e. For details see for instance [1].*

$$(\lambda x.e) : \tau \quad \Rightarrow \quad (\lambda y.e[x \mapsto y]) : \tau \qquad y \notin FV(e) \quad (\alpha)$$

$$((\lambda x.e_1) : \tau_\alpha \rightarrow \tau_\beta)\,(e_2 : \tau_\alpha) \quad \Rightarrow \quad e_1[x \mapsto e_2] : \tau_\beta \qquad (\beta)$$

$$(\lambda x.(e\,x)) : \tau \quad \Rightarrow \quad e : \tau \qquad x \notin FV(e) \quad (\eta)$$

$$f_j^0(e_1, \ldots, e_n)_{\circ j'} \quad \Rightarrow \quad f_{\widehat{j + j'}}^0(e_1, \ldots, e_n) \qquad \text{(FC1)}$$

$$f_j^k(e_1, \ldots e_n)_{\circ j'} \quad \Rightarrow \quad f_j^k(e_1, \ldots, e_{k \circ j'}, \ldots e_n) \tag{FC2}$$

$$\langle e_1, \ldots, e_n \rangle_{\circ j'} \quad \Rightarrow \quad \langle e_{1 \circ j'}, \ldots, e_{n \circ j'} \rangle \tag{SC}$$

$$(\lambda x.e)_{\circ j'} \quad \Rightarrow \quad \lambda x.(e_{\circ j'}) \tag{PC}$$

$$f_j^0(e_1, \ldots, e_n)_{\bullet j'} \quad \Rightarrow \quad f_{j \,\widehat{\cdot}\, j'}^0(e_1, \ldots, e_n) \tag{FS1}$$

$$f_j^k(e_1, \ldots e_n)_{\bullet j'} \quad \Rightarrow \quad f_j^k(e_1, \ldots, e_{k \bullet j'}, \ldots e_n) \tag{FS2}$$

$$\langle e_1, \ldots, e_n \rangle_{\bullet j'} \quad \Rightarrow \quad \langle e_{1 \bullet j'}, \ldots, e_{n \bullet j'} \rangle \tag{SS}$$

$$(\lambda x.e)_{\bullet j'} \quad \Rightarrow \quad \lambda x.(e_{\bullet j'}) \tag{PS}$$

$$f_j^k(e_1, \ldots e_n)^{\rightsquigarrow k'} \quad \Rightarrow \quad f_j^{k'}(e_1, \ldots e_n) \tag{IC}$$

It is assumed that the addition and multiplication operator, respectively $\widehat{+}$ and $\widehat{\cdot}$, always yields a result within $[-\omega; \omega]$ cf. Definition 4.

Definition 4. *The operators $\widehat{+}$ and $\widehat{\cdot}$ are defined cf. (2) and (3) such that they always yield a result in the range $[-\omega; \omega]$, even if the pure addition and multiplication might not be in this range.*

$$j \,\widehat{+}\, j' = \begin{cases} -\omega & \text{if } j + j' < -\omega \\ \omega & \text{if } j + j' > \omega \\ j + j' & \text{otherwise} \end{cases} \quad (2) \quad j \,\widehat{\cdot}\, j' = \begin{cases} -\omega & \text{if } j \cdot j' < -\omega \\ \omega & \text{if } j \cdot j' > \omega \\ j \cdot j' & \text{otherwise} \end{cases} \quad (3)$$

The presented definition of semantic expressions allows the binding between expressed sentiment and entities in the text to be analysed, given that each lexicon entry have associated the proper expression. Example 1 shows how to apply this for a the simple declarative sentence, while Example 2 considers an example with long distance dependencies.

Example 1. Figure 3 shows the deduction proof for the sentence "the hotel had an exceptional service" including semantics. The entity "service" is modified by the adjective "exceptional" which is immediately to the left of the entity. The semantic expression associated to "service" is simply the zero-argument functor, initial with a neutral sentiment value. The adjective has the "changed identity function" as expression with a change value of 40. Upon application of combinatorial rules, semantic expressions are reduced based on the rewrite rules given in Definition 3. The conclusion of the deduction proof is a sentence with a semantic expression preserving most of the surface structure, and includes the bounded sentiment values on the functors. Notice that nouns, verbs, etc. are reduced to their lemma for functor naming.

Example 2. Figure 4 shows the deduction proof for the sentence "the breakfast that the restaurant served daily was excellent" including semantics, and demonstrates variations of all combinator rules introduced. Most interesting is the

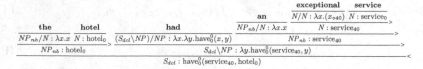

Fig. 3. Deduction of simple declarative sentence with semantics

correct binding between "breakfast" and "excellent", even though these are far from each other in the surface structure of the sentence. Furthermore the adverb "daily" correctly modifies the transitive verb "served", even though the verb is missing its object since it participates in a relative clause.

When the relative pronoun binds the dependent clause to the main clause, it "closes" it for further modification by changing the impact argument of the functor inflicted by the verb of the dependent clause, such that further modification will impact the subject of the main clause.

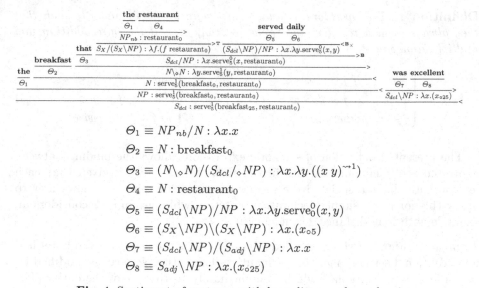

$$\Theta_1 \equiv NP_{nb}/N : \lambda x.x$$

$$\Theta_2 \equiv N : \text{breakfast}_0$$

$$\Theta_3 \equiv (N\backslash_\diamond N)/(S_{dcl}/_\diamond NP) : \lambda x.\lambda y.((x\ y)^{\leadsto 1})$$

$$\Theta_4 \equiv N : \text{restaurant}_0$$

$$\Theta_5 \equiv (S_{dcl}\backslash NP)/NP : \lambda x.\lambda y.\text{serve}_0^0(x, y)$$

$$\Theta_6 \equiv (S_X\backslash NP)\backslash (S_X\backslash NP) : \lambda x.(x_{\diamond 5})$$

$$\Theta_7 \equiv (S_{dcl}\backslash NP)/(S_{adj}\backslash NP) : \lambda x.x$$

$$\Theta_8 \equiv S_{adj}\backslash NP : \lambda x.(x_{\diamond 25})$$

Fig. 4. Sentiment of sentence with long distance dependencies

5   Results

The test data set chosen for evaluation of the case study was a subset of the *Opinosis data set* [8]. As the example texts might have hinted, the subset chosen was from the set of hotel and restaurant reviews. The *subject of interest* chosen for the analysis were *hotel rooms*, and the subset was thus randomly sampled from texts with high probability of containing this entity (i.e. containing any morphological form of the noun "room"). Two individuals were given a set of

35 review texts, and should mark each text as either positive, negative or unknown *with respect to the given subject of interest.*

5.1 Evaluation

An entity sentiment value was considered to *agree* with the human labeling if it had the correct sign (i.e. positive sentiment values agreed with positive labels, and negative values with negative labels). The baseline presented here is a sentence-level baseline, calculated by using the Naive Bayes Classifier available in the Natural Language Toolkit (NLTK) for Python. The *precision* and *recall* results for both the baseline and the presented method are shown in Table 2. As seen the recall is somewhat low for the proof of concept system, which is addressed below, while it is argued that precision of the system is indeed acceptable, since even humans will not reach a 100% agreement.

Table 2. Precision and recall results for proof of concept system

	Baseline	Presented method
Precision	71.5%	92.3%
Recall	44.1%	35.3%

The biggest issue was found to be the lack of correct syntactic tagging models. It is argued that models following a closer probability distribution of review texts than the one used would have improved the robustness of the system significantly. One might think that if syntactic labeled target data are needed then the presented logical method really suffers the same issue as machine learning approaches, i.e. *domain dependence*. However it is argued that exactly because the models needed are of *syntactic level*, and not of *sentiment level*, they really do not need to be *domain specific*, but only *genre specific*. This reduces the number of models needed, as a syntactic tagging model for reviews might cover several domains, and thus the *domain independence* of the presented method is intact.

6 Conclusions

This paper presents a formal logical method for deep structural analysis of the syntactical properties of texts using machine learning techniques for efficient syntactical tagging. The method is used to calculate entity level sentiment analysis, as an alternative to pure machine learning methods, which has been argued inadequate for capturing long distance dependencies between an entity and opinions, and of being highly dependent on the domain of the sentiment analysis. Empirical results show that while the correctness of the presented method seems acceptably high, its robustness is currently inadequate for most real world applications. However it is argued that it indeed is possible to improve the robustness significantly given further investment and development of the method.

References

1. Barendregt, H., Dekkers, W., Statman, R.: Lambda Calculus with Types. Cambridge University Press (2013)
2. Blitzer, J., Dredze, M., Pereira, F.: Biographies, Bollywood, Boomboxes and Blenders: Domain adaptation for sentiment classification. In: Annual Meeting of the Association of Computational Linguistics, pp. 440–447 (2007)
3. Bresnan, J., Kaplan, R.M., Peters, S., Zaenen, A.: Cross-serial dependencies in Dutch. Linguistic Inquiry 13(4), 613–635 (1982)
4. Cambria, E., Schuller, B., Liu, B., Wang, H., Havasi, C.: Statistical approaches to concept-level sentiment analysis. IEEE Intelligent Systems 28(3), 6–9 (2013)
5. Clark, S.: A supertagger for combinatory categorial grammar. In: International Workshop on Tree Adjoining Grammars and Related Frameworks, Venice, Italy, pp. 19–24 (2002)
6. Clark, S., Curran, J.R.: Wide-coverage efficient statistical parsing with CCG and log-linear models. Computational Linguistics 33(4), 493–552 (2007)
7. Feldman, R.: Techniques and applications for sentiment analysis. Commun. ACM 56(4), 82–89 (2013)
8. Ganesan, K., Zhai, C.X., Han, J.: Opinosis: A graph based approach to abstractive summarization of highly redundant opinions. In: International Conference on Computational Linguistics, pp. 340–348 (2010)
9. Hockenmaier, J.: Data and Models for Statistical Parsing with Combinatory Categorial Grammar. PhD thesis, University of Edinburgh (2003)
10. Hockenmaier, J., Steedman, M.: CCGbank: A corpus of CCG derivations and dependency structures extracted from the Penn treebank. Computational Linguistics 33(3), 355–396 (2007)
11. Joshi, A.K., Levy, L.S., Takahashi, M.: Tree adjunct grammars. Journal of Computer and System Sciences 10(1), 136–163 (1975)
12. Joshi, A.K., Shanker, K.V., Weir, D.: The convergence of mildly context-sensitive grammar formalisms. Technical Report, Department of Computer and Information Science, University of Pennsylvania (1990)
13. Liu, B.: Web Data Mining: Exploring Hyperlinks, Contents, and Usage Data. Springer (2007)
14. Marcus, M.P., Marcinkiewicz, M.A., Santorini, B.: Building a large annotated corpus of English: The Penn treebank. Computational Linguistics 19(2), 313–330 (1993)
15. Montague, R.: Formal Philosophy: Selected Papers of Richard Montague. Yale University Press (1974)
16. Pang, B., Lee, L.: Opinion mining and sentiment analysis. Foundations and Trends in Information Retrieval 2(1-2), 1–135 (2008)
17. Pollard, C.: Generalized Context-Free Grammars, Head Grammars and Natural Language. PhD thesis, Stanford University (1984)
18. Shieber, M.S.: Evidence against the context-freeness of natural language. Linguistics and Philosophy 8(3), 333–343 (1985)
19. Steedman, M.: Categorial grammar. In: The MIT Encyclopedia of Cognitive Sciences. The MIT Press (1999)
20. Steedman, M.: The Syntactic Process. The MIT Press (2000)
21. Vijay-Shanker, K., Weir, D.J.: The equivalence of four extensions of context-free grammars. Mathematical Systems Theory 27, 27–511 (1994)

Clustering View-Segmented Documents via Tensor Modeling

Salvatore Romeo[1], Andrea Tagarelli[1], and Dino Ienco[2]

[1] DIMES, University of Calabria, Italy
{sromeo,tagarelli}@dimes.unical.it
[2] IRSTEA, Montpellier, France
dino.ienco@teledetection.fr

Abstract. We propose a clustering framework for *view-segmented documents*, i.e., relatively long documents made up of smaller fragments that can be provided according to a target set of views or aspects. The framework is designed to exploit a view-based document segmentation into a third-order tensor model, whose decomposition result would enable any standard document clustering algorithm to better reflect the multi-faceted nature of the documents. Experimental results on document collections featuring paragraph-based, metadata-based, or user-driven views have shown the significance of the proposed approach, highlighting performance improvement in the document clustering task.

1 Introduction

Clustering has been long recognized as a useful tool for providing insight into the make-up of a document collection, and is often used as the initial step in the arduous task of knowledge discovery in text data. Document clustering research was initially focused on the development of general purpose strategies for grouping unstructured text data. Recent studies have started developing new methodologies and algorithms that take into account both linguistic and topical characteristics, where the former include the size of the text and the type of language used to express ideas, and the latter focus on the communicative function and targets of the documents. In particular, the length of documents and their topical variety are usually strongly interrelated, and in fact real-life collections are often comprised of very short or long documents. While short documents do not contain enough text and they can be very noisy, long documents often span multiple topics and this is an additional challenge to general purpose document clustering algorithms that tend to associate a document with a single topic. The key idea to solving this problem is to consider the document as being made up of smaller topically cohesive text blocks, named segments. When used as a base step in long document clustering, segmentation has indeed show significant performance improvements (e.g., [14]).

On the other hand, a useful mathematical tool to address the multi-faceted nature of real-world documents, along with the high dimensionality and sparseness in their text, is represented by tensor models and tensor decomposition

T. Andreasen et al. (Eds.): ISMIS 2014, LNAI 8502, pp. 385–394, 2014.
© Springer International Publishing Switzerland 2014

algorithms [5]. Tensors are considered as a multi-linear generalization of matrix factorizations, since all dimensions or modes are retained thanks to multi-linear models which can produce meaningful components. The applicability of tensor models has recently attracted growing attention in information retrieval and data mining related fields to solve problems such as link analysis [4], bibliographic data analysis [2], image clustering [15], document clustering [9,12].

In this paper we are interested in exploring the presumed benefits deriving from a combination of document segmentation and tensor models for clustering purposes. To this end, we deliberately keep the overall framework *simple* in terms of both document clustering scheme and tensor model/decomposition method, devoting our attention to evaluate clustering performance over different real-world scenarios. Important features of our proposed framework are *modularity*, since alternate methods for document clustering and tensor analysis can in principle be applied, and *domain versatility*. A key aspect of our approach is that segments can be provided according to a target set of *views* or aspects already defined over the data. These views should reflect an application target (e.g., different rating aspects in item reviews) or user-driven goals (e.g., microblogs posted by the same user). We stress this point in our experimental evaluation, using document collections from various domains and with different notions of "view". To the best of our knowledge, there has been no other study that explicitly treats multi-view documents in terms of their constituent, view-based text segments, while representing them under a multi-dimensional data structure.

Note that the problem we arise in this work is also related to the field of *multi-view clustering* [1], where the goal is to produce a partition of the instances exploiting all the different representations/views describing them. Multi-view clustering methods have also demonstrated to be effective on document clustering tasks (e.g., [11,3]). In particular, co-clustering approaches have shown to be a valuable tool to cluster sparse data. We will hence compare our approach with a recently proposed *multi-view co-clustering* approach which is specifically conceived to deal with text data [3].

2   View-Segmented Document Clustering

Given a collection of documents $\mathcal{D} = \{d_i\}_{i=1}^{N}$, we assume that each document is relatively long to be comprised of smaller textual units each of which can be considered cohesive w.r.t. a *view* over the document. The type of view is domain-dependent, and the way views are recognized in a document is supposed either to follow a topic detection approach or to reflect a metadata-level structuring of the documents. In the latter case, metadata can be of logical or descriptive type, such as, e.g., paragraph boundaries, specific subjects of discussion, or user-oriented aspects. We hereinafter refer to the view-oriented parts of a document with the term view-segment, or more simply *segment*.

The proposed view-segmented document clustering framework is shown in Fig. 1. The framework can be summarized by the following steps, which are discussed in detail next:

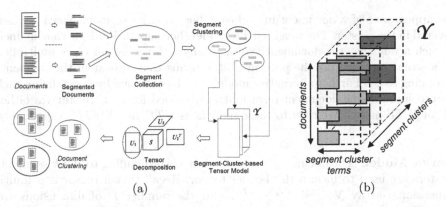

Fig. 1. (a) The view-segmented document clustering framework. (b) The third-order tensor model for the document collection representation based on the produced view-segment clusters.

1. Clustering the document segments, or exploiting meta-data to derive a grouping of the segments;
2. Computing a representation of the original document collection for each of the obtained clusters of segments;
3. Computing a third-order tensor for the document collection, upon the segment-cluster based representations.
4. Decomposing the tensor using a Truncated HOSVD;
5. Performing any document clustering algorithm on the mode-1 factor matrix to obtain the final document clustering solution.

Inducing view-segment Clusters. We are given a collection of segments $\mathcal{S} = \{s_j\}_{j=1}^n$ over $\mathcal{D}$, which can in principle be disjoint or not.

The first step of our framework is in charge of producing a clustering of the segments $\mathcal{C} = \{C_s\}_{s=1}^k$, by applying any document clustering algorithm over the segment collection $\mathcal{S}$. The obtained clusters of segments can be disjoint or overlapping.

Segment-cluster Based Representation. Upon the segment clustering, each document is represented by its segments assigned to possibly multiple segment clusters. Therefore, we derive a document-term matrix for each of the k segment clusters. For this purpose, we define four alternative approaches which are described next.

The basic approach is based on term-frequency information. Given a segment cluster C_s and the relating set of feature terms $\mathcal{F}(C_s)$, the representation of any document d_i in that cluster is defined as a vector of length $|\mathcal{F}(C_s)|$ that results from the sum of the feature vectors of the d_i's segments belonging to C_s; the feature vector of a segment is a vector of term-frequencies. We hereinafter refer to this approach as *TF*. An intuitive refinement of the TF model is to weight the appearance of a document in a cluster based on its segment-based portion covered in the cluster. The weighted TF model (henceforth WTF) is thus defined in such a way that the document vector of any d_i for a cluster C_s is multiplied by a scalar representing the portion of d_i's terms that appear in the segments belonging to C_s. Further alternative models can be obtained by normalizing each term-column in the document-term matrix obtained for each cluster via either TF or WTF model. We refer to these models as NTF and $NWTF$, respectively.

Tensor Model. The document-term matrices corresponding to the k segment-clusters are used to form a third-order tensor. Recall that a tensor is a multi-dimensional array $\mathcal{Y} \in \Re^{I_1 \times I_2 \times \cdots \times I_D}$, and the number D of dimensions (or modes) is called order of the tensor. A two-dimensional fragment of tensor defined by varying two indices and keeping the rest fixed is a 2-mode tensor called slice.

Our third-order tensor model is built by arranging as frontal slices the k segment-cluster matrices. However, since the segment clusters have possibly different feature subspaces and might cover different subsets of the document collection, the resulting matrices will have a different number of rows/columns. Therefore, in order to build the tensor, each matrix needs to be properly filled with as many zero-valued rows as the number of non-covered documents and with as many zero-valued columns as the number of non-covered feature terms. The resulting tensor will be $\mathcal{Y} \in \Re^{I_1 \times I_2 \times I_3}$, with $I_1 = |\mathcal{D}|$, $I_2 = \max\limits_{1 \leq s \leq k} |\mathcal{F}(C_s)|$, and $I_3 = k$. The proposed tensor model is sketched in Fig. 1(b).

Tensor Decomposition. The third-order tensor is decomposed through a Truncated Higher Order SVD (T-HOSVD) [6] in order to obtain a low-dimensional representation of the segment-cluster-based representation of the document collection; for document clustering purposes, we will consider the mode-1 factor matrix. Recall that T-HOSVD is a generalization of SVD [6], as it approximates a tensor into an orthogonal component matrix along each mode and a smaller all-orthogonal and ordered core tensor.

If we denote with r the number of output components for each mode required by T-HOSVD, the decomposed tensor is defined as $\mathcal{Y} \approx \mathcal{S} \times_1 \mathbf{U}_1 \times_2 \mathbf{U}_2 \times_3 \mathbf{U}_3$. It is worth also noting that the key idea of T-HOSVD is to capture the variation in each of the modes independently from other ones, which makes T-HOSVD particularly appropriate for clustering purposes.

Document Clustering. The mode-1 factor matrix is provided in input to a clustering method to obtain a final organization of the documents into K clusters. Note that there is no principled relation between the number K of final document clusters and k, however K is expected to reflect the number of views of interest for the document collection. Also, possibly but not necessarily, the same clustering algorithm used for the segment clustering step can be employed for this step.

Table 1. Evaluation datasets

Dataset	Type of Document	Segment granularity	Segment clue	# docs	# segs	fraction of nonzero entries*	avg # terms per doc	avg # terms per seg
RCV1	news article	paragraph	paragraph boundaries	15,813	128,031	4.6E-4	200	25
TripAdvisor	hotel review	rating aspect	metadata	170,867	810,314	1.1E-3	75	16
Twitter	user's tweets	tweets by hashtag	keywords	9,289	36,763	1.4E-3	147	37

* It refers to the segment-term matrix of the dataset.

3 Experimental Evaluation

3.1 Data

We used three collections of documents that fall into very different application scenarios (Table 1). The peculiarities of each dataset prompted us to identify segments at different granularity levels, exploiting different clues.

Scenario 1: Documents with paragraph-based views. We used a subset of the Reuters Corpus Volume 1 (RCV1) [7]. We filtered out very short news (i.e., original XML documents with size less than 3KB) and highly structured news (e.g., lists of stock prices), then we performed tokenization, stopword removal and word stemming. Each paragraph in a news article was regarded as a segment.[1] We exploited the availability of topic-labels associated with the news articles (i.e., values of Reuters TOPICS field) to sample the original dataset in order to select documents that satisfy certain requirements on the set of covered topics. For this purpose, we followed the lead of a methodology introduced in [10], whereby topic-sets are induced as sets of topic-labels that may overlap, whereas documents are kept organized in disjoint groups. Therefore, the assignment of topic-sets to documents results in a multi-topic, hard classification for the documents in the dataset. Moreover, we kept only the second-level topic-labels in order to ensure that there are no relations of containment between the topic-labels used to form the topic-sets. Upon the evaluation of frequency distribution of each possible set of second-level topic-labels occurring in the documents, we selected only topic-sets having at least two topic-labels and covering at least 1% of documents. Once the topic-sets were extracted, we collected all associated documents. The final dataset was composed of 15,813 documents belonging to 18 topic-sets.

Scenario 2: Documents with metadata-based views. We used the full DAIS Trip-Advisor dataset[2], which is a collection of 170,867 hotel reviews. A nice feature of this dataset is that all text reviews are already provided as segmented according to eight rating-aspects, namely "Overall rating", "Value", "Rooms", "Location", "Cleanliness", "Check in/front desk", "Service" and "Business Service".

---

[1] Note that clearly one can resort to text segmentation algorithms to induce segments at sentence level, and we indeed adopted this approach in previous work [14]; however, text segmentation typically requires the setting of several interrelated parameters, whose tuning is not an easy task.

[2] http://sifaka.cs.uiuc.edu/$\sim$wang296/Data/index.html.

Scenario 3: Documents with user-driven views. We used a collection of tweets from the Twitter UDI dataset [8]. Our key idea was to consider all tweets of a particular user as a document and to exploit the appearance of hashtags to group related tweets in the user's thread. However, many tweets can have a very few number of words, and the distribution of hashtags over the tweets can be very sparse. Therefore, we imposed constraints on the number of words per tweet and on the number of hashtags per user, in order to reflect the characteristic average length of tweet (around 5.9 in the original dataset) and to ensure high variability in the hashtags utilized by a user. We hence selected the users who posted at least 5 tweets having at least 6 content words, and for which at least 3 hashtags were used in his/her tweets. We then computed the user popularity of hashtags and we selected the top-25 most popular ones. Finally, we collected all tweets related to the users for which the above constraints are still valid. The dimensions of the final dataset are 161,623 tweets related to 9,289 users, with average number of tweets per user of 17.4, average number of hashtags per user of 3.96, and average number of tweets per segment of 4.8. Note also that we found a negligible average degree of segment overlapping (0.064 shared tweets).

Reference Classification. In RCV1, each of the 18 topic-sets is a document class. Similarly, in Twitter, each of the top-25 selected hashtags defines a class, and a user's tweet-document is assigned to the class that corresponds to the most frequently occurring hashtag in the tweets of that user. In TripAdvisor, the reviews are categorized according to the 8 rating aspects. Each review document is assigned to the class that corresponds to the most descriptive aspect of that review, i.e., the aspect described by the largest portion of terms.

3.2 Competing Methods

We compared our approach with two baseline clustering methods and a multi-view co-clustering algorithm.

In the first baseline method, dubbed *DocClust*, documents were represented by the conventional vector-space model equipped with the tf.idf term relevance weighting scheme. Clustering of the documents was performed by using the *Bisecting K-Means* [13] algorithm, which is widely known to produce high-quality (hard) clustering solutions in high-dimensional, large datasets.

The second baseline method, dubbed *SegClust*, utilizes the same text representation model and clustering scheme as the first baseline, however it applies on the collection of document segments. Once computed the clustering of segments, a document clustering solution is finally induced via majority voting. This approach was first explored in [14] and has shown to improve the final document clustering performance for relatively long multi-topic documents.

The multi-view co-clustering algorithm *CoStar* [3] searches for a solution that maximizes cross-association of the objects given the different views and viceversa. The clustering algorithm is formulated as a Pareto optimization problem and it does not require any parameter as input. Moreover, *CoStar* explores the search space choosing automatically the number of clusters.

Table 2. Best performance scores on RCV1. The best-performing setup of k and r is reported for each assessment criterion and tensor-slice representation model.

model	FM	k	r	E	k	r	Pty	k	r	NMI	k	r
TF	0.581	10	80	0.412	2	55	0.589	2	55	0.582	2	55
WTF	0.606	2	95	0.424	2	95	**0.602**	2	95	0.568	2	95
NTF	**0.608**	2	50	**0.394**	2	50	**0.602**	2	50	**0.601**	2	35
NWTF	0.566	4	75	0.418	12	70	0.579	4	75	0.582	12	70

3.3 Parameter Settings and Assessment Criteria

Our approach requires the setting of two parameters, namely the number k of view-segment clusters, and the number of output components r for each mode by T-HOSVD. We varied k from 2 to $\sqrt{n}$ (n is the total number of segments over $\mathcal{D}$) with increment of 2 and r from 5 to 100 with increment of 5.

We resorted to standard clustering validation criteria, namely F-Measure (FM), Entropy (E), Purity (Pty), and Normalized Mutual Information (NMI). We recall here that a larger (smaller) value is desirable for FM, Pty, and NMI (resp. E) to indicate better clustering quality. The interested reader is referred to [13] and [16] for details on the various assessment criteria.

4   Results

Tables 2–4 report on the best performance results by our method, with corresponding parameter settings, where the number of final document clusters was set equal to the number of dataset-specific reference classes. It can be noted that the tensor-slice representation models with normalization mostly led to higher quality scores. More precisely, NWTF was always the best-performing model on TripAdvisor and Twitter, while NTF led to the best results on RCV1. The latter would hint that in RCV1 the segments are more uniformly distributed along the segment clusters, thus reducing the weighting factor's influence.

A major remark is that, on all datasets, the best results were consistently achieved by using a quite small number of segment clusters; more precisely, k was mostly below 10, or even equal to the minimum value (i.e., 2). Moreover, in TripAdvisor, the best-performing results also occurred with very few tensor components (i.e., 5). This in general was not the case for the other datasets as well. In this regard, Figures 2–3 provide more insights by comparing the various tensor-slice representation models. In the figures, the distributions of performance scores are plotted over different numbers of tensor components. It can be noted that NWTF or NTF models generally corresponded to better performance on average. (Only results for FM and E criteria were shown due to space limits, but analogous remarks could be done for the other criteria.)

Comparison with Baselines. Table 5 compares the best performance obtained by our approach and the two baseline methods, DocClust and SegClust.

Table 3. Best performance scores on TripAdvisor. The best-performing setup of k and r is reported for each assessment criterion and tensor-slice representation model.

model	FM	k	r	E	k	r	Pty	k	r	NMI	k	r
TF	0.406	10	5	0.686	8	5	0.496	8	5	0.127	8	5
WTF	0.531	8	5	0.618	6	5	0.578	8	5	0.210	8	5
NTF	0.475	12	10	0.635	6	5	0.540	8	5	0.176	6	5
NWTF	**0.558**	8	5	**0.580**	6	5	**0.606**	8	5	**0.250**	6	5

Table 4. Best performance scores on Twitter. The best-performing setup of k and r is reported for each assessment criterion and tensor-slice representation model.

model	FM	k	r	E	k	r	Pty	k	r	NMI	k	r
TF	0.356	12	100	0.649	12	100	0.414	12	80	0.294	12	90
WTF	0.371	12	40	0.62	10	35	0.434	12	90	0.333	10	35
NTF	0.388	22	85	0.622	22	100	0.42	22	100	0.328	10	20
NWTF	**0.402**	22	55	**0.603**	20	95	**0.438**	26	80	**0.346**	10	15

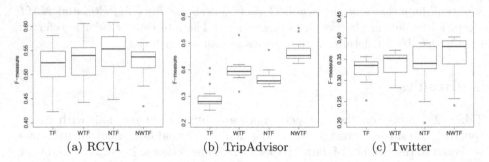

|(a) RCV1|(b) TripAdvisor|(c) Twitter|

Fig. 2. F-measure distribution over different numbers of tensor components

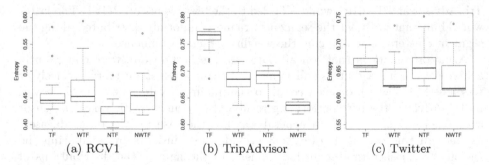

|(a) RCV1|(b) TripAdvisor|(c) Twitter|

Fig. 3. Entropy distribution over different numbers of tensor components

Our method outperformed the baselines according to all assessment criteria, with gains up to 0.256 FM, 0.198 E, 0.247 Pty, and 0.223 NMI. It should be emphasized that the maximum gains achieved by our method corresponded to TripAdvisor, where both the documents and segments are more than in the

Table 5. Best performance comparison with baseline algorithms

criteria	RCV1			TripAdvisor			Twitter		
	DocClust	*SegClust*	our method	*DocClust*	*SegClust*	our method	*DocClust*	*SegClust*	our method
FM	0.504	0.523	**0.608**	0.315	0.302	**0.558**	0.366	0.225	**0.402**
E	0.474	0.467	**0.394**	0.762	0.778	**0.580**	0.659	0.787	**0.603**
Pty	0.501	0.521	**0.602**	0.396	0.359	**0.606**	0.398	0.255	**0.438**
NMI	0.528	0.540	**0.601**	0.045	0.027	**0.250**	0.293	0.135	**0.346**

other datasets, but also shorter on average (cf. Table 1). From a qualitative viewpoint, by exploring the cluster descriptions in the form of top-ranked descriptive and discriminating terms [13], we observed an evident ability of our approach to detect clusters that better capture and separate the expected view-based classes. For instance, on TripAdvisor, DocClust and SegClust found clusters that mainly corresponded only to the classes "Overall rating" and "Rooms", while our method was able to discriminate also among the other classes.

Comparison with CoStar. As previously discussed in Section 3.2, CoStar automatically detects the number of clusters. When applied to our evaluation datasets, CoStar consistently obtained a larger number of clusters (e.g., 50 clusters on Twitter) than the dataset-specific reference classes. Therefore, in order to fairly compare our method with CoStar, we carried out an agglomerative hierarchical clustering method over the partition originally produced by CoStar, cutting the dendrogram at the level corresponding to the right number of reference classes. Moreover, since CoStar has a non-deterministic clustering behavior, we ran it multiple times (50) and hence evaluated each of the final CoStar clustering solutions w.r.t. the reference classification. Upon this, by comparing the best performance scores obtained by CoStar and by our method, we observed a similar behavior on Twitter and TripAdvisor, with marginal improvements by our method (i.e., order of 0.05 or less, for each assessment criterion). However, on RCV1, our method outperformed CoStar, with the following gains: 0.11 FM, 0.10 E, 0.09 Pty, and 0.09 NMI.

We conducted a further experimental session in which we constrained the number of document clusters to be produced by our method as equal to the number of clusters originally produced by CoStar. Using NMI (which is a symmetric evaluation index), we compared each of the multiple CoStar clusterings with those produced by our method, at varying parameter values. Results (here not shown due to space limits of this paper) have revealed a moderate alignment between the two methods, up to around 0.6 on average; also, the best alignment was again consistently achieved by very low values of our method's parameters, (from $k = 6$, $r = 10$ on RCV1 to $k = 8$, $r = 20$ on Twitter).

5   Conclusions

We have presented a tensor-based clustering framework for view-segmented documents. Experimental results have supported our intuition on the clustering improvement performance over different challenging datasets. As further work, it would be interesting to deepen our understanding of the influence each factor matrix has on the final document clustering solution. Another promising direction would be to combine the current framework with probabilistic models specifically tailored for view-segmented documents.

References

1. Bickel, S., Scheffer, T.: Multi-View Clustering. In: Proc. IEEE Int. Conf. on Data Mining (ICDM), pp. 19–26 (2004)
2. Dunlavy, D.M., Kolda, T.G., Kegelmeyer, W.P.: Multilinear algebra for analyzing data with multiple linkages. In: Graph Algorithms in the Language of Linear Algebra, Fundamentals of Algorithms, pp. 85–114. SIAM (2011)
3. Ienco, D., Robardet, C., Pensa, R.G., Meo, R.: Parameter-less co-clustering for star-structured heterogeneous data. Data Min. Knowl. Disc. 26(2), 217–254 (2013)
4. Kolda, T., Bader, B.: The TOPHITS model for higher-order web link analysis. In: Proc. Workshop on Link Analysis, Counterterrorism and Security (2006)
5. Kolda, T.G., Bader, B.W.: Tensor decompositions and applications. SIAM Review 51(3), 455–500 (2009)
6. Lathauwer, L.D., Moor, B.D., Vandewalle, J.: A Multilinear Singular Value Decomposition. SIAM J. Matrix Anal. Appl. 21(4), 1253–1278 (2000)
7. Lewis, D.D., Yang, Y., Rose, T., Li, F.: RCV1: A New Benchmark Collection for Text Categorization Research. J. Mach. Learn. Res. 5, 361–397 (2004)
8. Li, R., Wang, S., Deng, H., Wang, R., Chang, K.C.-C.: Towards social user profiling: unified and discriminative influence model for inferring home locations. In: Proc. ACM SIGKDD Int. Conf. on Knowledge Discovery and Data Mining (KDD), pp. 1023–1031 (2012)
9. Liu, X., Glänzel, W., Moor, B.D.: Hybrid clustering of multi-view data via Tucker-2 model and its application. Scientometrics 88(3), 819–839 (2011)
10. Ponti, G., Tagarelli, A.: Topic-Based Hard Clustering of Documents Using Generative Models. In: Rauch, J., Raś, Z.W., Berka, P., Elomaa, T. (eds.) ISMIS 2009. 5722, vol. 5722, pp. 231–240. Springer, Heidelberg (2009)
11. Ramage, D., Heymann, P., Manning, C.D., Garcia-Molina, H.: Clustering the tagged web. In: Proc. Int. Conf. on Web Search and Web Data Mining (WSDM), pp. 54–63 (2009)
12. Romeo, S., Tagarelli, A., Gullo, F., Greco, S.: A Tensor-based Clustering Approach for Multiple Document Classifications. In: Proc. Int. Conf. on Pattern Recognition Applications and Methods (ICPRAM), pp. 200–205 (2013)
13. Steinbach, M., Karypis, G., Kumar, V.: A Comparison of Document Clustering Techniques. In: Proc. KDD Workshop on Text Mining (2000)
14. Tagarelli, A., Karypis, G.: A segment-based approach to clustering multi-topic documents. Knowl. Inf. Syst. 34(3), 563–595 (2013)
15. Zhang, Z.-Y., Li, T., Ding, C.: Non-negative Tri-factor tensor decomposition with applications. Knowl. Inf. Syst. 34(2), 243–265 (2013)
16. Zhong, S., Ghosh, J.: A Unified Framework for Model-Based Clustering. J. Mach. Learn. Res. 4, 1001–1037 (2003)

Searching XML Element Using Terms
Propagation Method

Samia Berchiche-Fellag[1] and Mohamed Mezghiche[2]

[1] Université Mouloud Mammeri de Tizi-Ouzou, 15000 Tizi-Ouzou, Algérie
samfellag@yahoo.fr
[2] Université M'Hamed Bougara Boumerdes, Algérie
mohamed.mezghiche@yahoo.fr

Abstract. In this paper, we describe terms propagation method dealing with focussed XML component retrieval. Focussed XML component retrieval is one of the most important challenge in the XML IR field. The aim of the focussed retrieval approach is to find the most exhaustive and specific element that focus on the user need. These needs can be expressed through content queries composed of simple keyword. Our method provides a natural representation of document, its elements and its content, and allows an automatic selection of a combination of elements that better answers the user's query. In this paper we show the efficiency of the terms propagation method using a terms weighting formula that takes into account the size of the nodes and the size of the document. Our method has been evaluated on the «Focused» task of INEX 2006 and compared to XFIRM model which is based on relevance propagation method. Evaluations have shown a significant improvement in the retrieval process efficiency.

Keywords: Structured Information Retrieval (SIR), XML, terms propagation, CO query, terms weighting, element, INEX.

1   Introduction

XML documents are semi-structured documents which organize text through semantically meaningful elements labelled with tags. Hierarchical document structure can be used to return specific document components instead of whole documents to users.

Structural information of XML documents is exploited by Information Retrieval Systems (IRS) to return to users the most exhaustive[1] and specific[2] [1] documents parts(i.e. XML elements, also called nodes) answering to their needs. These needs can be expressed through Content queries (CO: Content Only) which contain simples keywords or through Content And Structure queries (CAS) which contain both keywords and structural information on the location of the needed text content. Most

---

[1] An element is exhaustive to a query if it contains all the required information.
[2] An element is specific to a query if all its content concerns the query.

T. Andreasen et al. (Eds.): ISMIS 2014, LNAI 8502, pp. 395–404, 2014.
© Springer International Publishing Switzerland 2014

of the retrieval models used for structured retrieval are adaptation of traditional retrieval models. The main problem is that the classical IR methods work at the document level. This does not perform well at the node level due to node nesting in XML as explained in [2] [3] [4].

The challenge in XML retrieval is to return the most relevant nodes that satisfy the user needs. Of most interest is the class of CO queries where the user doesn't know anything about the collection structure and issue her query in free text. The IRS exploits the XML structure to return the most relevant XML nodes that satisfy the user needs. Besides being relevant, retrieved nodes should be neither too large nor too small. In this aim we present our method which consists of searching the relevant nodes to a CO query composed of simple keywords in a large set of XML documents and taking into account the contextual relevance. The search process that we propose is based on a method of terms propagation.

Our method has already proved its effectiveness in [5] using the weighting formula usually used in IRS; In this paper we show that our method remains efficient using a weighting formula that takes into account the number of terms and the average number of terms in both the nodes and the document.

The rest of this paper is organized as follows: We present the state of the art in section 2. In section 3 we describe our baseline model, which uses a terms propagation method; we also present our weighting formula which uses node size and document size. Finally we present in section 4 results of our experimentations.

2   Related Work

The IR community has adapted traditional IR approaches to address the user information needs in XML collection. Some of these methods are based on the vector space model [6], [3] , [7] , or on the probabilistic model [8]. Language models are also adapted for XML retrieval [9], [10], as well as Bayesian networks in [11].

The aim of IRS dealing with XML documents is to retrieve the most relevant nodes the user need. For this purpose several approaches based on propagation methods were proposed by authors. Relevance propagation, terms propagation and weights propagation. In the relevance propagation approach, relevance score of leaf nodes in xml document tree is calculated and propagated to ancestors. Authors in [12] used linear combination of children's scores called "maximum-by-category » and « summation ».While the relevance propagation in [13] using XFIRM system is function of the distance that separates nodes in the tree. In [14], [15] authors used a method of weights propagation. For computing the weights of inner nodes, the weights from the most specific nodes in the document multiplied with an augmentation factor are propagated towards the inner nodes. Cui et al.[16], Benaouicha [17] , and Fellag[5] [18] used terms propagation method. In this case, textual content of leaf nodes in Xml document is propagated to their ancestor considering some conditions. In[16] and [5] [18] authors exploited both structural information and the statistics of term distributions in structured documents. In [5] [17] [18], a leaf node is represented by a set of weighted terms. These terms are propagated to their ancestors by reducing their weight depending on the distance that separates nodes in the tree. As a conclusion, whatever the considered approach, the relevance node's score strongly depends on its descendants' scores.

3   Proposed Approach

We consider that a structured XML document D is a tree, composed of simple nodes ni, leaf nodes lni and attributes ai. The textual information (terms) is at the leaf nodes lni. Other nodes only give indication on structure. Weights are assigned to terms in leaf nodes and inner nodes' weights are computed dynamically during the propagation.

Example of such document is given on figure 1 with its tree representation in figure 2.

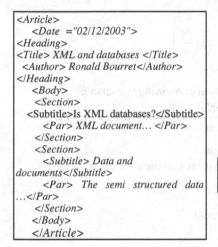

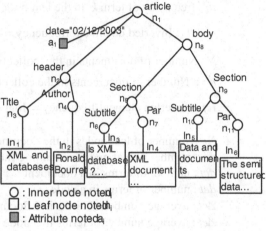

Fig. 1. Example of XML document

Fig. 2. Tree representation of the XML document in fig.1

3.1 Query Processing

The approach we propose for dealing with content queries is based on terms propagation method. The query processing is carried out as follows:

- assigning weights to terms in leaf nodes,
- pruning the document tree, retaining only informative nodes,
- Propagating the well distributed terms in the leaf nodes to their ancestor, in order to identify relevant and informative nodes
- evaluating the relevance score of the identified nodes and presenting the results descending scores.

Before presenting our terms propagation method, we considered useful to introduce first the term weighting formula and the matching score formula we use.

3.2 Weighting Terms in Leaf Nodes

The first step in query processing is to assign weights to terms in leaf nodes *ln*. In[5] the weighting formula we used is one of the adaptations from traditional IR to the granularity used in structured IR "the element" that is $w_k = tf_k \times ief_k$. In this paper we introduce the number of terms and the average number of terms in both the nodes and the document and proposed the following formula:

$$w_k = tf_k \times idf_k \times ief_k \times \frac{1}{h_1 + h_2 \times \frac{det}{\Delta et} + h_3 \times \frac{dct}{\Delta ct}} . \tag{1}$$

Where :

w_k: is the weight of term k in leaf node *ln*

tf_k : frequency of term k in the leaf node

Idf_k: Inverted document frequency $= \log\left(\dfrac{N}{n}\right)$

N : Number of documents in the collection

n : : Number of documents in the collection containing the term k

Ief_k: Inverted element frequency $= \log\left(\dfrac{Ne}{ne} + \alpha\right)$ with $0.5 \le \alpha \le 1$

Ne : number of leaf nodes in the document

ne : number of leaf nodes containing the term k in the document

dct : number of terms in a document

det : number of terms in a node

Δct : average number of terms in a document

Δet : average number of terms in a node

3.3 Evaluation Nodes Relevance Value

The final step in our query processing is to evaluate the relevance value of nodes n_i according to the query. Let $q = \{(t_1, w_{q1}), \ldots\ldots, (t_M, w_{qM})\}$ be a query composed of weighted keyword terms. t_k a query term, w_{qk} the weight of t_k in query q and M is the number of terms in the query. Relevance values are computed thanks to a similarity function called $RSV(q, n_i)$ of the vector space model (Inner product) as follows:

$$RSV(q, n_i) = \sum_{k=1}^{M} w_{qk} \times w_{nik} . \tag{2}$$

Where w_{qk} and w_{nik} evaluated with formula (1), are respectively the weight of the term k in the query q and in the node n_i.

3.4   Terms Propagation Method

In order to return the most relevant parts of XML documents that match the user query, we propose a terms propagation method, starting from leaf nodes to the document root.

The main issue here is: *what terms propagate?*

In this purpose, we introduce the concept of *informativeness* of the node. *Node is informative if it carries sufficient information to satisfy a user query.* The issue is: *how to measure it?* for this purpose we follow the intuition which guided us to take into account the node's size.

Indeed, a node that contains only the query terms, is specific to this query. However, not informative because it does not provide the required information to the user (eg a title node may be relevant to a query but is not informative). We define, for this purpose, a threshold that involves the minimum number of terms that a node must have to be considered informative. It is clear that we have no theoretical way to determine this threshold. We propose to fix it by experiment as is frequently the case in the IR area.

Two cases in the terms propagation, be considered:

    Case A: nodes whose number of terms is below the threshold.

    Case B: nodes whose number of terms is greater than the threshold.

Case A: Nodes Whose Number of Terms is Below the Threshold.

The document tree is traversed starting from leaf nodes. During the path, when the number of terms in the visited node is below the threshold, the node is removed from the tree and its content ascended to its parent node. This process is done recursively until reaching (or exceed) the threshold, or reach the root node of the document, or reach an inner node whose terms number of at least one of its child is greater or equal than threshold.

Case B: Nodes Whose Number of Terms is Greater than the Threshold.

We consider fundamental hypothesis which expresses that: *"terms of a node well distributed in its child may be representative terms for this node."*

Two cases can occur, a node can have several child nodes, or have only one (leaf node only has no child):

1. Case of Node with Several Child Nodes

Intuitively, we can think that a term of a node can be representative for its parent node if it appears at least on one sibling node. We consider this intuition insufficient and think that it is imperative to take into account the weight of terms in the nodes. Indeed, a term of a node may belong to all its child nodes, but if its weight is low compared to the weight of the other terms in these nodes. It cannot be discriminant for these nodes.

For this purpose we consider another hypothesis which consists to take into account only the terms which average weight in the child nodes where they appear is between the average and the maximum weight of all the terms of child nodes.

2. Case of Node with One Child Node

We consider the hypothesis which expresses that *"term of a node is representative for its parent. If its weight is between the average and the maximum weight of all the terms of the node.*

The term satisfying cases 1 or 2, is removed from its node and ascended to its parent node. Its weight in the parent node is equal to its average weight in the child nodes in case 1, or its weight in case 2.

These hypotheses are formalized as follows:

1. Let e be a node with several child nodes e'. Let t be a term of a child node e'. $w(t, e')$ the weight of term t in node e', calculated with the formula (1). t can be ascended to e, if t exists in at least one sibling node of e' and if the average weight of t in the child node of e where it appears, verify the following condition:

$$w_{avg} \leq \underset{e' \in chl(e)}{avg(w(t,e'))} \leq w_{max} \tag{3}$$

Where :

$$w_{avg} = \frac{\sum\limits_{e' \in chl(e)} \sum\limits_{i=1}^{Nte'} w(t_i,e')}{Nt}. \tag{4}$$

W_{avg} : average weight of terms in the nodes e' child of *node e*

chl(e): child of node e

$$avg_{e' \in cl(e)}(w(t,e')) = \frac{\sum\limits_{e' \in chl(e) / t \in e'} w(t,e')}{Ne'}. \tag{5}$$

Nte' : number of terms in the node e'

Nt : number of terms in all nodes e' child of node e

Ne' : number of nodes e' containing the term t

w_{max} : Maximum weight of terms in all nodes e' child of node e

The term t is removed from the child nodes e' and ascended to its parent node e its weight will be:

$$w(t,e) = \underset{e' \in chl(e)}{avg(w(t,e'))}. \tag{6}$$

2. Let e be a node with only one child node e'. Let t be a term of node e'. $w(t, e')$ the weight of term t in node e', calculated with the formula (1). t can be ascended to e, if it satisfies condition (7):

$$w_{avg} \leq w(t,e') \leq w_{max} \tag{7}$$

Where :

$$w_{avg} = \frac{\sum\limits_{i=1}^{Nte'} w(t_i,e')}{Nte'} \tag{8}$$

W_{avg} : average weight of terms in node e'

w_{max} : maximum weight of terms in node e'

Nte' : number of terms in the node e'

The term t is removed from the child node e' and ascended its parent node e with its weight:

$$w(t,e) = w(t, e')$$ (9)

Note that during the ascent of term t from child node e' to its parent node e,

It may previously be present. In this case, the term t is removed from child node(s) e', and its weight in node e is equal to the average weight in child node(s) e' and the parent node e as follow:

$$w(t,e) = \frac{w_{(6)/(9)}(t,e) + w_0(t,e)}{2}$$ (10)

where:

$w_0(t,e)$: initial weight of the term t in the node e

$w_{(6)/(9)}(t,e)$: weight should have (if it did not exist) the term t in node e calculated using the formula (6) or (9).

The propagation process runs recursively from leaves nodes to the document root.

At the end of the propagation process, the relevance score of the nodes represented by these terms according to the query terms is evaluated, the results are presented descending scores. The results nodes are relevant and informative.

4   Experimentations

The aim of this experiment is to test our model and to compare it to XFIRM model which uses relevance propagation.

4.1   INEX: Initiative for the Evaluation of XML Retrieval

We used for our experiments the INEX 2006 collection[19]. The test collection consists of a set XML documents, queries and relevance judgments. INEX consists of several tasks such as *"focused"* task, *"thorough"* task, *"Best in context"* task... We based our tests on the *"focused"* task.

4.2   Evaluation Protocol

The INEX 2006 collection contains about 659 388 documents and provides a set of 126 queries for evaluation. Our experiments are performed CO (Content Only) queries. We experimented 32 queries of INEX 2006 collection. We used the normalized cumulated gain $nxCG[i]$. With this measure, system performance was reported at several rank cutoff values (i).

$$nxCG[i] = \frac{xCG[i]}{xCI[i]} \ . \tag{11}$$

xCG[i] takes its values from the full recall-base of the given topic. *xCI[i]* takes its values from the ideal recall-base and i ranges from 0 and the number of relevant elements for the given topic in the ideal recall base.

4.3   Results

To assign value to threshold, we conducted preliminary experiments on small sample of INEX 2006 collection (15 queries and 100 documents). The various evaluations let us to set threshold equal to 50 terms.

Table 1. Results for the « Focused » Task with the nxCG metric at different cutoffs

	NXCG5	NXCG10	NXCG25	NXCG50
XFIRM	0,5333	0,2076	0,1215	0,0799
Our method	0,7188	0,2446	0,1399	0,0912
% improvement	25,810	15,099	13,132	12,486

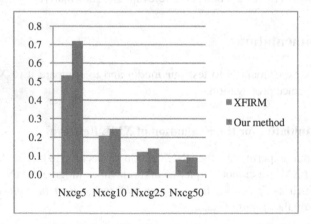

Fig. 3. Comparative graph the achieved results of our approach to that of XFIRM

As shown in [5], our terms propagation method also proved its efficiency through obtained results performance. Improvement in results is observed on different gain values. Even by applying a different formula other than the standard formula used in the IRS, our method remains effective and efficient than XFIRM which is relevance propagation. We conclude that considering the number of terms and the average number of terms in both the nodes and the document in the weighting formula, our method still effective..

5   Conclusion

We presented in this paper our contribution to XML element retrieval for retrieving the most relevant part of XML documents that the user needs. We proposed for this purpose terms propagation method. Which aim is to not only return the most exhaustive and specific nodes to a user query but mainly informative nodes are returned with the constraint about node's size we imposed. Our method has already proved its efficiency in [5]. Its performance is confirmed in this paper with another weighting formula, obtained results on the «Focused» task of INEX 2006 has shown a significant improvement in the retrieval process efficiency compared to XFIRM system.

References

1. Lalmas, M.: Dempster-Shafer's theory of evidence applied to structured documents: Modeling uncertainty. In: Proceedings of ACM-SIGIR, Philadelphia, pp. 110–118 (1997)
2. Mass, Y., Mandelbrod, M.: Retrieving the most relevant XML Component. In: Proceedings of the Second Workshop of the Initiative for the Evaluation of XML Retrieval (INEX), December 15-17 (2003)
3. Mass, Y., Mandelbrod, M.: Component Ranking and Automatic Query Refinement for XML Retrieval. In: Fuhr, N., Lalmas, M., Malik, S., Szlávik, Z. (eds.) INEX 2004. LNCS, vol. 3493, pp. 73–84. Springer, Heidelberg (2005)
4. Mass, Y., Mandelbrod, M.: Using the INEX Environment as a Test Bed for various User Models for XML Retrieval. In: Fuhr, N., Lalmas, M., Malik, S., Kazai, G. (eds.) INEX 2005. LNCS, vol. 3977, pp. 187–195. Springer, Heidelberg (2006)
5. Berchiche-Fellag, S., Mezghiche, M.: XML Element Retrieval using terms propagation. In: International Conference on Automation, Control, Engineering and Computer Science, ACECS 2014 (2014) (to be published)
6. Grabs, T., Scheck, H.J.: Flexible information retrieval from XML with Power DB XML. In: Proceedings of the First Annual Workshop of INEX, pp. 141–148 (December 2002)
7. Kakade, V., Raghavan, P.: Encoding XML in vector spaces. In: Losada, D.E., Fernández-Luna, J.M. (eds.) ECIR 2005. LNCS, vol. 3408, pp. 96–111. Springer, Heidelberg (2005)
8. Fuhr, N., Malik, S., Lalmas, M.: Overview of the initiative for the evaluation of XML retrieval (INEX) 2003. In: Proceedings of INEX 2003 Workshop, Dagstuhl, Germany (December 2003)
9. Ogilvie, P., Callan, J.: Using language models for flat text queries in XML retrieval. In: Proceedings of INEX 2003 Workshop, Dagstuhl, Germany, pp. 12–18 (December 2003)
10. Kamps, J., Rijke, M., Sigurbjornsson, B.: Length normalization in XML retrieval. In: Proceedings of SIGIR 2004, Sheffield, England, pp. 80–87 (2004)
11. Piwowarski, B., Faure, G.E., Gallinari, P.: Bayesian Networks and INEX. In: Proceeding in the First Annual Workshop for the Evaluation of Xml Retrieval, INEX (2002)
12. Anh, V.N., Moffat, A.: Compression and an IR approach to XML Retrieval. In: INEX 2002 Workshop Proceedings, Germany, pp. 100–104 (2002)
13. Sauvagnat, K.: Modèle flexible pour la recherche d'information dans des corpus de documents semi-structurés. Thèse Doctorat, Université Paul Sabatier de Toulouse (2005)
14. Fuhr, N., Grossjohann, K.: XIRQL, a query language for information retrieval in XML documents. In: Proceedings of SIGIR 2001, Toronto, Canada (2001)

15. Gövert, N., Abolhassanni, M., Fuhr, N., Grossjohann, K.: Content-Oriented XML Retrieval with HyreX. In: INEX 2002 Workshop Proceedings, Germany, pp. 26–32 (2002)
16. Cui, H., Wen, J.-R., Chua, J.-R.: Hierarchical indexing and flexible element retrieval for structured document (April 2003)
17. Ben Aouicha, M.: Une approche algébrique pour la recherche d'information structurée. Thèse de doctorat en informatique, Université Paul Sabatier, Toulouse (2009)
18. Berchiche-Fellag, S., Boughanem, M.: Traitement des requêtes CO (Content Only) sur un corpus de documents XML. In: Colloque sur l'Optimisation et les Systèmes d'Information (2010)
19. Denoyer, L., Gallinari, P.: The Wikipedia XML corpus. SIGIR Forum 40(1), 64–69 (2006)

AI Platform for Building University Research Knowledge Base[*]

Jakub Koperwas, Łukasz Skonieczny, Marek Kozłowski, Piotr Andruszkiewicz,
Henryk Rybiński, and Wacław Struk

Institute of Computer Science, Warsaw University of Technology, Nowowiejska 15/19,
00-665 Warszawa, Poland
{J.Koperwas,L.Skonieczny,M.Kozlowski,P.Andruszkiewicz,
H.Rybinski,W.Struk}@ii.pw.edu.pl

Abstract. This paper is devoted to the 3-years research performed at Warsaw
University of Technology, aimed at building of an advanced software for university research knowledge base. As a result, a text mining platform has been
built, enabling research in the areas of text mining and semantic information retrieval. In the paper some of the implemented methods are tested from the point
of view of their applicability in a real life system.

Keywords: digital library, artificial intelligence, knowledge base, scientific resources, repository.

1   Introduction

The last decade have shown an increased interest of the universities in the systems
concerning research data management and access to publicly funded research data. In
2010, a dedicated project, SYNAT, has been launched in order to address deficiencies
of scientific information infrastructure in Poland. The main SYNAT construction is
based on three levels of distributed knowledge bases, with a central database at the
highest level, and at the lower levels the domain oriented ones, as well as, the university ones. The ultimate goal of the knowledge base network is to ensure the dissemi-
nation of the Polish nation-wide scientific achievements and to improve integration
and communication of the scientific community, while leveraging existing infrastructure assets and distributed resources. In this paper we focus on research performed for
implementing the university level.

At the beginning, the university level system was assumed to be an institutional repository. There are many ready-to-use software solutions for building institutional
scientific information platforms, most of them have functionality well suited to the

---

[*] This work was supported by the National Centre for Research and Development (NCBiR)
under Grant No. SP/I/1/77065/10 devoted to the Strategic scientific research and experimental development program: "Interdisciplinary System for Interactive Scientific and Scientific-
Technical Information".

T. Andreasen et al. (Eds.): ISMIS 2014, LNAI 8502, pp. 405–414, 2014.
© Springer International Publishing Switzerland 2014

repository needs (like e.g. Fedora Commons, or DSpace, see e.g. [4]). However, having reconsidered the university needs, our goals have been expanded towards building around the repository an institutional knowledge base, concerning various types of research activities, e.g. publications, patents, supervised theses, participation in committees or projects.

In [7] we have presented an architecture of the knowledge base software, whereas in [8] we have shown extended functionality and some solutions of the knowledge base. In this paper we summarize our experience of research that was focused on the methods that can be practically applied within the knowledge base, reducing human effort in acquiring data, preparing them, and improving information retrieval. The research has been performed on a platform that has been implemented within the project, called Ω-Ψ^R.

The paper is organized as follows. In Section 2 we present a general architecture of the Ω-Ψ^R platform. In Section 3 we present some of the implemented tools, and show results of the performed experiments. Section 4 concludes the paper and presents our further plans.

2   Research Platform Ω-Ψ^R – General Architecture

As described in [7] the starting requirements for the university knowledge base were rather typical, focused on the repository functions. The main aim of the repository was to build institutional publication repository services, based on the open access idea to the most possible extent.

It has soon turned out that the solution being implemented must go far beyond a simple repository, mainly in order to make its crucial functionalities existent and usable. The key problem that occurred first, was to fill repository with data.

To this end, much effort was spent on assuring satisfactory starting point for filling the repository with the historical contents. Besides migrating legacy databases of various university units, the functionality that allows to complete the data directly by the authors themselves (crowdsourcing) was implemented. Nevertheless, expecting still a high incompleteness level of the database, it was decided that the tools for data acquisition from web should have been implemented. To this end, a specialized data acquisition module has been designed and implemented. It was given the name Ψ^R (for Platform for Scientific Information Retrieval). It is now integrated within the Knowledge Base system (Ω system), and it builds a complete platform, named Ω-Ψ^R.

The architecture of Ω-Ψ^R is presented on Fig 1. The need for using AI tools has arisen again when designing functionalities that are capable of utilizing collected repository to build a knowledge base for the overall university research activities. The key example is building author and faculty profile, that consists not only of simple historical statistics but also presents the most relevant research areas, allows finding and ranking of experts or teams that best conform to defined criteria, like best scientists in a given area, best PhD supervisor, etc. This need led to the introduction of knowledge base-focused data mining modules within Ω software that consist of semantic indexing, publication classifier and other algorithms.

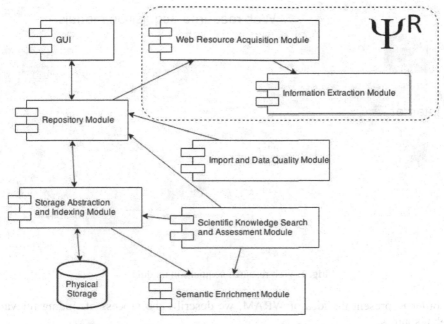

Fig. 1. The architecture of Ω-Ψ^R

3   Intelligent Tools in Ω-Ψ^R

3.1   Data Acquisition

There are many public information sources on the Internet that can be used to enrich University Research Knowledge Base. These sources are specific because they can be changed, removed, or appear at any time. Moreover, these sources could be unstructured or semi-structured. The examples are home pages of conferences that usually change for each separate event, e.g., in consecutive years. Furthermore, home pages, even these for the same conference in different years, could be different and have no defined structure. In order to find information sources on the Internet, we created Ψ^R (Platform for Scientific Information Retrieval), which is composed of Web Resource Acquisition Module (WRAM), and Information Extraction Module (IEM). Fig. 1 illustrates a general architecture of the integrated Ω-Ψ^R system.

The objective of WRAM   is to acquire addresses of web resources containing scientific information. The module has been implemented as a multi-agent system. As depicted in Fig. 2, it is divided into the following main sub-modules:

— Information searching, harvesting and information brokering sub-modules;
— Search definition and strategies sub-module;
— Task Agents sub-module;
— Resource Type Classification sub-module.

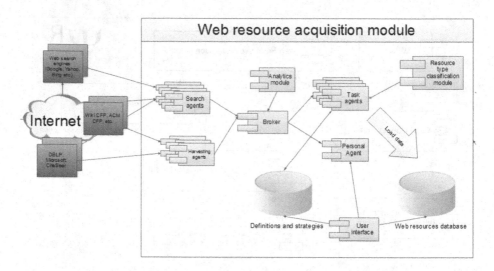

Fig. 1. Web resource acquisition module

In order to present the idea of WRAM, we describe the process of finding relevant web resources.

Strategies and definitions contain information on what should the module search for (type of object, e.g. *conferences, universities*), and how often. For instance, the definition contains information that the system should search for home pages of conferences, the strategy specifies how often the search should be performed, e.g., once a week.

Task agent is invoked by Strategy agent in order to execute a new search, according to a received task definition. It delivers a search query to Broker agent. The Broker agent executes the search (connects to Search agents and then to particular data sources). In the next step, it aggregates responses. Then Task agent receives the results (URLs of web resources) and transfers them to the Classifier module for a classification process. The most appropriate classifier is used to decide whether a given URL is of a desired type, e.g. *conference home page*. In the end, resources (URLs) are inserted into the web resources database, together with meta-information about classification and definition of a current search. Opposite to Task agent, Personal agent is designed to trigger on-demand searches that the user wants to perform.

We distinguish a variety of resources in the system, *inter alia person, university, conference, publisher*, etc. A query sent to Search agents is defined in terms of a key-value string, e.g., "conference: ICAART; year:2013"; then, depending on the data source interface, it will be converted to the most suitable query for a given data source (i.e., Google, Yahoo, Bing). The name or the short name of a conference is fetched from the repository (from the list of conferences), and having completed the process of searching for additional information, a given conference record is enriched with found information.

The main idea behind the classification module is to put all the tasks related to classification in one place. The module supplies simple interfaces for agents and classifiers implementations. Classifier implementation is generally not considered as a part of the module and can be developed separately as a brand new or library wrapper. The main function of the module is a website classification. Its interface is simple and consists of variants of invoking classifiers. In the simplest case, only a website URL and resource type are needed. There is no need for an outer system to care for algorithm selection and configuration. However, it is possible to indicate an algorithm and configuration parameters for a given object type. In the classification module we used SVM and Naive Bayes classifiers. However, the classification module is designed for easy incorporation of other classifiers to be used in the system, the ones existing in a library (and utilized as library wrappers), or the ones developed.

3.2 Information Extraction

Having web resources gathered, information extraction is used to enrich University Research Knowledge Base. The Information extraction module, which performs the task related to enrichment/extraction of attributes of objects defined in Ω-Ψ^R. For the specific types of objects very specific web resources should be analyzed. For instance, for acquiring information about researchers, the home pages of researchers can be used as information sources. In addition, introducing pages (web pages that introduce researchers) can be utilized as well. Information published on the page should be extracted and stored in the system. In Ψ^R we implemented information extraction tasks for researchers, conferences, and journals pages. For example, for researchers we extract the following information: position, affiliation, address, email, phone, fax, date, university and subject of bachelor, master and Ph.D. thesis, etc. The following methods were implemented: (1) regular expressions; (2) generalized sequence patterns in information extraction [1]; (3) SVM; (4) conditional random fields (linear and hierarchical); and (5) Markov logic networks.

Researchers' profiles can be also enriched by information extraction performed on web pages containing lists of workers associated with a given organization, especially when web pages are well structured. We implemented the solution for information extraction from structured web pages.

Research Papers Extraction
We use the Zotero software [17] to extract information about research papers from resources containing bibliographic entries. Original Zotero was developed as a browser extension, so it was not straightforward to build the application on top of it, as the module expects interacting with the user. Zotero is a free and open-source Firefox extension for managing researchers' bibliography. We implemented the Zotero-based web server and use it to extract metadata from websites containing bibliographic entries and convert into BibTex. Acquired BibTex entries are then imported into the system which merges and integrates every element with existing repository data, e.g. detects and links publication authors.

Person Name Disambiguation

In the case of extracting information about researchers, their activities, their publications, one of the crucial problems is the name disambiguation. Usually, we have a set of publications with a given text representing author's name and a set of researches (also with text strings representing their names) and we would like to assign each publication to one of the researchers. The problem is complicated because usually there are various forms of researchers' names, e.g., a name can be written with the first and last name of a researcher (with or without middle name), or with initials for the first name, with or without initials for the middle name. Often may happen that the authors' names are from the dictionary of first names, and family names are mixed up with first names (and then replaced by initials). Last but not least, for popular names very often happens that the various persons hold the same pairs *(first_name, family_name)*.

To deal with the disambiguation problem, we have implemented an approach based on first and middle names or initials matching. Having a set of publications of researchers with the same last name, we try to match first and middle names or initials and create groups of publications which authors' first and middle names match. We group publications in that way to obtain publications of one researcher in each group.

The proposed algorithm is used to perform basic name disambiguation. Moreover, it indicates publications that cannot be placed in the same group; that is, cannot have the same author or are highly probably written by the same author. The relations discovered by the proposed algorithm are used as attributes in other algorithms for the name disambiguation.

Having publications grouped, as a next step we perform clustering for each group separately. We have elaborated two clustering methods. The first method is to cluster publication based on their content, which reflects the subject of the publications. The idea behind this method is that a researcher usually publishes papers on a specific subject. This method enables the system to distinguish publications of authors with the same first and last names.

The second method is to cluster publications based on relations between them. As a starting point, we use the relations proposed in [2], namely co-authorship, citations, extended co-authorship, similarity of the titles, and users' restrictions. Each relation has a weight used in similarity calculation of publications. In order to choose the relation weights, we proposed a genetic algorithm. Both methods work complementary, and enable the system to assign a publication to the proper researcher.

3.3   Semantic Processing

Semantic processing aims at enriching the acquired objects by adding semantically meaningful descriptions and labels in order to improve information retrieval, as well as for discovering research areas of the researchers, and groups. The processing is performed on the repository documents (publications, theses, patents, etc). Two special semantic resources are used for this purpose Ontology for Scientific Journal [15] (in the sequel OSJ), and the Wikipedia resources. The process consists of:

1. Classifying the publications by assigning OSJ categories (domains, fields, sub-fields) to the publications;
2. Indexing with keywords – extracting from the documents the semantically meaningful descriptors;
3. Sense indexing – inducing senses of a given term and labeling text with them.

The target goals of this process are: (1) the retrieved OSJ publication categories are mainly used for building maps of research areas for individual researchers, and then, propagating the researchers interest to the affiliation-related university units; the steps (2) and (3) are crucial for enriching texts with semantic labels, which are also used for building researcher interest vectors (visualized in the form of word clouds), but mainly they are used for improving the search parameters, such as precision and recall. Below we describe the three modules in more details.

Publication Classifier

Scientific domain classification is the task consisting in providing a publication with one or more relevant tags, assigning the publication to one or more scientific classes. In Ω-Ψ^R we have decided to use the OSJ ontology as a classification schema. OSJ is a three level hierarchy, ended with the leaves on the last (fourth) level, which are simply scientific journal titles, so the path in OSJ from the root to a leaf (i.e., a journal title) assigns domain tags to the papers from the journal. The levels in the OSJ hierarchy are respectively domain, field, and subfield. Clearly, OSJ can be used straightforward for assigning tags to all the papers published in the journals that are contained in the OSJ list. The problem appears for the publications out of the OSJ journal lists, as well as theses, publications being conference papers, chapters in the books, etc. To this end, we have designed and implemented Bayesian classifier's model, which was trained on the OSJ papers. So, the science domain classifier works as follows for each document:

1. If the document is a paper from the OSJ list, take the tags assigned by OSJ to the journal;
2. Otherwise, use the model of bayesian classifiers on the available metadata, preferably including title, keywords and abstract, and use the result OSJ categories to classify the document.

We verified two solutions: one classifier for all the OSJ fields, or a tree of specific classifiers, each node representing a "specialized" classifier. The experiments have shown that the solution with the tree of "specialized" classifiers outperforms one common classifier. The tree of classifiers is a hierarchical structure with the depth 2, where each node represents a specialized classifier. The root is a classifier for the first OSJ level, its children are composed of 6 classifiers at level 2 (for each OSJ domain there is one fields classifier built). An average accuracy (10-fold cross validation) in a tree mode has reached 85%.

Keywords Extraction

Keywords extraction plays a crucial role in enhancing the intelligence of web, mainly by means of enterprise search. Nowadays, most of the semantic resources cover only specific domains. Bearing in mind that the whole University research do-main cannot be covered by one specific domain ontology, we have decided to apply Wikipedia (Polish and English) as a semantic knowledge resource and implement Wikipedia-based semantic indexing of documents in the $\Omega \Psi^R$ system.

Since a few years, both Wikipedia and DBpedia are used in many areas involving natural language processing, in particular for information retrieval and information extraction (see e.g. [5, 6, 11]). In our project we use Wikipedia in two ways – term oriented and text oriented. The first one is a module providing semantic information about an analyzed term. It is inspired by Milne, Medelyan, Witten [12], and Milne, Witten [13]. We process data in two steps. First, given a term extracted from the processed document, the module searches for an article with the title equal to or at least containing the term. Then the found article is processed in order to extract its labels, senses, translation, first paragraphs (summaries). Our contribution to this approach is that, while indexing the documents by the extracted keywords we additionally tag the keywords with a meaning, discovered by the SenseSearcher algorithm (SNS) [9], briefly described below.

Sense Indexing

For word sense induction we have used SnS (Sense Searcher) [9], which is based on closed frequent termsets. It provides as a result a tree of senses, and represents each sense as a context. The key feature of SnS is that it finds infrequent and dominated senses. It can be used on the fly by the end-user systems, and the results can be used to tag the keywords with the appropriate senses.

SnS consists of five phases. In Phase I, the index is built (i.e: full-text search index) using provided set of documents. In Phase II, we send a query with a given term to the index, and retrieve elements (paragraphs/snippets), which describe the mentioned term. Then the paragraphs/snippets are converted into a context representation (bag-of-words representation). In Phase III, significant contextual patterns are discovered in the contexts generated in the previous step. The Contextual patterns are closed frequent termsets occurring in the context space. In Phase IV, the contextual patterns are formed into sense frames, which build a hierarchical structure of senses. Finally, in Phase V sense frames are clustered in order to merge the similar frames referring to the same meaning. Clustered sense frames are represent senses.

An extensive set of experiments performed in [9] confirms that SnS provides significant improvements over existing methods be means of sense consistency, hierarchical representation, and readability. We tested SnS as a web search result clustering WSI-based algorithm. These experiments aimed at comparing SnS with the other WSI algorithms within the 2013 SemEval Task no 11 "Word Sense Induction within and End-User Applications" [14].

The clustering evaluation problem is a difficult issue, for which there is no unequivocal solution. Many evaluation measures have been proposed in the literature so, in order to get comparable results, we calculated four distinct measures: Rand Index (RI), Adjusted Rand Index (ARI), Jaccard Index (JI), and the F1 measure [9].

Table 1. Results for Rand Index (RI), Adjjusted RI (ARI), Jaccard Index (JI) and F1 in SemEval 2013 task 11

Type	System	RI	ARI	JI	F1
WSI	HDP-CLUSTERS-LEMMA	**65.22**	21.31	33.02	**68.30**
	HDP-CLUSTERS-NOLEMMA	64.86	**21.49**	33.75	68.03
	SATTY-APPROACH1	59.55	7.19	15.05	67.09
	DULUTH.SYS9.PK2	54.63	2.59	22.24	57.02
	DULUTH.SYS1.PK2	52.18	5.74	31.79	56.83
	DULUTH.SYS7.PK2	52.04	6.78	31.03	58.78
	UKP-WSI-WP-LLR2	51.09	3.77	31.77	58.64
	UKP-WSI-WP-PMI	50.50	3.64	**29.32**	60.48
SNS	SNS	**65.84**	**22.19**	**34.26**	**70.16**
WSD	RAKESH	58.76	8.11	30.52	39.49
BL	Singletons	60.09	0.00	0.00	100.00
	All-in-one	39.90	0.00	39.90	54.42

The SemEval 2013 task 11 is measured by a diversified number of indicators: RI, ARI, JI, F1. We show the results for those four measures in Table 1. As one can see, SnS outperforms the best systems that took part in SemEval - HDP based methods. The SnS-based system reports considerably higher values in RI and ARI. It achieves significantly better results in terms of F1. In the case of JI the best values of SnS and UKP-WSI-WACKY-LLR are similar. Generally, SnS obtains the best results in all measures. To get more insights into the performance of the various systems, we calculated the average number of clusters and the average cluster size per clustering produced by each system, and compared it with the gold standard average. The best performing system in the case of all above mentioned categories has clustering size and clusters size similar to the gold standard.

4   Conclusions and Future Work

One of the lessons learned was that with building an information system, for the first glance looking as a fairly typical one, we have encountered many interesting real life research problems in such areas like knowledge acquisition and discovery, text mining, or information retrieval.

Although the research that was briefly described in this paper has been in most cases applied in the working knowledge base, it is far from being completed. While the practical goals of the SYNAT project have been achieved, i.e. the Ω-Ψ^R platform has been successfully implemented at WUT as the university knowledge base, and is subject of implementing at other universities in Poland, we will be extending the research part of the platform. The already built repository of scientific publications, mostly in English, is quite heterogeneous in terms of the covered research areas, and as such, it provides a lot of challenges. Special emphasis will be put on semantic cross-lingual search, giving rise to a more symmetric retrieval for English and Polish, i.e., giving similar results for queries regardless of the language. In addition, some web mining tools aimed at discovering knowledge about journals and conferences are still under way.

References

1. Hazan, R., Andruszkiewicz, P.: Home Pages Identification and Information Extraction in Researcher Profiling. In: Bembenik, R., et al. (eds.) Intelligent Tools for Building a Scientific Information Platform: Advanced Architectures and Solutions, pp. 41–51 (2013)
2. Tang, J., Yao, L., Zhang, D., Zhang, J.: A combination approach to web user profiling. ACM Transactions on Knowledge Discovery from Data, TKDD 5(1), 2 (2010)
3. Bembenik, R., et al. (eds.): Intelligent Tools for Building a Scientific Information Platform. SCI, vol. 390. Springer, Heidelberg (2012)
4. Berman, F.: Got Data? A Guide to Data Preservation in the Information Age. CACM 51(12) (2008)
5. Gabrilowich, E., Markovitch, S.: Overcoming the brittleness bottleneck using Wikipedia: Enhancing text categorization with encyclopedic knowledge. AAAI (2006)
6. Gabrilowich, E., Markovitch, S.: Wikipedia-based semantic interpretation for natural language processing. Journal of Artificial Intelligence Research 34, 443–498 (2009)
7. Koperwas, J., Skonieczny, Ł., Rybiński, H., Struk, W.: Development of a University Knowledge Base. In: Bembenik, R., Skonieczny, Ł., Rybiński, H., Kryszkiewicz, M., Niezgódka, M. (eds.) Intell. Tools for Building a Scientific Information. SCI, vol. 467, pp. 97–110. Springer, Heidelberg (2013)
8. Koperwas, J., Skonieczny, Ł., Kozłowski, M., Rybiński, H., Struk, W.: University Knowledge Base – Two Years of Experience. In: Bembenik, R., Skonieczny, Ł., Rybin´ski, H., Kryszkiewicz, M., Niezgódka, M. (eds.) Intelligent Tools for Building a Scientific Information Platform - From Research to Implementation. SCI, vol. 541, pp. 257–274. Springer, Heidelberg (2014)
9. Kozłowski, M.: Word sense discovery using frequent termsets, PhD Thesis, Warsaw University of Technology (2014)
10. Di Marco, A., Navigli, R.: Clustering and Diversifying Web Search Results with Graph-Based Word Sense Induction, Computational Linguistics, vol. 39(3), pp. 709–754. MIT Press (2013)
11. Medelyan, O., Milne, D., Legg, C., Witten Ian, H.: Mining meaning from Wikipedia. Int'l. J. Hum.-Comput. Stud. 67(9), 716–754 (2009)
12. Milne, D., Medelyan, O., Witten, I.H.: Mining domain-specific thesauri from Wikipedia: A case study. In: IEEE/WIC/ACM International Conference on Web Intelligence, Hong Kong, China, pp. 442–448 (2006)
13. Milne, D., Witten, I.H.: An effective, low-cost measure of semantic relatedness obtained from Wikipedia links. In: Wikipedia and Artificial Intelligence: An Evolving Synergy, Chicago, IL, pp. 25–30 (2008)
14. Navigli, R., Vannella, D.: SemEval-2013 Task 11: Word Sense Induction & Disambiguation within an End-User Applications. In: Proc. of 7th Int'l Workshop on Semantic Evaluation, 2nd Joint Conf. on Lexical and Computational Semantics, pp. 193–201 (2013)
15. Ontology of Scientific Journal, classification of scientific journals, http://www.science-metrix.com/eng/tools.htm
16. Omelczuk, A., Andruszkiewicz, P.: Agent-based Web Resource Retrieval System for Scientific Knowledge Base (2013)
17. Zotero, http://www.zotero.org/

A Seed Based Method for Dictionary Translation*

Robert Krajewski, Henryk Rybiński, and Marek Kozłowski

Warsaw University of Technology
Warsaw, Poland
{R.Krajewski,H.Rybinski,M.Kozlowski}@ii.pw.edu.pl
http://www.ii.pw.edu.pl

Abstract. The paper refers to the topic of automatic machine translation. The proposed method enables translating a dictionary by means of mining repositories in the source and target repository, without any directly given relationships connecting two languages. It consists of two stages: (1) translation by lexical similarity, where words are compared graphically, and (2) translation by semantic similarity, where contexts are compared. Polish and English version of Wikipedia were used as multilingual corpora. The method and its stages are thoroughly analyzed. The results allow implementing this method in human-in-the-middle systems.

Keywords: Machine translation, dictionary translation, semantic similarity, multilingual corpus.

1 Introduction

Knowledge resources, such as thesauri, taxonomies, and recently ontologies are of high importance in the applications of nowadays information technologies, especially for information retrieval, information extraction, or knowledge discovery methods. Especially with the enormous development of Web the role of knowledge resources has increased drastically. However, one of the real barriers in wide use of them is multilinguality of information resources - according to [1], about 65 % of the Internet is a non-English content.

In the context of information retrieval the attempts towards solving the multilinguality problems go back to the Salton's works in early 70-ties of the previous century (see e.g. [2]). Since then, several methods of using multilingual dictionaries for improving information retrieval in multilingual text databases have been developed ([3,4,5]). On the other hand, a lot of time and effort has been invested to build and maintain multilingual thesauri and/or flat dictionaries, to be used for enhancing information retrieval in multilingual databases. Many of

---

* This work was supported by the National Centre for Research and Development (NCBiR) under Grant No. SP/I/1/77065/10 devoted to the Strategic scientific research and experimental development program: "Interdisciplinary System for Interactive Scientific and Scientific-Technical Information".

T. Andreasen et al. (Eds.): ISMIS 2014, LNAI 8502, pp. 415–424, 2014.
© Springer International Publishing Switzerland 2014

them (e.g. Eurovoc, GEMET, Agrovoc, INIS thesaurus) are multilingual and domain oriented. Their translation possibilities are very limited, restricted to the concepts (main descriptors in the thesaurus languages), nevertheless they are extensively used for information retrieval in the international, usually multilingual, databases.

In order to cope with the multilinguality problems, multilingual components of knowledge resources become of highest importance. Although recently novel methods based on Wikipedia cross-language links have been presented [6], it seems that specialized multilingual domain-oriented knowledge resources will still play important role.

In particular, the main important research areas, where the translation quality between languages has essential importance is the field of multilingual information retrieval (MLIR) [1] , where the main objective is to perform search within a multilingual set of documents and collect relevant multilingual documents.

In this paper we present an approach for translating a dictionary based on mining subject-similar repositories in the source and target language. We call this method Seed Based Dictionary Builder (SBDB). It works without any explicitly predefined relationships between the source and target languages. Instead, it uses two raw text repositories in the source and target languages respectively. The paper is organized as follows: In Section 2 we discuss related work. Section 4 presents the algorithm. Then we provide experimental results of the method in Section 4. Section 5 concerns future works and the final sections concludes the paper.

2   Related Work

Automatic translation from one language to another, better known as Machine Translation, is a long-standing aim of AI. Usually, two approaches to the MT problems are distinguished: knowledge-rich or knowledge poor. The first ones have a language knowledge embedded within the algorithms processing texts, and/or are based on using advanced predefined knowledge bases (ontologies, thesauri, semantic dictionaries, etc.). Opposite to the knowledge rich approaches, the latter ones do not use semantic knowledge bases that are difficult and costly to build, and the algorithms used for text processing do not have embedded deep language dependent knowledge. In the research, several directions can be distinguished, but the main goal to reduce the human involvement in the translation process, and speed up this process, remains unchanged.

As a matter of fact, there are many domain-oriented multilingual thesauri (such as Agrovoc, or Eurovoc), which could be used for multilingual information retrieval ([8]). They are rich in concepts specific for the domains they are used for, however, there are serious limitations e.g: the high cost of maintenance, insufficient lexical granularity, lack of polysemy.

---

[1] Actually two problems are distinguished, namely a more general one, known as multilingual information retrieval (MLIR), or more specific one, cross-lingual information retrieval (CLIR).

Statistical machine translation (SMT) treats the translation of natural language as a machine learning problem. By examining many samples of human-produced translation, SMT algorithms automatically learn how to translate. SMT has made tremendous strides in less than two decades, and new ideas are constantly introduced. One of the first systems of this kind was IBM Model 1, presented in [9]. It was based on the EM algorithm and provided word-to-word translation. Then the approach has been extended by [10] for word-to-phrase translating. The idea has been also developed by [12] for word-to-phrase translation, and with the use of the Hidden Markov Models. Based on the features of IBM Model 4, Deng [12] improves essentially the translation quality by using the HMM models. Phrase-based statistical machine translation [10] has emerged as the dominant paradigm in machine translation research. It is based on heuristic learning of phrase translations from word-based alignments and lexical weighting of phrase translations. In [11], an open source toolkit Mosese, inspired by Koehn previous works, has been proposed.

An interesting approach, called CL-ESA (*Cross Lingual - Explicit Semantic Analysis*), has been presented in [6]. The approach is an extension of Explicit Semantic Analysis (ESA), the idea presented in [14]. Both, ESA and CL-ESA can be definitely classified as knowledge rich, as they are based on extensive use of Wikipedia as a provider of semantic knowledge (ESA) and structured cross-lingual relationships between Wikipedia articles. Given a document, CL-ESA uses a document-aligned cross-lingual reference collection in Wikipedia to represent the document as a language-independent concept vector (usually expressed by tf-idf). The human-generated translations provided in Wikipedia (inter-language links) are also exploited by BabelNet [15].

Other semantic support in translating the dictionaries can be taken from existing ontologies or thesauri. A method combining statistical and semantic approaches for translating thesauri and ontologies is presented in [16]. Unfortunately, these methods have limitations. Namely, they cannot be used when the semantic resources (Wikipedia, thesauri or ontologies) are missing or insufficient.

The idea of lexicon translation with nonparallel corpora has been presented in [17]. The presented method is based on the observation that if the words a and b collocate often in one language then their translations should collocate in the repository in the target language too. Additionally, for large corpora the frequencies of the collocations should be similar. However conceptually correct, this solution turns to be computationally very expensive. There is a strong need for initial seed inventory. Unfortunately, the problem of how to obtain a starting seed lexicon is not considered in the paper.

An interesting solution for obtaining the seed lexicon has been proposed by [19]. Namely the authors postulate building the seed translation dictionary from the source and target repositories, based on the existence of some words in both languages in the same form, or similar in terms of spelling of the corresponding source and target words. It seems therefore to be straightforward to apply the similarities for building seed dictionaries not only for the languages belonging to the same family.

Our approach for building bilingual dictionaries has been influenced by [18] and [19]. Also in our approach we build a seed dictionary within the first phase of building bilingual dictionary, but opposite to [19], we show that knowledge poor data mining methods can be used successfully even for the languages belonging to various families (English and Polish). Having built the seed, we also build vectors for the source terms, and translate them with the seed dictionary, but the way the vectors are constructed is different. In the sections 3 and 4 we describe algorithm and some preliminary experiments.

3   The SBDB Method

In our approach, we focus on automating the process of seed lexicon building. The proposed method uses knowledge-poor text mining algorithms. In the sequel we denote by R_S and R_T the monolingual source and target repositories, and by D_S and D_T the dictionaries extracted from R_S and R_T respectively. The SBDB method consists of the following three phases:

1. First, for the repositories R_S and R_T the monolingual dictionaries D_S and D_T are extracted;
2. Given D_S and D_T we mine translation rules and build a seed dictionary;
3. R_S repository is used to build *aggregated context vectors* for each $s \in D_S$, additionally for R_T the context vectors are built for each $t \in D_T$; then, with the use of the seed dictionary we look for the most similar translate candidates.

The first phase (building dictionaries) is a standard procedure, where we reject stop words and select only specific parts of speech.

For the second phase the idea is that the algorithm compares words from two dictionaries source and target, and builds the similarity translation function (bijection) $\gamma : D'_S \rightarrow D'_T$, $D'_S \subseteq D_S$, and $D'_T \subseteq D_T$ by using an edit distance measure. Additionally, during the comparison process, the algorithm mines the transformation rules that are specific for the source and target languages, S and T respectively. The rules are dynamically added to the set of rules, which are then reused in the continued comparison process. We call this phase *syntactic translation*, as the pairs are identified by syntactic similarities. As one can notice this phase does not use any extra language-dependent knowledge about the rules for transcription from the source language to the target one.

In Step 3, the results of the previous steps are used for building the final translation function $\Gamma : D_S \rightarrow 2^{D_T}$. In order to reach this goal, for each $s \in D_S$ and $t \in D_T$ aggregated context vectors are built. Now, the semantic relatedness can be measured with the seed dictionary. Namely, the target terms vectors are translated, and the vectors are limited only to the seed translations. Having this done, for each $s \in D_S$ the k semantically closest target language vectors are identified. This phase is called *semantic translation*, as it mines contexts in the source and target languages and looks for the semantic similarities.

The conceptual diagram of the whole process is presented in Figure 1. In the consecutive subsections the particular phases of the algorithm are discussed in detail.

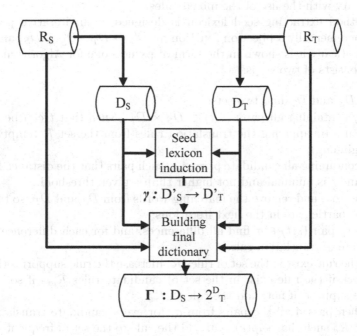

Fig. 1. Conceptual diagram of the method

3.1   Lexical Similarity

The idea of using lexical similarity is based on the assumption that some words exist in a similar form in the source and target languages. Linguists indicate various reasons of similarities between the dictionaries of various languages. For the languages belonging to the same group, to large extent the common roots in the past decide on the similarities. Additionally, an influence of one language to another one may result from various factors, such as, e.g. technology transfer, cultural influence, etc. This may also refer to the languages belonging to various groups, like e.g. German and Polish.

The words having similar form (and meaning) are named by linguists cognates, and it turns out that they appear quite often across the modern languages.

To measure similarity between words an edit distance computation algorithm can be employed. We will use a slightly modified Levenshtein edit distance measure. Usually the distance measure between the terms p and q is defined as follows: $d(p,q) = n/max\{l(p), l(q)\}$, where n is the number of changes (in characters), which have to be performed to achieve q from p, and $l(t)$ is the length (in characters) of term t.

The proposed algorithm differs from the ones known from the literature that along with the process of looking for similar words it mines additionally the rules. So, if a given type of differences between similar words appears often enough, the algorithm adds new translation rules to the set of rules and continues building the dictionary with the use of the mined rules.

The method extracting seed lexicon is designed as an iterative process of building incrementally a bijection function $\gamma: D'_S \to D'_T$, $D'_S \subseteq D_S$, and $D'_T \subseteq D_T$. Schematically it is shown in the form of pseudocode for Algorithm 1. Each iteration consists of two steps:

1. Given D_S and D_T find two sets:
   (a) $\Delta\gamma$ containing all pairs $(s,t) \in D_S \times D_T$, such that t can be reached from s by applying the translation rules from the set $\mathcal{R}$ (empty at the beginning);
   (b) σ containing all candidate pairs, i.e. such pairs that the distance between s and t is minimal and not higher than a given threshold.
   Add $\Delta\gamma$ to γ and remove the matching words from D_S and D_T, so that they will not participate in the next iterations;
2. For every pair $(s,t) \in \sigma$ find all differences, and for each difference build a candidate of translation rule.
   (a) If the rule exist in the set of rules $\mathcal{R}$, increase the rule support, otherwise, check if the rule exist in the set of candidate rules $\mathcal{R}_c$ - if so, increase the support, if not, add the rule to $\mathcal{R}_c$.
   (b) When passed all the pairs from σ, for every candidate translation rule $r \in \mathcal{R}_c$ such that $sup(r) > \delta$, add the rule to the set of frequent rules $\mathcal{R}$.

First, let us define the notion of transformation rule r. The rule has the form $x \to y$, $x, y \in \Sigma_S^*$, and $y \in \Sigma_T^*$, where Σ_S, and $y \in \Sigma_T$ is the alphabet of the languages S and T [2].

Given a set of rules ρ, and the terms p and q we say that p can be transformed to q with ρ iff every rule $x \to y$ from ρ is applied at least once in substituting the substring x of p into y, and the final term obtained is q. For this we write $p \xrightarrow{\rho} q$, and say that ρ is the edit difference set of rules for the pair (p,q). In order to indicate that ρ refers to the pair (p,q), whenever needed we will write ρ_{pq} For example we say that

1. *orthography* can be transformed to *ortografia* with the set of rules $\rho = \{th \to t, ph \to f\}$
2. *comic* can be transformed to *komik* with the set of rules $\rho = \{c \to k\}$

We call the rule r frequent if in the process of translating from D_S to D_T, it is applied more than ϵ times, where ϵ is a user-predefined threshold, i.e. $sup(r) > \epsilon$. At i-th iteration of computing the translation function γ, the set $\mathcal{R}$ of frequent

---

[2] We presume that a large part of the alphabets of the source and target languages is shared by both languages, which is the case when e.g. both languages are based on the Latin alphabet.

transformation rules is available. Now, the translation procedure uses a modified edit distance measure between terms t and s. Actually, we define the distance measure as a function of terms t, s, and a set of rules $\mathcal{R}$. Given a pair of terms (s,t) and the edit difference set of rules ρ for this pair, we define the distance between s and t as $d_{\mathcal{R}}(s,t) = d(s,t) - ||\rho \cap \mathcal{R}||$, where $d(s,t)$ is the standard Levenshtein's measure, and $||\rho \cap \mathcal{R}|| = |\rho \cap \mathcal{R}|/max\{l(t), l(s)\}$.

Given $s \in D_S$ and $\mathcal{R}$, the translation procedure looks for $t \in D_T$ such that $d_{\mathcal{R}}(s,t) = 0$, i.e. $\rho_{st} \subseteq \mathcal{R}$. If there are more possible translations of s than one, the procedure selects t_0, which maximizes the total support of ρ in $\mathcal{R}$, i.e:

$$\forall t \in D_T \sum_{r \in \rho_{st}} sup(r) \leq \sum_{r \in \rho_{st_0}} sup(r) \qquad (1)$$

As we can see, for each $s \in D_S$ we look for $t_0 \in D_T$ such that the distance $d_{\mathcal{R}}(s,t_0) = 0$, and the condition (3) is satisfied for t_0. If t_0 is found, the pair (s,t_0) is added to the result function γ and t and s_0 are removed from D_S and D_T respectively, which guarantees that γ is a bijection.

As a result, we receive a seed dictionary in a form of the translation function γ. In the next subsection we will present how with a seed dictionary the semantic translation can be performed.

3.2   Semantic Similarity

The second phase of the method performs an analysis based on semantic similarity, and extends the previously created seed. The algorithm is based on a context found out for each term extracted from the source and destination repositories, R_S and R_T respectively. In order to find semantically similar terms we construct vectors for the terms in D_S and D_T reflecting similar relationships against the seed dictionary γ. A general idea is that given contexts (termsets) for all the terms $x \in D_S$ we limit them only to those termsets that contain $t \in Dom(\gamma)$ and replace t by $\gamma(t)$ in these termsets. Then limiting the termsets only to the terms from $Dom(\gamma)$ we can build for each x the aggregated context vectors, where the i-th position is the frequency of t_i in all contexts of x. In a similar way we proceed with the terms $y \in D_T$, limiting the aggregated context vectors to the terms $\gamma(t), t \in Dom(\gamma)$. Now, having vectors for the source and target languages, we perform a trimming procedure reducing the vector space. The trimming is based on the k top frequent terms for each vector.

Finally, for each x we search for the closest y by means of the cosine pseudo-metrics (justified by the fact that R_s and R_t are different in size), and obtain the translation candidates.

4   Experiments

The experiments have been performed for English (source) and Polish (target). We used two corpora, built from the English and Polish Wikipedia respectively.

The wikilinks between the articles, descriptive labels and interlingual links have been ignored.

Referring to the first phase (lexical similarity), the achieved results outperformed the method based on manual definition of rules in [19], in spite of the fact that the source and target languages are in different language families. The lexical similarity experiments are summarized on Fig. 2 and 3. Fig. 2 shows the most frequent examples of the rules induced from D_S and D_T. We have selected 5000 most frequent words from each corpus. As one can see (Fig. 3), from these starting dictionaries, 1507 translations have been properly detected, giving the precision 66,9% (Fig. 3.a). The precision can be increased to 78,1% by eliminating the rules $s \to \varnothing$, reducing slightly the recall (Fig. 3.b).

regua			ilo wystpie
c	→	k	213
l	→	∅	103
e	→	∅	69
v	→	w	59
y	→	ia	38
ph	→	f	35
∅	→	ny	30
all	→	zn	30
tion	→	cja	28
s	→	z	22

Fig. 2. Top frequent rules generated by lexical similarity translation

Fig. 3. Precision and recall of lexical similarity translation

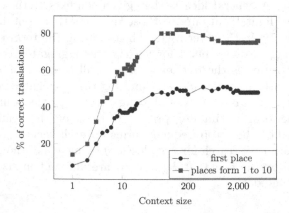

Fig. 4. Dependency between the context size and number of correctly translated terms

The experiments concerning semantic similarity were performed on the whole corpora, based on the obtained seed. The results are shown on Fig. 4, where the dependency between the dimensionality of the aggregated vector space and the number of proper translations is illustrated. The square-dotted line shows the cases where the proper translation appears within top 10 candidates. The circle-dotted line shows the number of properly translated words, i.e. they are on the first position among the top ten candidate translations. One can also see from the figure that for finding the proper translations on the first position the optimal vector size is about 400, whereas for finding the proper translation in the top 10 candidates is about 120.

5 Conclusions and Future Work

In this work we proposed a novel approach to identify word translations from non-parallel or even unrelated texts. This task is quite difficult, because most of the statistical information, which is useful in the processing of parallel texts, cannot be applied in this case. The novel element introduced in SBDB is the phase of inducing lexicographical rules of translations. As shown, the proposed method works well even for the languages from different language families. We also proved that trimming of aggregated context improves computational efficiency without significant loss of precision. Most of well-known evaluations [18], [19] were performed on small datasets (100-1000 words) extracted from a radio news corpus, or some existing lexicons. The achieved results vary from 40 to 72 % of correctness. Opposite to that, we performed evaluation on all words from wikipedia articles. Our method reports 50% of correct translations found as the first candidates, and 80% among the top 10 candidates.

The proposed technique can be essentially improved by incorporating word sense induction. Differentiations of meanings could be applied within the semantic similarity translation algorithm to produce more accurate translations. To take into account the ambiguity problem, first we need to know possible meanings of particular terms. Having large and raw textual corpora, unsupervised approach to discover meanings can be applied. We plan using the methods described in [20] and in [21].

References

1. Pimienta, D., Prado, D., Blanco, A.: Twelve years of measuring linguistic diversity in the Internet: Balance and perspectives. United Nations Educational, Scientific and Cultural Organization (2009)
2. Salton, G.: Automatic processing of foreign language documents. Journal of the American Society for Information Science 21(3) (1970)
3. Hull, D., Grefenstette, G.: Querying across languages: a dictionary-based approach to multilingual information retrieval. In: Proceedings of the 19th Annual International ACM SIGIR Conference on Research and Development in Information Retrieval. ACM (1996)
4. Ballesteros, L., Croft, W.: QPhrasal translation and query expansion techniques for crosslanguage information retrieval. In: ACM SIGIR Forum, vol. 31. ACM (1997)

5. Pirkola, A.: The effects of query structure and dictionary setups in dictionary-based cross-language information retrieval. In: Proceedings of the 21st Annual International ACM SIGIR Conference on Research and Development in Information Retrieval. ACM (1998)
6. Sorg, P., Cimiano, P.: Cross-lingual information retrieval with explicit semantic analysis. In: Working Notes for the CLEF Workshop (2008)
7. Sorg, P., Cimiano, P.: Enriching the crosslingual link structure of wikipedia classification-based approach. In: Proceedings of the AAAI 2008 Workshop on Wikipedia and Artifical Intelligence (2008)
8. Soergel, D.: Multilingual thesauri in cross-language text and speech retrieval. In: AAAI Symposium on Cross-Language Text and Speech Retrieval (1997)
9. Brown, P., et al.: A statistical approach to machine translation. Computational linguistics 16(2) (1990)
10. Koehn, P., Och, F., Marcu, D.: Statistical phrase-based translation. In: Proceedings of the 2003 Conference of the North American Chapter of the Association for Computational Linguistics on Human Language Technology, vol. 1. Association for Computational Linguistics (2003)
11. Koehn, P., et al.: Moses: Open source toolkit for statistical machine translation. In: Proceedings of the Annual Meeting of the Association for Computational Linguistics (2007)
12. Deng, Y., Byrne, W.: Hmm word and phrase alignment for statistical machine translation. IEEE Transactions Audio, Speech, and Language Processing (2008)
13. Dumais, S., Letsche, T., Littman, M., Landauer, T.: Automatic cross-language retrieval using latent semantic indexing. In: AAAI Spring Symposium on Cross-Language Text and Speech Retrieval (1997)
14. Gabrilovich, E., Markovitch, S.: Computing semantic relatedness using wikipedia-based explicit semantic analysis. In: IJCAI (2007)
15. Navigli, R., Ponzetto, S.: BabelNet: Building a very large multilingual semantic network. In: 48th Annual Meeting of the Association for Computational Linguistics, Uppsala, Sweden (2010)
16. McCrae, J., Espinoza, M., Montiel-Ponsoda, E., Aguado de Cea, G., Cimiano, P.: Combining statistical and semantic approaches to the translation of ontologies and taxonomies. In: Proceedings of the Fifth Workshop on Syntax, Structure and Semantics in Statistical Translation, Uppsala, Sweden (2010)
17. Rapp, R.: Identifying word translations in non-parallel texts. In: Proceedings of the 33rd Annual Meeting on Association for Computational Linguistics. Association for Computational Linguistics (1995)
18. Rapp, R.: Automatic identification of word translations from unrelated english and german corpora. In: Proceedings of the 37th Annual Meeting of the Association for Computationald Linguistics on Computational Linguistics. Association for Computational Linguistics (1999)
19. Koehn, P., Knight, K.: Learning a translation lexicon from monolingual corpora. In: Proceedings of the ACL 2002 Workshop on Unsupervised Lexical Acquisition, vol. 9. Association for Computational Linguistics (2002)
20. Rybiński, H., Kryszkiewicz, M., Protaziuk, G., Kontkiewicz, A., Marcinkowska, K., Delteil, A.: Discovering Word Meanings Based on Frequent Termsets. In: Raś, Z.W., Tsumoto, S., Zighed, D.A. (eds.) MCD 2007. LNCS (LNAI), vol. 4944, pp. 82–92. Springer, Heidelberg (2008)
21. Kozlowski, M.: Word sense discovery using frequent termsets. PhD Thesis, Warsaw University of Technology (2014)

SAUText — A System for Analysis of Unstructured Textual Data*

Grzegorz Protaziuk, Jacek Lewandowski, and Robert Bembenik

Institute of Computer Science, Warsaw University of Technology,
Nowowiejska 15/19, 00-665 Warsaw, Poland
{G.Protaziuk,J.Lewandowski,R.Bembenik}@ii.pw.edu.pl

Abstract. Nowadays semantic lexical resources, like ontologies, are becoming increasingly important in many systems, in particular those providing access to structured textual data. Typically such resources are built based on already existing repositories and by analyzing available texts. In practice, however, building new or enriching existing resources of such type cannot be accomplished without using an appropriate tool. In this paper we present SAUText – a new system which provides infrastructure for carrying out research involving usage of semantic resources and analyzing unstructured textual data. In the system we use dedicated repository for storing various kinds of text data and take advantage of parallelization in order to speed up the analysis.

Keywords: Text mining, text analysis system, ontology enrichment.

1 Introduction

Text mining and ontology building/enriching are related tasks. Text mining is the discovery and extraction of interesting, non-trivial knowledge from free or unstructured text [4]. The techniques used in the mining process are essentially the same as techniques used in the mining of non-textual data. The difference lies in the preparation of unstructured textual resources and their representation, so that well-known data mining techniques can be used [10]. The preparation step typically consists in converting all processed documents to a standard format, tokenization, lemmatization, stemming, dictionary generation, sentence boundary determination, part of speech tagging, word sense disambiguation, phrase recognition, named entity recognition, and feature generation. There exists a plethora of text mining algorithms and methods developed to date.

Ontology formally represents knowledge as a set of concepts within a do-main, using a shared vocabulary to denote the types, properties and interrelationships of those concepts. Specialized tools which support the task of ontology engineering are necessary to reduce the costs associated with the engineering and

* This work is supported by the National Centre for Research and Development (NCBiR) under Grant No. SP/I/1/77065/10 by the Strategic scientific research and experimental development program: Interdisciplinary System for Interactive Scientific and Scientific-Technical Information.

T. Andreasen et al. (Eds.): ISMIS 2014, LNAI 8502, pp. 425–434, 2014.
© Springer International Publishing Switzerland 2014

maintenance of ontologies [2]. As data in various forms, especially textual, is massively available, many methods have been developed in order to support ontology engineering basing on text mining, among other approaches. These data-driven techniques of ontology engineering are known as ontology learning. Noteworthy approaches to learning ontologies include TextToOnto [5], Text2Onto [3], OntoLearn [9], OntoLT [1], SPRAT [6], and OntoUSP [7].

In the following we present SAUText – a system thought as a platform capable of enriching ontologies basing on a repository of documents using text mining methods. The system will ultimately be composed of text mining as well as ontology enriching parts. In this paper we present the part of the system that is responsible for knowledge discovery from textual data that can further be used in ontology enriching.

The rest of the paper is organized as follows. Chapter 2 contains assumptions concerning the SAUText system and its architecture. Chapter 3 describes the key solutions applied in the system, whereas chapter 4 presents tests of the system. Finally, chapter 5 contains concluding remarks.

2   The SAUText System

SAUText - a System for Analysis of Unstructured Textual Data is thought as an environment allowing a single user, as well as a team, to carry out research in the widely understood area of discovering knowledge from textual data focusing on text mining and generating candidate entries to ontologies using text repositories. In our opinion such platform should have the following properties:

- provide libraries of NLP and text mining methods,
- allow users for sharing configurations of experiments and their results, especially those obtained by applying NLP methods,
- use open architecture facilitating easiness of development, especially the possibility of adding new components being implementations of NLP or text mining algorithms at the platform configuration level rather than at the source code level.

In a system designed with carrying out multiple, oftentimes repetitive experiments in mind efficiency is a very important factor. It seems that efficiency depends foremost on the following three factors: (i) algorithms, (ii) implementation quality, and (iii) level of parallelism. As the first two factors are difficult to control in an open architecture, in the design of SAUText we focused on the parallelization of knowledge discovery processes.

2.1   Architecture

The general idea of the system is presented in Fig. 1. The system consists of:

- Client application - a client program which enables a user to specify parameters of a knowledge discovery process and obtain its results.

- Server application – a service providing access to core functionality of the system responsible for managing users and their profiles. An auxiliary database is used for keeping data required for tasks assigned to a server.
- Ontologies – access to ontologies is realized with the use of an external component also responsible for storing ontologies, implemented foremost in OWL language.
- Knowledge discovery process manager – the main component responsible for creating and executing knowledge discovery process (KD process).
- Ontology manager - a library providing access to ontologies stored in the Ontologies module to be used with NLP library and text mining methods.
- NLP library – a module offering various NLP methods.
- Text mining methods library – a library of methods for analyzing text documents focused on widely understood text mining algorithms.
- Text document repository – an autonomous module offering functionality concerning storing text documents and products obtained from processing those documents using NLP methods. It also provides possibility of searching for data of interest (documents or products).

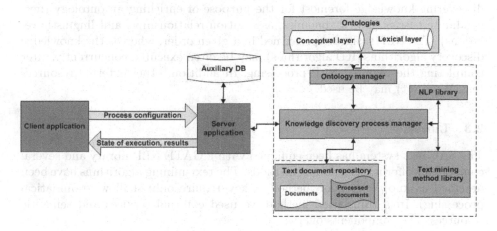

Fig. 1. SAUText system architecture

2.2 Knowledge Discovery Process

Knowledge discovery process (KD process) can be seen as a sequence of steps which should be performed in a given order. These steps are: collecting input data, preprocessing and performing the actual analysis. The important thing is that in each step (apart from the last) some data is generated which constitutes input for the next step. In SAUText we applied such a model of knowledge discovery process. It is illustrated in Fig. 2.

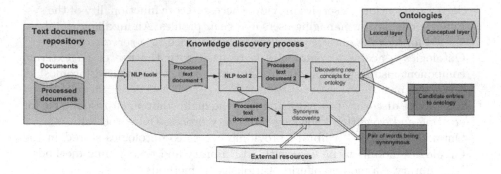

Fig. 2. Schema of the knowledge discovery process

Input data for this process consists of a set of text documents (original or pre-processed), whereas the output depends on algorithms applied during a given realization of the process. For example, it may be a set of candidate entries to an ontology or a set of pairs of words being synonyms. In general, the process is built of several steps, which are composed of NLP tasks and algorithms discovering knowledge foremost for the purpose of enriching an ontology (new candidate entries, new taxonomies, association relationships, and linguistic relations). The NLP tasks are performed in a given order, whereas the knowledge discovery algorithms (KD algorithms) may be often executed concurrently, after completing the required NLP processing. In addition, other external resources (e.g. WordNet) may be used.

2.3 Libraries

The SAUText system has been equipped with a GATE NLP library and several components offering text-mining methods. The text-mining algorithms have been designed as knowledge-poor methods (they require only shallow grammatical processing). In a translation method we used external services and semantic resources. The components offer:

- T-GPS algorithm for discovering non-taxonomical relations and proper nouns. This is the new version of an algorithm introduced in [8] and implemented in the multithread manner in Scala language.
- Synonyms identification. The new method is based on association rules and disjoint contexts of single words.
- The translation method of domain vocabulary. In the introduced method machine translation services are used for obtaining candidate translations and a domain document repository is used for verification and providing translations taken from natural texts, also for multiword expressions.

3   Implementation of the SAUText System

In this section we describe the selected solutions applied in the design and implementation of the SAUText system, which concern the main components of the system and are important in understanding the way the system is built.

3.1   Asynchronous Processing

In the applied general model of a knowledge discovery system a given step is executed after the previous step has been completed. However, in many cases the entire input data is not necessary for starting execution of a consecutive step. For example, NLP processing required in a KD process can often be started after receiving a single text document from a repository; there is no need to wait for all documents. In order to execute the steps in a parallel fashion the transfer of partial data between consecutive steps should be provided. Asynchronous communication seems to be a very good solution to that problem. In the system we developed, asynchronous and autonomous queues are widely used for exchanging data between components used to implement a KD process as well as between components and the repository.

The queue works in its own thread. It requires a producer to be defined and a fixed number of outputs to be connected to it. The aim of the queue is to maintain its internal buffer so that elements can be retrieved without any lags, and with the use of limited amount of memory. This goal is achieved by keeping the buffer filling ratio between fixed bounds. When the ratio falls below the lower bound, the queue thread starts to call the producer continuously to fetch new elements. This process is paused when the buffer filling ratio reaches the upper bound. The queue remains alive until the producer has more elements to provide. The producer informs the queue about being exhausted by returning no elements.

In order to retrieve elements from the queue, one or more individual outputs have to be created. Each output works independently (has its own buffer) and acts like a plain blocking queue – one has to call a blocking method to get the next element. When the speeds of producing and consuming elements are balanced, the queue keeps calling its producer all the time and the data flows fluently from the input to the output.

The queues, are used to connect processing units (which will be described later), so that they form a processing graph. The processing units can be fed by the queues, while the queues can supply transferred elements to other processing units. The execution is performed by pulling elements from the global outputs (we assume the graph is directed and acyclic, so that the roots are the system inputs, whereas the leaves are the system outputs). Once we want an element to be processed, we *pull* it from the system output and this pull request is propagated down the graph.

3.2 Text Repository

The text repository is a main data store for SAUText. The main purpose of this component is to manage all documents which were loaded into or generated by the system. At first glance, it serves as a plain, indexed documents store. However, it also provides the ability to manage the 'made-of' relation between stored objects. Therefore the text repository contains a set of document trees, each of which is rooted in the object that represents a document loaded into the system from the outside.

The text repository provides rich search capabilities. Since the indexing is handled by Solr server[1], it is possible to use all Solr specific search operators. Nevertheless, we designed a higher abstraction layer over the plain Solr REST API. We implemented special classes to represent document-related data, as well as the new search API.

Client API. The model available for a client comprises two main entities - Document and DocumentProduct. Document entity represents roots of document trees in the repository, that is, documents loaded into the system from the outside. On the other hand, DocumentProduct entity represents all the results of the processing. As far as it is possible to run multiple transformations on the same DocumentProduct, a number of derived objects can be related to it. This is the way a single document tree is created. Documents and products differ in one more detail - for documents, we are enabled to upload the original binary file and store it along with the document, whereas for document products we may attach additional information about the transformation process. The process identifier and execution parameters can be used as additional search constrains then.

The client API supports creating new document trees, adding and removing document products, modifying metadata associated with each object and searching with various criteria. All these operations can be performed with the use of buffered streams, which were employed as the main communication mechanism in the SAUText system.

We also designed a DSL with the use of Scala, which allows to construct queries to the repository. Queries may refer to fields in the aforementioned entities and may express join clauses between documents and document products. Each query can be serialized into and deserialized from an XML node. It facilitates making the query a part of the process definition.

Internal Architecture. Internally, the text repository uses Solr server and Cassandra database[2]. Solr is responsible for providing the indexing functionality, whereas Cassandra stores all the data. The whole infrastructure is based on Akka[3] actors in order to provide horizontal scalability and required isolation of

---

[1] Apache Solr, https://lucene.apache.org/solr/
[2] Apache Cassandra http://cassandra.apache.org/
[3] Akka http://akka.io/

operations. This is achieved by partitioning the workload into several entities - actors. The workload is partitioned according to the hashes of document keys.

Once a query is received by the text repository, it is translated into the form understandable by Solr and executed. Solr returns a collection of document identifiers, and a new buffered stream is initiated between the repository and SAUText. The stream is based on the object producer which takes an identifier and retrieves the corresponding object from the database. It is done when the client (SAUText) requests the objects to process. By loading the documents lazily, we achieved a few advantages:

- we do not waste memory on the server side - we do not have to keep documents buffered before sending them
- we do not waste memory on SAUText side - we do not have to keep documents buffered before processing them
- processing of document can be started once the first document arrives - we do not have to wait for all the results

We also eliminated lags that would occur when subsequent documents are requested. In a simple scenario, when we ask the repository for the next document, the repository loads the document from a database and sends it back to a client. The client has to wait for these procedure to finish before it can continue. We addressed this problem by attaching autonomous buffers on both ends of the stream. The buffer on the server side is responsible for preloading some number of documents before the are requested, whereas the buffer on the client side is responsible for pre-fetching some number of documents from the server. Ultimately, while some processing units are working, the subsequent documents are automatically loaded from the database and fetched by the client so that they can be immediately retrieved and processed.

3.3   Components of a KD Process

One of the aspects which facilitates development of the SAUText system is a well-defined, simple interface which is used for steering the execution of a discovery process. The hierarchy of those interfaces is presented in Fig. 3.

In the design we distinguished objects which generate input data for other elements (transformation objects) of a KD process from objects providing

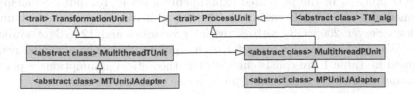

Fig. 3. Interface hierarchy for elements of a knowledge discovery process

knowledge discovery methods (processing objects). The interface includes parameterized Scala traits (*ProcessUnit* and *TransformationUnit*) for defining basic functions of objects that implement a step in a KD process and abstract classes (*MultithreadPUnit*, *MultithreadTUnit*) delivering a frame for multithread processing or transformation of input data. Algorithms in which input data cannot be divided into elements and processed in separate threads may be implemented by using the *TM_alg* abstract class. These classes provide functionality required by the system such as not-blocking execution of the main function or steering of an execution. In the interface functions for controlling level of parallelism are also provided. The interface introduces functions helpful for logging or efficiency analysis and provides adapters for objects implemented in the java language (*MPUnitJAdapter*, *MTUnitJAdapter*).

3.4 Implementation and Execution of a KD Process

Knowledge discovery process manager module provides functionality for building a KD process from XML specification, carrying out that process and saving results of transformations of text documents. In the implementation of the applied model of the KD process we assumed that steps involving text mining methods (TM steps) occur in a process after the last NLP step. First, all runtime objects providing functionality for steps of a KD process are created based on an XML specification and by applying the reflection mechanism (which essentially facilitates using external libraries in the SAUText system). Each object is an autonomous element and communication between them is limited to data exchange. As we applied asynchronous queues for data exchange, all necessary producers and outputs for them are created before starting execution of a process. Each step is executed in its own thread as well as the procedure of saving data into the repository. Application of asynchronous queues with many outputs makes it possible to save data into the repository without suspending execution of the main flow of a process.

4   Tests

In this section we present a selection of experimental results concerning NLP processing tasks. Typically, a KD process includes NLP processing and many NLP tasks involve processing only one document at a time. Carrying out these tasks in a parallel manner may greatly shorten time needed for preprocessing in a KD process. In the presented experiments a set of 137 pdf documents of combined size of 64.4 MB was used. The experiments were carried out under Windows Server 2008 OS, with 8 virtual processors and 12 GB of available RAM. We repeated tests 3 times for each configuration. The obtained results presented in Table 1 are consistent with the theoretical assumptions – parallel processing allowed for significant gain in terms of execution time.

Table 1. NLP processing time depending on concurrent number of threads

Number of threads	Avg. execution time	% of single-threaded
execution time		
1	7966 s	100%
2	5582 s	70%
4	5018 s	63%
7	4901 s	62%

5 Conclusions

The SAUText system briefly presented in this paper is still under development. While the core of the system has been implemented and the tests gave legitimacy to the applied solutions focusing on multithreaded processing, the number of text-mining methods available in the system as well as its ontology enriching part are still not entirely satisfactory. Therefore we are planning to continue development of the SAUText system and carrying out research focusing on building ontologies from text repositories and available semantic text resources.

References

1. Buitelaar, P., Olejnik, D., Sintek, M.: A protege plug-in for ontology extraction from text based on linguistic analysis. In: Proceedings of the 1st European Semantic Web Symposium (ESWS), Heraklion, Greece (2004)
2. Cimiano, P., Mdche, A., Staab, S., Völker, J.: Ontology learning. In: Staab, S., Studer, R. (eds.) Handbook on Ontologies, International Handbooks on Information Systems, pp. 245–267. Springer, Heidelberg (2009)
3. Cimiano, P., Völker, J.: Text2onto - a framework for ontology learning and data-driven change discovery. In: Montoyo, A., Muñoz, R., Métais, E. (eds.) NLDB 2005. LNCS, vol. 3513, pp. 227–238. Springer, Heidelberg (2005)
4. Kao, A., Poteet, S.R.: Natural Language Processing and Text Mining. Springer (2007)
5. Maedche, A., Volz, R.: The Text-To-Onto Ontology Extraction and Maintenance System. In: Workshop on Integrating Data Mining and Knowledge Management Co-Located with the 1st International Conference on Data Mining, San Jose, California, USA (November 2001)
6. Maynard, D., Funk, A., Peters, W.: Sprat: a tool for automatic semantic pattern-based ontology population. In: International Conference for Digital Libraries and Semantic Web (2009)
7. Poon, H., Domingos, P.: Unsupervised ontology induction from text. In: Hajic, J., Carberry, S., Clark, S. (eds.) ACL, pp. 296–305. The Association for Computer Linguistics (2010)
8. Protaziuk, G., Kryszkiewicz, M., Rybiński, H., Delteil, A.: Discovering compound and proper nouns. In: Kryszkiewicz, M., Peters, J.F., Rybiński, H., Skowron, A. (eds.) RSEISP 2007. LNCS (LNAI), vol. 4585, pp. 505–515. Springer, Heidelberg (2007)

9. Velardi, P., Navigli, R., Cucchiarelli, A., Neri, F.: Evaluation of OntoLearn, a methodology for automatic population of domain ontologies. In: Buitelaar, P., Cimiano, P., Magnini, B. (eds.) Ontology Learning from Text: Methods, Applications and Evaluation. IOS Press (2006)
10. Weiss, S.M., Indurkhya, N., Zhang, T.: Fundamentals of Predictive Text Mining. Texts in Computer Science. Springer (2010)

Evaluation of Path Based Methods
for Conceptual Representation of the Text

Łukasz Kucharczyk and Julian Szymański

Department of Computer Systems Architecture,
Gdańsk University of Technology, Poland
{lukasz.kucharczyk,julian.szymanski}@eti.pg.gda.pl

Abstract. Typical text clustering methods use the *bag of words* (BoW) representation to describe content of documents. However, this method is known to have several limitations. Employing Wikipedia as the lexical knowledge base has shown an improvement of the text representation for data-mining purposes. Promising extensions of that trend employ hierarchical organization of Wikipedia category system. In this paper we propose three path-based measures for calculating document relatedness in such conceptual space and compare them with the *Path Length* widely used approach. We perform their evaluation using the OPTICS clustering algorithm for categorization of keyword-based search results. The results have shown that our method outperforms the Path-Length approach.

Keywords: text representation, documents categorization, information retrieval.

1 Introduction

One of the most important factors for achieving good results in data mining tasks is the selection of appropriate features as a document representation. Widely used approach is based on the *bag of words* (**BoW**) method, where each document is represented as a normalized vector of weighted term frequencies [1]. This method has a number of known drawbacks, specifically, it ignores important relations between terms. It also operates on very high dimensional spaces, which make accurate computation of distances difficult due to the *curse of dimensionality* [2]. Recent research addresses these problems by enriching or replacing the standard *bag of words* features with *conceptual-based* lexical information, some of the most promising using Wikipedia for that purpose [3].

Some of these solutions successfully employ Wikipedia Category Graph (**WCG**) as a source of additional text features, although they take relatively little advantage of categories hierarchical organization [3]. It seems as a promising direction to build an abstract document representation based on WCG hierarchical concepts. Intuitively, the WCG feature space should perform better than the standard BoW, because it already introduces a basic form of concept classification.

Creation of such representation is not an easy task and requires successful implementation of at least two major steps: a) automatic tagging of arbitrary documents with Wikipedia categories, preferably through deployment of a multi-label classifier; b) efficient extraction of relations between categories from Wikipedia Category Graph. This paper describes the first part of our research, which focuses on the latter of these tasks.

T. Andreasen et al. (Eds.): ISMIS 2014, LNAI 8502, pp. 435–444, 2014.
© Springer International Publishing Switzerland 2014

Our aim is to ultimately replace the standard BoW with the conceptual representation, derived from Wikipedia category system. As the first step towards this direction, we investigate usefulness of *path based* measures for clustering a priori tagged documents[1]. We propose several methods for computation of relatedness in such conceptual space, and compare them with the standard *Path Length* measure, which has been already shown to be successfully adapted for Wikipedia Category Graph [4]. We expect that clustering should especially benefit from usage of representation based on WCG, so we employ the OPTICS algorithm for evaluation of our approaches.

2   Related Works

Previous works that employ WCG traverse usually a small part of category hierarchy and use relatively simple weighting schemes. Both [5] and [6] traverse the hierarchy only up to the depth of three levels, while others use only categories directly related to articles [7][8]. Gabrilovich has completely resigned from the usage of Wikipedia categories after unsatisfying results obtained with Open Directory Project (which is also a hierarchical category system) [9]. He points out that hierarchical organization of data violates the orthogonality requirement, which is crucial for computing concepts relatedness [10], and proposes for extending representation the *Explicit Semantic Analysis* method [10]. His observations seem to be in term with those of other researchers, who noticed that limiting WCG search depth improves received results and helps to avoid various anomalies [11] [5] [6]. Through the depth limit, they reduce the size of processed hierarchy and thus limit the impact of the orthogonality problem.

On the other hand, Zesch has shown that Wikipedia Category Graph shares many important properties with other semantic networks like WordNet [4] or Roget's thesaurus [4]. He has also adopted classical *relatedness measures*[2] to WCG and concluded, that WCG can be effectively used for natural language processing tasks [4]. Zesch also compares efficiency of WCG with that of the GermanNet and finds out, that Wikipedia outperforms GermaNet in regard to computing *semantic relatedness*. Strube performs similar experiments on WordNet and also gets promising results [11].

Moreover, despite mentioned difficulties, researchers who employed limited usage of WCG reported consistent improvement in their results. Specifically [7], [6], [8] and [13] adopted Wikipedia categories for enhancing document clustering, while [14][15] used them for cross-lingual and multilingual information retrieval.

Those various controversies regarding WCG point several important directions for future studies. Although Gabrilovich did not perform any empirical evaluation of WCG to support his claims (and admits that Wikipedia possesses much less noise than ODP [10]) his observation regarding orthogonality seems valid enough to be taken into consideration. Because the orthogonality problem applies mainly to the standard *cosine* (or

---

[1] For evaluation purposes we use documents which are a priori tagged with Wikipedia categories. However, for general applicability of described methods, introduction of classifier is necessary.

[2] *Semantic similarity* is typically defined via the lexical relations of synonymy and hypernymy, while *semantic relatedness* is defined to cover any kind of lexical or functional association that may exists between two words [12].

euclidean) measure, we decided to analyze a more graph oriented approach presented by Zesch.

After the analysis of several relatedness measures, Zesch established that *path based* ones are the most appropriate for Wikipedia Category Graph [4] [16]. Thus, in this paper, we propose three different methods for improving those measures and evaluate their usability for document clustering.

3 Path Based Relatedness Measures

In order to perform clustering of documents represented as WCG concepts, we need to establish a method for computing semantic relatedness. Path based techniques define relatedness as distance between nodes in concept graph [4].

In the next subsection, we first adapt path based measures to document clustering. To achieve this, we introduce a method for easy computation of path lengths between any two documents, so that we can avoid traversal of entire concept graph for each distance calculation.

Then we proceed to the main topic of this paper. Firstly, we define our baseline method, and describe observations we made during our attempts to employ it for clustering purposes. Then, we present our proposed methods of computing relatedness, which we developed in an attempt to overcome weaknesses of the baseline.

Adaptation of Path Based Measures for Clustering. Given the set of articles $A = \{a_0, a_1, ..., a_N\}$, for each article a_i we perform *Breadth First Search* of its associated category branches. For each traversed category c we assign weight equal to its distance to article a_i:

$$w_{PL}(a_i, c_j) = length(a_i, c_j) \tag{1}$$

Then *Path Length* between any two documents can be easily calculated using the following procedure:

1. For any article a_i, its *document vector* contains distances to all of its connected categories. Categories which are not connected have their distance represented as *infinity*:

$$V_i = V(a_i) = \{w_{PL}(a_1, c_1), ..., w_{PL}(a_1, c_K)\} \tag{2}$$

2. For any two *document vectors* V_1 and V_2, calculate the *distance vector* V_d as their sum:

$$V_d(a_1, a_2) = V_1 + V_2 = \{V_d[k] : 1 \leq k \leq K; V_d[k] = V_1[k] + V_2[k]\}$$
$$V_d(a_1, a_2) = \{w_{PL}(a_1, c_1) + w_{PL}(a_2, c_1), ..., w_{PL}(a_1, c_K) + w_{PL}(a_2, c_K)\} \tag{3}$$

3. We notice that for any two *document vectors* V_1, V_2 and their *distance vector* V_d:

(a) if $V_1[i]$ is the length of the shortest path $p_{(a_1,c_i)} = \{a_1, ..., c_i\}$ connecting article a_1 and category c_i:

$$V_1[i] = |p_{(a_1,c_i)}| = |\{a_1, ..., c_i\}| \tag{4}$$

(b) and if $V_2[i]$ is the length of the shortest path $p_{(a_2,c_i)} = \{a_2, ..., c_i\}$ connecting article a_2 and category c_i:

$$V_2[i] = |p_{(a_2,c_i)}| = |\{a_2, ..., c_i\}| \tag{5}$$

(c) then $V_d[i]$ is the length of the shortest path $p_{(a_1,c_i,a_2)} = \{a_1, ..., c_i, ..., a_2\}$ connecting articles a_1 and a_2 through the common category[3] c_i:

$$\begin{aligned} V_d[i] &= V_1[i] + V_2[i] \\ &= |\{a_1, ..., c_i\}| + |\{a_2, ..., c_i\}| \\ &= |\{a_1, ..., c_i, ..., a_2\}| \end{aligned} \tag{6}$$

(d) then the *distance vector* V_d contains distances for possible paths $p = \{a_1, ..., c_i, ..., a_2\}$ between articles a_1 and a_2:

$$V_d = \{|p_{(a_1,c_1,a_2)}|, ..., |p_{(a_1,c_K,a_2)}|\} \tag{7}$$

4. Using the above observation, we can find the minimum distance between documents a_1 and a_2 by taking the minimum value of the *distance vector* V_d:

$$dist_{PL}(a_1, a_2) = \min_{1 \le k \le K} V_d[k] \tag{8}$$

3.1   Baseline Method: Path Length (PL)

Path Length is one of the standard WordNet relatedness measures adapted for Wikipedia Category Graph by Zesch, which has been also shown to be most successful[16][4]. Thus we employ this method as a *baseline* for comparison of our proposed approaches. The measure is defined as path length between two nodes (measured in number of edges):

$$dist_{PL}(c_i, c_j) = length(c_i, c_j) \tag{9}$$

Because each Wikipedia article can be assigned to multiple categories Zesch proposes two methods of aggregation [4]. We define C_1 and C_2 as the set of categories assigned to articles a_1 and a_2, respectively. We compute PL distance for each category

---

[3] Note that WCG is in fact a *directed graph* describing a generalization relation, so c_i is a *common ancestor* of articles a_1 and a_2. Given distances $w_{PL}(a_1, c_i)$, $w_{PL}(a_2, c_i)$ between articles and their common ancestor, Equation 6 calculates the total path length through c_i. Note it implies, that article nodes are only valid as start or end points for path $p_{(a_1,c_i,a_2)}$. All in between points must be category nodes.

pair (c_k, c_l) with $c_k \in C_1$ and $c_l \in C_2$. Then we use either the minimum value or average over all calculated pairs:

$$dist_{PL+Min}(a_1, a_2) = \min_{c_i \in C_1, c_j \in C_2} dist_{PL}(c_i, c_j) \tag{10}$$

$$dist_{PL+Avg}(a_1, a_2) = \frac{\sum_{c_i \in C_1} \sum_{c_j \in C_2} dist_{PL}(c_i, c_j)}{|C_1||C_2|} \tag{11}$$

Both adaptation schemes posses serious drawbacks, which make them unsuitable for document clustering. $PL + Min$ ignores all information provided by categories laying outside the shortest path. This way a lot of information about documents similarity (or dissimilarity) is lost. Let us consider three documents a_1, a_2, a_3 and their assigned categories $C_1 = \{History, Technology\}$, $C_2 = \{History, Technology\}$, $C_3 = \{History, Art\}$. In such setting $PL + Min$ would give results $dist(a_1, a_2) = dist(a_1, a_3) = dist(a_2, a_3)$ instead of $dist(a_1, a_2) < dist(a_1, a_3) = dist(a_2, a_3)$. If clustered document set contains many articles sharing a common category, $PL + Min$ will assign the same distance value to all of them, effectively preventing construction of meaningful clusters.

On the other hand, $PL + Avg$ suffers from quite an opposite weakness. Averaging over all possible category pairs gives excessive weight to redundant or unusual ones, which is especially problematic if documents posses different number of categories. Lets consider two documents a_1, a_2 and their assigned categories $C_1 = \{Technology, Engines, Hi-TechIndustry, AmericanCompany\}$, $C_2 = \{AmericanCompany\}$. Clearly both documents should be clustered together (assuming there are no other technical oriented articles), however their distance value will be artificially increased by the fact, that categories in C_1 are mostly redundant.

So each of these adaptation schemes pose its own problems when applied to the field of document clustering. They either discard too much information (PL+Min) or give excessive weight to redundant categories and noise (PL+Avg). In consequence, both become sensitive to data distribution within clustered collection, which makes extraction of clusters difficult.

3.2 Method 1: Semi-Average Path Length (PL+Avg*)

In this method we attempt to overcome the weakness of *Path Length* by introducing a new adaptation scheme. Instead of calculating average over all category pairs we apply the following procedure. Given documents a_1, a_2 and their assigned categories C_1, C_2 we calculate distances between each category and its *opposite document* i.e. for each category in C_1 we calculate its distance to document a_2, and for each category in C_2 we calculate distance to a_1. Then we calculate the average over all these distances:

$$dist_{PL+Avg*}(a_1, a_2) = \frac{\sum_{c_i \in C_1} dist_{PL}(c_i, a_2) + \sum_{c_j \in C_2} dist_{PL}(c_j, a_1)}{|C_1| + |C_2|} \tag{12}$$

The major difference between this method and $PL + Avg$ is that in this setting we are computing distances between category and article, so only several most sensible category pairs are calculated. This way redundant categories do impose much less penalty on final distance score.

3.3 Method 2: Semi-Average Path Length with Frequency Reduction (PL+Avg*+df)

This is an extension to the previous method $PL + Avg*$, where we additionally ignore nodes whose *Document Frequency*[1] value is below the given threshold. This way we expect to remove unnecessary noise in the same fashion as it is done for the *bag of words* representation [17].

During traversal of Category Graph, if node is found to be below the defined document frequency threshold, it is assigned an $infinite$ distance (it is effectively treated as unreachable). However, its descendants are normally processed with the only difference, that the cost of travel through reduced node is equal to zero. In effect, all descendants of the removed node are considered as being one level closer to its source document:

$$dist_{PL+df}(a, c) = dist_{PL}(a, c) - |R_{a,c}| \tag{13}$$

$$R_{a,c} = \{r : r \in p_{a,c} \wedge df(r) < THRESHOLD\} \tag{14}$$

where $p_{a,c} = \{a, ..., ..., c\}$ is a set of nodes lying on path between article a and category c, and $R_{a,c}$ is a subset of $p_{a,c}$ consisting of those nodes whose document frequency is below $df(c) < THRESHOLD$.

Then, we can compute $dist_{PL+Avg*+df}$ in the same way as defined in section 3.2. We simply substitute values of $dist_{PL}$ in Equation 12 with values of $dist_{PL+df}$.

3.4 Method 3: Minimum Weighted Path Length (PL+Min+Idf)

This method is similar to our baseline $PL+Min$ in that it uses only the shortest path for distance calculation. However $PL + Min + Idf$ does not use the simple *Breadth First Search* approach, but instead performs shortest path search with usage of the *Dijkstra* algorithm. In this method WCG is interpreted as a *weighted* graph where travel cost through node c_i is equal to the inverse of its *Idf* statistic [1]:

$$idf(c_i) = log \frac{N}{df(c_i)} \tag{15}$$

$$cost(c_i) = \frac{1}{idf(c_i)} \tag{16}$$

where N is a number of documents in clustered collection. This way travel through common terms should be more expensive and we expect it to generate greater differences in documents distance values.

4 Results and Evaluation

For evaluation of the proposed methods presented in section 3, we have used Wikipedia articles as a priori tagged document set. In our future research, we are going to employ a multi-label classifier to automatically assign Wikipedia categories, so clustering of arbitrary documents could be performed. At this stage, we use Wikipedia articles for testing purposes, as the usage of classifier could introduce additional noise to measurements.

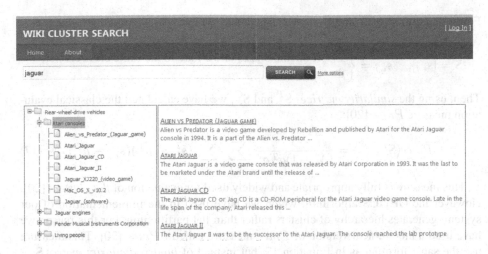

Fig. 1. User interface for WikiClusterSearch engine

To make our evaluation procedure more close to real world applications, we have implemented a clustering search engine (Figure 1), which groups keyword-based search results within Wikipedia in a similar fashion as Clusty[4] or Carrot[5] does for WebPages. For document clustering we employed the *OPTICS* [18] algorithm with parameters $MinPts = 3$ and $\epsilon = infinity$. Because original method for extracting clusters from *Reachability Plots* is very sensitive to input parameters [19] (and so unapprioriate for evaluation purposes), we have used the *Cluster Tree* algorithm [19], which in turn is essentially parameterless (its only input argument is strongly recommended to be set to a fixed value of 0.75 [19]).

4.1 Evaluation Procedure

For evaluation of the proposed methods we have decided to use *external validation measures* which are known to be more accurate than *internal* ones [20]. For this purpose we have manually prepared sets of reference groupings using the following procedure:

1. Keyword search has been performed to select a set of articles containing selected phrase
2. Returned results have been manually grouped into hierarchy of clusters. That hierarchy was then saved as reference partitioning $C^T = \{C_1^T, ..., C_{K^T}^T\}$ (we call C_k^T sets as *classes*, K^T is the number of classes). We call such manually prepared partitioning C^T as *Golden Standard* [1].
3. Returned results have been used again as an input for automatic clustering, using methods described in section 3. That hierarchy was then saved as partitioning $C^C = \{C_1^C, ..., C_{K^C}^C\}$ for comparison (we call C_i^C sets as *clusters*, K^C is the number of clusters)

---

[4] http://clusty.com/
[5] http://search.carrot2.org

For both partitions, C^T and C^C, we can calculate a binary $N x N$ *similarity matrix* $\mathbf{S}$:

$$\mathbf{S} = [s_{i,j}], \quad s_{i,j} = \begin{cases} 1 & \text{if documents } x_i \text{ and } x_j \text{ are in the same class/cluster,} \\ 0 & \text{otherwise.} \end{cases}$$

Then, using the *similarity matrices* $\mathbf{S}^T$ and $\mathbf{S}^C$, we have calculated the classical evaluation measure $Pmcc$ [20]:

$$Pmcc(\mathbf{S}^T, \mathbf{S}^C) = \frac{1}{(M-1)\sigma^T \sigma^C} \sum_{i=1}^{N-1} \sum_{j=i+1}^{N} (s_{i,j}^T - \mu^T)(s_{i,j}^C - \mu^C) \quad (17)$$

This measure is fully appropriate and widely used for evaluation of *flat clusters* [20]. However the OPTICS algorithm, which we have used in the implementation of our system, generates hierarchy of clusters rather than flat partitioning. For this reason we have used a hierarchical version of $Pmcc$ measure called $hPmcc$ [20]. This measure use the same formula as in Equation 17, but instead of *binary similarity matrix* $\mathbf{S}$, it uses *non-binary similarity matrix* $\mathbf{S}_{ca}^*$. Non-binary similarity matrices are described in detail in [20].

4.2 Results

We have used $PL + Min$ and $PL + Avg$ (described in section 3.1) as the two baselines for the evaluation, both of which have been already reported to provide good results for the computation of category relatedness [4] [16]. We have expected that modifications which we introduced to those measures should remove problems mentioned in section 3.1 and make them suitable for document clustering. Tests performed on several prepared *Golden Standards* show that one of our proposed methods, $PL + Avg*$, has achieved significantly better results than both baselines (Table 1).

It can also be seen that $PL + Min$ introduces the highest variance of the results, just as expected. Analysis of generated Reachability Plots has shown, that this method generates very well formed plots if data distribution is favorable (as can be seen for golden standard *Relation*), although it generally fails to capture significant inter-document differences and so extracts less clusters. $PL + Avg$ also behaves according to expectations-its Reachability Plots are significantly flattened, due to exceeding averaging of distance values .

Surprisingly, the $PL + Min + Idf$ method presents a very low score. We hoped that use of collection level information will allow to improve results, but in effect we received Reachability Plots with very shallow dents. High density of Wikipedia Category Graph turned out to provide many alternative routes for the traversal algorithm, so introduction of additional weights has made the distance score to approach its average value. $PL + Avg* + Df$ have also performed below expectations. We have assumed that removal of apparently noisy nodes from the concept graph should improve clustering results. Better scores in *Game* and *Kernel* golden standards partially confirm this assumption, but unfortunately this solution turned out to be not globally optimal. Deleted nodes proved often to hold additional information about distances and they removal had generally a negative impact on the performance of otherwise successful method $PL + Avg*$.

Table 1. Methods evaluation using hPMCC

	[PL+Min]	[PL+Avg]	[PL+Avg*]	[PL+Avg*+Df]	[PL+Min+Idf]
Class	0,43	0,42	**0,56**	0,47	0,52
Game	0,31	0,25	0,38	**0,48**	0,15
Jaguar	0,7	**0,73**	**0,73**	0,56	0,57
Kernel	0,31	0,5	0,57	**0,68**	0,42
Relation	**0,65**	0,38	0,45	0,43	0,1
Sphere	0,24	0,36	**0,52**	0,33	0,35
Element	0,29	0,39	**0,52**	0,43	0,22
Feature	0,15	0,29	**0,42**	0,25	0,19
Part	0,41	0,35	**0,6**	0,33	0,34
Average	0,39	0,41	**0,53**	0,44	0,31

5   Future Directions

In this paper we proposed a method to use Wikipedia category hierarchy as an alternative document representation for clustering. Then we introduced three new, path based methods for computation of distance in such defined concept space. To test our assumptions we implemented the clustering search engine and performed their empirical evaluation. Our experiments has shown that one of the proposed techniques, $PL + Avg*$, achieves better scores than both baseline methods. In our further research we plan:

1. Employ a *large scale multi-label text classifier* [21] for automatically tagging with Wikipedia categories raw texts. This should allow us to apply the method based on a conceptual representation of a text for a wider scale rather than only for Wikipedia articles.
2. Improvement in system performance, especially the *OPTICS* algorithm that could be increased through implementation of methods similar to those described in [22].
3. Support for different document repositories other than Wikipedia. In particular, system could be adopted to group search results within the *MEDLINE* repository, which has been shown to respond well to category-based enhancements [6].

Acknowledgements. This work has been partially supported by the National Center for Research and Development (NCBiR) under research Grant No. PBS2/A3/17/2013.

References

1. Manning, C.D., Raghavan, P., Schütze, H.: An Introduction to Information Retrieval. Cambridge University Press, Cambridge (2009)
2. Rajaraman, A., Ullman, J.D.: Mining of Massive Datasets. Cambridge University Press (2011)
3. Medelyan, O., Milne, D., Legg, C., Witten, I.H.: Mining meaning from wikipedia. Int. J. Hum.-Comput. Stud. 67, 716–754 (2009)
4. Zesch, T., Gurevych, I.: Analysis of the Wikipedia Category Graph for NLP Applications. In: Proceedings of the TextGraphs-2 Workshop (NAACL-HLT) (2007)

5. Syed, Z.S., Finin, T., Joshi, A.: Wikipedia as an ontology for describing documents. In: ICWSM (2008)
6. Hu, J., Fang, L., Cao, Y., Zeng, H.J., Li, H., Yang, Q., Chen, Z.: Enhancing text clustering by leveraging Wikipedia semantics. In: Proceedings of the 31st Annual International ACM SIGIR Conference on Research and Development in Information Retrieval, SIGIR 2008, pp. 179–186. ACM, New York (2008)
7. Banerjee, S., Ramanathan, K., Gupta, A.: Clustering short texts using wikipedia. In: Proceedings of the 30th Annual International ACM SIGIR Conference on Research and Development in Information Retrieval, SIGIR 2007, pp. 787–788. ACM, New York (2007)
8. Hu, X., Zhang, X., Lu, C., Park, E.K., Zhou, X.: Exploiting wikipedia as external knowledge for document clustering. In: Proceedings of the 15th ACM SIGKDD International Conference on Knowledge Discovery and Data Mining, pp. 389–396. ACM (2009)
9. Gabrilovich, E., Markovitch, S.: Overcoming the brittleness bottleneck using wikipedia: Enhancing text categorization with encyclopedic knowledge. In: Proceedings of the 21st National Conference on Artificial Intelligence, AAAI 2006, vol. 2, pp. 1301–1306. AAAI Press (2006)
10. Gabrilovich, E., Markovitch, S.: Computing semantic relatedness using wikipedia-based explicit semantic analysis. IJCAI 7, 1606–1611 (2007)
11. Strube, M., Ponzetto, S.P.: Wikirelate! computing semantic relatedness using wikipedia. In: Proceedings of the 21st National Conference on Artificial Intelligence, pp. 1419–1424. AAAI Press (2006)
12. Budanitsky, A., Hirst, G.: Evaluating wordnet-based measures of lexical semantic relatedness. Computational Linguistics 32, 13–47 (2006)
13. Yazdani, M., Popescu-Belis, A.: Using a wikipedia-based semantic relatedness measure for document clustering. In: Proceedings of TextGraphs-6: Graph-based Methods for Natural Language Processing, TextGraphs-6, pp. 29–36. Association for Computational Linguistics, Stroudsburg (2011)
14. Sorg, P., Cimiano, P.: Exploiting Wikipedia for cross-lingual and multilingual information retrieval. Data & Knowledge Engineering 74, 26–45 (2012)
15. McCrae, J.P., Cimiano, P., Klinger, R.: Orthonormal explicit topic analysis for cross-lingual document matching. In: Proceedings of the 2013 Conference on Empirical Methods in Natural Language Processing, pp. 1732–1740. Association for Computational Linguistics, Seattle (2013)
16. Zesch, T., Gurevych, I., Mühlhäuser, M.: Comparing Wikipedia and German Wordnet by Evaluating Semantic Relatedness on Multiple Datasets. In: Proceedings of Human Language Technologies: The Annual Conference of the North American Chapter of the Association for Computational Linguistics (NAACL-HLT) (2007)
17. Liu, T., Liu, S., Chen, Z.: An evaluation on feature selection for text clustering. In: ICML, pp. 488–495 (2003)
18. Ankerst, M., Breunig, M.M., Peter Kriegel, H., Sander, J.: Optics: Ordering points to identify the clustering structure, pp. 49–60. ACM Press (1999)
19. Sander, J., Qin, X., Lu, Z., Niu, N., Kovarsky, A.: Automatic extraction of clusters from hierarchical clustering representations. In: Whang, K.-Y., Jeon, J., Shim, K., Srivatava, J. (eds.) PAKDD 2003. LNCS (LNAI), vol. 2637, pp. 75–87. Springer, Heidelberg (2003)
20. Draszawka, K., Szymanski, J.: External validation measures for nested clustering of text documents. In: ISMIS Industrial Session, pp. 207–225 (2011)
21. Draszawka, K., Szymanski, J.: Thresholding strategies for large scale multi-label text classifier. In: 2013 The 6th International Conference on Human System Interaction (HSI), pp. 350–355 (2013)
22. Kryszkiewicz, M., Lasek, P.: Ti-dbscan: Clustering with dbscan by means of the triangle inequality. In: Szczuka, M., Kryszkiewicz, M., Ramanna, S., Jensen, R., Hu, Q. (eds.) RSCTC 2010. LNCS, vol. 6086, pp. 60–69. Springer, Heidelberg (2010)

Restructuring Dynamically Analytical Dashboards Based on Usage Profiles

Orlando Belo[1], Paulo Rodrigues[1], Rui Barros[1], and Helena Correia[2]

[1] ALGORITMI R&D Centre, School of Engineering, University of Minho
Campus de Gualtar, 4710-057 Braga, Portugal
[2] Business Optimization Division, Portugal Telecom Inovação
Rua Eng. José Ferreira Pinto Basto, 3810-106 Aveiro, Portugal

Abstract. Today, analytical dashboards assume a very important role in the daily life of any company. For some, they could be seeing as simple "cosmetic" software artefacts presenting analytical data in a more pleasant way. However, for others, they are very important analysis instruments, quite indispensable for current decision-making tasks. Decision-makers use to defend strongly their use. They are simple to interpret, easy to deal, and fast showing data. However, a regular dashboard is not capable to adapt itself to new user needs, having not the ability to personalize themselves dynamically during a regular OLAP session. In this paper, we present the structure, components and services of an analytical system that has the ability to restructure dynamically the organization and contents of its dashboards, following usage patterns established previously on specific users' OLAP sessions.

Keywords: On-Line Analytical Processing, Adaptive Dashboards Systems, OLAP Personalization, Private Data Clouds, and Multi-agent Systems.

1 Introduction

From executive information systems to enterprise performance management tools it is a long distance to run. However in every single step we find dashboards helping decision-makers facing daily challenges, conciliating in simple views a set of business indicators supporting decision-making at a glance [3] [4]. They are quite useful tools for monitoring and control of all kind of business activities, being today a clear and strong element of differentiation in the world of companies. Rarely, when available, their use is dispensable by any decision-maker. Dashboards are recognized as tools that reduce significantly the process of information analysis once that they are prepared previously for monitoring and alerting when necessary critical business cases and events using simple visual structures quite simple to interpret and understand. The saying "simple is beautiful" never had a greater practical application as with dashboards. Dashboards have been a clear factor of differentiation for decision-makers that want to have "the right decision at the right time", being alerted about critical business indicators in useful time. A right set of dashboards could have an extreme value to the success of a company. However, frequently they still have doing things for what they were prepared and configured. This is not bad at all, but

T. Andreasen et al. (Eds.): ISMIS 2014, LNAI 8502, pp. 445–455, 2014.
© Springer International Publishing Switzerland 2014

they are static components and tools in the sense that they are not able to evolve them selves, to adapt automatically to new business requirements or user needs and preferences. So, we felt that integrating some kind of "evolution" mechanisms in the structure of a dashboard it would be a clear advantage factor, especially in terms of acquiring and reflecting user preferences – system personalization [2] [8]. Although the implementation of such system require very demanding computational platforms in part due to the great volume and aggregation levels of data they use. Thus, the possibility of distributing the systems' workload for several entities capable of acting on different platforms (even with heterogeneous natures) is very interesting, allowing not only accelerate the data processing system and ensure a very robust and scalable system. Multi-agent systems [15] [16] perfectly satisfy operational requirements of this type, fitting well in the scenario of a naturally distributed corporate data warehousing system. This paper presents an automatic personalization system for dashboarding platforms. All systems' functionalities and services are described, especially the aspects related to the agent based community (organized as a multi-agent system) that supports the means to restructure in terms of data and metadata a set of analytical dashboards based on the usage that users gave them. We organized this paper as the follow: section 2 presents and describes the overall system's structure, giving particular emphasis to the adaptive functionalities of dashboarding platforms; and section 3 presents some conclusions and future work.

2   A Dashboard Personalization System

2.1   General Overview

Over the last few years, decision support systems have shown some development trends that point to build components increasingly unconventional. One of these aspects of development has been outlined in order to accommodate preferences of users in monitoring processes and data querying. Following this trend, we designed and implemented a multi-agent system [16] [15] [7] with the ability to incorporate and reflect preferences of users in an adaptive dashboarding system. Although no longer unusual the use of multi-agent systems in many areas – e.g. decision support systems [5], e-commerce applications [6], or product analysis processes [9] - as far as we know, their applications are rare in the context of dashboarding systems. The advantages that arise from the adoption of a multi-agent system are various. Just to name a few, we can underline the possibility of modelling a system according to some individual software components with the ability to interact with each other, the better distribution of computational resources (and their capabilities) that we can get in a network of software agents, or the improvement of the performance of a system through the provision of distributed processing entities more robust, reliable, maintainable, and reusable. The system was designed taking into consideration several distinct aspects that we assumed as very relevant for the goals we pursued, especially the ones related to the natural distribution of information sources that occurs frequently in a data warehousing system, the versatility of a data cloud system   [11] [13] for ensuring the right resource allocation in cases when the system scales, and, of course, the importance of gathering usage information in order to reflect posteriorly in the format and contents of a dashboard. Basically, these aspects drove us to a system configuration organized in to three distinct services layers (figure 1), namely:

1) *Layer 1* - Gathering and conciliating, where are located all the mechanisms and structures responsible for collecting the information we need across the several instances of a corporate data warehousing system distributed along the company facilities, and conciliating it latter in a specific data staging area where data are treated and filtered if necessary.

2) *Layer 2* - Global storage and management, that was defined over a private data cloud structure receiving all the multidimensional structures designed for providing data required by users' dashboarding platforms.

3) *Layer 3* - Providing, exploring and restructuring, which has querying mechanisms and structures to get on system's data cloud the information that dashboards need, caching it locally, and restructuring it latter accordingly the usage user did when used the several dashboards that he it has at its disposal.

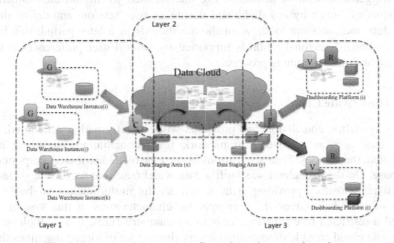

Fig. 1. System architecture configuration – layers and services

2.2 Agents and Services

Based on the characteristics of many agent-based applications (e.g. [5] or [7]), we designed services of Layers 1 and 3 as specialized software agents. These were organized into communities, where each agent has the necessary skills and knowledge to perform tasks of its own competence inside the system. Agents have also task-oriented acquaintance structures and mechanisms provided for communication and cooperation with others agents inside the system. We can find in the system five different communities of agents. To know:

1) *Gatherers* (G), which are responsible to collect data required to feed user dashboards in the several information sources (in this case, instances of a corporate data warehouse) referred in their local agendas, and send it to the respective conciliator agent.

2) *Conciliators* (C), these agents that conciliate data that comes from gatherers, transform it according predefined directives that instruct conciliators about the multidimensional structures located in the system's data cloud.

3) *Providers* (P), similar to conciliators, but doing opposite things, providers collects data required by visualizers in the system's data cloud, storing it locally in a multidimensional data base supported by Mondrian; providers maintain as long as they are valid instances the data that support all the active OLAP sessions triggered by one or more visualizers.

4) *Visualizers* (V), when a dashboard platform is activated a visualizer is launched for providing the structure and data that each dashboard needs; visualizers have locally predefined agendas that contain timely tagged all multidimensional queries that will be send to a provider agent in order to feed dashboards.

5) *Restructurers* (R) that are the most critical system's components; they are responsible to analyse regularly – a time interval is defined previously to trigger restructurers actions – log data related to all the multidimensional queries launch by any dashboard during an user session, and define the new data and metadata to show on the dashboards in a later period; this kind of dashboard personalization is supported by several user preferences detection algorithms and mining processes.

2.3   The Private Cloud

As we saw earlier, conciliators and providers work directly with a data cloud, which in this case acts as an overall framework for reconciliation, sheltering all the information that comes from the various instances of a distributed corporate data warehouse. On the date cloud was built a data warehouse that serves the system as a global data repository providing data to refresh the multidimensional bases under control of providers. Given the very specific characteristics of this system, which imposed a restricted or exclusive access to all data structures, it was implemented a private data cloud capable of sustaining a very diverse set of virtual machines that can be instantiated with the resources necessary for the normal execution of all system activities. The adoption of a private cloud-based system allows us to obtain the computational resources we need in a simple and rapid manner, and at reduced prices, without worrying about the infrastructure or the process of provisioning those resources. Private clouds are recommended for this type of environments, whether for implementing transactional systems or data warehousing systems [10] [12]. For the system we developed, this particular kind of cloud bring us the advantages of public clouds but without some of their problems, especially when we are dealing with clouds owned by the organization itself or managed internally. Private clouds are more controlled environments, providing dedicated computing resources and offering better conditions to face unpredictable workloads that use to occur in a conventional data warehousing system. Moreover, being a dedicated environment, a private cloud simply ends the multi-tenant model that is used in public clouds, often requiring the implementation of additional practices in order to promote confidentiality, privacy and security of the system and its information.

2.4 System's Life Cycle

In spite of being strong autonomous entities, agents behave following a well-defined system life cycle, which can be seen as a typical process of refreshment of a data warehouse. Agents have high independence when they are performing some specific task. However, globally, they act accordingly with some set of instructions that was previously recorded in system's agendas. These scheduling structures were configured accordingly the refreshment cycle of the dimensional structures maintained in the system's data cloud (layer 2). Thus, the system life cycle begins with one or more gatherer that was triggered by its agenda, starting collecting information in the sources, and acting at the same time cooperating with one or more conciliators in order to prepare the data collected and posted it after in the system's data cloud.

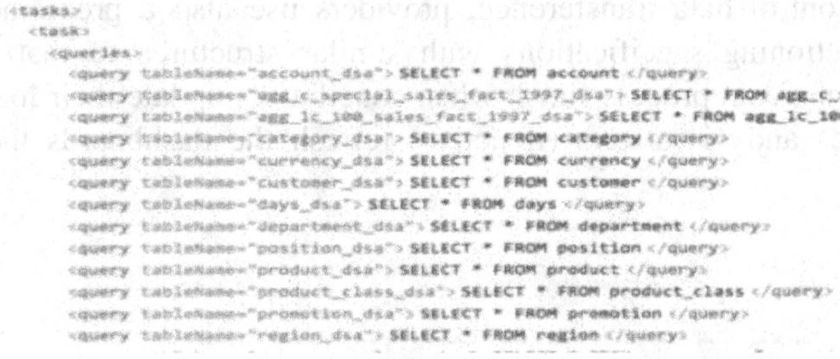

```
<tasks>
  <task>
    <queries>
      <query tableName="account_dsa"> SELECT * FROM account </query>
      <query tableName="agg_c_special_sales_fact_1997_dsa"> SELECT * FROM agg_c_special_sales_fact_1997 </query>
      <query tableName="agg_lc_100_sales_fact_1997_dsa"> SELECT * FROM agg_lc_100_sales_fact_1997 </query>
      <query tableName="category_dsa"> SELECT * FROM category </query>
      <query tableName="currency_dsa"> SELECT * FROM currency </query>
      <query tableName="customer_dsa"> SELECT * FROM customer </query>
      <query tableName="days_dsa"> SELECT * FROM days </query>
      <query tableName="department_dsa"> SELECT * FROM department </query>
      <query tableName="position_dsa"> SELECT * FROM position </query>
      <query tableName="product_dsa"> SELECT * FROM product </query>
      <query tableName="product_class_dsa"> SELECT * FROM product_class </query>
      <query tableName="promotion_dsa"> SELECT * FROM promotion </query>
      <query tableName="region_dsa"> SELECT * FROM region </query>
```

Fig. 2. An excerpt of a gatherer's agenda

The agenda of a gatherer (figure 2) contains a predefined list of queries that is expected to be launched over the data sources associated with the agent, as well as indicates the periodicity and the frequency of the queries. All results obtained from such queries are filtered according to some gathering specifications (expressed through filtering tags) that were associated with each single query. Data is extracted in blocks (e.g. 10,000 records chunks at a time) as a way to avoid very time consuming data transferences. At the same time data is extracted from the sources they are sent to the respective conciliator agent. Each gathering process ends when the corresponding conciliator agent receives all records collected. From this point on, conciliators have the necessary conditions to perform the tasks for which they were scheduled. These agents work directly with the cloud system, populating the data warehouse inside it with the data sent by gatherers. The conciliators do not need to acquaint the existence of other system's agents to accomplish a given task, once are the recruiters who start processes related to the integration of data in system's structures.

However, at the booting phase, conciliators have to register themselves in the system as data providers, so that gatherers have the possibility to know which agents are active and available to receive data that they collected on their information sources. At that time, they also receive a configuration file indicating them the location of their agenda, and the credentials they needed to access the system's cloud and its local database. This last repository act as a staging area and is used to support some transformation processes over the records received from gatherers, namely adapting metadata operations, or cleaning, standardizing and reconciling system's data. All these operations are performed in accordance to agenda's specifications of the agent. The transference of data to the multidimensional structures stored in the cloud only begins when the referred data transformation processes are finished. This last task ends what we call by the upstream process of system's data cloud refreshment.

The second part – the downstream process – starts when one or more providers launch over the system's data cloud the multidimensional queries that were defined previously in its own populating agenda. After receiving results, conciliators act together with one or more visualizers refreshing data and metadata of the dashboards under their control. Conciliators are clients of the system's cloud. They use it to populate their own multidimensional database, which stores every single view (typical fragments of data cubes) selected by users through visualizers during some data exploration process. To control data transference, providers use also a predefined agenda and a set of functioning specifications with similar structures to that of gathering agents. The downstream process finish when conciliators update their local multidimensional databases and visualizers (if active) refresh the dashboards they supervise with new data.

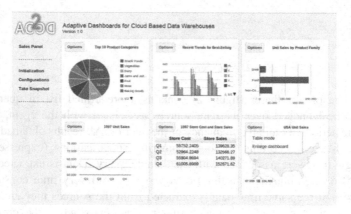

Fig. 3. A system's dashboard screenshot

2.5   The Adaptive Dashboards

Accessing the system, a user has available one or more analytical dashboards, each one including six different segments of analysis (figure 3) reflecting a set of user's preferences for using the system at a given time. The access to each panel is restricted, requiring the user authentication with proper credentials, before making any action of data exploration. When a session starts, a preconfigured dashboard is loaded into the user's desktop. After this, the user has its dashboarding environment prepared according to a specific configuration for data analysis reflecting their initial preferences for multidimensional querying. During a working session the user interacts (when permitted) with each of the segments that he has on the dashboard –

he can also submit new queries through a specific interface available in each segment. All actions that he performs over dashboards are recorded in a specific log file (figure 4). Later, a restructurer that is associated with the visualizer supporting the user dashboard operationally will work this log. Based on a few algorithms for identifying preferences, the restructurer suggests a new configuration for the user's dashboard, which can be or not accepted by the user.

id integer	user_id integer	date time date time without time zone	panel_number integer	element_number integer	dashboard_type_id integer	query text	initialization boolean	query_title character varying(100)
1	1	201110:40:51	1	1	1	select {[Uni	t	Top 10 Product Categori
2	1	201110:40:51	1	2	3	with set [Te	t	Recent Trends for Best
3	1	201110:40:51	1	3	2	select {[Mea	t	Unit Sales by Product F
4	1	201110:40:51	1	4	4	select {[Mea	t	1997 Unit Sales
5	1	201110:40:51	1	5	5	select {[Mea	t	1997 Store Cost and Sto
6	1	201110:40:51	1	6	6	select {[Sto	t	USA Unit Sales
7	1	201110:41:25	1	2	3	SELECT Measu	f	Profit of top 5 cities
8	1	201110:41:30	1	2	1	SELECT Measu	f	Profit of top 5 cities
9	1	201110:41:34	1	2	6	SELECT Measu	f	Profit of top 5 cities
10	1	201110:41:37	1	6	7	select {[Sto	f	USA Unit Sales
11	1	201110:41:45	1	1	1	select {[Uni	t	Top 10 Product Categori
12	1	201110:41:45	1	2	1	SELECT Measu	t	Profit of top 5 cities
13	1	201110:41:45	1	3	2	select {[Mea	t	Unit Sales by Product F
14	1	201110:41:45	1	4	4	select {[Mea	t	1997 Unit Sales
15	1	201110:41:45	1	5	5	select {[Mea	t	1997 Store Cost and Sto

Fig. 4. A dashboard usage log excerpt

The service provided by the restructuring agents it's done jointly with the visualizers and the providers agents. It aims to generate new settings for system's dashboards, reflecting usage preferences users revealed during their past working sessions with the several elements of a panel. Users can customize their dashboards in two different ways: by changing the data display mode of a dashboard's element, or changing the data that are usually presented in a dashboard's element. Both operations are recorded in specific log files. After an appropriate treatment, the information of these files is stored together with other information relative to the settings defined for each of the available dashboards in a database (figure 5) specifically designed for this purpose.

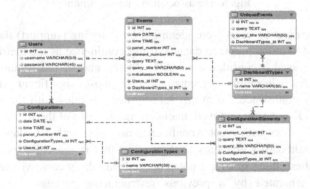

Fig. 5. System's usage preferences database

The customization and automatic restructuring of a dashboard can be made through the use of a process of discovering association rules. This data mining technique was used to find associations among distinct events - data exploration dashboards control tasks - that occurred in one or more elements of a panel, enabling its restructuring based on pairs of events more frequent occurred in the dashboard's environment.

Thus, it is possible to represent the sequences of actions that a user has developed during a data exploration process. Before running the association rules algorithm it was necessary to prepare the data coming from the logs of actions, assigning unique identifiers to each data exploration action - every action refers to an interrogation MDX, a title, and a data visualization mode. This preparation task serves exclusively to associate multiple identical entries performed on different periods of time to a same identifier - a configuration of a dashboard's element has always a unique identifier. This new dataset is then used to a relation of pairs of configurations of elements that will be used as input to the association rules algorithm. The algorithm that was selected for discovering the association rules was the Apriori [1] that is available in Weka [14]. Figure 6 presents the pseudo-code of the association rules algorithm.

```
Input: uid = the id of the authenticated user in the system,
       panelNumber = the panel to restructure
Output: Conf = The new configuration calculated according to the association
        rules algorithm

1: Let uniqueEvents be a mapping between a unique id and the
       corresponding user action associated to it;
2: Let dataPairs be a list that holds pairs of events that have happened in
       sequence;
3: previousActionId = -1; currentElementProcessing = -1;
4: userActions = GetEvents(uid, panelNumber);
5: foreach action in userActions do
6:     put action in uniqueEvents map;
7:     if currentElementProcessing != action.elementNumber then
8:         currentElementProcessing = action.elementNumber;
9:         previousActionId = -1;
10:    else
11:        if previousActionId != -1 and previousEventId != action.eventId then
12:            put {previousEventId, action.eventId} in dataPairs;
13:        end if
14:    end if
15: end foreach
16: rules = executeAprioriAlgorithm(dataPairs);
17: Conf = createConfiguration(rules, uniqueEvents);
```

Fig. 6. The association rules algorithm

Basically, association rules were identified by counting (support) the various data instances included in the prepared dataset, using confidence as the metric of interest - in our case we used a confidence of 60%. At the end of the mining process, the association rules algorithm presents the 10 most significant rules, ordered decreasingly by confidence. From this set of rules, all rules with the same meaning (e.g. A-> B and B-> A) are removed, maintaining only one of them.

Finally, the 3 rules with greater confidence are presented – 3 rules correspond to six different dashboard's elements configuration that is the capacity of a system's dashboard. If these three rules are not proposed, the dashboard will receive the last configuration generated by a previous restructuring process. To represent the sequence of actions determined by a set of association rules that was generated, we used the first row of elements of a dashboard to house the antecedents of the rules and the second row for the correspondent consequents. Just as a demonstration example, figure 7 presents a possible instantiation of a dashboard organized according to a set of association rules that integrates the 'A-> B', 'C-> D', and 'E-> F' rules.

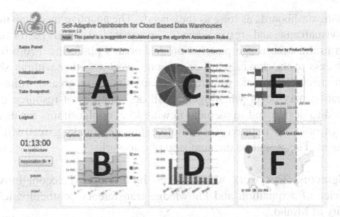

Fig. 7. Reflecting association rules in a system's dashboard

2.6   Implementation Aspects

During the project design were made some assumptions about the implementation means and tools. We intended to conceive a multi-agent system for dashboarding personalization entirely using products coming from open source initiatives. The reasons are pretty obvious. From the software agents' development platforms and components, to the private data cloud environment we used open-source software. To implement and support the private data cloud environment, provably the most critical system's component, we used *OpenNebula* (opennebula.org), an *Infrastructure as a Service* tool, which supported the implementation of the data warehouse in the system's data cloud as well provided the means to receive data form conciliators and feed the providers' multidimensional databases. The *Analytical Server Mondrian* (mondrian.pentaho.com), from Pentaho Corporation, ensured the maintenance and functioning of these last data structures. All system's dashboards were implemented appealing to regular Web technology, like JSP, HTML, JavaScript, or CSS, being the graphical layer developed with *Google Chart Tools* (developers.google.com/chart). System's software agents were developed in Java and their community environment for data sharing communication and collaboration supported by the framework JADE (jade.tilab.com) that follows the standard specifications of FIPA (www.fipa.org).

3   Conclusions and Future Work

In this work, we designed and implemented an automatic personalization system for analytical dashboards. Using a specialized community of agents, the system monitors and controls all the actions (multidimensional queries) that users do over a set of dashboards, establish when possible a set of usage patterns – a ranking querying selection, sequences of queries, "pairs" of queries, and so on -, and according them restructure automatically all the dashboards referred in usage patterns. But this is not a permanent mandatory action. Users may to inhibit this adaptive feature remaining with their static dashboards as long as they need. The adaptive mechanisms – a group of agents, visualizers and restructurers – presented a good performance level, reflecting in useful time usage patterns on dashboards platforms. From the system's upstream process to the downstream one, agents work together without showing any kind of bottleneck. However, we are conscious that with greater volumes of data conflict situation may occur when agents will act on cases of resource overlapping. Also, we need to design and implement new and better algorithms for usage patterns recognition in order to improve the adaptive features, making restructurers agents more intelligent. This it will be the next step of our work.

Acknowledgments. This work was financed by Portugal Telecom Inovação, S.A. under a service of acquisition and knowledge transference protocol celebrated with the University of Minho.

References

[1]   Agrawal, R., Srikant, R.: Fast algorithms for mining association rules in large databases. In: Bocca, J.B., Jarke, M., Zaniolo, C. (eds.) Proceedings of the 20th International Conference on Very Large Data Bases, VLDB, Santiago, Chile, pp. 487–499 (September 1994)

[2]   Bellatreche, L., Giacometti, A., Marcel, P., Mouloudi, H., Laurent, D.: A Personalization Framework for OLAP Queries. In: Proceedings of DOLAP 2005, Bremen, Germany, November 4-5 (2005)

[3]   Eckerson, W.: Performance Dashboards: Measuring, Monitoring, and Managing Your Business. Wiley (2010)

[4]   Few, S.: Dashboard Confusion. Intelligent Enterprise (March 20, 2004)

[5]   Lavbiş, D., Rupnik, R.: Multi-Agent System for Decision Support in Enterprises. Journal of Information and Organizational Sciences 33(2) (2009)

[6]   Kang, N., Han, S.: Agent-based e-marketplace system for more fair and efficient transaction. Decision Support Systems 34(2), 157–165 (2003)

[7]   Kishore, R., Zhang, H., Ramesh, R.: Enterprise integration using the agent paradigm: foundations of multi-agent-based integrative business information systems. Decision Support Systems 42(1), 48–78 (2006)

[8]   Kozmina, N., Niedrite, L.: OLAP Personalization with User-Describing Profiles. In: Forbrig, P., Günther, H. (eds.) MFCS 1978. LNBIP, vol. 64, pp. 188–202. Springer, Heidelberg (1978)

[9]   Lee, W.P.: Applying domain knowledge and social information to product analysis and recommendations: an agent-based decision support system. Expert Systems 21(3), 138–148 (2004)

[10]   Madsen, M.: Cloud Computing Models for Data Warehousing. Technology White Paper. Third Nature Inc. (2012)

[11]   The NIST Definition of Cloud Computing. National Institute of Standards and Technology (July 2011),
       http://csrc.nist.gov/publications/nistpubs/800-145/
       SP800-145.pdf (accessed April 14, 2014)

[12]   Russom, P.: TDWI Checklist Report - Consolidating Data Warehousing on a Private Cloud (2011),
       http://i.zdnet.com/whitepapers/
       Oracle_DW_US_EN_WP_Checklist_2.pdf (accessed April 14, 2014)
[13]   Voorsluys, W., Broberg, J., Rajkumar, B.: Introduction to Cloud Computing. In: Buyya, R., Broberg, J., Goscinski, A. (eds.) Cloud Computing: Principles and Paradigms, pp. 1–44. Wiley Press, New York (2011)
[14]   Weka, 2013. Weka Documentation (2013),
       http://www.cs.waikato.ac.nz/ml/weka/documentation.html
       (accessed April 14, 2014)
[15]   Wooldridge, M.: An Introduction Multi-Agent Systems. Jonh Wiley & Sons, Ltd. (2002)
[16]   Wooldridge, M., Jennings, N.: Intelligent agents: Theory and practice. The Knowledge Engineering Review 10(2), 115–152 (1995)

Enhancing Traditional Data Warehousing Architectures with Real-Time Capabilities

Alfredo Cuzzocrea[1], Nickerson Ferreira[2], and Pedro Furtado[2]

[1] ICAR-CNR and University of Calabria, Italy
[2] University of Coimbra, Portugal
cuzzocrea@si.deis.unical.it,
{nickerson,pnf}@dei.uc.pt

Abstract. In this paper we explore the possibility of taking a data warehouse with a traditional architecture and making it real-time-capable. Real-time in warehousing concerns data freshness, the capacity to integrate data constantly, or at a desired rate, without requiring the warehouse to be taken offline. We discuss the approach and show experimental results that prove the validity of the solution.

1 Introduction

The traditional data warehouse architecture model assumes that new data loading occurs only at certain times, when the warehouse is taken offline and the data is integrated during a more or less lengthy time interval. This offline procedure is required for three main reasons: there should be no interference between the loading process and the query sessions running on the data warehouse, so that there is no significant slowdown; the warehouse is typically a set of interconnected data marts, schemas (stars), with constraints (e.g. foreign keys, not null constraints, primary keys), lots of indexes (b-tree, bitmap indexes, others), materialized views and other summary or derived data, which are created to speedup query answering. From the point of view of data integration, constraints and indexes considerably slow the process down, as well as the refreshing of all those structures with the new data. The appropriate solution for these problems in traditional warehouses is to take the whole warehouse offline, disable/drop the constraints and indexes that cause loading slowdown, load the whole data and refresh the datasets, and then rebuild the auxiliary structures and constraints.

It would be convenient if, instead of using completely different architecture and database engine solutions, it would be possible to transform a warehouse to make it real-time, by adding some mechanisms to it. We explore this possibility in this paper, propose an approach and show that it achieves the desired objectives.

Our approach is based on a real-time integration component that is added to the traditional data warehouse and provides real-time capabilities without modifications to the existing setup. We concentrate on the warehouse itself, which includes loading and refreshing. This assumes that extraction and transformation phases of the ETL

T. Andreasen et al. (Eds.): ISMIS 2014, LNAI 8502, pp. 456–465, 2014.
© Springer International Publishing Switzerland 2014

processes are real-time-capable, or modified to be so. In short, the frequency of extraction should increase to provide the desired freshness (e.g. instead of once every day it could be once every 5 minutes), using Change-Data-Capture approaches and producing mini-batches, and transformation procedures should be guaranteed to run efficiently for the small mini-batches.

The paper is organized as follows: Section 2 reviews related work on the subject. Section 3 discusses the basic structures of a traditional data warehouse, then sections 4 and 5 proposes adding the real-time data warehouse component to provide real-time capabilities and how to provide 24/7 operation. Section 6 contains experimental results and section 7 concludes the paper.

2   Related Work

In [1] the authors discuss the two contexts for a DW, both the traditional and the real-time one. Theirs is an important work, in that it defines requirements and provides some indicative solutions for near-real-time data warehouse architecture. They suggest an architecture with some mechanisms that guarantee constant refresh of data in the DW. The architecture has five levels: 1) the level where the data extraction is done into data holders; 2) a level that synchronizes data from the data holders and transfers the data, periodically or push-based, into the intermediate level, known as the Data Processing Area (DPA). This level does the transformations; 3) the level that synchronizes the DPA and the DW. Two issues are left out of their work: the authors do not consider, during the loading and refreshing phase, the existence of indexes, materialized summaries and data marts. The performance problems introduced by the simultaneity between online querying and continuous data loading, and the difficulties in real-time loading introduced by having indexes and summary tables are not considered or evaluated.

The authors of [2,3,4] all propose a similar, temporary-tables based, approach to deal with real-time. For instance, in [3] the authors add time-interval granularity partitions, e.g. one for the last hour and one for the last day. The main limitation of those solutions is that there is not an adequate decoupling between the data loading and querying services, since they are done in the same database instance in the same machine.

In [5] the authors propose to build a tool based on Data Streams that allows data freshness and continuous integration, besides other criteria also evaluated by the authors. However, these stream-based solutions return approximate results, due to the large cost that would be incurred to have exact results using continuous streams.

The authors of [6] suggest a solution based in SOA (Service Oriented Architecture). This approach does data extraction initially, through a web service, and stores the results in caches. There are various levels of cache concerning different update levels, for instance, 5, 10, 30 and 60 minutes. After the data passes through all caches, it is stored in the DW. This solution also has a component that joins information from the caches. This proposal modifies the architecture and the structures of the data warehouse.

Finally, there are also commercial systems that propose real-time and freshness in DW, such as the ones in [8,9].

3   ETL and Typical Data Warehouse Architecture

Figure 1 illustrates the traditional data warehouse architecture. Data sources (DSx) are typically operational systems, most often those would be distributed business unit sources, such as the stores in a supermarket chain, that produce transaction data.

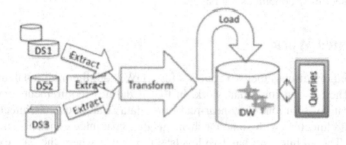

Fig. 1. Elements of ETL and Warehouse Architecture

As transactions happen in those data source operational systems, extract sets are created for posterior shipping into the data warehouse. Change Data Capture mechanisms (CDC) provide a convenient way to efficiently generate those extract sets, as seen in [9,10,11,12,13], identifying the data that is to be extracted and creating the log for posterior integration into the DW. The extracted data goes into a staging area, the place where transformation and loading is supposed to happen. In order to achieve near-real-time extraction, CDC mechanisms are used and mini-batches are created to be integrated more frequently than in the traditional non-real-time warehouse.

Transformation refers to the modifications that data suffers before being incorporated into the data warehouse. Transformation speedup for near-real-time data integration can be achieved by parallelization [1], lookups can be done in-memory, and certain transformations can be pushed to the data sources.

The data warehouse itself is typically made of one or more database schemas, with a possibly large number of stars, which contain business data viewed from different perspectives and in a format that is easy to query and analyze. A star schema contains a central fact table and several dimension tables. The fact table contains foreign keys to the dimension tables, since a fact row reports business measure data, such as sales, for a specific combination of dimensions rows. Figure 2 shows a typical star schema of a data warehouse. Data to be loaded at a certain instant typically include fact data and dimension data. The data warehouse also includes a number of query speedup structures, such as indexes, materialized views and other forms of data summaries. The purpose of the summary data is to speedup queries that can be answered using the summarized data, such as sales per month and per store. Since there may be several data marts and summaries, and those are specialized in answering certain types of queries, there may be a large number of such structures in a data warehouse. The star schema is typically indexed for fast access, with both b-tree indexes for fast lookup in dimension tables or in fact tables, bitmap indexes, which encode attribute or predicate values per row for faster query execution, or even more complex structures such as join bitmap indexes that avoid join processing for some queries. Column constraints such as NOT NULL, FOREIGN KEYS, UNIQUE, PRIMARY KEYS are also common in these as in any other database schema.

All the structures that we described in the previous paragraph are created in order to speedup query processing significantly. This is important, since the data warehouse can be huge and still require users to have their analysis queries answered in few seconds. In particular, users analyzing data charts and modifying their perspective online require interactive response times of few seconds.

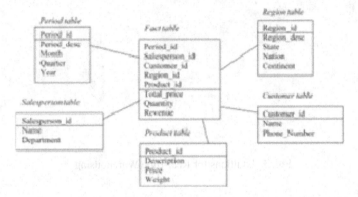

Fig. 2. Typical Star Schema

In the traditional data warehouse, the output of the data transformation is put into a load log. When the transformation ends and the log is filled-up, a bulk-load mechanism is called to load and refresh the data warehouse. When transforming the traditional data warehouse into a real-time one, the frequency of runs of the batch loading have to be increased (e.g. instead of once per day or once per week, they would run once every 5 minutes or once every hour), or in the limit, every new data item would be integrated as soon as it would be generated from the operational data sources.

But this is typically not possible for two main reasons: due to the need for the data warehouse to remain online during the "in-use hours", and due to the significant loading slowdown if indexes, constraints and summaries are kept during the loading. The alternative, to disable the indexes and other structures, load and then rebuild those structures, is also extremely heavy.

4   The Real-Time Warehouse

The purpose of the Real-time Warehouse is to allow data to be integrated in much shorter time than the traditional periodic loading intervals. The requirements are that it be an add-on to the traditional architecture, requiring no or few modifications to the existing infrastructure. It is based on two components: the static and the dynamic data warehouse components.

Data warehouses have multiple structures, including star schemas, multiple indexes, summary tables. Loading and refreshing all this redundant data slows the process down significantly if there are queries running simultaneously. If a schema has many indexes, insertion of new data will be slowed due to index updates. But dropping indexes before loading the data and rebuilding them after the loads is a heavy process if its done online, with queries running simultaneously. In our architecture those problems are eliminated by completely separating the light real-time data integration component, the D-DW, from the component holding the bulk of the data and all the indexing and summarizing elements on the huge less recent data, the S-DW.

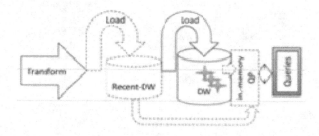

Fig. 3. Additions for Real-time Warehousing

4.1   Static Data Warehouse (S-DW) – The DW Component

The S-DW follows the typical data warehouse organization and holds the bulk of the data. First of all, it is expected to have multiple star schemas. Indexes such as B-trees (I) or Bitmap indexes (BI) (possibly many) are expected there, to speedup accesses. The cost of reorganizing those indexes can be quite high. Another existing element in the S-DW are any number of summary tables, which summarize the data to allow very fast answers to queries that may access those them instead of base data.

4.2   Dynamic Data Warehouse (D-DW) – The Real-Time Component

Only recent data will live in the D-DW (e.g. data from the current day), resulting in a comparatively small amount of data. Queries over the D-DW are always answered fast in absolute terms, since the quantity of data that needs to be processed is small. This is an important point, since it also means that this component can work efficiently even without indexes and aggregated views, at least comparatively to the S-DW. The fact that it dispenses indexes and materialized views altogether means that loading of mini-batches can be online, with extreme efficiency, in real-time and with no or very little noticeable performance degradation for the queries. This is shown in Figure 3.

Comparing the S-DW to the D-DW, the base star schemas have the same fact and dimension tables. Materialized aggregates (MV) are replaced by non-materialized views (V), there are no constraints (e.g. no explicit links between facts and dimensions), and both B-tree (I) and Bitmap (BI) indexes are absent. These differences and the size of the data (the D-DW contains a much smaller amount of data than the S-DW) makes the system load and run much faster. There is no need to update indexes or views.

4.3   Merger

Since in this design the data is spread over two different components, and those should reside in different nodes for faster processing, the approach features a merger, which submits the query in each of the two independent components simultaneously (S-DW and D-DW), which send the results back to the merger. This merger then computes the final answer. The merger is simply another database instance, in our prototype implementation it was an in-memory database for faster merging operation (we used H2). Figure 4 illustrates the merger component.

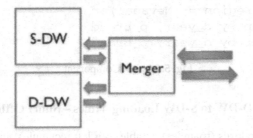

Fig. 4. Merger Component

4.4   Merger – Query Processing

Query processing in the merger is very similar to the typical parallel query processing approaches. We describe the approach using a simple SQL-based example. In the example a query is broken down into two queries and the results are merged using an UNION operator.

4.5   Same Instance D-DW versus Independent Instance D-DW

The dynamic data warehouse component (D-DW) can also be implemented in the same instance as the S-DW, as temporary tables. The problem then is that this alternative does not eliminate the simultaneity of querying and loading activity, therefore the performance is still degraded.

```
select sum(l_revenue), d_year, p_brand
from (select l_revenue, d_year, p_brand
  from lineorder, date_dim, part, supplier
  where
  l_orderdate = d_datekey and
  l_partkey = p_partkey and
  l_suppkey = s_suppkey and
  p_brand = 'MFGR#2221' and
  s_region = 'Nevada')
UNION ALL
(select l_revenue, d_year, p_brand
  from lineorder@ddw, date_dim@ddw, part@ddw,
supplier@ddw
  where
  l_orderdate = d_datekey and
  l_partkey = p_partkey and
  l_suppkey = s_suppkey and
  p_brand = 'MFGR#2221' and
  s_region = 'Nevada')
group by d_year, p_brand
order by d_year, p_brand;
```

Fig. 5. Merger Component

4.6   Minimizing D-DW to S-DW Loading Times – Short Offline Periods

Figure 6 illustrates indexes (triangles), a table (big left rectangle), and a time-partitioned table (right rectangles). While in the left organization a whole index has to be dropped and rebuilt when loading data, using the right organization only the index for the last partition that is the one loaded needs to be rebuilt. This saves a large amount of time.

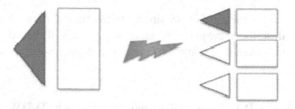

Fig. 6. Indexing a table versus partitions

5   Experiments

The following experiments show that it is possible to add the RWC component to achieve real-time. We test the real-time loading limitations of the DW for the setup scenario, compare it with the real-time loading capability when the RWC is added.

We also show that query response times do not suffer significantly. The experimental setup and results, taken from [7], are based on the SSB (Star Schema Benchmark) [15], augmented to include: A) the Orders star, representing sales, and the corresponding dimension tables (Time_dim, Date_dim and Customer); B) a star representing LineOrder, which contains order items, with 5 dimension tables (Time_dim, Date_dim, Customer, Suppllier and Part). Figure 7 shows the star schemas. Aggregated views were also added, measuring sales by Supplier, Part and Customer over granularities (hour, day, month and year). The schema was indexes - 8 B-tree indexes for the star schemas (4+4), and 6 bitmap indexes. The size of the data warehouse was 30 GB. The same query workload of the original SSB benchmark was used, consisting of 13 queries that target data with different granularity [15]. Experiments included loading time, from log files in the operating system. Additional details on the modified benchmark (SSB-RT) are available in [7]. The database Server used in the experiments was an Intel(R) Core(TM) i5 3.40GHz with 16GB of RAM memory, running Oracle version 11g.

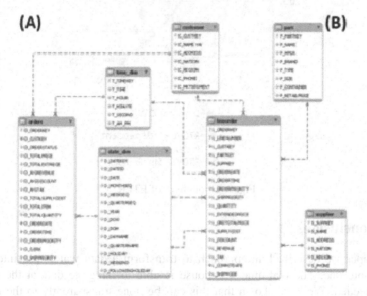

Fig. 7. RT-Benchmark Schema

The SSB-RT experimental setup includes the ETL. The extracted data was simulated by means of TPC-H dbgen generated orders and lineitem log files (for facts), and customer, part and supplier (for dimensions). The transformation process started by reading those files, and included selection of columns to load, translation of coded values (5 fields), encoding of lineorder, orderpriority and ship-priority fields, computation of the total price from lineorder prices, generation of surrogate keys, generation of data and time rows and surrogate keys for those. Details of the SSB-RT are available in [7].

In our experiments, we assess how much query sessions impact the ETL process, that is, the throughput of the ETLR process. The setup places 10 simultaneous query sessions always running queries that they choose randomly form the SSB-RT query set. The throughput is measured as the number of rows of the batch that our system is able to process per second. Three alternatives are compared: traditional DW (S-DW), same-instance D-DW, and the solution of independent instance D-DW and merging in a third component. Figure 8 shows the results. It clearly states the benefits of our research.

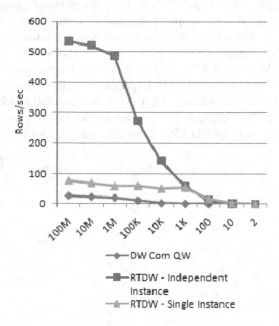

Fig. 8. Throughput of ETLR

6   Conclusions

In this paper we described an approach to transform a data warehouse into a real-time-capable one. The real-time warehouse is able to integrate data at the rate that may be needed. We have shown that this can be done transparently to the user and existing data warehouse by adding a Real-time Warehouse component, and small middleware component that rewrites client queries into queries to two components and merges results in a single result to be returned to clients. We also discussed how to minimize D-DW to S-DW loading times. Experimental results have shown the validity of the proposed approach.

References

1. Vassiliadis, P., Simitsis, A.: Near Real Time ETL. New Trends in Data Warehousing and Data Analysis. Annals of Information Systems 3, 1–31 (2009)
2. Jain, T., Rajasree, S., Saluja, S.: Refreshing Datawarehouse in Near Real-Time. International Journal of Computer Applications 46(18), 24–29 (2012)

3. Zuters, J.: Near Real-Time Data Warehousing with Multi-stage Trickle and Flip. In: Grabis, J., Kirikova, M. (eds.) BIR 2011. LNBIP, vol. 90, pp. 73–82. Springer, Heidelberg (2011)
4. Santos, R.J., Bernardino, J.: Real-Time Data Warehouse Loading Methodology. In: Proceedings of ACM IDEAS, pp. 49–58. ACM Press, New York (2008)
5. Nguyen, M., Tjoav, A.M.: Zero-Latency Data Warehousing for Heterogeneous Data Sources and Continuous Data Streams. In: Kotsis, G., Bressan, S., Ibrahim, I.K. (eds.) iiWAS, vol. 170. Austrian Computer Society (2003)
6. Zhu, Y., An, L., Liu, S.: Data Updating and Query in Real-time Data Warehouse System. In: Proceedings of IEEE CSSE, vol. 5, pp. 1295–1297. IEEE Computer Society, Washington, DC (2008)
7. Ferreira, N.: Realtime Warehouses: Architecture and Evaluation, MSc Thesis, U. Coimbra (June 2013)
8. Vertica, http://www.vertica.com/the-analytics-platform/real-time-loading-querying/
9. Oracle, Best Pratices for Real-time Data Warehousing, White Paper (August 2012)
10. Kim, N., Moon, S.: Concurrent View Maintenance Scheme for Soft Real-time Data Warehouse Systems. Journal of Information Science and Engineering 23(3), 725–741 (2007)
11. Jain, T., Rajasree, S., Saluja, S.: Refreshing Datawarehouse in Near Real-Time. International Journal of Computer Applications 46(18), 24–29 (2012)
12. Shi, J., Bao, Y., Leng, F., Yu, G.: Study on Log-based Change Data Capture and Handling Mechanism in Real-time Data Warehouse. In: Proceedings of CSSE, vol. 4, pp. 478–481. IEEE Computer Society, Washington, DC (2008)
13. Ram, P., Do, L.: Extracting Delta for Incremental Data Warehouse Maintenance. In: Proceedings of ICDE, pp. 220–229. IEEE Computer Society, Washington, DC (2000)
14. Furtado, P.: Efficiently Processing Query-Intensive Databases over a Non-Dedicated Local Network. In: Proceedings of IPDPS, vol. 1, p. 72. IEEE Computer Society, Washington, DC (2005)
15. O'Neil, P., O'Neil, E., Chen, X., Revilak, S.: The Star Schema Benchmark and Augmented Fact Table Indexing. In: Nambiar, R., Poess, M. (eds.) TPCTC 2009. LNCS, vol. 5895, pp. 237–252. Springer, Heidelberg (2009)

Inference on Semantic Trajectory Data Warehouse Using an Ontological Approach

Thouraya Sakouhi[1], Jalel Akaichi[1], Jamal Malki[2], Alain Bouju[2],
and Rouaa Wannous[2]

[1] Institut Supérieur de Gestion, Tunis, Tunisia
{thouraya.sakouhi,j.akaichi}@gmail.com
[2] University of La Rochelle, La Rochelle, France
{jmalki,alain.bouju,rouaa.wannous}@univ-lr.fr

Abstract. Using location aware devices is getting more and more spread, generating then a huge quantity of mobility data. The latter describes the movement of mobile objects and is called as well *Trajectory* data. In fact, these raw trajectories lack contextual information about the moving object goals and his activity during the travel. Therefore, the former must be enhanced with semantic information to be called then *Semantic Trajectory*. The semantic models proposed in the literature are in many cases ontology-based, and are composed of thematic, temporal and spatial ontologies and rules to support inference and reasoning tasks on data. Thus, calculating inference on moving objects trajectories considering all thematic, spatial, and temporal rules can be very long depending on the amount of data involved in this process. On the other side, TDW is an efficient tool for analyzing and extracting valuable information from raw mobility data. For that we propose throughout this work a TDW design, inspired from an ontology model. We will emphasis the trajectory to be seen as a first class semantic concept. Then we apply the inference on the proposed model to see if we can enhance it and make the complexity of this mechanism manageable.

Keywords: Trajectory data, semantic modeling, ontology, trajectory data warehouse, inference.

1 Introduction

Nowadays, using location aware devices is getting more and more spread, generating then, a huge quantity of mobility data. Since that, tracing mobile objects movement has become possible due to the enhanced accuracy of these technologies. Thanks to this evolution, the management of big amounts of spatio-temporal data derived from those devices is expected to extract useful and novel knowledge about the moving objects and facilitate, then, the understanding of their behavior according to different domains. Thus, this has given rise to a large number of applications dedicated to this end such as: land planning, animals tracking, mobile marketing, etc. The captured data are called *Trajectory* and are in the heart of these applications. While this raw data is stored in databases,

T. Andreasen et al. (Eds.): ISMIS 2014, LNAI 8502, pp. 466–475, 2014.
© Springer International Publishing Switzerland 2014

this provides a limited support for analyzing and understanding the behavior of moving objects. In fact, these raw trajectories doesn't contain contextual information about the moving object goals neither his activity during the travel which does not satisfy the applications requirements. Therefore, the raw data must be enhanced with semantic information. Data with the additional semantic layer will be called then *Semantic Trajectory*. The state of the art semantic models that are built around trajectory data are in many cases ontology-based. A trajectory ontology is generally composed of thematic, spatial and temporal ontologies, respectively with thematic, temporal and spatial rules to support inference and reasoning tasks on data. Thus, calculating inference on moving objects trajectories considering all thematic, spatial and temporal rules can be very long depending on the amount of data involved in this process. So that's what explains the huge time and storage space taken for the inference process when considering an important mass of data. Inference difficulty was shown during the experiments performed in [1] by measuring the complexity of executing the entailment. The latter increases incredibly when increasing the number of loaded data. Thereby, others mechanisms are still needed to get manageable complexity and reason readily on the whole ontology. The aforementioned problem can be solved using novel analytical techniques that can exploit the available rich mobility semantics and provide performance features at the same time. In this work we tend to extend the traditional analytical techniques so as to take into consideration these limitations and handle efficiently raw trajectory data using the best of both *Data Warehousing* and *Semantic Modeling* worlds. Indicatively, a TDW (Trajectory Data Warehouse) is an efficient tool for analyzing and extracting valuable information from raw mobility data. For that we propose throughout this work, in a first level, a STrDW (Semantic Trajectory Data Warehouse) model, extracted from an ontology. This emphasis the trajectory to be seen as a first class semantic concept, not only a spatio-temporal path, providing then a semantic multidimensional model which is meant to be more than a spatio-temporal data repository for storing and querying raw movement. In a second level, we apply the inference over the proposed model to see if we can enhance it and make the complexity of this mechanism manageable due to the increased performance of the DW technology structure. The outline of this paper is as follows. In section 2, we review the main works on spatio-temporal data modeling approaches. Section 3 describes the design of the proposed STrDW and the inference process to be applied over it. In section 4, we introduce the implementation of our model and the evaluation of the inference results. Finally, section 5 summarizes and concludes this work and discusses interesting open issues.

2   State of the Art

Since that representing trajectory data and asking queries about them are important issues to understand the behavior of the moving object, many works in the literature have focused on these issues. Trajectory data were first stored and manipulated by traditional DBMS, which presents limitations on managing moving objects positions that change their values continuously. It's that they only consider attributes that are constant in time. Even if updating operations can be made in transactional systems, they are always considered as infrequent [2]. Due the conducted researches in this field dating from the second half of the 90's, many approaches to model trajectories considered as a set of spatio-temporal data were defined [2], [3]. Thereby, Spatio-Temporal Databases (STDB) and Moving Object Databases (MOD) have emerged. However, current DBMS ability, even with its extensions, is limited only on storing raw mobility data which is not bearing any semantic information. To make efficient exploitation of this data, there are attempts to enrich it with semantic annotations to make it more comprehensible, manageable and understand moving object behavior. For that, the latest years a big interest has been shown to the semantic approach for modeling spatio-temporal data. The first works were based on conceptual models such as [4] and [5]. Then recently, likely semantic models were in many cases ontology-based as in the works of [1] and [6].

There are many ways for efficiently analyzing mobility data. Warehousing and mining techniques are, among others, supporting the extraction of valuable information from raw moving object data. TDW is the application of DW techniques on trajectory data. A TDW, and generally Spatio-Temporal Data Warehouse (STDW), is then an extension to the MOD or any other data structure storing moving objects footsteps. Indeed, the latter can only handle simple transactional queries with not a lot of history, neither data aggregation capabilities compared to the complex analysis queries required for the decision making process. There have been various proposals of multidimensional models for STDW [7], [8]. The aim behind these proposals was generally the integration of various data sources into a STDW. These problems could be linked to spatial, temporal or semantic heterogeneity, seeing that requirements on data may change from one period of time to another. A trajectory is defined as a line described by an object during his movement resulting from the change of his spatial location in time [9]. Indeed, trajectory data are special types of spatio-temporal data. Then, a TDW is obviously a special case of STDW.

In fact, DW (Data Warehouse) technology is nowadays standard for complex analytical capabilities with a high performance and a rapid execution time, even when dealing with spatio-temporal data that is considered as a recent research issue. Nevertheless, current DW models present some limitations, mainly when dealing with the semantic aspect of data. Adding a semantic layer in the above of a DW structure would be an interesting idea to enhance its flexibility and leap ahead of low-level details. Also, many works in the literature believe that the trajectory concept exceeds a raw set of time stamped positions, to be considered as a semantic entity related to a layer of thematic information. By exploring the literature, we find wide concrete work on semantic multidimensional modeling, [10], [11] and [12] are the most recent. Taking a brief overview on similar works, the closest and the only true design of TDW using semantic concepts that can be recalled, is the solution discussed in [12]. In this paper, a framework for the design and the implementation of semantic DW supporting spatio-temporal data sets, notably, a semantic TDW is proposed. The latter is tailored around the *Move, Stop* and *Episode* components of a *Semantic Trajectory* detailed in [4] and [5]. Firstly, this proposal uses generic semantic concepts to model the semantic TDW. That's what prevents the resultant model from satisfying the application-specific semantic requirements. Secondly, the semantic TDW is inspired from a conceptual model, not an ontology model, for describing trajectory data semantics. Actually, ontology model are similar to conceptual models, but in many aspects of knowledge representation the former exceeds the latter [13]. Among which we mention here the ones that are most relevant for our STrDW model are:

- *Consensuality*: to offer the STrDW designer a global and universal view of the application domain.
- *Reasoning*: apply inference on the STrDW model to deduce new information from existing facts.

Also, the aforementioned work didn't propose a methodology to extract the semantic TDW model from the semantic trajectory model. So, the need to emphasis the trajectory to be seen as a first class concept motivates us to propose throughout this work a design for a STrDW. Our work, on the contrary of [12] is describing an approach used for designing the STrDW whose multidimensional concepts (fact, dimension, measure, etc) are becoming from a domain, spatial and temporal ontologies.

3   STrDW Model

In the above, we mentioned most of the works in the literature related to modeling trajectory data. Inhere, we are mostly interested in TDW modeling. We think that, modeling a TDW related to our experimental data sets (seal trajectories), could help making inference on them and optimizing then the time taken through this mechanism. Considering this, our contributions are twofold:

- A semantic model for TDW (STrDW) related to seal trajectories using an ontological approach.
- Inference on the STrDW, using thematic, spatial and temporal rules becoming from the source ontology.

To the best of our knowledge, none of the proposed multidimensional models in the literature brought together the aforementioned features (spatial/temporal data types, semantics, reasoning, performance). Even more, the problem of inference on DW was never discussed, neither in the previous works on TDWs, nor on works dedicated to classical DWs.

3.1   Application Scenario

This subsection is aimed to illustrate the application scenario used to exemplify this work. The hereinafter conducted researches are motivated by the scenario from the marine mammals tracking application domain, related to the biology research field, regarding especially the seal animals trajectories, and are inspired by the research works for modeling semantic trajectories [1]. The source trajectory datasets are time-stamped locations collected from the GPS devices tied to seals during their travel. Additional informations related to the seal, and its activity during the trip, are provided too. The main components of the dataset are:

- Ref: the seal's reference;
- Dive-dur, Sur-dur and Max-depth: are dive duration, surface duration and maximum depth of a dive, respectively;
- TAD: is Time Allocation at Depth which defines the shape of a seal's dive;
- Long and Lat: are respectively longitude and latitude, the spatial coordinates of the seal's position;
- B-date and E-date: are the temporal coordinates of the start and the end positions of a seal's trajectory.

Actually, the movement of the seals population is still relatively unknown. The research team involved in this work is interested in collecting and analyzing data becoming from these animals in their natural environment, to understand their behavior. Domain experts find it hard to exploit raw trajectory data to that end. For that, a semantic layer was added to data and three prominent semantic components were revealed:

- *Domain Ontology*: representing the real facts of the trajectory (Activities)
- *Spatial Ontology*: localization of the facts in the space
- *Temporal Ontology*: localization of the facts in time

The embedded system continuously captures data and maintains a model of three states: *Haul-out*, *Cruise* or *Dive*. In this work, researchers are more interested on the seal's activities conducted during dives (instances of the Dive concept). According to the experts, this comes to four main activities: resting, traveling, foraging and traveling-foraging. The seal activity is determined based on the dive duration, surface duration, maximum depth and TAD parameters describing the dive state of the seal and included in the source datasets. The DW design we developed is tailored around the main concepts from this ontological model, and that's what makes this former support spatio-temporal semantic concepts.

3.2   The Proposed Model

Many works in the literature, as mentioned before, proposed ontology-based methods for the design of semantic DWs, [10] and [11] are the most recent. In the proposal of [10] threshold values must be set by the designer for the annotation process of the DW's concepts, and no indications were shown on how to fix their values, which complicates this task for the designer. For that, we adopted the [11] method for our proposal due to its clarity and completeness.

Such as any DW model, the design of the STrDW is expected to pass through the requirements analysis phase. We expressed requirements by the means of a goal-driven approach. As its name indicates, this approach identifies goals that guide decisions of users and it is based on a *Goal Model*. There is two main co-ordinates describing a goal: a *Result* to analyze that can be quantified by given metrics measuring the satisfaction of the goal, and *Criteria* influencing this re-sult. Requirements analysis is ontology-based, as the STrDW model is derived from the already existing thematic, spatial and temporal ontologies denoted as the GO (Global Ontology). After identifying users' goals and their coordinates from the Goal Model, a connection is made between coordinates of each goal (Result and Criteria) from the one side and the resources (Concepts and Roles) of the GO from the other side. This projection allows then the extraction of sub-ontology from the GO called TDWO (Trajectory Data Warehouse Ontology). This step is of paramount importance. Firstly, it will permit, later, the definition of the STrDW conceptual model based on ontological concepts that express as much as possible effective user's goals. Secondly, it permits the identification of data sources, presented in the GO, that are the most relevant to be integrated in the STrDW. That's why this DW design methodology follows a mixed approach [11], seeing that it appeals at the same time user requirements and data sources, giving them then the same role in the modeling of the resultant STrDW. Users goals are also used for the annotation of the TDWO by multidimensional con-cepts such as fact, dimension, measures and dimension attributes, to result on the STrDW conceptual model.

Here is a Goal example "Analyze seal activities in a given time interval in a specific area". The result to analyze is the rate of different seal activities in specific place and time. The aforementioned result is quantified by some metrics which are in this case dive duration, surface duration, TAD and maximum depth. And finally the criteria influencing this result are the time, the space and the seal characteristics (gender and age). A first possible design model for the application scenario is given in figure 1. It appeals numeric measures (dive-dur, surf-dur, max-depth, tad) and 3 dimensions:

- *Time-Dim* organized following the hierarchy: second, minute, hour, day, month and year.
- *Space-Dim* with the hierarchy: position, geo-sequence,
- *Seal-Dim* represented by the seals' attributes: gender, age and reference.

Space-Dim is the spatial dimension of the model, and contains 2 levels related to a spatial hierarchy. Those levels reference geometric objects. The aggrega-tion function applied against the measures *Activity Rate* is actually the rate of seal activities (foraging/resting/traveling/traveling-foraging) calculated using the following formula: $Activity - Rate = \frac{Foraging-Sum}{All-Activities-Sum}$. This custom ag-gregation function is implemented to take into consideration the requirements inflicted by our model and its aims. The fact table is composed of dimensions' keys at their lower level that form the symbolic coordinates for the value of the measure. In this model, activities are the subject of the multidimensional analysis, so the user (a biologist) can deduce information about the activity of the seal during a special period of time and area in the space.

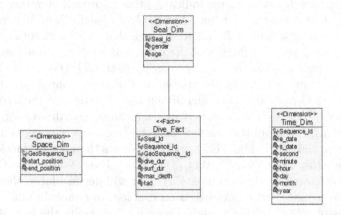

Fig. 1. The Proposed Model of the STrDW

3.3   Inference on the STrDW

The inference mechanism is the base for the reasoning on ontologies or any other knowledge representation model. In the case of ontologies, it's needed to respond semantic queries asked by users. Inference is defined as the automatic process of discovering new relationships using big data sets and auxiliary informations, mostly a set of rules. This process is the main targeted task behind the semantic web. Whether the new informations are explicitly added to the set of data, or are returned at query time, is an implementation issue. In our case, the big data sets and the auxiliary informations, needed to discover new relationships by the means of the inference mechanism, are respectively, the STrDW data and the domain, spatial, temporal rules. Therefore, inference will be applied here on the data residing in the STrDW. The inference process is expected to be more efficient due to the enhanced structure of the DW technology. Using facts and dimensions improves query performance because users often analyze data by drilling down on known hierarchies, according to which dimensions are organized. It's is that the single-column primary key populated with values reduce the size of cubes. Also, dimension tables are de-normalized, permitting the improvement of the query performance, since it reduces the number of joins. Actually, throughout this work the inference mechanism is launched by asking semantic queries to the STrDW. This can be done seeing that our STrDW has already been established on the head of an ontological model, inheriting then the domain concepts, relationships between them and rules to permit the generation of new relationships. Indeed, our model gathers at a time the performance and query optimization of the relational models and the high-level concepts and relationships of the semantic models. The imperative part of the ontological model of [1] is presented by a set of rules which are thematic, spatial or temporal. The thematic rules are application-specific and are specified by the domain experts. For the choice of the spatial and temporal rules, a comparative study between the available ones was proposed in [14]. For that, the OGC and the Allen algebra, respectively spatial and temporal rules were considered. Once the user asks a query on the DW model, for each record of the query result the inference algorithm appeals the domain, spatial and temporal rules. If those rules return a result, then the latter is saved in the knowledge repository and added to the query result, else the next rule is applied.

4 Experimentations

The dataset we used to feed our STrDW are collected from GPS devices tied to a selected list under test of seals of the English Channel who are distributed in many islands of the Brittany Coast and tripping until the Britannic islands in 2011. The dataset is provided by the AMARE research team, from the LIENSs research unity, working on the capacity of marine mammals to face their environment modifications. The sets of data are firstly stored in Excel files and are extracted using the ETL process defined for the given STrDW. To implement our STrDW we used the *Oracle 11.2*. We opted for the Oracle database due to its performance and extensibility features, in addition to that, starting from the 11 version, Oracle support spatial data types (*SDO-GEOMETRY*). In this model, the spatial dimension is reduced to its geometric part: *geo-sequence* (*Line* type) and *position* (*Point* type) hierarchy. Relationship between its levels is then a topological operator (spatial inclusion). We choose the ROLAP implementation, which stores the fact and dimension definitions and its data in a relational form in the database. ROLAP is preferable to store detailed, high volume data or you have high refresh rates combined with high volumes of data, and this is the case for our data.

For the inference mechanism, we implemented functions for the thematic, spatial and temporal already defined for the source ontology using *Java* language. To manipulate spatial data types on Java we made use of the *JGeometry* class that maps the *Oracle Spatial* type *SDO-GEOMETRY*. The domain and spatial inference execution time results are shown in the following figure 2. In fact, domain inference is of linear complexity and did not present difficulties. In contrast, spatial inference is of polynomial complexity (n^2), seeing that, for each couple of dives in a dataset of n dives, it appeals spatial relationships. The ontology model makes use of the spatial rules that call in their turn the spatial Oracle's operations what takes a lot of execution time. Since we are using the Oracle Spatial operations directly (relational implementation) in our model, the spatial inference execution time was reduced notably. It's useful for the biologist to visualize his data and query results in a map, particularly as we manage here

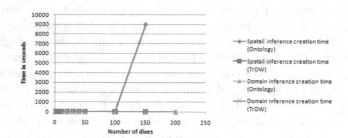

Fig. 2. Calculating Spatial and Domain Inference Creation Time

Fig. 3. Visualization of Spatial Data and Query Results

spatial data. For that, we developed a Java application to extract data from the source files or those becoming from query results, and create then a KML (Keyhole Markup Language) file. The generated KML files then permit the plotting of the data on Google Earth (figure 3).

5   Conclusion and Future Work

Throughout this work, we have been motivated by the need to visualize trajectory data models from a different perspective to respond to limitations in the foregoing models. To meet this need, we presented inhere a STrDW ontology-based model. We also applied the inference process on this model to manage its complexity, beforehand important. Research on these and related topics is crucial for expanding the usefulness of multidimensional models to nontraditional applications. At first, we aim in future works to offer designers the possibility to express their requirements in natural-language-like syntax. Indeed, STrDW contain huge amounts of data, so optimization issues are of paramount importance either for the data storage and retrieval issues or concerning the performance of the inference process. Moreover, we are highly interested in implementing a case tool to support user on the STrDW design. This work leads to consider also the transition to the logical and physical models. Considering the uncertain aspect of trajectory data rather than its unbounded and unpredictable streams are issues that should also be explored in future works.

References

1. Wannous, R., Malki, J., Bouju, A., Vincent, C.: Modelling mobile object activities based on trajectory ontology rules considering spatial relationship rules. In: Amine, A., Mohamed, O.A., Bellatreche, L. (eds.) Modeling Approaches and Algorithms. SCI, vol. 488, pp. 249–258. Springer, Heidelberg (2013)
2. Wolfson, O., Sistla, P., Xu, B., Zhou, J., Chamberlain, S.: Domino: Databases for moving objects tracking. In: ACM SIGMOD, pp. 547–549 (1999)
3. Güting, R.H., Böhlen, M.H., Erwig, M., Jensen, C.S., Lorentzos, N.A., Schneider, M., Vazirgiannis, M.: A foundation for representing and querying moving objects. ACM Trans. Database Syst. 25, 1–42 (2000)
4. Spaccapietra, S., Parent, C., Damiani, M., Demacedo, J., Porto, F., Vangenot, C.: A conceptual view on trajectories. Data & Knowledge Engineering 65(1), 126–146 (2008)
5. Yan, Z., Parent, C., Spaccapietra, S., Chakraborty, D.: A hybrid model and computing platform for spatio-semantic trajectories. In: Aroyo, L., Antoniou, G., Hyvönen, E., ten Teije, A., Stuckenschmidt, H., Cabral, L., Tudorache, T. (eds.) ESWC 2010, Part I. LNCS, vol. 6088, pp. 60–75. Springer, Heidelberg (2010)
6. Yan, Z., Macedo, J., Parent, C., Spaccapietra, S.: Trajectory ontologies and queries. Transactions in GIS 12(s1), 75–91 (2008)
7. Moreno, F., Arias, J.A.E., Losada, B.: A conceptual spatio-temporal multidimensional model. Revista IngenierÃas Universidad de MedellÃn 9, 175–183 (2010)
8. Zhou, L., Bao, M., Yang, N., Lao, Y., Zhang, Y., Tian, Y.: Spatio-temporal analysis of weibo check-in data based on spatial data warehouse. In: Bian, F., Xie, Y., Cui, X., Zeng, Y. (eds.) GRMSE 2013 Part II. CCIS, vol. 399, pp. 466–479. Springer, Heidelberg (2013)
9. Marketos, G.D.: Data warehousing and mining techniques for moving object databases (2009)
10. Thenmozhi, M., Vivekanandan, K.: An ontology based hybrid approach to derive multidimensional schema for data warehouse. International Journal of Computer Applications 54, 36–42 (2012)
11. Bellatreche, L., Khouri, S., Berkani, N.: Semantic data warehouse design: From ETL to deployment à la carte. In: Meng, W., Feng, L., Bressan, S., Winiwarter, W., Song, W. (eds.) DASFAA 2013, Part II. LNCS, vol. 7826, pp. 64–83. Springer, Heidelberg (2013)
12. Campora, S., Fernandes, J., Spinsanti, L.: St-toolkit: A framework for trajectory data warehousing. In: AGILE Conf. Lecture Notes in Geoinformation and Cartography. Springer (2011)
13. Khouri, S., Boukhari, I., Bellatreche, L., Sardet, E., Jean, S., Baron, M.: Ontology-based structured web data warehouses for sustainable interoperability: requirement modeling, design methodology and tool. Computers in Industry 63(8), 799–812 (2012)
14. Malki, J., Bouju, A., Mefteh, W.: An ontological approach for modeling and reasoning on trajectories taking into account thematic, temporal and spatial rules. Technique et Science Informatiques 31(1), 71–96 (2012)

Combining Stream Processing Engines and Big Data Storages for Data Analysis

Thomas Steinmaurer, Patrick Traxler, Michael Zwick,
Reinhard Stumptner, and Christian Lettner

Software Competence Center Hagenberg, Austria
{thomas.steinmaurer,patrick.traxler,michael.zwick,
reinhard.stumptner,christian.lettner}@scch.at

Abstract. We propose a system combining stream processing engines and big data storages for analyzing large amounts of data streams. It allows us to analyze data online and to store data for later offline analysis. An emphasis is laid on designing a system to facilitate simple implementations of data analysis algorithms.

1 Introduction

Advances in sensor technology, wireless communication, the advent of mobile platforms, and other developments of recent years brought new challenges to areas such as databases, operating systems, data mining, and machine learning.

A common pattern in practice is that a large amount of devices send data to a data center. Making use of data is crucial for many applications and services. The shear amount of data makes it however difficult to solve data analysis problems which are considered simple otherwise. Such problems include computing means and medians. More complex problems include linear and median (quantile) regression.

Besides the amount of data to be processed the programming models are crucial for applications. We distinguish between processing data as soon as it arrives, i.e. online processing, and processing historical data, i.e. offline processing. One example which requires online data analysis is fault detection of devices. The goal is usually to detect a malfunctioning device as soon as possible. This may require an interaction between online and offline processing: We use some algorithm to learn parameters of a model of the devices behavior offline and check online if the device works according to the model.

New systems for storing and processing large amounts of data emerged over the last decade [12,7,3,9]. A common system to process and store large amounts of data is MapReduce together with a distributed file system [7,12] or a key-value storage [3]. We refer to this kind of systems as *big data storages*, without mentioning explicitly the capability to process data in parallel.

In the context of data streams, *stream processing engines* are common [13,2,1,14]. A data stream is a sequence of data items, each associated with a timestamp. Data streams arise naturally in electrical engineering as signals or

T. Andreasen et al. (Eds.): ISMIS 2014, LNAI 8502, pp. 476–485, 2014.
© Springer International Publishing Switzerland 2014

in environmental sciences as measurements of temperature or other quantities. Another example is the geographic location (φ_t, λ_t) of a mobile device at time t.

Combining both kinds of systems seems to be necessary at the current state of art as there seems to be no system which has all the benefits of a combination. MapReduce together with a distributed file system is well-suited for distributing data among machines and parallel batch processing. Stream processing requires on the other hand on-going or continuous computations. Conversely, stream processing engines are well-suited for their purpose since, for example, they work in-memory only.

Another problem is to design and implement a system which enables simple adaption of new models and algorithms. This requirement is crucial for data analysis in practice. Models and algorithms improve and change over time. There is a need to adapt and test them easily.

We are facing a situation that was partially considered in [11]. The system HiFi is designed for high fan-in. In our situation we have many data streams coming in. Another work which is close to ours is [5,6]. Hadoop Online Processing is a system to process data in MapReduce similar to data streams. On a technical level, [5,6] integrate pipelines into the MapReduce system. Both approaches try to build a system from scratch. Our intention is to combine existing systems. The reason is to have some choice in the implementation and to set-up a running system easily. The goal of this work in comparison to [11,5,6] is thus to identify the problems in combining SPEs and BDSs and solve them appropriately. Machine learning algorithm for the MapReduce framework are considered for example in [4,10].

1.1 Results

Summarizing our results, we designed a system and adapted algorithms for fast analysis of *data streams*, both *online* and *offline*. The implementation of the system and algorithms were simple, the effort reasonable. We needed to implement a single system component. It deals with *multiplexing* many data streams into a single data stream and with the problem of delayed data items. Additionally we implemented for convenience a component we named *replay*. It generates a data stream from stored data. We can thus apply continuous queries to historical data.

2   System Design and Data Model

We want a system for the purpose of analyzing data streams. We have two requirements. We want to process data online. For this purpose we use a stream processing engine (SPE). And we want to store data for later offline processing. For this purpose we use a big data storage (BDS) with MapReduce.

The reason for this choice is that the amount of data in online processing is moderate since we usually process only most recent data online. An SPE builds on this fact. For example, it does only in-memory computations. Storing time dependent data leads however to a massive amount of data usually. A BDS has the capability of distributing data across many machines.

The system design is depicted in Fig. 1. SPE and BDS work in parallel. An additional system component is denoted by MUX in Fig. 1. It functions as a multiplexer, i.e. it transforms many data streams into a single data stream. The result is a coherent data model for SPE and BDS. The multiplexer has another function too. It synchronizes data w.r.t. time. We discuss the data model first and data consistency and synchronization issues thereafter.

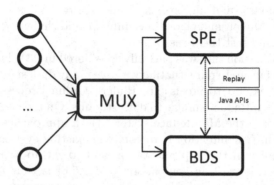

Fig. 1. System design

2.1   Data Model

We have a single data stream $d_1, d_2, \ldots$ We denote by $\mathsf{key}(d_i)$ the key, by $\mathsf{val}(d_i)$ the value, by $\mathsf{id}(d_i)$ the identifier, and by $\mathsf{ts}(d_i)$ the timestamp of d_i. The identifier $\mathsf{id}(d_i)$ and timestamp $\mathsf{ts}(d_i)$ constitute $\mathsf{key}(d_i)$. We assume that $\mathsf{key}(d_i)$ uniquely identifies $\mathsf{val}(d_i)$ and that $\mathsf{ts}(d_i) \leq \mathsf{ts}(d_j)$ for $i < j$.

Example. We get temperature values from s data sources. The timestamp ts and identifier id are both fixed-size strings. For example, $\mathsf{ts}(d) = \mathtt{20070611T172151}$ is the timestamp of the measurement and $\mathsf{id}(d) = \mathtt{000011}$ is the identifier of the data source, the weather station for example. The key $\mathsf{key}(d)$ is the concatenation of $\mathsf{id}(d)$ and $\mathsf{ts}(d)$.

The data stream $d_1, d_2, \ldots$ is generated by MUX. MUX sends the data items to SPE and BDS. Our data model is appropriate for many implementations of SPEs and especially for key-value storages, a special kind of BDSs. We may also just write the data to a distributed file system. We decided however to use a key-value storage and in particular HBase [9], a reimplementation of BigTable [3]. A reason is that $\mathsf{val}(\cdot)$ may be semi-structured. This means in our situation that $\mathsf{val}(\cdot)$ follows no data scheme. A second reason is that it allows key-value queries. A third reason is locality of data.

Locality of Data. Locality of data is important to avoid network traffic. Different BDSs also provide different mechanisms to distribute the data. A distribution mechanism has to decide which data items stay on the same or nearby machine. In the context of key-value storages it is the key which a mechanism for distributing the data gets as input. Range-based mechanisms divide the keys into ranges. A range is a list of consecutive keys, for example, in lexicographical order if the keys are strings. Ranges are then stored locally. BigTable [3] and HBase [9] support range-based mechanisms.

Example cont. If we order our data items lexicographically per key, they are ordered per id first, i.e. data sources, and then by ts, i.e. time. Computing the average temperature at the 11. of June, 2007 of data source number 11 favors computations with local data due to the definition of the key.

2.2 Data Consistency

Running two systems with partially the same data in parallel leads to data redundancy and thus to possible data inconsistency. The problem is more subtle than it seems at first sight. To understand the problem we need to introduce the concept of *application time* and *system time* of an SPE. The application time in our data model is $\mathsf{ts}(d)$ of some data item d. The system time of d is the time it enters the SPE. We also need to understand how the SPE organizes its sliding windows w.r.t. application time. One possibility is the following. (It is implemented in the SPE we are going to use [8].)

Sliding Window Rule. A new data item d arrives for a sliding window. While $|\mathsf{ts}(d) - \mathsf{ts}(d_{\mathrm{oldest}})| > \Delta$ remove some oldest data item d_{oldest} until the condition is satisfied. Insert d.

We can conclude eventual consistency w.r.t. to updates in the BDS for many situations. First, we assume there is no interaction between SPE and BDS. Second, we assume that a new data item d' with $\mathsf{ts}(d') > \mathsf{ts}(d_{\mathrm{oldest}})$ arrives continuously. Assume we change the value of some data item d in the BDS. The SPE operates on sliding windows. It only adds and removes elements from the sliding window. The value of a sliding window depends only on the elements in it. The sliding window rule implies that the data item remains in the sliding window iff no new data item d' with $|\mathsf{ts}(d') - \mathsf{ts}(d)| > \Delta$ arrives. By our assumption d gets removed from any sliding window at some point and our system gets thus consistent w.r.t. d.

Basic Time Synchronization. We remarked above that the multiplexer MUX also does time synchronization. The reason for this is that some SPEs (such as [8]) implement the sliding window rule: It is possible that $\mathsf{ts}(d) < \mathsf{ts}(d_{\mathrm{oldest}})$ for a new item d and some oldest item d_{oldest}, i.e. d arrives too late. Although there are situations where such behavior makes sense we decided to exclude this behavior for reasons of simplicity. MUX sends a data item d from source $\mathsf{id}(d)$

to the SPE if it is not too late. For this purpose MUX stores a timestamp t for every data source. If $\mathsf{ts}(d) \geq t$ then d is sent to the SPE and t is set to $\mathsf{ts}(d)$. MUX thus guarantees the relative order per data stream w.r.t. application time. This holds just for the SPE. MUX sends any data item, independent of its time of arrival, to the BDS.

2.3 Replay and Query Language

We also included a replay capability. Data stored in the BDS can be fed into the SPE. Combining the BDS with an SPE in this way, the query language of the SPE can be utilized to function as a query language for the BDS. This is convenient because many BDSs and in particular key-value storages do not have a declarative query language available for processing data streams or even no query language at all. Query languages for stream processing are available for the stream processing engines STREAM [13], Aurora [2], Borealis [1]. For an example of a query in the language of Esper [8]. We also note here that not every system for processing streams has a query language. An example is S4 [14].

3 Implementation

3.1 Components and Interfaces

As a stream processing engine we deployed Esper [8]. The Apache project HBase, which is a reimplementation of BigTable [3], was used as a BDS. HBase uses Hadoops distributed filesystem (HDFS) to store its data. It provides redundant storage to achieve fault tolerance and in conjunction with the MapReduce framework parallel processing capabilities.

BDS and SPE can interact in a number of ways. For example, results from offline analysis such as aggregations can be sent directly to the SPE via Java APIs or forwarded to a Relational Database Management System (RDBMS). Continuous queries running inside Esper can then access data from the RDBMS.

3.2 Experiments

We conducted three different experiments during the evaluation of the HBase cluster.

A *test data generator* generated random measurement values according to our data model to simulate data sources. A *test data scanner* executed a query which reads all measurement values of one day and one particular data source. This is the ad-hoc query most commonly used in our application scenario. Finally, a *MapReduce Aggregator* job which computes aggregation values (avg, min, max) on a day level for all data sources in the cluster.

The overall goal was to verify that a Hadoop cluster can handle the amount of data and queries used in our application scenario and also if our data model fits our requirements.

We used a virtualized cluster running on 2 physical servers with a cumulated 24 cores, 96 GB RAM and 4.8 TB hard disk space (16 * 300 GB). Using the physical servers we configured a virtualized HBase cluster with 8 data nodes. The resources were assigned evenly to the virtual machines.

Test Data Generator. The test data generator ran on 4 virtualized client machines to simulate parallel inserts of measurement values. Each client executed 100 threads simulating data sources. Each thread then inserted measurement values comprising approximately 3 years of data. In total 4.4 billion measurement values or about 550 million HBase rows where inserted into the cluster.

As the full row key is stored with each HBase cell, i.e. measurement value, table compression has to be used to prevent exhaustive use of storage space. Several compression formats are available, each having a different trade-off between compression and speed (see Table 1). In our application scenario we use Snappy, because it offers the highest decompression speed.

Table 1. Test data generator results

	GZ	LZO	Snappy
storage space (GB)	73.45	154.35	161.56
HBase rows written per sec.	4314	4287	6743
measurement values written per sec.	34510	34296	53941

Test Data Scanner. The goal of the test data scanner was to test the response time of our cluster for ad-hoc queries. As an evaluation query we used the most common ad-hoc query in our application scenario, "Select all measurement values from (e.g.) temperature for one particular day and data source", which yields 288 data rows. We used our 4 virtualized test machines, each simulating 8 clients for a total of 32 clients accessing the cluster simultaneously. The average response time was <0.015 seconds on all machines (see Table 2).

Table 2. Test data client results

	VM1	VM2	VM3	VM4
avg. response time in sec.	0.0139	0.0144	0.0136	0.0138
avg. HBase rows retrieved per sec.	1823	1752	1846	1831

Aggregator. In order to evaluate the batch processing performance of our HBase cluster, we calculated aggregates (avg, max, min) for each day and all data sources in the cluster using a MapReduce job. Gradually increasing the amount of data, we calculated the aggregates using different job settings. In the end, the MapReduce job processed 1.8 billion rows which equals approximately 14.5 billion measurement values. Table 3 shows the execution times of 3 independent runs using local combiners and 8 reduce tasks (one on each data node).

Table 3. Aggregation test results

	run 1	run 2	run 3
execution time (hh:mm:ss)	6:45:00	6:42:47	6:45:54
HBase rows processed per sec.	74409	74818	74244

4   Data Analysis

An important question which remained open so far is whether we can use our system for data analysis. The answer to this question depends however strongly on the application.

We described our data model above. We recall that a data item has a key and a value. In what follows the value is either $x \in \mathbf{R}$ or $(x, y) \in \mathbf{R}^2$. Our restriction has the mere purpose to simplify the presentation since many SPEs allow to project, filter, and join different streams. We can thus generate streams with complex structured values from streams with simple values.

Sliding Window. A *sliding window* at time t and with size Δ are all data items d such that $t - \Delta \leq \mathsf{ts}(d) \leq t$. Stream processing engines implement this model. An algorithm in this model gets updates of data items to be added and removed. The resulting data items after a series of updates form a sliding window at some later time. The algorithm needs to process these updates (add or remove). Most important it should provide quick access to the value of the sliding window. The value of a sliding window is usually a statistic, e.g. the sample mean or sample median.

A perhaps subtle detail of this model is that updates have elements to be added *and* removed. This enables highly efficient aggregations, e.g. counting n elements with memory space $O(\log(n))$ bits. In this case the algorithm does not need to store the elements explicitly.

MapReduce. The MapReduce system [7] distributes the data, a list of n key-value pairs, from the storage to mapper instances, to combiner instances, to reducer instances in the following way:

Mapper. A mapper gets a single key-value pair and outputs a list of key-value pairs.

Combiner. The input of a combiner is a key-list pair $(k, (v_1, \ldots, v_m))$. Every pair $(k, v_1), \ldots, (k, v_m)$ emerged from a mapper instance. Note that every key-value pair has the same key. The output is another key-list pair. We also note that mapper and combiner instances are executed on the same machine the input data is stored [7].

Reducer. The input of a reducer is a key-list pair $(k', (v'_1, \ldots, v'_m))$. It results from *all* combiner outputs with the key k'.

Example. Let $d_1, d_2, \ldots$ be a data stream. As an example we consider summing values $v \in \mathbf{R}$. Let $v = \mathsf{val}(d)$ and $k = \mathsf{key}(d)$ for some d. We map (k, v) to $(0, (v))$. The combiner gets $(0, (v_1, \ldots, v_m))$. It sums up all values $v := v_1 + \ldots + v_m$ and outputs the key-list pair $(0, (v))$. The reducer does the same as the combiner. The combiner computes partial sums and the reducer the total sum.

The following definition is intended to help us to decide if an algorithm fits well our needs. It is motivated by [4] which consider MapReduce for parallel computing. Distributing data among machines and locality of data is not an issue in [4].

Definition 1. *We define n_c as the maximum size of a list the combiner gets as input. The maximum is over all input instances of n key-value pairs and instances of the combiner. Moreover, let $M \geq 1$ be the number of combiner instances.*

Example Cont. Let n be the number of input key-value pairs. Assume that $n_c = \Theta(\frac{n}{M})$. The mapper works in time $O(1)$, the combiner in time $O(\frac{n}{M})$, and the reducer in time $O(M)$.

The example shows that summation fits well to MapReduce. This was for example already observed in [4]. The assumption $n_c = \Theta(\frac{n}{M})$ can be seen as an optimal case. It may happen contrary to this optimal case that one out of the M combiner instances gets a list with all n input values pairs, i.e. $n_c = n$. We depicted the optimal case in Fig. 2 for Ex. 4. At the bottom level we have n key-value pairs split into 6 chunks. There is one mapper instance per chunk (second level). There are $M = 3$ combiner instances (third level). Ideally, every combiner instance gets a list of $\lfloor \frac{n}{3} \rfloor$ or $\lceil \frac{n}{3} \rceil$ elements as input. Finally, there is a single reducer instance (fourth level).

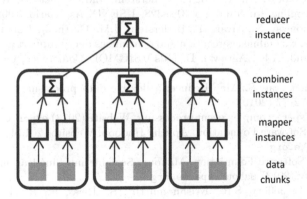

Fig. 2. Summation example in MapReduce

5   Future Research

Our system design has the benefit that it can be implemented by current technologies without much effort. Running two systems in parallel leads to problems of data consistency and time synchronization that we addressed.

An open problem is the extra effort for implementing algorithms for both programming models. We partially solved this problem with a replay function: It takes data from the BDS and puts it into a *single* SPE. It can thus deal only with data streams of moderate size. It remains open how to deal with a large amount of data streams. One research direction is to have an SPE per machine or combiner instance.

Acknowledgements. This work has been supported by the COMET-Program of the Austrian Research Promotion Agency (FFG).

References

1. Abadi, D.J., Ahmad, Y., Balazinska, M., Cetintemel, U., Cherniack, M., Hwang, J.-H., Lindner, W., Maskey, A.S., Rasin, A., Ryvkina, E., et al.: The design of the borealis stream processing engine. In: CIDR (2005)
2. Abadi, D.J., Carney, D., Çetintemel, U., Cherniack, M., Convey, C., Lee, S., Stonebraker, M., Tatbul, N., Zdonik, S.: Aurora: a new model and architecture for data stream management. The VLDB Journal 12(2), 120–139 (2003)
3. Chang, F., Dean, J., Ghemawat, S., Hsieh, W.C., Wallach, D.A., Burrows, M., Chandra, T., Fikes, A., Gruber, R.E.: Bigtable: A distributed storage system for structured data. ACM Trans. Comput. Syst. 26(2) (2008)
4. Chu, C.-T., Kim, S.K., Lin, Y.-A., Yu, Y., Bradski, G.R., Ng, A.Y., Olukotun, K.: Map-Reduce for machine learning on multicore. In: Schölkopf, B., Platt, J.C., Hoffman, T. (eds.) NIPS, pp. 281–288. MIT Press (2006)
5. Condie, T., Conway, N., Alvaro, P., Hellerstein, J.M., Elmeleegy, K., Sears, R.: Map-Reduce online. In: NSDI, pp. 313–328. USENIX Association (2010)
6. Condie, T., Conway, N., Alvaro, P., Hellerstein, J.M., Gerth, J., Talbot, J., Elmeleegy, K., Sears, R.: Online aggregation and continuous query support in mapReduce. In: Elmagarmid, A.K., Agrawal, D. (eds.) SIGMOD Conference, pp. 1115–1118. ACM (2010)
7. Dean, J., Ghemawat, S.: Map-Reduce: a flexible data processing tool. Commun. ACM 53(1), 72–77 (2010)
8. EsperTech. Esper – complex event processing. Website (2013) esper.codehaus.org
9. The Apache Software Foundation. Apache Hadoop. Website (2013), hadoop.apache.org
10. The Apache Software Foundation. Mahout: Scalable machine-learning and data-mining library (2013) mahout.apache.org
11. Franklin, M.J., Jeffery, S.R., Krishnamurthy, S., Reiss, F., Rizvi, S., Wu, E., Cooper, O., Edakkunni, A., Hong, W.: Design considerations for high fan-in systems: The HiFi approach. In: CIDR (2005)
12. Ghemawat, S., Gobioff, H., Leung, S.-T.: The Google file system. In: Scott, M.L., Peterson, L.L. (eds.) SOSP, pp. 29–43. ACM (2003)

13. Motwani, R., Widom, J., Arasu, A., Babcock, B., Babu, S., Datar, M., Manku, G., Olston, C., Rosenstein, J., Varma, R.: Query processing, resource management, and approximation in a data stream management system. In: CIDR (2003)
14. Neumeyer, L., Robbins, B., Nair, A., Kesari, A.: S4: Distributed stream computing platform. In: Fan, W., Hsu, W., Webb, G.I., Liu, B., Zhang, C., Gunopulos, D., Wu, X. (eds.) ICDM Workshops, pp. 170–177. IEEE Computer Society (2010)

Representation and Evolution of User Profile in Information Retrieval Based on Bayesian Approach

Farida Achemoukh and Rachid Ahmed-Ouamer

LARI Laboratory, Mouloud Mammeri University, Tizi-Ouzou, Algeria
{achemoukh.farida,ahm_r}@yahoo.fr

Abstract. In the web personalization how to represent user profile is one of the key issues. The user profile refers to his/her interests which change over time. This paper, presents a personalized search approach for representation and evolution of the user profile, based on dynamic bayesian network. The theoretical framework provided by these networks allows to infer and to evolve the user profile from his /her interactions with the search system. An experimental evaluation was designed to appraise the exploitation impact of the user profile defined by his/her interests on the search results relevance.

Keywords: Personalized Search, Short term user profile, long term user profile, Dynamic Bayesian Network.

1 Introduction

To find information adapted to the user's needs, he/she used web search, which is considered as a privileged information source. In both ways how to acquire the user's interest and how to represent it is the key issues. Various personalization strategies were proposed in literature [Pitkow et al 2002], [Teevan et al 2011], [Tan et al 2006]. The user profile refers to his/her interests which change over time [Shen et al 2005],[Tan et al 2006]. User profile can be short-term and long-term types. Short term user profile limited to a single search session, which contains a consecutive interactions with the search system with a coherent user interests during a short period of time. Long term user profile hold persistent user interests generally stable for a long time.

The main contributions in this paper are as follows: First, we proposed a bayesian network modeling interaction between the user and the search system, which combines a query and relevant documents used as a source of information to infer the user interest, and then we modeled the user interests changing over time, defining the short term and long term user profile. The profile will be exploited in the search process in order to re-rank search results. Finally, we evaluated the proposed approach on clueweb09 (clueweb09_English1) test collection and we compared it with the baseline classical search model, noted BM25 Model [Robertson et al 1998] provided by the Terrier-3.5 platform (http://ir.dcs.gla.ac.uk/terrier/). Experimental results showed that our approach gives an improvement over the baseline model.

T. Andreasen et al. (Eds.): ISMIS 2014, LNAI 8502, pp. 486–492, 2014.
© Springer International Publishing Switzerland 2014

The rest of the paper is structured as follows: the next section gives an overview of related works. In Section 3, we describe our approach to represent and evolve the user profile in search session based on bayesian networks. Section 4 presents the experiments on performance evaluation of our approach. In the last section, we present our conclusion and the future works.

2 Related Work

A user model allows representing data that characterize him/her. Without user profile, an information retrieval system behaves in the same way with all users, who have different knowledge's, preferences, goals and interests.

The personalization of search consists of the user modeling way, which is a process at different steps, namely, the user's profile representation, construction and then its exploitation in the search process [Gowan 2003], [Shen et al 2005],[Tan et al 2006] represent the user profile as one or more vectors of terms. In others, a user profile is represented as a hierarchical concepts structure representing the interest's domains [Micarelli et al 2004], [Chirita et al 2005] or with a structured model of predefined dimensions (personal data, interests, preferences... etc) [Kostadinov et al 2005]. Others studies use external domain ontology as an additional evidence to model the user profile as a set of concepts issued from predefined ontology[Daoud et al 2009].

The construction of the user profile consists of collecting information that represent the user, it can be done in two ways: that users specified explicitly [Chirita et al 2005], or implicitly, from the consulted documents and the user behavior [Shen et al 2005].

User's interest defining the user profile could change with time; this generated the problem of the user profile evolution. In most of personalization search systems the profile evolution consists to adapt its content to the variations of the user information needs. In [Chen and al 2002] the concept of the life cycle of the interest is introduced and the user profile structure can be modulated while the user interests change.

User profiles can be short-term ones and long-term ones. Short-term user profile is related to a single search session, which contains a sequence of user interactions with a coherent information need during a short period of time. Long- term user profile holds persistent user interests generally stable for a long time [White et al 2010], [Cao et al 2009]. [Mihalkova et al 2009], [Shen et al 2005], [Daoud et al 2009], [Xiang et al 2010].

The work presented in this paper differs from previous works in several important ways. First, we present a probabilistic formalism for modeling user's interests exploitation and we present a user profile evolution modeling with dynamic bayesian network, and the impact of short term user profile exploitation on search results.

3 User Profile Representation and Evolution Approach

The user profile evolution is modeled by a dynamic Bayesian networks. It represents temporal dependencies between interactions at different times, this means that if we have a sequence of user interactions modeled by bayesian networks models for different time slices $t \in [0..T-1]$, i.e. $I_0, I_1, I_2,.....I_{T-1}$, we have used temporal links to connect nodes in the bayesian networks over different time slices.

The model components consist of the sequences of random variables S, I, and Du for T consecutive time slices $t \in [0..T-1]$. With S, I, and Du denote respectively the search session, the user interaction and the remaining stay-time at time t. Each user interaction with the system is modeled as a directed acyclic graph composed on two sets, nodes set V and arcs set E. Each arc represents the relationship between two nodes.

3.1   Probabilistic Description

In this section we present the description of probabilities involved in the model

3.1.1   Search Session
Let $S = \{1,...m,...,M\}$ be the set of M search sessions, where the random variable S_t takes value. The session probability is defined for each $m \in S$ by

$$P (S_t = m) = v_{t,m} \tag{1}$$

Where $v_{t,m}$ is the probability that the session m is being at time t . The probability distribution v_t is then a sum-to-one vector of M elements (M=number of sessions).

3.1.2   Initial Interaction Probability
We consider a set of N user interactions, where the random variable I_t takes value at $I = \{1, 2...N\}$.

Each user interaction with the system at time t, regroups the query q_t and the inferred user's interest noted $c_{k,t}$. and is represented by a triplet < Q, C, F> where $Q=q_t$ represents the query submitted at the interaction, $C=c_k$ represents the user profile defined by his/her interest and F is a relevance search function that estimates the relevance of document d_j with   respect to the   user's interest c_k and   query q, expressed by:

$$F(q, d_j, c_k) = P(d_j / q) \times P(c_k / q) \times P(q) \tag{2}$$

The resulted probabilities for each query q with instantiations of each document dj and interest ck are represented in a matrix $X_{n,m}$ of dimension (number of documents× number of user's interests) [Achemoukh et al 2012].

However the probability of an initial user interaction $P(I_0=n)$ regrouping query q_0 submitted at time t=0 and the user's interest $c_{k,0}$ inferred at time t=0 according to the search session is defined by:

$$P(I_0=n|S_0=m) = P(q_0, c_{k,0}|S_0=s)=A_{0,m,n} \tag{3}$$

Where $A_{0,m,n}$ gives the probability to start   with user interaction ($I_0=n$)given the initial   search session ($S_0=m$) . The probability distribution A_0 is then a matrix of M (number of sessions) rows and N (number of Interactions) columns.

3.1.3   Initial Stay Time Probability
The stay time probability gives the distribution of the time spent in each possible (session, user interaction) configuration. Let $Du = \{1...d,...,D\}$ where the random

variable Du_t takes value, with D denotes the maximum number of time units, that can spend   any user interaction and session configurations.

For the initial user interaction $I_0=n$ and session $S_0=m$, the initial stay time probability distribution is given by:

$$P (Du_0=d|I_0=n, S_0=m)=L_{0,n,m,d} \tag{4}$$

Where $L_{0,n,m,d}$ is the probability to spend d time units in the initial interaction and the session configuration. It is represented as a matrix of NM (number of interactions× number of sessions) rows and D columns (number of time units).

3.1.4   Interaction Transition Probability

To define the user profile evolution indicated by the user's interest changing, throughout the various user interactions with the system, we are interested at the interactions transition probabilities $P(I_t|I_{t-1},Du_{t-1}, S_t)$, $t >0$; this corresponds to:

$$P(It|I_{t-1},Du_{t-1}, S_t)= P(q_t, c_{k,t}| q_{t-1} , c \acute{k}_{,t-1}, Du_{t-1}, S_t) \tag{5}$$

Where $c_{k,t}$ and q_t are the user's interest and the query of the current user interaction $(I_t=n)$, $c\acute{k}_{,t-1}$ is the user's interest of the past user interaction $(I_{t-1}=n')$ and Du_{t-1} is the remaining stay time to be spent in the   past user interaction I_{t-1} .
Two cases have to be dealt with:

1. Du t-1 =1: indicates that the stay time in the past user interaction I_{t-1} is up. Therefore a user interaction   transition occurs at time t, according to the following distribution:

$$P(I_t=n|I_{t-1} =n',Du_{t-1}=1, S_t =m)=A_{n',m,n} \tag{6}$$

Where $A_{n',m,n}$ is a matrix of   NM (number of user interactions× number of sessions) rows and N (number of user interactions)columns giving the probability of user interaction transition from I_{t-1} to I_t in the session S_t.

2. $Du_{t-1}>1$: indicates that it remains some time to be spent in the past user interaction I_{t-1}. This means that there is no the user interaction transition, which is defined by:

$$P(I_t=n|I_{t-1} =n',Du_{t-1}>1, s_t =m)=Id_{n',n} = \begin{cases} 1 \text{ if } n = n' \\ 0 \text{ otherwise} \end{cases} \tag{7}$$

3.1.5   Stay Time Transition Probability

The stay time transition probability distribution aims to update the remaining stay time in the user interaction and to provide a new stay time when a transition occurs. Two cases have to be considered:

1. Du $_{t-1}$ =1: indicates that a user interaction transition occurs and a new stay time is selected according to the following distribution:

$$P (Du_t=d|I_{t-1}=n', Du_{t-1}=1 , S_t=m, I_t=n)=L_{n',m,n,d} \tag{8}$$

Where $L_{n',m,n,d}$ is a matrix of N^2M ((number of user interactions)$^2 \times$ number of sessions)rows and D (number of time units) columns specifying the stay time probability distribution given the session S_t, the current user interaction I_t and then the past   user interaction I_{t-1}.

2. Du $_{t-1}$=d', with d' >1:indicates that the stay in the user interaction I_{t-1} is not over and the remaining stay time has to be deterministically counted down as follows:

$$P (Du_t=d | I_{t-1}=n', Du_{t-1}=d', S_t=m, I_t=n) = C_{d',d} = \begin{cases} 1 & \text{if} \quad d=d'-1 \\ 0 & \text{otherwise} \end{cases} \qquad (9)$$

Where $C_{d',d}$ is square matrix of D (number of time units) ,defined as the identity matrix with   D-1 rows concatenated with a zero-valued column on the right.

In order to define the user profile evolution, we consider a set of user interactions trajectories, denoted by R= ($R_{0,n,d}$, $R_{1,n,d}$,...,$R_{T-1,n,d}$)   .

Therefore, the probability of any user interactions trajectories of length t $P(R_{t,n,d})$, $0<=t<=T-1$ is represented by: $P(R_{t,n,d})=P(R_{0,n,d},R_{1,n,d},..., R_{T-1,n,d})$, calculated according to the formula (10).

$$P(R_{t,n,d})= \begin{cases} P(R_{0,n,d}) = P(S_0 = m, Ac_0 = n, Du_0 = d) = \sum_{m \in S} v_{0,m} A_{0,m,n} L_{0,m,n,d} & \text{if } t=0 \\ \sum_{n' \in R_{t-1}} \sum_{d'=1}^{D} P(R_{t-1,n',d'}) \sum_{m \in S} v_{t,m} A_{n',m,n} L_{n',m,n,d} & \text{if } t>0 \end{cases} \qquad (10)$$

4   Experiments and Results

To evaluate the performance of our approach we used a clueweb09_English1 test collection   contains a set of documents, queries issued from the TREC Session Track2011 collection, includes 61 main queries (topics), 202 interactions queries and 75 currents queries, relevance judgments and sessions. Each session includes a main query (topic), different interactions queries and a current query, and we compared our approach with the baseline classical search model, noted BM25 Model [Robertson et al 1998] provided by the Terrier-3.5 platform. Table 1 shows the results obtained with the currents queries by our approach and the BM25 classical search model [Robertson et al 1998] at P5, P10 and mean average precision (MAP).

Table 1. Performance Comparison of our approach and BM25 model

Measure	BM25	Our approach	Our approach % BM25
P5	0,5290	0,9354	43,44%
P10	0,4096	0,8064	49,21%
MAP	0.1350	0,1030	-31,06%

We notice that our approach gains a statistically significant improvement over the BM25 Model of the Terrier-3.5 search system over P5, P10. More particularly, our approach brings an average improvement of 43.44% in P5 and 49.21% in P10, but there is a decrease in the mean average precision (MAP). However, these results are acceptable given the values of P5 and P10.

To get a better understand the difference between our approach and the BM25 model, figure (fig.1) shows the comparison of average improvement between the two models for each current query.

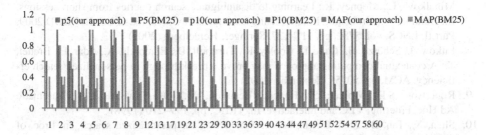

Fig. 1. Query by query comparison between the two models

5   Conclusion

In this paper we described the user profile construction and evolution model in the personalized search. The user profile evolution is modeled by a dynamic Bayesian network which represents temporal dependencies between interactions at different times. To evaluate the performance of our approach, and the user profile evolution impact on the search results, we conducted experiments on the clueweb09 (clueweb09_English1) test collection and we compared our approach with the BM25 Model of the Terrier-3.5 search system. The obtained results show the ability of our approach to infer short term user profile in search session. Its inclusion in the matching model provides a good results ranking. As perspectives of this work, we plan to test and compare our approach to other personalized search approaches and to show the impact of exploitation of user profile in improving the search quality for a recurring query.

References

1. Cao, H., Hu, D.H., Shen, D., Jiang, D., Sun, J., Chen, E., Yang, Q.: Context aware query classification. In: Proc. SIGIR, pp. 3–10 (2009)
2. Chen, C., Chen, M., Sun, Y.: A self-adaptive personal view agent. Journal of Intelligent Information Systems, 173–194 (2002)
3. Chirita, P.-A., Nejdl, W., Paiu, R., Kohlschutter, C.: Using ODP Metadata to Personalize Search. In: Proc. 28th Ann. Int'l ACM SIGIR Conf. Research and Development in Information Retrieval (SIGIR 2005), pp. 178–185 (2005)
4. Daoud, M.L., Tamine-Lechani, M.: A session based personalized search using an ontological user profile. In: SAC 2009: Proceedings of the ACM Symposium on Applied Computing, New York, pp. 1732–1736 (2009)

5. Li, L., Yang, Z., Wang, B., Kitsuregawa, M.: Dynamic adaptation strategies for long term and short-term user profile to personalize search. In: Dong, G., Lin, X., Wang, W., Yang, Y., Yu, J.X. (eds.) APWeb/WAIM 2007. LNCS, vol. 4505, pp. 228–240. Springer, Heidelberg (2007)
6. Micarelli, A., Gasparetti, F., Sciarrone, F., Gauch, S.: Personalized Search on the World Wide Web. In: Brusilovsky, P., Kobsa, A., Nejdl, W. (eds.) Adaptive Web 2007. LNCS, vol. 4321, pp. 195–230. Springer, Heidelberg (2007)
7. Mihalkova, L., Mooney, R.: Learning to disambiguate search queries from short sessions. In: Buntine, W., Grobelnik, M., Mladenić, D., Shawe-Taylor, J. (eds.) ECML PKDD 2009, Part II. LNCS, vol. 5782, pp. 111–127. Springer, Heidelberg (2009)
8. Pitkow, J., Schütze, H., Cass, T., Cooley, R., Turnbull, D., Edmonds, A., Adar, E., Breuel, T.: A contextual computing approach may prove a breakthrough in personalized search efficiency. ACM, pp. 50–55 (September 2002)
9. Robertson, S.E., Walker, S., Hancock-Beaulieu, M.: Okapi at TREC-7: Automatic Ad Hoc, Filtering, VLC and Interactive. In: TREC, pp. 199–210 (1998)
10. Shen, X., Tan, B., Zhai, C.: Implicit user modeling for personalized search. In: Proc. of CIKM, pp. 43–50 (2005)
11. Tan, B., Shen, X., Zhai, C.: Mining long-term search history to improve search accuracy. In: KDD 2006: Proceedings of the 12th ACM SIGKDD International Conference on Knowledge Discovery and Data Mining, pp. 718–723. ACM, New York (2006)
12. Teevan, J., Liebling, D., Geetha, G.R.: Understanding and prediction personal navigation. In: Proc.WSDM, pp. 85–94 (2011)
13. White, R.W., Bennett, P.N.: Precting short term interests using activity based search context. In: Proc. CIKM, pp. 1009–1018 (2010)
14. Xiang, B., Jiang, D., Pei, J., Sun, X., Chen, E., Li, H.: Context aware ranking in web search. In: Proc. SIGIR, pp. 451–458 (2010)

Creating Polygon Models for Spatial Clusters

Fatih Akdag, Christoph F. Eick, and Guoning Chen

University of Houston, Department of Computer Science, USA
{fatihak,ceick,chengu}@cs.uh.edu

Abstract. This paper proposes a novel methodology for creating efficient polygon models for spatial datasets. A comprehensive analysis framework is proposed that takes a spatial cluster as an input and generates a polygon model for the cluster as an output. The framework creates a visually appealing, simple, and smooth polygon for the cluster by minimizing a fitness function. We propose a novel polygon fitness function for this task. Moreover, a novel emptiness measure is introduced for quantifying the presence of empty spaces inside polygons.

Keywords: Spatial data mining, Polygon Models for Point Sets, Spatial Clustering, Polygon Fitness Function, Polygon Emptiness Measure.

1   Introduction

Polygons serve an important role in the analysis of spatial data. In particular, polygons can be used as a higher order representation for spatial clusters, such as for defining the habitat of a particular type of animal, for describing the location of a military convoy consisting of a set of vehicles, or for defining the boundaries between neighborhoods of a city consisting of sets of buildings. It is computationally much cheaper to perform certain calculations on polygons than on sets of objects. For example, polygons have been used to describe the functional regions of a city [1]. A given location can be assigned to one of those functional regions efficiently by checking in which polygon the location is included. Moreover, relationships and changes between spatial clusters can be studied more efficiently and quantitatively by representing each spatial cluster as a polygon. Polygon analysis is particularly useful to mine relationships between multiple related datasets, as it provides a useful tool to analyze discrepancies, progression, change, and emergent events [2].

However, there is not an established procedure in the literature on how to derive polygonal models from spatial clusters. The objective of the research described in this paper is to find an optimal set of polygons for two dimensional spatial clusters. The input of this process is a spatial cluster containing a set of points and its output is a polygon—the model of the cluster. As shown in Figure 1, many different polygon models (or a set of polygons as in Figure 1e) can be generated for the same set of points. Therefore, it is desirable to define application specific criteria for evaluating different polygon models. Coming up with such criteria and evaluation measures is the focus of this paper.

T. Andreasen et al. (Eds.): ISMIS 2014, LNAI 8502, pp. 493–499, 2014.
© Springer International Publishing Switzerland 2014

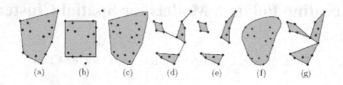

Fig. 1. Different shapes generated for the same set of points (taken from [3])

Main contributions of this paper include:

- A novel quantitative polygon fitness function is introduced to guide the generation of polygons from point clouds, alleviating the parameter selection problem when using existing polygon generation methods.
- A novel emptiness measure is introduced that quantifies the presence of empty areas in a polygon.

The rest of the paper is organized as follows. In Section 2, we compare the existing methods for creating polygon models. Section 3 provides a detailed discussion of our methodology. We present the experimental evaluation in Section 4, and Section 5 concludes the paper.

2   Related Work

Convex hulls are the simplest way to enclose a set of points in a polygon. However, convex hulls may contain large empty areas that are not desirable. Creating polygon models based on Voronoi diagrams or Delaunay triangulations is another commonly used approach. Matt Duckham et al. [4] propose a *"simple, flexible, and efficient algorithm for constructing a possibly non-convex, simple polygon that characterizes the shape of a set of input points in the plane, termed a Characteristic shape"*. The algorithm firstly creates the Delaunay triangulation of the point set—which actually is the convex hull of the point set—and then reduces it to a non-convex hull by replacing the longest outside edges of the current polygons by inner edges of the Delaunay triangulation until a termination condition is met.

The Alpha shapes algorithm, introduced by Edelsbrunner et al. [5] also uses Delaunay triangulation as the starting step and generates a hull of polylines, enclosing the point set and this hull is not necessarily a closed polygon. Thus, the Alpha shapes algorithm requires post-processing for creating polygons out of the polylines. Besides, there is no easy way of determining the proper parameter for Alpha shapes algorithm.

Chaudhuri et al. [6] introduce s-shapes and r-shapes; the proposed algorithm firstly generates a staircase like shape called s-shape, which is determined using an s parameter and then reduces it to a smoother shape using the r parameter. Authors state that *"to get a perceptually acceptable shape, a suitable value of r should be chosen, and there is no closed form solution to this problem"*.

A commercial algorithm, Concave Hull [7], generates polygons by using a method that is similar to the "gift-wrapping algorithm" used for generating convex hulls. It employs a k-nearest neighbors approach to find the next point in the polygon and creates a simple connected polygon unless the smoothness parameter k is too large and the points are not collinear. A density-based clustering algorithm, DContour [8] uses density contouring for generating polygonal boundaries. However, selecting the proper kernel width for the density estimation approach is non-trivial.

3   Methodology

In this section, we firstly discuss desirable polygon models and then propose a methodology that addresses the shortcomings of the existing algorithms.

3.1   Polygon Models

Figure 2 depicts three polygons that were created for the same cluster. The generated polygon in Figure 2a covers the largest area, and has the smallest perimeter, the least number of edges and the smoothest shape. However, it is obviously not a good model for the cluster because it includes large empty areas that are not relevant for the cluster. On the other hand, the polygon in Figure 2b has the largest perimeter, the most number of edges and covers the smallest area. Yet, it is also not a good representation for the cluster due to its ruggedness. Additionally, this polygon has a potential overfitting problem as it is quite complex and therefore more sensitive to noise. The polygon in Figure 2c balances the two objectives as it does not include large empty areas and has a low degree of ruggedness.

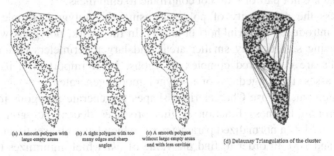

(a) A smooth polygon with large empty areas   (b) A tight polygon with too many edges and sharp angles   (c) A smooth polygon without large empty areas and with less cavities   (d) Delaunay Triangulation of the cluster

Fig. 2. Three polygons (a-c) generated for a cluster and its Delaunay Triangulation(d)

In the following, a polygon generation framework will be introduced which fits a polygon P to a set of spatial objects D minimizing the two objectives, we introduced earlier; namely, generating smooth polygons that have *a low emptiness with respect to D* and a low complexity. Additionally, we require that all objects in D are inside the polygon P. More formally, we define the problem of fitting a polygon P to a set of spatial objects as follows: Let D be a set of spatial objects in the cluster. Our goal is to find a polygon P that minimizes the following fitness function:

$$\phi(P,D)= Emptiness(P,D) + C * Complexity(P) \tag{1}$$

subject to the following constraint:

$$\forall o \in D: inside(o,P) \tag{2}$$

where C is a parameter which assesses the relative importance of polygon complexity with respect to polygon emptiness; e.g. if we assign a large value of C, smooth polygons will be preferred. *Emptiness(P,D)* is a quantitative emptiness measure that assesses the degree to which P contains empty regions with respect to D. *Complexity(P)* measures the complexity of polygon P.

3.2   Measuring the Emptiness and Complexity of P with Respect to D

In this paper we use Delaunay Triangulation DT(D) of a point set D, to define emptiness of polygon P which is supposed to model D. In general, as can be seen in Fig. 2d, areas with very low density can be identified as large triangles in the Delaunay triangulation; that is, triangles whose area is above a certain size θ. Let $P_{CONV}=(\cup_{t \in DT(D)}\, t)$ be the outer polygon of the DT(D) which is the convex hull of D. We define emptiness of a polygon P with respect to a point cloud D as follows:

$$Emptiness(P,D):= (\Sigma_{t \in DT(D) \wedge area(t) > \theta \wedge inside(t,P)}\ (area(t)-\theta)\ /area(P_{CONV}) \tag{3}$$

When assessing emptiness, we go through the triangles inside P and add the differences between θ and the area they cover, but only if the size of their area is above θ, and divide this sum by the area of the convex hull of D; be aware that p_{CONV} is not the area P covers, but a usually larger polygon which is the union of all triangles of Delaunay triangulation. It should be noted that when measuring emptiness, triangles that are not part of P do not contribute to emptiness.

We assess the complexity of polygons using the polygon complexity measure which was introduced by Brinkhoff et al. [9]. In this work, polygons with too many notches, having significantly smaller areas and larger perimeters compared to their convex hulls are considered complex polygons. Most importantly, it is a suitable measure to assess the ruggedness of a polygon model generated.

At the moment, we use Characteristic shapes to generate polygons in conjunction with the proposed fitness function as this produces decent polygon models. The algorithm itself has a normalized parameter *chi* which has to be set to an integer value between 1 and 100. In order to find the value of *chi* which minimizes the employed fitness function we exhaustively test all 100 *chi* values, and return the fittest polygon.

4   Experimental Evaluation

In this section, we present the experimental results using the fitness function ϕ defined in equation (1) for spatial clusters in a dataset called Complex8 [10]. Figures 3b-3d depict the polygons generated for the cluster in Fig. 3a using different C parameters and Table 1 reports the area, perimeter, emptiness, and complexity for polygons in these figures along with the optimal *chi* parameter values selected by the fitness function to create these polygons. We observe that, setting C=0.35 gives quite reasonable results. We also observe that setting C to higher values generates larger polygons. For very low C values, the generated polygons are quite complex having many edges and cavities.

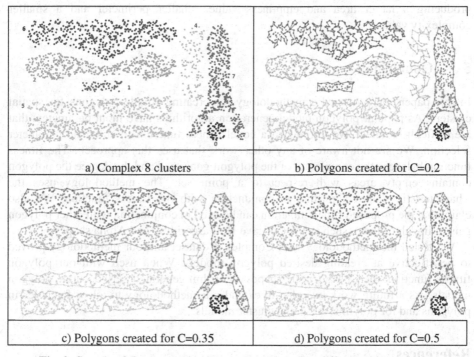

a) Complex 8 clusters	b) Polygons created for C=0.2
c) Polygons created for C=0.35	d) Polygons created for C=0.5

Fig. 3. Complex 8 Dataset and polygons generated using different chi parameters

Table 1. Statistics for polygons in Figures 3b-3d separated by comma in respective order. P0-P7 represent polygons for clusters 0-7 in the dataset and colored respectively.

	area	perimeter	emptiness	complexity	chi
P0	1088, 2030, 2030	328, 173, 173	0.077, 0.219, 0.219	0.49, 0.02, 0.02	37, 70, 70
P1	2697, 2697, 2741	287, 287, 286	0.144, 0.144, 0.148	0.052, 0.052, 0.046	34, 34, 37
P2	21492, 23107, 23107	1052, 997, 997	0.084, 0.096, 0.096	0.125, 0.072, 0.072	6, 13, 13
P3	9477, 18057, 20146	2465, 1058, 954	0.072, 0.192, 0.233	0.589, 0.118, 0.02	2, 5, 10
P4	4829, 8408, 11246	1171, 751, 561	0.057, 0.113, 0.197	0.562, 0.319, 0.089	5, 10, 16
P5	9007, 19122, 20413	2460, 968, 947	0.063, 0.19, 0.211	0.606, 0.043, 0.015	2, 8, 18
P6	17560, 34719, 35061	3019, 1018, 1003	0.044, 0.162, 0.168	0.632, 0.054, 0.04	4, 13, 14
P7	19759, 19759, 20807	1003, 1003, 984	0.042, 0.042, 0.049	0.188, 0.188, 0.17	8, 8, 14

It can be seen that quite different chi values are chosen for different spatial clusters by our approach. The polygon P4 (cyan-colored) best illustrates the effect of changing the C parameter. The generated polygon for P4 in Fig. 3b is very tight and rugged having a smaller area, larger perimeter, smaller emptiness and larger complexity values compared to polygons generated with larger C values. On the other hand, the generated polygon for P4 in Fig. 3d is smoother; it has fewer edges and empty areas producing a larger area and emptiness value, smaller perimeter and a smaller complexity value.

5   Conclusion

In this paper, we proposed a methodology for creating simple polygons for spatial clusters. As popular polygon model generation algorithms have input parameters that are difficult to select, we introduced a novel fitness function to automate parameter selection. We are not aware of any other work that uses this approach. The fitness function balances the complexity of the polygon generated and the degree the polygon contains empty areas with respect to a point set. The methodology uses the Characteristic shapes algorithm in conjunction with the fitness function. We also claim that the proposed fitness function can be used in conjunction with other polygon generating algorithms, such as the Concave Hull algorithm, and Alpha shapes.

We tested the methodology with Complex 8 dataset and our methodology proved to be effective at creating desired polygon models. When used with our polygon fitness function, the Characteristic shapes algorithm generated very accurate polygon models. As a future work, we plan to extend our methodology to allow for holes in polygons and for polylines in spatial cluster models.

References

1. Cao, Z., Wang, S., Forestier, G., Puissant, A., Eick, C.F.: Analyzing the Composition of Cities Using Spatial Clustering. In: Proc. 2nd ACM SIGKDD International Workshop on Urban Computing, Chicago, Illinois (2013)
2. Wang, S., Chen, C.S., Rinsurongkawong, V., Akdag, F., Eick, C.F.: A Polygon-based Methodology for Mining Related Spatial Datasets. In: Proc. of ACM SIGSPATIAL International Workshop on Data Mining for Geoinformatics (DMG), San Jose (2010)
3. Galton, A., Duckham, M.: What is the region occupied by a set of points? In: Raubal, M., Miller, H.J., Frank, A.U., Goodchild, M.F. (eds.) GIScience 2006. LNCS, vol. 4197, pp. 81–98. Springer, Heidelberg (2006)
4. Duckham, M., Kulik, L., Worboys, M., Galton, A.: Efficient generation of simple polygons for characterizing the shape of a set of points in the plane. Pattern Recognition 41, 3224–3236 (2008)
5. Edelsbrunner, H., Kirkpatrick, D.G., Seidel, R.: On the shape of a set of points in the plane. IEEE Transactions on Information Theory 29, 551–559 (1983)
6. Chaudhuri, A.R., Chaudhuri, B.B., Parui, S.K.: A novel approach to computation of the shape of a dot pattern and extraction of its perceptual border. Computer Vision and Image Understanding 68, 57–275 (1997)

7. Moreira, A., Santos, M.Y.: Concave hull: a k-nearest neighbours approach for the computation of the region occupied by a set of points. In: International Conference on Computer Graphics Theory and Applications GRAPP (2007)
8. Chen, C.S., Rinsurongkawong, V., Eick, C.F., Twa, M.D.: Change Analysis in Spatial Data by Combining Contouring Algorithms with Supervised Density Functions. In: Theeramunkong, T., Kijsirikul, B., Cercone, N., Ho, T.-B. (eds.) PAKDD 2009. LNCS, vol. 5476, pp. 907–914. Springer, Heidelberg (2009)
9. Brinkhoff, T., Kriegel, H.-P., Schneider, R., Braun, A.: Measuring the Complexity of Polygonal Objects. In: Proc. of the Third ACM International Workshop on Advances in Geographical Information Systems, pp. 109–117 (1995)
10. Salvador, S., Chan, P.: Determining the Number of Clusters/Segments in Hierarchical clustering/Segmentation Algorithm. In: ICTAI, pp. 576–584 (2004)

Skeleton Clustering by Autonomous Mobile Robots for Subtle Fall Risk Discovery

Yutaka Deguchi and Einoshin Suzuki*

Department of Informatics, ISEE, Kyushu University, Japan

Abstract. In this paper, we propose two new instability features, a data pre-processing method, and a new evaluation method for skeleton clustering by autonomous mobile robots for subtle fall risk discovery. We had proposed an autonomous mobile robot which clusters skeletons of a monitored person for distinct fall risk discovery and achieved promising results. A more natural setting posed us problems such as ambiguities in class labels and low discrimination power of our original instability features between safe/unsafe skeletons. We validate our three new proposals through evaluation by experiments.

1 Introduction

Among the various problems the elderly face in their daily lives, accidental falls are critical [3], e.g., in the U.S., falls occur 30-60% of older adults each year, and 10-20% of these falls result in injury, hospitalization, and/or death [3]. Several large-scale European projects use mobile robots for detecting such falls at home, e.g., GraffPlus, HOBBIT. We have constructed autonomous mobile robots that apply incremental clustering to human skeleton data of a monitored person for distinct fall risk discovery [4,5][1]

Although our previous method [5] yielded promising experimental results, it is not so effective in discovering subtle fall risk discovery, which is a more significant problem in terms of fall prevention. Moreover, Normalized Mutual Information (NMI) is inappropriate as an evaluation measure for this type of clustering, since the safe and unsafe skeletons are hard to be distinguished.

2 Skeleton Clustering with Instability Features for Fall Risk Discovery

We have constructed a monitoring system with two autonomous mobile robots to observe human $\mathbf{H}$ [2,4]. Robot k is equipped with a Kinect sensor, which can

---

* A part of this research was supported by JSPS, under Bilateral Joint Research Project between Japan and France, and KAKENHI 24650070 and 25280085.

[1] We previously formalized our monitoring system as conducting a positioning problem of two robots [2]. We then tackled the problem of clustering skeletons for distinct fall risk discovery [5] with one robot. Note that the two applications were mentioned very briefly in our introduction of our projects [4].

T. Andreasen et al. (Eds.): ISMIS 2014, LNAI 8502, pp. 500–505, 2014.
© Springer International Publishing Switzerland 2014

Table 1. ID i and weight w_i of joint $\boldsymbol{p}_{k,t,i}$, where L, C, and R represent left, center, and right, respectively

ID	0	1	2	3	4	5	6
Joint	Hip C	Spine	Shoulder C	Head	Shoulder L	Elbow L	Wrist L
Weight	6.90	6.90	10.45	3.55	8.50	2.50	1.30
ID	7	8	9	10	11	12	13
Joint	Hand L	Shoulder R	Elbow R	Wrist R	Hand R	Hip L	Knee L
Weight	0.40	8.50	2.50	1.30	0.40	12.50	7.80
ID	14	15	16	17	18	19	
Joint	Ankle L	Foot L	Hip R	Knee R	Ankle R	Foot R	
Weight	3.10	0.90	12.50	7.80	3.10	0.90	

be used to observe the skeleton $\mathcal{S}(k,t)$ of the monitored person $\mathbf{H}$ at time t. A skeleton $\mathcal{S}(k,t)$ consists of 20 joints $\{\boldsymbol{p}_{k,t,0}, \cdots, \boldsymbol{p}_{k,t,19}\}$ as shown in Table 1, where $\boldsymbol{p}_{k,t,i}$ represents the joint i observed by robot k at time t. Kinect outputs $\boldsymbol{p}_{k,t,i}$ as a point $(\boldsymbol{p}_{k,t,i}.x, \boldsymbol{p}_{k,t,i}.y, \boldsymbol{p}_{k,t,i}.z)$ in the 3D space with the position of the sensor as its origin. We use shift and rotation transforms so that $\boldsymbol{p}_{k,t,0}$ becomes the point of origin, and x and z axes become parallel to the ground, respectively.

We define our target problem as a clustering problem of skeleton data $\{\mathcal{S}(k,1), \ldots, \mathcal{S}(k,T)\}$ $(k = 1,2)$ observed by two robots, where T represents the number of the skeletons. As the clustering method, we use BIRCH [8], which is standard in data stream mining, as we did in [5]. BIRCH constructs micro-clusters from $\{\mathcal{S}(k,1), \ldots, \mathcal{S}(k,T)\}$ $(k = 1,2)$ then obtains a set of clusters $C = \{c_1, \ldots, c_{K(\mathcal{F},\Gamma,\theta_{\text{leaf}})}\}$ by merging them. Here $\mathcal{F}, \Gamma, \theta_{\text{leaf}}$ represent the set of instability features, the clustering method, and the absorption threshold of BIRCH, respectively. The latter is used to judge whether an example is absorbed in a micro-cluster.

Takayama et al. defined two features $\delta_{\text{HC}}(t), \alpha_{\text{W}}(t)$ to represent the instability of the skeleton [5]. $\delta_{\text{HC}}(t)$ is defined as the horizontal deviation of the center $\boldsymbol{g}_{k,t}$ of gravity of the body from the 0th joint $\boldsymbol{p}_{k,t,0}$ observed by robot k.

$$\delta_{\text{HC}}(t) = f(\boldsymbol{g}_{k,t}, \boldsymbol{p}_{k,t,0}) \tag{1}$$

$$f(\boldsymbol{a}, \boldsymbol{b}) = \sqrt{(\boldsymbol{a}.x - \boldsymbol{b}.x)^2 + (\boldsymbol{a}.z - \boldsymbol{b}.z)^2}, \qquad \boldsymbol{g}_{k,t} = \frac{\sum_{i=0}^{19} w_i \boldsymbol{p}_{k,t,i}}{\sum_{i=0}^{19} w_i}, \tag{2}$$

where w_i as shown in Table 1 represents the weight ratio of each joint [5]. $\alpha_{\text{W}}(t)$ is defined as the larger angle along the y axis of the neck - ankle line.

$$\alpha_{\text{W}}(t) = \max_{i=14,18} \left[\arctan \frac{f(\boldsymbol{p}_{k,t,i}, \boldsymbol{p}_{k,t,2})}{|\boldsymbol{p}_{k,t,i}.y - \boldsymbol{p}_{k,t,2}.y|} \right] \quad \left(0 < \alpha_{\text{W}}(t) < \frac{\pi}{2} \right) \tag{3}$$

3   Proposed Method

3.1   Focus-on Transformation for an Instability Feature

For the subtle fall risk discovery problem, the instability feature values for safe and unsafe skeletons can be similar, which deteriorates clustering performance. To circumvent this problem, we propose a method which transforms instability feature x to a new feature x'. The methods "focuses on" a range of large values in x to distinguish such skeletons in x'.

We define the "middle" x_b of the range as a large value in safe skeletons.

$$0.01 = \frac{1}{2} \int_{x_b}^{\infty} \frac{1}{\sqrt{2\pi\sigma^2}} \exp\left[-\frac{(t-\mu)^2}{2\sigma^2}\right] dt, \tag{4}$$

where μ and σ represent the mean and the standard deviation of dataset X_{post} of safe skeletons, respectively. The values 0.01 and $1/2$ were chosen to represent a rare case and reflect the one-side long tail distribution of x, respectively. A normal distribution $N(x_b, \sigma_b^2)$ is assumed to define the range, where $\sigma_b = 2\sigma$. The transformation is nonlinear, emphasizing differences around x_b.

$$x' = \int_{-\infty}^{x} \frac{1}{\sqrt{2\pi\sigma_b^2}} \exp\left\{-\frac{(t-x_b)^2}{2\sigma_b^2}\right\} dt \tag{5}$$

3.2   Additional Instability Features

To capture subtle fall risk, we define $\delta_{UL}(t)$ and $\alpha_U(t)$ as new instability features. Since the lower body sustains the upper body, their deviation signifies instability. $\delta_{UL}(t)$ is the deviation of the center $g_{k,t}^{up}$ of gravity of the upper body (joints 0-11) and that $g_{k,t}^{low}$ of the lower body (joints 12-19).

$$\delta_{UL}(t) = f(g_{k,t}^{up}, g_{k,t}^{low}) \tag{6}$$

$$g_{k,t}^{up} = \frac{\sum_{j=1}^{11} w_j p_{k,t,j}}{\sum_{i=1}^{11} w_i}, \qquad g_{k,t}^{low} = \frac{\sum_{j=12}^{19} w_j p_{k,t,j}}{\sum_{i=12}^{19} w_i}, \tag{7}$$

In the analysis by Tinetti et al. [6], the 24 of 25 elderly who are recurrent fallers were unable or unwilling to extend their backs. When the back is curved, the center of gravity of the human moves forward and becomes unstable, so the probability that the human falls down is relatively high. We introduce $\alpha_U(t)$, which is the angle of the upper body to represent the curvature of the back.

$$\alpha_U(t) = \arctan \frac{f(p_{k,t,1}, p_{k,t,2})}{|p_{k,t,1}\cdot y - p_{k,t,2}\cdot y|} \qquad \left(0 < \alpha_U(t) < \frac{\pi}{2}\right) \tag{8}$$

We inspected the distributions of the values of the instability features and found that two of them follow normal distributions while the remaining two follow exponential distributions. The latter features rarely indicate a high value because a human takes a balance to prevent the risk of fall unconsciously. For each instability feature, we judge whether it follows a normal distribution or an exponential distribution. In the latter case, we then transform it to follow a normal distribution via a chi-squared distribution [1].

Table 2. Datasets used in the experiments, where an abnormal part refer to the body part(s) with tools for simulating the physical situation of the elderly

$\mathcal{D}$	num. of examples	person	robots	abnormal parts
Base	790	A	R_1, R_2	
A0	1136	A	R_2, R_3	none
A1	1551	A	R_2, R_3	eyes
A2	1362	A	R_2, R_3	one leg
A3	1131	A	R_2, R_3	both legs
B0	827	B	R_2, R_3	none
B1	580	B	R_2	eyes
B2	1525	B	R_2	one leg
B3	1164	B	R_2	both legs

3.3 Evaluation for Clustering Results with Stable Skeletons

As we explained, NMI, the standard evaluation measure for clustering, is inappropriate for our problem. We propose a new evaluation method which exploits a set $\mathcal{D}_{\text{stable}}$ of stable skeletons of the subject, a normal data set $\mathcal{D}_{\text{normal}}$, and an abnormal data set $\mathcal{D}_{\text{abnormal}}$. We formalize the question as to measure how close $\mathcal{D}_{\text{normal}}$ is to $\mathcal{D}_{\text{stable}}$ and how far $\mathcal{D}_{\text{abnormal}}$ is to $\mathcal{D}_{\text{stable}}$, given Γ. Intuitively, the former can be accomplished by clustering $\mathcal{D}_{\text{normal}}$ and $\mathcal{D}_{\text{stable}}$ together and checking the degree of overlaps between them. The latter can be done in a similar way but by negating the degree.

We first propose the degree $\Delta_\Gamma(\mathcal{D}_1, \mathcal{D}_2)$ of the similarity between two datasets $\mathcal{D}_1 = \{d_1, \ldots, d_{M_1}\}$ and $\mathcal{D}_2 = \{d_{M_1+1}, \ldots, d_{M_1+M_2}\}$ with respect to the obtained clusters $C = \{c_1, \ldots, c_K\}$ from the merged dataset $\mathcal{D} = \{d_1, \ldots, d_{M_1+M_2}\}$.

$$\Delta_\Gamma(\mathcal{D}_1, \mathcal{D}_2) = \sum_{i=1}^{K} \sqrt{H_1(i) H_2(i)} \qquad \text{where } H_j(i) = \frac{N_{c_i, \mathcal{D}_j}}{M_j} \qquad (9)$$

$N_{c_i, \mathcal{D}_j}$ is the number of examples of $\mathcal{D}_j$ included in c_i. If we view $H_j(i)$ as a ratio of a histogram of which bin corresponds to a cluster, $\Delta_\Gamma(\mathcal{D}_1, \mathcal{D}_2)$ corresponds to Bhattacharyya Coefficient, which is widely used for representing the similarity between two histograms. Our new evaluation measure $E(\Gamma)$ of a clustering method Γ is defined as the difference of the above degrees.

$$E(\Gamma) = \Delta_\Gamma(\mathcal{D}_{\text{normal}}, \mathcal{D}_{\text{stable}}) - \Delta_\Gamma(\mathcal{D}_{\text{abnormal}}, \mathcal{D}_{\text{stable}}) \qquad (10)$$

4 Experiments

As preliminary experiments for our objective, we used eight datasets $\mathcal{D}$ of two persons shown in Table 2. **Base** data set contains skeletons observed by robots R_1 and R_2 while each of other data sets R_2 and R_3. Due to the special usage of the former, we believe the influence is small. Fig. 1 shows a snapshot of the

Fig. 1. Snapshot of the experiments (**A3**)

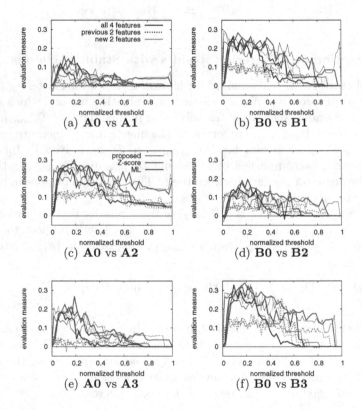

Fig. 2. Results of the experiments. For the plots, the left and right ones are for person **A** and **B** while the top, middle, and bottom ones simulate deficiencies in the eyes, one leg, and both legs, respectively. For the polylines, the red, green, and blue ones represent the proposed method, the Z-score transformation method, and the distance learning method while the bold, normal, and dashed lines represent all, the new 2, and the previous 2 features, respectively.

experiments for **A3**. We compared our method against a simplified method which applies the Z-score transformation instead of our focus-on transformation and its variant which, instead of the Euclidean distance, uses the Mahalanobis distance learned from 1603 safe skeletons and 92 unsafe skeletons by a metric learning method [7]. For each of the three methods we tested three kinds of instability features: all 4, previous 2 only, and new 2 only.

For a combination of $(\mathcal{D}, \mathcal{F}, \Gamma)$, one value for threshold θ_{leaf} of BIRCH yields different numbers of clusters. Thus we define normalized thresholds θ'_{leaf}.

$$\theta'_{\text{leaf}} = \frac{\theta_{\text{leaf}}}{\max_{\mathcal{D}} \Theta_{\text{leaf}}(\mathcal{F}, \Gamma | \mathcal{D})}, \tag{11}$$

where $\Theta_{\text{leaf}}(\mathcal{F}, \Gamma | \mathcal{D})$ represents the minimum value of θ_{leaf} when a single cluster is obtained from $\mathcal{D}$ with $(\mathcal{F}, \Gamma)$. $\max_{\mathcal{D}}$ represent the maximum value $\mathcal{D}$ except $\mathcal{D} = A0 \cup A2$, which possibly indicates an exceptional situation as $\Theta_{\text{leaf}}(\mathcal{F}, \Gamma | \mathbf{A0} \cup \mathbf{A2})$ is about three times larger than in other cases.

The results of experiments are shown in Figure 2. The six plots show that the dashed lines exhibit low evaluation measures, which validates our two new features. Evaluation measures are generally high in plots (b), (c), (f) but low in other plots because persons **A** and **B** were affected in 1-leg deficiency case and in the other two cases, respectively. In the latter plots, our method with 4 features is stable and exhibits relatively high evaluation measure values. In plots (b), (c), (f), which are more adequate for fall risk discovery, the Z-score transformation with 4 features exhibits the highest evaluation measure value but in a narrow range of the threshold. We see that our method with 4 features is stable and outperforms other methods in wide ranges of the threshold, which validates our proposals in these cases.

References

1. Deguchi, Y.: Construction of a Human Monitoring System Based on Incremental Discovery by Multiple Mobile Robots. Master thesis, Graduate School of ISEE, Kyushu University (2014) (in Japanese)
2. Deguchi, Y., Takayama, D., Takano, S., Scuturici, V.-M., Petit, J.-M., Suzuki, E.: Multiple-Robot Monitoring System Based on a Service-Oriented DBMS. In: Proc. PETRA (2014) (accepted for publication)
3. Rubenstein, L.Z.: Falls in Older People: Epidemiology, Risk Factors and Strategies for Prevention. Age and Ageing, 35(suppl. 2), ii37–ii41 (2006)
4. Suzuki, E., Deguchi, Y., Takayama, D., Takano, S., Scuturici, V.-M., Petit, J.-M.: Towards Facilitating the Development of a Monitoring System with Autonomous Mobile Robots. In: Post-Proc. ISIP 2013. Springer (2013) (accepted for publication)
5. Takayama, D., Deguchi, Y., Takano, S., Scuturici, V.-M., Petit, J.-M., Suzuki, E.: Online Onboard Clustering of Skeleton Data for Fall Risk Discovery (submitted)
6. Tinetti, M.E., Williams, T.F., Mayewski, R.: Fall Risk Index for Elderly Patients Based on Number of Chronic Disabilities. The American Journal of Medicine 80(3), 429–434 (1986)
7. Weinberger, K.Q., Saul, L.K.: Distance Metric Learning for Large Margin Nearest Neighbor Classification. JMLR 10, 207–244 (2009)
8. Zhang, T., Ramakrishnan, R., Livny, M.: BIRCH: A New Data Clustering Algorithm and its Applications. Data Min. Knowl. Discov. 1(2), 141–182 (1997)

Sonar Method of Distinguishing Objects Based on Reflected Signal Specifics

Teodora Dimitrova-Grekow and Marcin Jarczewski

Faculty of Computer Science, Bialystok University of Technology,
Wiejska 45A, Bialystok 15-351, Poland
t.grekow@pb.edu.pl, marcin.jarczewski@o2.pl

Abstract. This paper presents a method of pattern recognition based on sonar signal specificity. Environment data is collected by a Lego Mindstorms NXT mobile robot using a static sonar sensor. The primary stage of research includes offline data processing. As a result, a set of object features enabling effective pattern recognition was established. The most essential features, reflected into object parameters are described. The set of objects consists of two types of solids: parallelepipeds and cylinders. The main objective is to set clear and simple rules of distinguishing the objects and implement them in a real-time system: NXT robot. The tests proved the offline calculations and assumptions. The object recognition system presents an average accuracy of 86%. The experimental results are presented. Further work aims to implement in mobile robot localization: building a relative confidence degree map to define vehicle location.

Keywords: Object recognition, sonar signal, features extraction, intelligent information retrieval, mobile robotics, navigation.

1 Introduction

The accuracy of a distance measured by sonar depends mainly on the applied sensor, the send signal, and the method of processing the received echo-information. In real conditions, a single sonar measurement is quite irrelevant. Much more can be expected from a data sequence. This article describes an experiment which checks how effectively a simple sonar can recognize if an object is cylinder or parallelepiped. The main objective was to set clear and simple rules of distinguishing the objects and implement them in a real-time system: NXT robot. The experiment confirmed expectations.

1.1 Previous Work

Ultrasonic signal is mostly used as a source distance information or obstacle detection in mobile robotics [1], [2]. Nevertheless, its sequences contain much about the environment [3]. In fact, even too much: interferential echo signals from all possible reflection surfaces in the actual workspace [4], [5].

T. Andreasen et al. (Eds.): ISMIS 2014, LNAI 8502, pp. 506–511, 2014.
© Springer International Publishing Switzerland 2014

Sonar advantages are commonly known: it is an easily accessible, inexpensive sensor with satisfactory precision [6], [7]. The main disadvantage of ultrasonic sensors are connected with the conical emission area of the signal. And hence, many additional complicating and disturbing echo receptions arise. That complicates too much for direct use of sonar data for localization [8]. The real-time work of sonar-navigated systems at times significantly slows down. Also a powerful calculation engine is required for information processing. However, there are examples of sonar use for mapping [7], localization [9] and SLAM [10].

1.2   Structure of the System

The presented method of pattern recognition is based on the specificity of sonar signals. A mobile Lego Mindstorms NXT robot was used as a data collecting vehicle. The first stage of the research involved data processing. Consequent data observations showed a number of interesting dependencies between the received reflections and the scanned object. As a result, a set of extracted features enabling effective object recognition was fixed. The most essential features, are described in the next section.

The study uses a simple set of objects: two types of solids - parallelepipeds and cylinders. The tests shown in the third section were done on 3 different sizes of each type. The last stage of the system was developing recognition rules, which had been applied to the real-time test vehicle and proved the primary idea - to get more than the usual noise using only very simple tools. The object recognition system showed an accuracy of 86%. The experimental results are shown in the third section. The further work aims an implementation in mobile robot localization.

2   The Essence of the Method

The main objective of the article was to develop an object recognition method. The study consisted of following steps: (1) environment data collection and data processing, (2) features extraction, (3) building object recognition rules.

2.1   Environment Data Collection and Data Processing

Raw sonar measurements were saved into a file. The raw data was double processed: on the one hand it was smoothed so that the registered shapes became sharper and more clear, and on the other, its exceptions were exactly counted, related, and compared. Preliminary data processing removed all individual extreme, irregular measurements (Fig. 1). Among the registered values are very much 'glitters' - reading exceptions observed by comparison with neighboring ones. Their quick smoothing facilitates further processing of the received signal.

After scanning the environment, a secondary correction was also done. It focused on the larger 'exceptions' and smoothing of the discovered shapes. The so-called 'exceptions' were a result of overlapping echoes from various objects

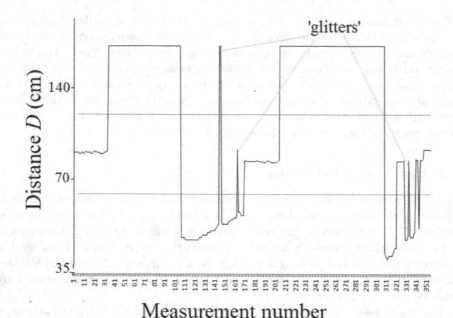

Fig. 1. Raw measurements of the sonar

and measurements made in sequence. This will be especially useful in the future development of the system in the navigation area.

The adjustment has been empirically determined during the experiments and is part of continued refinement-cast. The criterion for the selection is effectiveness of the extracted features for further classification of objects, especially in the right end of the vehicle localization. After many observations, several important dependencies were found.

2.2   Features Extraction

The criterion for selection was the effectiveness of the extracted features in the classification of objects. The extracted features were used by building the object recognition rules. Every file with the data saved from the sonar was interpreted as a vector. The three most important selected characteristics were:

Wall-greater Distances: D_{OV} - percentage of measurements greater than the distance to the wall N_W

The source of such values can be different: NXT sonars use value '255' as an error measurement. The total value D_{OV} is calculated as follows:

$$D_{OV} = \frac{(N_{255} + N_{GW})}{N} \tag{1}$$

Where:

N_{255} - number of '255' values,
N_{GW} - number of values greater than the distance to the wall,
N - total number of measurements.

Alien Echos: D_{NR} - percentage of measured distances less than N_W although no object is displaced at the spot

Obviously, this is result of delayed or side reflections:

$$D_{NR} = \frac{N_{LLW}}{N} \tag{2}$$

Where:

N_{LLW} - number of values less than the distance to the wall, cached in the object's free space.

2.3   Building Object Recognition Rules or Rapid Distinguishing of Parallelepipeds and Cylinders

For fulfilling this goal, a set of $D_{NR} = f(D_{OV})$ characteristics for ca. several hundred vectors were built. An unambiguous linear dependence (Fig. 2), which was implemented and loaded on the robot for online tests, resulted from the characteristic families. Our approach of object recognition was based on the concept of similarity or s'template matching'. To discriminate parallelepipeds and cylinders with minimum calculations, we studied the features - theirs behavior and relations to each other. A basic optimization of these observations is proof of the simple reliance between D_{OV} and D_{NR}.

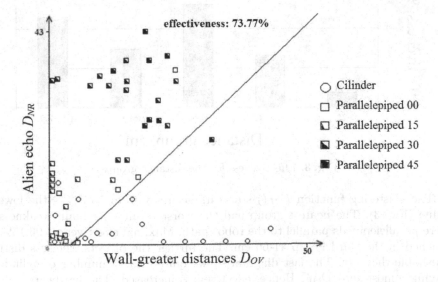

Fig. 2. Distinguishing parallelepipeds and cylinders

Example of the dependency $D_{NR} = f(D_{OV})$ is visualized in Fig. 2. The objects become splitted into two groups: cylinders and parallelepipeds. The objects clustering was determined by linear function $F_{CL}(x) = ax + b$. Points located on the same side of the line belong to the same cluster (cylinders or parallelepipeds). Derivation of this function can be described as follows: defining points (objects) in a certain distance from each other on the x and y axes, appointing a collection of lines for each of these points and finily selecting the best of the lines by evaluating the test collection.

3   Experimental Results

In order to facilitate the presentation of results, as well as to simplify the algorithms, data were split according to the minimum read distance. This division was determined empirically on the basis of the difference in readings. Groups, to which we assign individual points are:8 - 25 cm; 26 - 40 cm; 41 - 60 cm; 61 - 80 cm; 81 - 120 cm.

In the realization of this study we conducted over 300 tests divided into two sets: Teaching set - used exclusively for deriving proprer clustering for the objects and testing set - serving only for evaluation of clustering effectiveness.

All measurements were conducted with the same movement speed and frequency of scanning. The results of the clustering function derive are shown as a set of graphs. Interpreting them you should take into account that for the largest distances ($> 100cm$) most of the test objects were hardly 'visible' thus parameters D_{OV} and D_{NR} were almost the same.

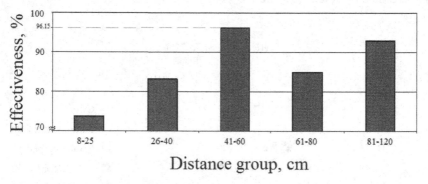

Fig. 3. Effectiveness for the distance groups

The clustering function $F_{CL}()$ starts its accuracy from 73.77% - the lowest value (Fig. 3). The nearest group had the worst result - its main weaknesses were parallelepipeds parallel to the robot path. Maximal effectiveness 96.15% is reached in the third group 41-60 cm. The further the objects, the less distinguishable they are. The last diagram shows quite a large number of cylinders having almost zero D_{OV}. Hence, effectiveness increased. The hardware unfortunately disallows work with longer distances. All experiments were conducted in real-time conditions implementing $F_{CL}()$ on the robot. The tests, repeated several hundred times, proved the calculated accuracy.

4 Conclusions

This paper presents an original approach of object distinguishing using a sonar sensor and minimized calculations for real-time implementation. Studies found a simple distinguishing rule: linear function $F_{CL}()$ expressing the dependencies between the sonar signal features in the described conditions. The average accuracy obtained was 86.21%, which seem very optimistic. The tests on the robot Lego Mindstoms NXT proved the effectiveness of function $F_{CL}()$. The issue shows an intelligent information retrieval based on sonar signal specificity.

The determination of the distinguishing rule can be additionally improved in several directions: exploring new features through additional dependencies, varying the reading frequency, bringing in movements to the reading sensor, and increasing the number of ultrasonic receivers. Further work also aims to implement in mobile robot localization: building a relative confidence degree map to define vehicle location.

Acknowledgments. This paper is supported by the S/WI/1/2013.

References

1. Ronaldo, V.: Representational information: a new general notion and measure of information. Information Sciences 181, 4847–4859 (2011)
2. Lenser, S., Veloso, M.: Visual sonar: Fast obstacle avoidance using monocular vision. Intelligent Robots and Systems 1, 886–891 (2003)
3. Worth, P., McKerrow, P.: An approach to object recognition using CTFM sensing. Sensors and Actuators A: Physical 179, 319–327 (2012)
4. Blaschke, T.: Object based image analysis for remote sensing. ISPRS Journal of Photogrammetry and Remote Sensing 65, 2–16 (2010)
5. Kiyoshi, O., Masamichi, M., Hiroyuki, T., Keihachiro, T.: Obstacle arrangement detection using multichannel ultrasonic sonar for indoor mobile robots. Journal Artificial Life and Robotics 15, 229–233 (2010)
6. Joong-Tae, P., Jae-Bok, S., Se-Jin Lee, M.K.: Sonar Sensor-Based Efficient Exploration Method Using Sonar Salient Features and Several Gains. Journal of Intelligent and Robotic Systems 63, 465–480 (2011)
7. Santana, A.M., Aires, K.R.T., Veras, R.M.S.: An Approach for 2D Visual Occupancy Grid Map Using Monocular Vision. Electronic Notes in Theoretical Computer Science 281, 175–191 (2011)
8. Barshan, B., Kuc, R.: A bat-like sonar system for obstacle localization. Pattern Analysis and Machine Intelligence 12, 686–690 (1992)
9. Frommberger, L.: Representing and Selecting Landmarks in Autonomous Learning of Robot Navigation. In: Xiong, C.-H., Liu, H., Huang, Y., Xiong, Y.L. (eds.) ICIRA 2008, Part I. LNCS (LNAI), vol. 5314, pp. 488–497. Springer, Heidelberg (2008)
10. Yap, T.N., Shelton, C.R: SLAM in large indoor environments with low-cost, noisy, and sparse so-nars. In: Proceedings of IEEE International Conference on Robotics and Automation, pp. 1395–1401 (2009)

Endowing Semantic Query Languages with Advanced Relaxation Capabilities

Géraud Fokou, Stéphane Jean, and Allel Hadjali

LIAS/ENSMA-University of Poitiers
1, Avenue Clement Ader, 86960 Futuroscope Cedex, France
{geraud.fokou,stephane.jean,allel.hadjali}@ensma.fr

Abstract. Most of studies on relaxing Semantic Web Database ($\mathcal{SWDB}$) queries focus on developing new relaxation techniques or on optimizing the top-k query processing. However, only few works have been conducted to provide a fine and declarative control of query relaxation using an $\mathcal{SWDB}$ query language. In this paper we first define a set of requirements for an $\mathcal{SWDB}$ cooperative query language($\mathcal{CQL}$). Then, based on these requirements, we propose an extension of query language with a new clause to use and combine the relaxation operators we introduce. A similarity function is associated with these operators to rank the alternative answers retrieved.

1 Introduction

With the widespread adoption of RDF, specialized databases called Semantic Web Databases ($\mathcal{SWDB}$) have been developed to manage large amounts of RDF data during the last decade (e.g, RDF-3X[8] or OntoDB[5]). Unlike relational databases where the schema is fixed, $\mathcal{SWDB}$s use a generic schema (a triple table or one of its variants) that can be used to store a set of diverse data, ranging from structured to unstructured data (valuation or not valuation of several properties of the concepts). A consequence of this flexibility is the difficulty for users to formulate queries in a correct and complete way. This can often lead to the problem of empty/unsatisfactory answers.

Several works have been proposed to relax queries in the $\mathcal{SWDB}$s context [1,4,2,7]. They mainly focus either on introducing new relaxation operators or on the efficient processing of the top-k approximate answers. However, only very few of them have addressed the need of defining a cooperative query language ($\mathcal{CQL}$) for $\mathcal{SWDB}$s that is capable of expressing most meaningful relaxation operators in a convenient and declarative way. Starting from our paper [9] where a set of relaxation operators has been proposed: *GEN* (super classes/properties), *SIB* (sibling classes/properties) and *PRED* (predicates) which are associated with a ranking function. The following novel contributions are made in this paper:

- a set of requirements for $\mathcal{SWDB}$'s $\mathcal{CQL}$ are defined and a critical review of existing works according to these requirements is then provided;
- a clear and complete formalization of the proposed relaxation operators, their formal properties and their combination are discussed explicitly.

T. Andreasen et al. (Eds.): ISMIS 2014, LNAI 8502, pp. 512–517, 2014.
© Springer International Publishing Switzerland 2014

The paper is structured as follows. First, Section 2 presents the main requirements we have defined for an $SWDB$'s CQL and the limitations of existing works. Section 3 describes an extension of $SWDB$ languages with relaxation operators and discusses their mixed use as well. Finally, Section 4 concludes the paper and outlines some future work.

2 CQL Requirements and Limitations of Previous Works

As a first step for designing an CQL for $SWDB$s, we define below a set of requirements which an CQL must fulfill.

R_1: **Relaxation operators combination.** For more efficiency of the CQL, it must allow specification of combination of relaxation operators.

R_2: **Ranking function.** The CQL should be able to provide the user with a discriminated set of answers. The rank-ordering should leverage the similarity queries and the satisfaction scores of the retrieved answers.

R_3: **Guide and control of the relaxation process.** The CQL should support a guiding and controlling of the relaxation process.

R_4: **Implementation.** The CQL language should be implemented in an $SWDB$.

In Table 1, we provide a summary of existing works w.r.t. the previous requirements. One can observe that there is no work that fulfills all the five requirements for designing an (CQL) in the $SWDB$ context. In particular, the integration of different relaxation operators in an CQL as well as the controlling of the relaxation process have not been addressed by most of these works.

Table 1. Characterization of existing works w.r.t. the four requirements

	Requirements			
	R_1	R_2	R_3	R_4
Dolog et al.[1]	n/a	Based on users preferences	trigger rules (no tuning)	Sesame
Hurtado et al.[7]	RELAX clause	Distance-based	Top-k (no tuning)	n/a
Huang et al.[6]	n/a	Distance-based Content-based	Top-k (no tuning)	Jena (LUBM)
Elbassuoni et al.[2]	n/a	Content-based	Domain target approach	Tests on real data set
Hogan et al.[4]	n/a	Content-based	Domain target approach	Generic framework for EADS

3 Extending of SWDB Query Languages

In this section, we present an extension of the language OntoQL [5] with operators and clauses of relaxation. Precisely, we introduce the following clauses and operators: *Approx* clauses, *PRED*, *GEN* and *SIB* operators.

3.1 Approx Clause

The clause *APPROX* shows explicitly the operators of relaxation and its parameters. The main operators of relaxation we use are: *PRED, GEN* and *SIB*
which will be presented in the next section. The syntax of *APPROX* clause is
defined as:
$\prec$ *approx clause* $\succ ::=$ APPROX $\prec$ *approx exp* $\succ\prec$ [TOP $\prec$ *integer* $\succ$]
$\prec$ *approx exp* $\succ ::=\prec$ *relax operator* $\succ|\prec$ *approx exp* $\succ$ AND $\prec$ *relax operator* $\succ$
$\prec$ *relax operator* $\succ ::=\prec$ *pred operator* $\succ|\prec$ *gen operator* $\succ|\prec$ *sib operator* $\succ$
    If the query's results is empty, the clause *APPROX* triggers the process of
the relaxation. An example of query with APPROX clause writes:
select *Name, Price* from *Motel* where *Price* $\geq$ 113 and *Price* $\leq$ 114
Approx(*PRED*(*Price*, 2) *AND GEN*(*Motel*), *TOP* 15).

3.2 Relaxation Operators

Now, we discuss a set of primitives operators for query relaxation. Each operator
has a precise action on the query to relax. This action will be performed according
to some given parameters.

PRED: Operator *PRED* applies on conditions expressed in the form of predicates. The *PRED* operator is repeated until the top-k answers are obtained (if
the K answers is desired by the user) or the predicate can't be relaxed (the stopping condition holds). The *PRED* operator computes approximate answers as
well as their satisfiability degrees and then rank-orders the answers. The syntax
of *PRED* operator is defined as follows:
    $\prec$ *pred operator* $\succ ::= PRED$ ($\prec$ *var* $\succ$ [, *tol*][, *interval*]), where *var* is the prop-
erty to relax, the constant *tol* is the tolerance value and *interval* is the validity interval, both are optional. So, the signature of PRED is: $\mathbb{Q}\times\mathbb{P}\times Real\times Interval \to$
$\mathbb{Q}$ where $\mathbb{Q}$ is the set of queries, $\mathbb{P}$ the set of properties.

Example 1. A query Q: select *Name, Price* from *Motel* where *Price* $\geq$ 113 and *Price* $\leq$ 114,
can be relaxed using the *PRED* as follows: $Q' = PRED(Q, Price, 2)$; with $Q'=$
select *Name, Price* from *Motel* where *Price* $\geq$ 113 and *Price* $\leq$ 114 *Approx*(*Pred*(*Price*, 2),
*TOP*15).

GEN: This operator uses entailment of ontology $\mathbb{O}$ to guide the relaxation. With
the operator *GEN* super-classes are using for relaxing one class by substitution.
The syntax of the operator *GEN*: $\prec$ *gen operator* $\succ ::= GEN$ ($\prec$ *var* $\succ$ [, *Integer*]),
var is the property to relax and the constant *Integer* is the level of generalization
given by the user, its default value is 1. The signature of *GEN* writes then:
$\mathbb{Q} \times \mathbb{C} \times Integer \to \mathbb{Q}$ where $\mathbb{C}$ is the set of classes. In $GEN(Q, C, i)$, Q is
the query to relax, C is the class to generalize, it must be in the set of classes
on which the selection is operated (noted $Dom(Q)$), and i is the maximal level
allowed for the superclass C' which will generalize C. The reference level 0 refers
to the class C itself.

Example 2. Let us consider again the query Q of the example 1:
If we have $Q' = GEN(Q, Motel, 2)$ then Q' writes:

select *Name, Price* from *Hotel* where *Price* $\geq$ 113 and *Price* $\leq$ 114

union select *Name, Price* from *Lodging* where *Price* $\geq$ 113 and *Price* $\leq$ 114,

where *Hotel* is the superclass of *Motel* at *level 1* and *Lodging* at *level 2*.

SIB: Sibling is another form of substitution, it uses entailment but in a different way than *GEN* operator. Sibling uses the sibling classes of the class to relax for replacing it. The syntax of the operator *SIB* is defined as follows:
$\prec$ *sib operator* $\succ ::= SIB$ ($\prec$ *var* $\succ$ [, var_1[(, var_2)*]]) where the first variable is the class to relax, the second is the sibling class used for the relaxation and the optional other variables denote other sibling classes, in the case where the relaxation process is applied more than one time. The signature of *SIB* is:
$\mathbb{Q} \times \mathbb{C} \times 2^{\mathbb{C}} \to \mathbb{Q}$. In $SIB(Q, C, [C_1, C_2, ..., C_n])$, Q is the query to relax, C is the class in $Dom(Q)$ to replace and $C_1, C_2, ..., C_n$ a list of sibling classes of C.

Example 3. Considering the query Q of the example 1:
If we have $Q' = SIB(Q, Motel, [Inn, Resort, Retreat])$ then Q' writes:

select *Name, Price* from *Inn* where *Price* $\geq$ 113 and *Price* $\leq$ 114

union select *Name, Price* from *Resort* where *Price* $\geq$ 113 and *Price* $\leq$ 114

union select *Name, Price* from *Retreat* where *Price* $\geq$ 113 and *Price* $\leq$ 114.

3.3   Combining Relaxation with and Logic Operator

The operator *GEN, SIB* and *PRED* can be associated with a logical connective *AND* for extending the relaxation operation. But for using this logic operator one needs to define its syntax and semantics. Since *PRED* extends the values selected and *GEN* or *SIB* change the domain, one can check that these two kinds of operators can be handled separately.

Conjunction of PRED: Let Q be a query with condition clauses on properties p_1 and p_2, we have only the following case:
1. $Q' = PRED(Q, p_1, \epsilon_1, I_1)$ AND $PRED(Q, p_2, \epsilon_2, I_2)$ since the two operators act on two different properties, each property can be relaxed independently of the other before execution of Q', $Q' = (Q_1, Q_2)$ where $Q_1 = PRED(Q, p_1, \epsilon_1, I_1)$ and $Q_2 = PRED(Q, p_2, \epsilon_1, I_2)$ and (Q_1, Q_2) means simultaneous relaxation.

Conjunction of GEN and/or SIB: For the conjunction of the same operator on different attribute classes:
1. $Q' = GEN(Q, c_1, level_1)$ AND $GEN(Q, c_2, level_2)$ the relaxation is applied independently on C_1 and C_2. The relaxation is also simultaneous, which means if $Q_1 = GEN(Q, c_1, level_1)$ and $Q_2 = GEN(Q, c_2, level_2)$ we will have $Q' = (Q_1, Q_2)$.
2. $Q' = SIB(Q, c_1, [c_1, .., c_n])$ AND $SIB(Q, c_2, [c'_1, ...; c'_m])$ can be written under the form $Q' = \bigcup\bigcup_{c_i c'_j}(Q_{c_i}, Q_{c'_j})$, with $Q_{c_i} = SIB(Q, c_1, [c_i])$ and

$Q_{c'_j} = SIB(Q, c_2, [c'_j])$.

GEN and *SIB* can be combined on the same classes or on different classes:

1. $Q' = GEN(Q, c, level_1)$ AND $SIB(Q, c, [c_1, .., c_n])$, since $level_1 \geq 1$ the generalization of c will use a superclass of c, as $Q_1 = GEN(Q, c, level_1)$ will already include all the sibling of c, so $Q_2 = (Q, c, [c_1, .., c_n]) \subset Q_1$. Hence $Q' = GEN(Q, c, level_1)$.

2. $Q' = GEN(Q, c, level_1)$ AND $SIB(Q, c', [c'_1, .., c'_n])$ the class c is substituted in all the sub-queries of the *Union* clause of *SIB*. So,

$$Q' = \bigcup_{i \in \{1...level_1\}} (SIB(GEN(Q, c, i), c', [c'_1, .., c'_n])).$$

As it can be seen, all these operators allow users to control the relaxation process via their underlying parameters.

3.4 Computation of Similarity and Satisfiability

We describes now how compute the similarity for each relaxation operator and for a combination of operators. The similarity can be defined as the quality of the answer, with 1 for the best answer and 0 for the worst. We make use of two measures, $Sim(Q, Q')$: similarity between queries and $SatQ(h_r)$: satisfiability of an alternative answer.

For $GEN(Q, C, i)$ we have: $Sim(Q, Q') = \min_{j=1..i} Sim(C, C_j)^1$ with

$Sim(C, C') = IC(msca(C, C'))/(IC(C) + IC(C') - IC(msca(C, C')))$.

Where $IC(C) = -log(Pr(C))^2$ is the information content of the class C, $msca(C, C')$ is the first concept which subsumes the two concepts C and C'. For an instance we have the satisfaction degree: $SatQ'(h_r) = \max_{T \in directType(h_r)}$

$Sim(T, C')$ where $directType(\mathbb{T})$ is the set[3] of most specific classes of an instance.

For $SIB(Q, C, [C_1, C_2, ..., C_n])$ we have: $Sim(Q, Q') = \min_{j=1..n} Sim(C, C_j)$. The satisfaction degree $SatQ'(h_r)$ for *SIB* is computed as *GEN*.

For $PRED(Q, Prop, tol)$ we have: $Sim(Q, Q') = Sim(Prop, Prop')$ where $Prop'$ is the relaxed property and $Sim(Prop, Prop') = 1/(1 + Dist(Prop, Prop'))$ with $Dist = distance of Hausdorff[3]$. The satisfaction degree for *PRED* is $SatQ'(h_r) = \mu_{Prop}(h_r.Prop)$.

For combination of operators, we take the worst case. So, for expressing this pessimistic attitude we use the operator *min* for aggregating the two measures. Then for the final satisfaction degree we have $SatQ(h_r) = \min(Sim(Q', Q), SatQ'(h_r))$.

4 Concluding Remarks

We have done experimentation on real dataset "HotelBase"[9] and we have observed that the *PRED* operator gives more alternative answers with a regular

---

[1] C_j generalize class of C at level j.

[2] $(Pr(C))$ is the probability of getting an instance of the class C in the ontology $\mathbb{O}$.

[3] It is a set due to possibility of multiple instantiation.

repartition values of the instances in the classes. While *GEN* and *SIB* give more alternative answers with a homogeneous repartition of instances between classes. So, the distribution of data affects the efficiency of relaxation operators and the similarity measure which decreases fast or slowly.

In this paper, we have addressed the issue of query relaxation in the $SWDB$ context. We have proposed a set of relaxation operators and shown how these operators can be integrated in a query language. A set of experiments has been conducted to demonstrate the feasibility of the approach. The analysis of experiment results reveals that the structure of the ontology and the data repartition in classes impact the performance and quality of each operator.

For future, we will conduct end-users experimentation in order to evaluate the similarity measure used. We will also study some optimization techniques for computing the *topk* answers desired by using the notion of selectivity to estimate the appropriate relaxation step and integrate multi-query optimization notions for executing all the relaxed queries. So, we will be able to propose a set of complete process of relaxation using relaxation operators according to the requirements of performance and quality of end-users.

References

1. Dolog, P., Stuckenschmidt, H., Wache, H., Diederich, J.: Relaxing rdf queries based on user and domain preferences. IJIIS 33(3), 239–260 (2009)
2. Elbassuoni, S., Ramanath, M., Weikum, G.: Query relaxation for entity-relationship search. In: Antoniou, G., Grobelnik, M., Simperl, E., Parsia, B., Plexousakis, D., De Leenheer, P., Pan, J. (eds.) ESWC 2011, Part II. LNCS, vol. 6644, pp. 62–76. Springer, Heidelberg (2011)
3. Fokou, G., Jean, S., Hadjali, A.: Endowing Semantic Query Languages with Advanced Relaxation Capabilities. Technical Report 14618, http://www.lias-lab.fr/members/geraudfokou
4. Hogan, A., Mellotte, M., Powell, G., Stampouli, D.: Towards fuzzy query-relaxation for rdf. In: Simperl, E., Cimiano, P., Polleres, A., Corcho, O., Presutti, V. (eds.) ESWC 2012. LNCS, vol. 7295, pp. 687–702. Springer, Heidelberg (2012)
5. Dehainsala, H., Pierra, G., Bellatreche, L.: Ontodb: An ontology-based database for data intensive applications. In: Kotagiri, R., Radha Krishna, P., Mohania, M., Nantajeewarawat, E. (eds.) DASFAA 2007. LNCS, vol. 4443, pp. 497–508. Springer, Heidelberg (2007)
6. Huang, H., Liu, C., Zhou, X.: Approximating query answering on rdf databases. World Wide Web 15(1), 89–114 (2012)
7. Hurtado, C.A., Poulovassilis, A., Wood, P.T.: Query relaxation in RDF. In: Spaccapietra, S. (ed.) Journal on Data Semantics X. LNCS, vol. 4900, pp. 31–61. Springer, Heidelberg (2008)
8. Neumann, T., Weikum, G.: Rdf-3x: A risc-style engine for rdf. In: Proceedings of the 34th International Conference on VLDB (August 2008)
9. Jean, S., Hadjali, A., Mars, A.: Towards a cooperative query language for semantic web database queries. In: Meersman, R., Panetto, H., Dillon, T., Eder, J., Bellahsene, Z., Ritter, N., De Leenheer, P., Dou, D. (eds.) OTM 2013. LNCS, vol. 8185, pp. 519–526. Springer, Heidelberg (2013)

A Business Intelligence Solution for Monitoring Efficiency of Photovoltaic Power Plants

Fabio Fumarola, Annalisa Appice, and Donato Malerba

Dipartimento di Informatica, Università degli Studi di Bari Aldo Moro
via Orabona, 4 - 70126 Bari - Italy
{fabio.fumarola,annalisa.appice,donato.malerba}@uniba.it

Abstract. Photovoltaics (PV) is the field of technology and research related to the application of solar cells, in order to convert sunlight directly into electricity. In the last decade, PV plants have become ubiquitous in several countries of the European Union (EU). This paves the way for marketing new smart systems, designed to monitor the energy production of a PV park grid and supply intelligent services for customer and production applications. In this paper, we describe a new business intelligence system developed to monitor the efficiency of the energy production of a PV park. The system includes services for data collection, summarization (based on trend cluster discovery), synthetic data generation, supervisory monitoring, report building and visualization.

1 Introduction

The global photovoltaic (PV) market has continued to expand despite the economic and financial crisis [3]. With the large-scale PV development taking shape in the EU, as well as outside the EU, maximizing the output of PV plants is vital for developers and utilities alike. Comprehensive performance, yield monitoring and ability to combat field under-performance have become crucial challenges, in order to optimize overall output and deliver reliable energy to the park.

In this paper, we describe a novel web-based Business Intelligence architecture for on-line monitoring a PV park. The proposed system, called Sun Inspector[1], is developed according to the "Software as Service" (SaaS) paradigm. It centralizes and monitors energy production data of PV portfolios in a unified, coherent manner, regardless of system configuration and inverter or sensor manufacture. This idea moves away from the plethora of monitoring systems [8,7,6,5] already developed from the PV community. The system includes services for data collection, summarization, synthetic data generation, supervisory monitoring, report building and visualization. On-site energy production data, which are backed up with panel strings, inverters and transformers, as well as irradiation measures, are summarized using the trend cluster discovery [1] and saved in a data warehouse for any future queries and analysis. The end users can query and display

---

[1] A demo of Sun Inspector is available at www.kdde.uniba.it/suninspector

T. Andreasen et al. (Eds.): ISMIS 2014, LNAI 8502, pp. 518–523, 2014.
© Springer International Publishing Switzerland 2014

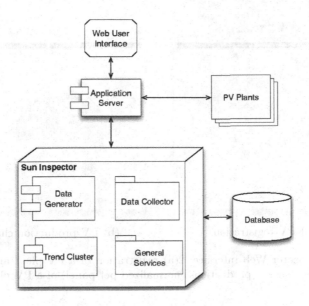

Fig. 1. Sun Inspector Architecture

in charts both energy production data and trend cluster summaries, with the purpose of monitoring performances and view anomalies.

The paper is organized as follows. In Section 2, we illustrate the Sun Inspector architecture that integrates the trend cluster discovery for creating log-book of historical data, as well us performing innovative explorative analysis of PV efficiency at spatio-temporal levels. In Section 3 we illustrate an example case. Finally, conclusions are drawn.

2   Sun Inspector Architecture

The architecture of the proposed system consists of several components, each of which is in charge of performing specific tasks as shown in Figure 1. A brief description of each component is given below.

2.1   General Services

The General Services Component is in charge of administering the database and providing authorized access to the saved data. It enables the execution of services scheduled by the Sun Inspector administrators through the web interface. Services include registering a PV plant, obtaining information about a PV plant, displaying energy production data and accessing the Business Intelligence services. For example, Figure 2(a) displays the web page to register a new PV plant in Sun Inspector. Figure 2(b) displays the bar chart of daily energy productions of a selected PV plant. This chart can allow the end users to monitor PV plant performances and check for daily production anomalies.

   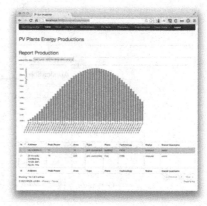

(a) PV registration   (b) PV production chart

Fig. 2. Sun Inspector Web Interface: 2(a) registration of a PV plant and 2(b) graph chart of the day's energy productions (normalized per panel) of a PV plant

2.2   Data Collector

The data collector component acquires the measurements of energy productions of registered PV plants. Production data are extracted by using the protocol MODBUS [2] every 15 minutes and send to Sun Inspector via a REST Web Service [4] that accepts the data formatted as tab-separated values or in a json format. Measurement setups (based on timing and latency properties) are used for synchronizing data acquisition from multiple plants. In order to use the data collection service, data loggers or micro-controllers measure acquire the signals from the PV plants and transmit them to Sun Inspector through the web. Each transmission contains the identifier of the PV plant stored in Sun Inspector, the timestamp, and the measures of the actual energy production plus additional parameters. After receiving these data, Sun Inspector stores them in the database. The energy production of a plant is normalized on the number of panel strings before storage, where the number of panel strings is a technical characteristic of the PV plant registered.

2.3   Trend Cluster-Based Summarizer

The summarizer component wraps the system SUMATRA, described in [1]. It segments the energy productions, collected from a starting time point, into windows, computes summaries window by-window and stores these summaries in a data warehouse, in place of actual data which can be discarded. Trend clusters are discovered as summaries of each window. They are clusters of plants which measure production values, whose temporal variation (called trend polyline) is similar over the time horizon of the window. A trend polyline is a time series that can be drawn as the sequence of straight-line segments. They fit clustered

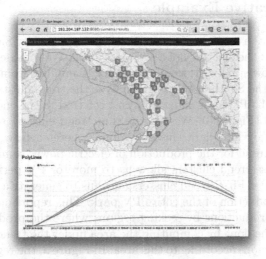

Fig. 3. Sun Inspector Web Interface: the web page to view the trend cluster base summarization of the production data collected on July 5, 2013. Plants grouped in a cluster are associated to the cluster identifier (top figure). For each cluster region, the trend of the energy production, aggregated over the associated region, is plotted (bottom figure).

values as they are transmitted along the window time horizon. The trend time series is the summary over time of this cluster of data. In SunInspector, trend clusters are made available to end users through the graphical web interface. Users can visualize trend clusters and check trend of energy productions (see Figure 3).

2.4 Data Generator

The Data Generator component is a web service to simulate the energy production of a PV plant. It is implemented by wrapping an extension of the web application *Photovoltaic Geographical Information System - Interactive Maps (PVGIS-IM)* implemented by the European Commission. PVGIS-IM[2] is a radiation database, which can be used to estimate the solar electricity produced by a PV plant over the year as well as the monthly/daily solar radiation energy, which hits one square meter in a horizontal plane in one day. It can be queried by filling in a form with several parameters related to the geographic position, the inclination and the orientation of the PV plant. The Data Generator component wraps the PVGIS-IM by offering a Rest Web Service interface, which can be queried by automatic services. Moreover, it offers a new service that, given the characteristics of a PV plant as input, simulates its day-by-day energy productions.

---

[2] http://re.jrc.ec.europa.eu/pvgis/apps4/pvest.php

3 An Illustrative Example

We illustrate how the analysis of trend clusters, supported by Sun Inspector, can be used to monitor the efficiency of a PV plant park. We consider energy productions (electrical power in kw/h) simulated every 15 minutes from 52 PhotoVoltaic (PV) plants over the South of Italy. We analyze the trend cluster representation of the history of the PV park under analysis, referred to the energy productions on July 5th, 2013 (Figure 3). We observe that trend clusters segment the map of the South of Italy into eight cluster regions, where each region is characterized by a similar trend for the production of electric power summarized over the selected day. These trend clusters allow us to monitor PV performances of the park in several ways. First, trend clusters allow us to generate a summary report for the energy production of the entire PV park. This report depicts the spatial variation of the productions by seeking regions where the production potential is high. This can be done by computing the area under the trend time series and ranking cluster regions according to this area. In Figure 3, the region labeled with the cluster identifier equal to 4 is that which exhibits the maximum production over one day (see the purple colored trend time series in Figure 3). This cluster covers the regions of Apulia and Calabria in Italy, which are solar regions with a production potential that is very high. This analysis is confirmed by statistics provided by Gestore dei Servizi Elettrici (GSE), which indicates that Apulia was the region with the higher installed capacity (2.186 MW) in 2011 (details at http://www.eniscuola.net/en/energy/contenuti/solar/). Second, trend clusters allow us to identify suspicious productions and detect possible faults. The presence of a plant, that is installed in dense area of plants, but is clustered alone, can indicate faulty panels or wrong system configurations that are degrading the performances. In this case, an alarm can be raised. Based on this alarm, the installer can promptly arrange repair activities by guaranteeing high performance of the plant and minimizing the loss of income. This case is displayed in Figure 3, since we forced the configuration of the two plants clustered alone in cluster 1 and cluster 2, respectively to under-produce electrical power. The plant clustered alone in cluster 0 was switched-off. In all these cases, plants are clustered alone, and surrounded by plants belonging to a dense cluster. The energy productions of trend clusters associated to these plants were lower than the energy productions of surrounding plants. Finally, trend clusters can be used for comparative studies of performance. The day's raw productions of a plant can be compared to some past production trends of this plant in the region of reference. This permits to monitor possible decay of efficiency of a plant over time.

4 Conclusion

We have described the general architecture of Sun Inspector, a Business Intelligence solution that offers services for taking a census of PV plants as well as managing all data of the efficiency of a PV park grid. It provides reliable services of PV data acquisition and storage, data aggregation, synthetic data generation, supervisory monitoring and report building. It integrates a data mining system, which is deployed for the discovery of trend clusters from historical efficiency data. Trend clusters allow us to create a compact, knowledgeable log-book of historical data. When trend clusters are discovered, they are stored in a data warehouse for any future analyses, while original data are discarded. Users can visualize trend cluster to perform a cross-grid analysis of efficiency of PV power plants and check for trend productions. As future work, we plan to further extend Sun Inspector by integrating the data mining systems which use trend clusters for forecasting analysis, interpolation and fault detection.

Acknowledgment. This work fulfills the research objectives of the Startup project "VIPOC: Virtual Power Operation Center" funded by the Italian Ministry of University and Research (MIUR).

References

1. Appice, A., Ciampi, A., Malerba, D.: Summarizing numeric spatial data streams by trend cluster discovery. Data Mining and Knowledge Discovery, 1–53 (2013)
2. Drury, B.: Control Techniques Drives and Controls Handbook, 2nd edn. (2009)
3. EurObserv'ER. Photovoltaic energy barometer. Le Journal du Photovoltaique 7 (2012)
4. Pautasso, C., Zimmermann, O., Leymann, F.: Restful web services vs. "big"' web services: making the right architectural decision. In: Proceedings of the 17th International Conference on World Wide Web, WWW 2008, pp. 805–814. ACM, New York (2008)
5. Sugiura, T., Yamada, T., Nakamura, H., Umeya, M., Sakuta, K., Kurokawa, K.: Measurements, analyses and evaluation of residential pv systems by japanese monitoring program. Solar Energy Materials and Solar Cells 75(3-4), 767–779 (2003)
6. Ulieru, V.D., Cepisca, C., Ivanovici, T.D., Pohoata, A., Husu, A., Pascale, L.: Measurement and analysis in pv systems. In: International Conference on Circuits, ICC 2010. WSEAS, pp. 137–142 (2010)
7. Zahran, M., Atia, Y., Al-Hussain, A., El-Sayed, I.: Labview based monitoring system applied for pv power station. In: International Conference on Automatic Control, Modeling and Simulation, ACMOS 2010, pp. 65–70. WSEAS (2010)
8. Zahran, M., Atia, Y., Alhosseen, A., El-Sayed, I.: Wired and wireless remote control of pv system. WSEAS Transactions on Systems and Control 5, 656–666 (2010)

WBPL: An Open-Source Library for Predicting Web Surfing Behaviors

Ted Gueniche[1], Philippe Fournier-Viger[1], Roger Nkambou[2],
and Vincent S. Tseng[3]

[1] Dept. of computer science, University of Moncton, Canada
[2] Dept. d'informatique, Université du Québec á Montréal, Canada
[3] Dept. of computer science and inf. eng., National Cheng Kung University, Taiwan
{etg8697,philippe.fournier-viger}@umoncton.ca, tsengsm@mail.ncku.edu.tw

Abstract. We present WBPL (Web users Behavior Prediction Library),
a cross-platform open-source library for predicting the behavior of web
users. WBPL allows training prediction models from server logs. The
proposed library offers support for three of the most used webservers
(Apache, Nginx and Lighttpd). Models can then be used to predict the
next resources fetched by users and can be updated with new logs efficiently. WBPL offers multiple state-of-the-art prediction models such as
PPM, All-K-Order-Markov and DG and a novel prediction model CPT
(Compact Prediction Tree). Experiments on various web click-stream
datasets shows that the library can be used to predict web surfing or
buying behaviors with a very high overall accuracy (up to 38 %) and is
very efficient (up to 1000 predictions /s).

Keywords: Web behavior prediction, sequence prediction, accuracy.

1 Introduction

Sequences prediction algorithms have been widely integrated in webservers for
numerous applications such as reducing latency by prefetching content, predicting buying behavior and recommending products [2,7,5]. The prediction task
usually consists in predicting the next x resources or pages that a user will fetch
using information contained in sequences from previous users. The prediction
task is challenging since it is expected that prediction models can be trained
efficiently and incrementally, and that fast and accurate predictions can be performed. Although there is an important need for predicting the next objects that
a user will prefetch, there generally exists few public libraries that can be used
for this task. Moreover, to our knowledge all existing libraries suffer from one or
more of the following limitations. Several libraries offering sequence prediction
models do not offer web-specific functionalities such as reading HTTP logs and
filtering unwanted information [3]. Other libraries are specialized for application
domains such as biological sequence prediction [6]. Therefore, these libraries
cannot be directly used in the Web context without performing time-consuming

T. Andreasen et al. (Eds.): ISMIS 2014, LNAI 8502, pp. 524–529, 2014.
© Springer International Publishing Switzerland 2014

modifications, and their efficiency has not been demonstrated for this task. Another important problem is that prediction libraries offer prediction models that cannot be trained incrementally [3].

In this paper, we address these limitations by proposing an open source library for the prediction of web users behavior, named WBPL (Web users Behavior Prediction Library). The library offers implementation of state-of-the-art prediction models such as Prediction by Partial Matching [1], Dependency Graph [7], All-K-Order Markov [9] and our own approach; Compact Prediction Tree [4]. The library is cross-platform. Prediction models and other core components are implemented in Java. Furthermore, we provide a web GUI implemented in Javascript to quickly test and compare prediction models on user uploaded weblogs, using various parameters. The library can be easily integrated in web applications since it is open-source, has no external dependencies and it provides important web-specific functionalities. Source code is available at http://goo.gl/QvVXQZ.

The rest of this paper is organized as follows. Section 2 briefly describes the prediction models offered in our software. Section 3 explains how to use the GUI to quickly test prediction models and compare their performance. Section 4 explains how the library can be integrated in web applications to perform prediction. Finally, section 5 draws a conclusion.

2   Presented Models

The Problem of Sequence Prediction. Given a finite alphabet of items (resources) $I = \{i_1, i_2, ..., i_m\}$, an individual sequence is defined as $S = \langle s_1, s_2, ..., s_n \rangle$, a list of ordered resources where $s_i \in I$ ($1 \leq i \leq m$). Let $T = \{s_1, s_2, ..., s_t\}$ be a set of training sequences used to build a prediction model M. The problem of sequence prediction consists in predicting the next item s_{n+1} of a given sequence $\langle s_1, s_2, ..., s_n \rangle$ by using the prediction model M.

Prediction by Partial Matching. [1] (PPM) considers the last K items of a sequence to perform a prediction, where K is the order of the model. A K-order PPM model can be represented as a graph where nodes are subsequences and arcs indicate the probabilities that a subsequence will be followed by another subsequence. In a K-order PPM, arcs are outgoing from sequences of K consecutive items to nodes containing a single item. Predicting a sequence is performed by identifying the node containing the sequence's last K items and then following the arc having the highest probability. This approach has been proven to yield good results in certain areas such as web sequence predictions [1,2]. However, this approach is very sensitive to noise. The smallest variation in a sequence will affect the prediction outcome. Because of this, prediction accuracy can greatly deteriorate for noisy datasets.

The All-K-Order Markov Model. (AKOM) is a variation of the previous approach, consisting of training PPM predictors of order $1, 2...K$ to perform predictions. AKOM yields higher accuracy than fixed order PPM models in most cases [2]. But it suffers from a much higher space complexity. A lot of research has been done to improve the speed and memory requirement, for example by pruning nodes [2,9,8], but few to improve accuracy.

The Dependency Graph. (DG) [7] model is a graph where each node represents an item $i \in I$. A directional arc connects a node A to a node B if and only if B appears within x items from A in training sequences, where x is the *lookahead window length*. The weight of the arc is $P(B|A)/P(A)$ for the training sequences. Predicting a sequence consists of finding the node containing its last item and then following the arc with the highest weight.

Compact Prediction Tree. (CPT) is our proposed prediction model [4]. It is built on the hypothesis that a lossless prediction model would yield higher accuracy than lossy models such as DG, PPM and AKOM because all relevant information from training sequences could be used to perform each prediction. The CPT consists of three data structures. The Prediction Tree (PT) is a tree-like data structure that stores and compresses training sequences. The Inverted Index (II) is a structure designed to efficiently identify in which sequences a set of items appears. Finally, the Lookup Table (LT) is an index that provides an efficient way to find a specific sequence in the prediction tree using its id. A CPT can be built and updated incrementally.

Predicting the next item of a sequence S using a CPT is performed as follows. Let x be an integer named the prefix length. The prediction algorithm finds all sequences containing the last x items of S in any order and in any position, using the inverted index. Next, these sequences are traversed in the prediction tree to count the number of occurrences of each item appearing after the last x items of S. The item having the largest occurrence count is the predicted item. The prediction algorithm is tolerant to noise because it does not look for items appearing in a fixed order. Furthermore, it incorporates a strategy named the *recursive divider* to remove potentially noisy items (cf. [4] for more details).

3   Using the Web GUI

In this section, we describe the four steps to use the web GUI. The web GUI (cf. Fig. 1) is useful to compare prediction models on web log data with various parameters, to evaluate their accuracy for a given web application.

Step1 1: Choosing a Data Source. The user should first upload their own raw log file in the default format of one of the three main supported webserver platforms (Nginx, Lighttpd or Apache) or select one of three well known public web datasets namely BMS, FIFA and Kosarak. BMS contains 15,806 sequences (average length of 6 items) and 495 unique items. FIFA contains 28,978 sequences (average length of 32 items) and 3,301 unique items. Kosarak contains 638,811 sequences (average length of 12 items) and 39,998 unique items. Uploaded log files are converted into sequences of items that can be used by the prediction models. Converting a log file into a set of sequences is done as follows. Each sequence represents a web session of a user. A web session is a sequence of

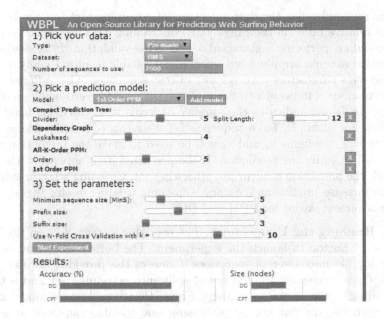

Fig. 1. WSBL user interface

HTTP requests found in the logs. A user is identified by its IP address and user-agent if it is provided. Two consecutive HTTP request from a user belongs to the same web session if the time difference between the two is smaller than the time window. The time window is a parameter that must be specified in the user interface. Note that multiple individual sessions for a given user are not grouped together because of the limited information provided in log files. For more accurate identification of users, a webserver could use a user authentication mechanism or a permanent tracking cookie.

Step 2. Choosing Prediction Model(s). The user next has to select one or more prediction models for comparison. Each model has its own set of parameters that needs to be set (cf. Section 2 for a description of each parameter). All the predictors do not perform equally well in all situations. Therefore, it is recommended to test them all for a given dataset.

Step 3. Tuning Parameters for Running an Experiment. The user should next select parameters to indicate how to compare the selected prediction models. These parameters are independent of the choice of prediction models. The *MinS* parameter indicates the minimum sequence length. Any sequence shorter than *MinS* will not be used for the experiment (either for training or testing). Setting the right value for this parameter may enhance the accuracy of prediction models because some sequences are too short to generate accurate predictions. Note that PPM and DG are not affected by this parameter. The user should also select a sampling strategy to determine how to divide the dataset into training and test sets. Two sampling methods are offered. The *standard random* *sampling* randomly split the data into two sets, training and testing, by specifying the relative ratio (in the range $[0, 1]$) of training data desired. The *K-Fold cross validation* performs a standard k-fold cross-validation. It is slower than the standard random sampling but offers more reliable results. Finally, the user should set two parameters with respect to test sequences. The *suffix size* indicates the number of items of a test sequence to be used for verifying a prediction, while the *prefix size* indicates the number of items to be used for making the prediction. For example, for a sequence $\langle s_1, s_2, s_3, s_4, s_5 \rangle$, suffix size = 1 and prefix size = 2, the items s_3 and s_4 will be used to make the prediction and s_5 will be used to verify the prediction. If the predicted item appears in the suffix (here, s_5), the prediction is accurate, otherwise not. Choosing a larger suffix size generally increase prediction accuracy. Choosing a larger prefix size may also increase accuracy, except for PPM and DG.

Step 4. Running the Experiment. The user should next click on the "Start Experiment" button to launch the experiment. The library loads and converts the raw log file into a set of sequences if one of the provided dataset is not selected. Once converted, the dataset is split into a training set and a testing set according to the sampling strategy. Each selected prediction model is then trained with the training set by processing one training sequence at a time. Testing sequences are then used to evaluate accuracy, coverage and time of each prediction model.

Experimental Results. To evaluate the performance of the library, we compared the prediction models on the BMS dataset. Models were tuned with their best parameters (DG lookahead window = 4, AKOM order = 5). We ran the experiment on a 3rd generation Core i3 processor running Windows 7. We found that the library can predict next fetched resources with a high accuracy. For BMS the most and second most accurate models were CPT (38.4 %) and DG (36 %). Results also shows that the library is very efficient. For BMS the best and second best models in terms of predictions per second was PPM (1000 /s) and DG (250 /s). We also found that the library is also very efficient in terms of training time. The fastest and second fastest models for inserting a sequence for BMS are PPM (0.01 s) and CPT (0.02 s).

4 Using the Library

WBPL is composed of two parts, a web GUI and the library. The web GUI consists of a small HTML, CSS and Javascript website. The interface can be uploaded to a server and linked to the backend framework using the provided PHP bridge. Since running experiments using the web GUI can be resource intensive, we recommend using a dedicated environment for testing purposes. The library is written in Java and the source code is available at `http://goo.gl/QvVXQZ`. The source code is highly modular and easy to customize for specific needs. The library provides an interface to add new prediction models or edit existing ones. Just like when using the web GUI, it is possible to load custom log files such as Apache logs, Nginx logs or Lighttpd logs, and use them as datasets. The library is designed to be used alongside a web application. In this case, the application has to first train the desired prediction model(s) with training data and then the application can ask the library to perform sequence predictions. We provide a simple PHP bridge to communicate back and forth between the library and a web application.

5 Conclusion

We presented WBPL, a cross-platform open-source library for predicting the next ressources fetched by web users. The library offers implementations of state-of-the-art prediction models DG, PPM, AKOM and CPT. Furthermore, if offers support for reading logs files from the three main webservers (Apache, Nginx and Lighttpd). We also presented a web interface to quickly test the prediction models offered in the library with various parameters on user-given web datasets. Experiments on real-life web click-stream datasets shows that the library can achieve a very high overall accuracy (up to 38 %) and can perform very fast prediction (up to 1000 predictions /s). For future work, we will work on further enhancing the performances of the proposed prediction models [2,9,8] proposing additional models such as Context Tree Weighting and Neural Networks and providing ways to use the library with other web frameworks like Ruby on Rails and .NET.

Acknowledgement. This project was supported by an NSERC grant from the Government of Canada.

References

1. Cleary, J., Witten, I.: Data compression using adaptive coding and partial string matching. IEEE Trans. on Inform. Theory 24(4), 413–421 (1984)
2. Deshpande, M., Karypis, G.: Selective Markov models for predicting Web page accesses. ACM Transactions on Internet Technology 4(2), 163–184 (2004)
3. Google Prediction API, https://developers.google.com/prediction (accessed: February 15, 2014)
4. Gueniche, T., Fournier-Viger, P., Tseng, V.-S.: Compact Prediction Tree: A Lossless Model for Accurate Sequence Prediction. In: Motoda, H., Wu, Z., Cao, L., Zaiane, O., Yao, M., Wang, W. (eds.) ADMA 2013, Part II. LNCS (LNAI), vol. 8347, pp. 177–188. Springer, Heidelberg (2013)
5. Hassan, M.T., Junejo, K.N., Karim, A.: Learning and Predicting Key Web Navigation Patterns Using Bayesian Models. In: Gervasi, O., Taniar, D., Murgante, B., Laganà, A., Mun, Y., Gavrilova, M.L. (eds.) ICCSA 2009, Part II. LNCS, vol. 5593, pp. 877–887. Springer, Heidelberg (2009)
6. HMMgene (v. 1.1), http://www.cbs.dtu.dk/services/HMMgene (accessed: February 15, 2014)
7. Padmanabhan, V.N., Mogul, J.C.: Using Prefetching to Improve World Wide Web Latency. Computer Communications 16, 358–368 (1998)
8. Domenech, J., de la Ossa, B., Sahuquillo, J., Gil, J.A., Pont, A.: A taxonomy of web prediction algorithms. Expert Systems with Applications (9) (2012)
9. Pitkow, J., Pirolli, P.: Mining longest repeating subsequence to predict world wide web surng. In: Proc. 2nd USENIX Symposium on Internet Technologies and Systems, Boulder, CO, pp. 13–25 (1999)

Data-Quality-Aware Skyline Queries

Hélène Jaudoin, Olivier Pivert, Grégory Smits, and Virginie Thion

Université de Rennes 1, Irisa, Lannion, France
{helene.jaudoin,olivier.pivert,
gregory.smits,virginie.thion}@univ-rennes1.fr

Abstract. This paper deals with skyline queries in the context of "dirty databases", i.e., databases that may contain bad quality or suspect data. We assume that each tuple or attribute value of a given dataset is associated with a quality level and we define several extensions of skyline queries that make it possible to take data quality into account when checking whether a tuple is dominated by another. This leads to the computation of different types of gradual (fuzzy) skylines.

1 Introduction

As is well-known, databases may suffer from quality issues, in particular when the data have been gathered from different sources. For instance, data may be incomplete, inaccurate, or out-of-date [1]. Two main approaches can be considered for managing low-quality data: i) the *data cleaning* approach [9], which deals with detecting and removing errors and inconsistencies from data, or ii) the approach that consists in living with the "dirty data" (known as the *manage* approach). However, data cannot always be repaired and cleaning may lead to losing useful information. In this paper, we consider the second data management approach and focus on a particular type of queries, i.e. Skyline queries [2] that aim to retrieve the points of a dataset that are not Pareto-dominated (in the sense of a set of preference criteria) by any other. Our objective is to take data quality into account during the evaluation of a skyline query, so as to avoid that a bad quality point hides more interesting, better quality ones.

In the approach we propose, we assume that a level of quality attached to each tuple (or suspect attribute value) is available (different models have been proposed in the literature for assessing data quality, see for instance [7,4]).

To the best of our knowledge, there exists very few work about the impact of data quality on skyline queries in the literature. In [6], Lofi *et al.* propose an approach based on crowdsourcing for eliciting missing values during runtime, which corresponds to a data cleaning type of technique, whereas in the approach we propose, the very definition of skyline queries is extended in order to incorporate the data quality aspect.

The remainder of the paper is structured as follows. Section 2 provides a refresher about skyline queries. Section 3 shows how data quality can be taken into account while evaluating a skyline query. Two cases are considered: that where data quality is assessed at a tuple level, and that where any attribute value may be assigned a confidence level. Section 4 recalls the main contributions and outlines perspectives for future work.

T. Andreasen et al. (Eds.): ISMIS 2014, LNAI 8502, pp. 530–535, 2014.
© Springer International Publishing Switzerland 2014

2 Refresher About Skyline Queries

The notion of a skyline in a set of tuples is easy to state (since it amounts to exhibit non dominated points in the sense of Pareto ordering). Assume we have:

- a given set of criteria $C = \{c_1, \ldots, c_n\}(n \geq 2)$ associated respectively with a set of attributes A_i, $i = 1, \ldots, n$;
- a complete ordering $\succcurlyeq_i$ given for each criterion i expressing preference between attribute values[1] (the case of non comparable values is left aside).

A tuple $u = (u_1, \cdots, u_n)$ in a database D *dominates* (in the sense of Pareto) another tuple $u' = (u'_1, \cdots, u'_n)$ in D, denoted by $u \succ_C u'$, iff u is at least as good as u' in all dimensions and strictly better than u' in at least one dimension:

$$u \succ_C u' \Leftrightarrow \forall i \in \{1, \ldots, n\}, \; u_i \succcurlyeq_i u'_i \text{ and} \atop \exists i \in \{1, \ldots, n\} \text{ such that } u_i \succ_i u'_i. \tag{1}$$

A tuple $u = (u_1, \cdots, u_n)$ in a database D belongs to the skyline S, denoted by $u \in S$, if there is no other tuple $u' = (u'_1, \cdots, u'_n)$ in D which dominates it:

$$u \in S \Leftrightarrow \forall u', \; \neg(u' \succ_C u). \tag{2}$$

Table 1. An extension of relation *car*

	make	category	price	color	mileage
t_1	Opel	roadster	4500	blue	20,000
t_2	Ford	SUV	4000	red	20,000
t_3	VW	roadster	5000	red	10,000
t_4	Opel	roadster	5000	red	8000
t_5	Fiat	roadster	4500	red	16,000
t_6	Renault	coupe	5500	blue	24,000
t_7	Seat	sedan	4000	green	12,000

Example 1. Let us consider the relation *car* of schema (*make, category, price, color, mileage*) (cf. Table 1), and the query involving four preference criteria:
select * from *car* **preferring**
(*make* = 'VW' **else** *make* = 'Seat' **else** *make* = 'Opel' **else** *make* = 'Ford') **and**
(*category* = 'sedan' **else** *category* = 'roadster' **else** *category* = 'coupe') **and**
(**least** *price*) **and** (**least** *mileage*);
In this query, "$A_i = v_{1,1}$ *else* $A_i = v_{1,2}$" means that value $v_{1,1}$ is strictly preferred to value $v_{1,2}$ for attribute A_i. It is assumed that any domain value which is absent from a preference clause is less preferred than any value explicitly specified in the clause (but it is not absolutely rejected). Here, the tuples that are not dominated

---

[1] $u \succ v$ means u is preferred to v. $u \succcurlyeq v$ means u is at least as good as v, i.e., $u \succcurlyeq v \Leftrightarrow u \succ v \vee u \approx v$, where $\approx$ denotes indifference.

in the sense of the *preferring* clause are $\{t_3, t_4, t_7\}$. Indeed, t_7 dominates t_1, t_2, and t_5, whereas every tuple dominates t_6 except t_2.

Notice that if we add the preference criterion (*color* = 'blue' **else** *color* = 'red' **else** *color* = 'green') to the query, then the skyline is $\{t_1, t_2, t_3, t_4, t_5, t_7\}$, i.e., almost all of the tuples are incomparable.   ◇

3   Taking Data Quality into Account

We are concerned with the possible presence of *bad quality* points in the dataset over which the skyline is computed. The impact of such points on the skyline may obviously be important if they dominate some other, better quality ones. Our goal is to revisit the definition of the skyline so as to take into account the quality of the points in the database, in order to control the impact of "suspect" elements. If data quality is defined as a Boolean notion (modeled by a predicate $quality(t) = 1$ if t is reliable, 0 otherwise) , one may define the skyline as follows:

$$p \in S_Q \Leftrightarrow \forall q \in D, \; (quality(q) \Rightarrow \neg(q \succ_C p)) \tag{3}$$

which expresses that a point belongs to the skyline iff it is not dominated by any good quality point.

However, it is of course much more natural to view data quality as a gradual notion. We thus assume that each tuple t from D is associated with a quality degree in $(0, 1]$, denoted by $\mu_{qual}(t)$ (a refinement would be to associate a quality degree with each attribute value, cf. Subsection 3.2). This amounts to viewing the relation over which the skyline is computed as a *fuzzy relation*, i.e., a relation where each tuple is associated with a membership degree in $(0, 1]$.

3.1   Tuple-Level Quality

Taking data quality into account leads to making the skyline *gradual*, i.e., to view it as a fuzzy set of points. We consider that a point totally belongs to the skyline (membership degree equal to 1) if it is dominated by no other point. A point does not belong at all to the skyline (membership degree equal to 0) if it is dominated by at least one point of perfect quality. In general, the degree associated with a point p depends on the highest quality degree among those associated with the points that dominate p. The definition stems from Formula (3), interpreting $\forall$ by the triangular norm minimum (as is classical in fuzzy set theory), and $\Rightarrow$ by Kleene-Dienes implication: $x \to_{KD} y = \max(1 - x, y)$:

$$S_Q = \{\mu/p \mid p \in D \wedge \mu = \min_{q \in D} (\max(1 - \mu_{qual}(q), 1 - \mu_{\succ_C}(q, p)))\} \tag{4}$$

where $\mu_{\succ_C}(q, p)) = 1$ if q dominates p (i.e., $q \succ_C p$ is true), 0 otherwise. In Formula (4), the notation μ/p means that point p belongs to S_Q to degree μ. Equation (4) may be rewritten as follows:

$$S_Q = \{\mu/p \mid p \in D \wedge \mu = 1 - \max_{q \in D \mid q \succ_C p} (\mu_{qual}(q))\}. \tag{5}$$

Remark 1. We straightforwardly have: $core(S_Q) = S$ since every point that is not dominated by any other gets degree 1 through Equation (5) (it is implicitly assumed that $\max_{q \in D \mid q \succ_C p} (\mu_{qual}(q)) = 0$ when $\nexists q \in D$ such that $q \succ_C p$).

Table 2. Relation *car* with quality degrees attached to the tuples

	make	*category*	*price*	*color*	*mileage*	μ_{qual}
t_1	Opel	roadster	4500	blue	20,000	0.6
t_2	Ford	SUV	4000	red	20,000	1
t_3	VW	roadster	5000	red	10,000	0.4
t_4	Opel	roadster	5000	red	8000	0.9
t_5	Fiat	roadster	4500	red	16,000	0.7
t_6	Renault	coupe	5500	blue	24,000	0.3
t_7	Seat	sedan	4000	green	12,000	0.2

Example 2. Let us consider the relation and query from Example 1. Table 2 shows the quality degrees associated with the tuples of *car*. We get the skyline:

$$\{0.8/t_1, \; 0.8/t_2, \; 1/t_3, \; 1/t_4, \; 0.8/t_5, \; 0.1/t_6, \; 1/t_7\}.$$

In this result, the degree 0.8 obtained by t_1, t_2 and t_5 corresponds to $1 - \mu_{qual}(t_7)$ since t_7 is the only tuple that dominates each of them, whereas the degree 0.1 associated with t_6 equals $1 - \max_{t \in \{t_1, t_3, t_4, t_5, t_7\}}(\mu_{qual}(t))$. ◇

Notice that every element of the result is in fact associated with two scores: its degree of membership to the skyline, and its original quality degree, which makes it possible for the user to precisely assess each answer. For rank-ordering the skyline points, he/she may give priority either to the degree of membership to the skyline or to the original quality degree, using lexicographic order.

3.2   Value-Level Quality

In order to handle "suspect values", a solution is to use an uncertain database model, and to model such attribute values as probability or possibility distributions. The issue of computing skyline queries in such frameworks has been tackled notably in [8] (probabilistic case) and in [3] (possibilistic case). However, this implies having available information about the possible candidate values.

On the other hand, if all we have to represent the quality of a value is a degree of confidence — which can be related to the freshness of the value, for instance, or the reliability of the source it comes from — $\rho \in (0, 1]$ attached to it (assuming by default that $\rho = 1$), it has to be taken into account in the comparison between u_i and u'_i, whose result itself gets a confidence degree. We interpret confidence as certainty in the framework of possibility theory [5] and denote by ρ_i (resp. ρ'_i) the confidence degree associated with u_i (resp. u'_i).

Let us consider the event $(u.A_i\,\theta_i\,u'.A_i)$ where θ_i is $\succ_i$, $\succcurlyeq_i$ or $\approx_i$, $u.A_i = (u_i, \rho_i)$ and $u'.A_i = (u'_i, \rho'_i)$. According to the axioms of possibility theory, we have:

$$conf(u.A_i\,\theta_i\,u'.A_i) = \begin{cases} 0 \text{ if } \neg(u_i\,\theta_i\,u'_i), \\ \min(\alpha,\,\beta,\,\gamma) \text{ otherwise.} \end{cases}$$

where

$\alpha = \rho'_i$ if $\exists y \in dom(A_i)\backslash\{u'_i\}$ such that $\neg(u_i\,\theta_i\,y), 1$ otherwise,

$\beta = \rho_i$ if $\exists x \in dom(A_i)\backslash\{u_i\}$ such that $\neg(x\,\theta_i\,u'_i), 1$ otherwise,

$\gamma = \begin{cases} \max(\rho_i,\,\rho'_i) \text{ if } \exists(x,\,y) \in (dom(A_i)\backslash\{u_i\} \times dom(A_i)\backslash\{u'_i\}) \text{ s.t. } \neg(x\,\theta_i\,y), \\ 1 \text{ otherwise.} \end{cases}$

As to the overall tuple comparison, we have, $\forall(u,\,u'), u \neq u'$:

$$conf(u \succ_C u') = \min(\min_i\ conf(u_i \succcurlyeq_i u'_i),\ \max_i\ conf(u_i \succ_i u'_i)) \qquad (6)$$

$$conf(u \approx_C u') = \max(\ \min(\max_i\ conf(u_i \succ_i u'_i),\ \max_i\ conf(u'_i \succ_i u_i)),$$
$$\min_i\ conf(u_i \approx_i u'_i)). \qquad (7)$$

Formula (6) is the "translation" of (1) whereas Expression (7) stems from:

$$u \approx_C u' \Leftrightarrow ((\exists i \text{ such that } u_i \succ_i u'_i) \wedge (\exists i \text{ such that } u'_i \succ_i u_i))$$
$$\vee (\forall i,\ u_i \approx_i u'_i). \qquad (8)$$

The skyline obtained is then represented as a fuzzy set of points, let us denote it by S'. The degree of certainty of the event "u belongs to S'" is defined as:

$$\mu_{S'}(u) = \min_{u' \in D,\,u' \neq u}\ \max(conf(u \succ_C u'),\ conf(u \approx_C u')) \qquad (9)$$

Remark 2. We have: $support(S') = S$ since as soon as $\exists u'$ such that $u' \succ_C u$, both $conf(u \succ_C u')$ and $conf(u \approx_C u')$ equal 0.

Example 3. Let us consider the query from Example 1 along with the data from Table 3 (where the values in brackets correspond to confidence degrees). The skyline obtained is: $\{0.4/t_3,\ 0.3/t_4,\ 0.6/t_7\}$.   ◇

Table 3. Relation *car* with suspect attribute values

	make	category	price	color	mileage
t_1	Opel (1)	roadster (0.8)	4500 (1)	blue (0.6)	20,000 (0.8)
t_2	Ford (0.9)	SUV (1)	4000 (0.7)	red (1)	20,000 (0.6)
t_3	VW (1)	roadster (0.4)	5000 (0.9)	red (1)	10,000 (0.8)
t_4	Opel (0.7)	roadster (1)	5000 (0.3)	red (0.6)	8000 (1)
t_5	Fiat (0.6)	roadster (1)	4500 (0.8)	red (1)	16,000 (0.9)
t_6	Renault (1)	coupe (1)	5500 (1)	blue (1)	24,000 (1)
t_7	Seat (1)	sedan (1)	4000 (0.8)	green (1)	12,000 (0.7)

4   Conclusion

In this paper, we have proposed different extensions of skyline queries with the objective of taking into account data quality when computing the points that are not dominated by any other. Two cases have been considered: that where a quality level is associated with every tuple of the dataset considered, and that where data quality is assessed at the level of attribute values. In each case, the skyline obtained is represented as a fuzzy set of points, where each answer is associated with a degree in the unit interval, which may have the meaning of a membership degree or of a certainty degree depending on the type of data quality information available.

The main perspective concerns query optimization. A preliminary study — that could not be included here due to lack of space — shows that it is possible to extend classical algorithms for computing extended skyline queries without any significant additional cost. However, this preliminary study needs to be pursued. In particular, the computation of extended skylines by means of parallel algorithms is a topic that needs to be investigated. It would also be worth studying the case where *several* degrees of quality are attached to a given tuple (or attribute value), corresponding to different dimensions of quality (freshness, accuracy, etc).

References

1. Batini, C., Scannapieco, M.: Data Quality: Concepts, Methodologies and Techniques (Data-Centric Systems and Applications). Springer-Verlag New York, Inc. (2006)
2. Börzsönyi, S., Kossmann, D., Stocker, K.: The skyline operator. In: Georgakopoulos, D., Buchmann, A. (eds.) ICDE, pp. 421–430. IEEE Computer Society (2001)
3. Bosc, P., Hadjali, A., Pivert, O.: On possibilistic skyline queries. In: Christiansen, H., De Tré, G., Yazici, A., Zadrozny, S., Andreasen, T., Larsen, H.L. (eds.) FQAS 2011. LNCS, vol. 7022, pp. 412–423. Springer, Heidelberg (2011)
4. Dai, B.T., Koudas, N., Ooi, B.C., Srivastava, D., Venkatasubramanian, S.: Column heterogeneity as a measure of data quality. In: CleanDB (2006)
5. Dubois, D., Prade, H.: Possibility Theory: An Approach to Computerized Processing of Uncertainty. Plenum Press, New York (1988); with the collaboration of Farreny, H., Martin-Clouaire, R., Testemale, C.
6. Lofi, C., Maarry, K.E., Balke, W.-T.: Skyline queries over incomplete data – error models for focused crowd-sourcing. In: Ng, W., Storey, V.C., Trujillo, J. (eds.) ER 2013. LNCS, vol. 8217, pp. 298–312. Springer, Heidelberg (2013)
7. Mihaila, G.A., Raschid, L., Vidal, M.E.: Using quality of data metadata for source selection and ranking. In: WebDB (Informal Proceedings), pp. 93–98 (2000)
8. Pei, J., Jiang, B., Lin, X., Yuan, Y.: Probabilistic skylines on uncertain data. In: Koch, C., Gehrke, J., Garofalakis, M.N., Srivastava, D., Aberer, K., Deshpande, A., Florescu, D., Chan, C.Y., Ganti, V., Kanne, C.C., Klas, W., Neuhold, E.J. (eds.) VLDB, pp. 15–26. ACM (2007)
9. Rahm, E., Do, H.H.: Data cleaning: Problems and current approaches. IEEE Data Eng. Bull. 23(4), 3–13 (2000)

Neuroscience Rough Set Approach for Credit Analysis of Branchless Banking

Rory Lewis

Department of Computer Science, University of Colorado at Colorado Springs,
Colorado Springs, CO 80918, USA

Abstract. This paper focuses on mobile banking; very often referred
to as "branchless banking" which presents a platform wherein rough set
theory algorithms can enhance autonomous machine learning to analyze credit for a purely mobile banking platform. First, the terms "mobile banking" and " branchless banking" are defined. Next, it reviews
the huge impact branchless banking with credit analysis will have on
the world and the traditional banking models as it becomes a reality
in Africa. Credit Analysis techniques of current branchless banks such
as Wonga are then explained and an improvement on their techniques
is presented. Finally, experiments taken implementing the author's neuroscience algorithms and applied with rough SVMs, Variable Precision
Rough Set Models and Variable Consistency Dominance-based Rough
Set Approach models are performed on financial data sets and their
results are presented.

1 Introduction

Every time a mobile phone invokes a financial transaction against a bank account, it is recognized as "branchless banking" because a conventional "brick and
mortar" bank was not part of the equation [14]. Branchless banking is exploding
in third world countries because it is far more convenient for poor "unbanked"
people to use their mobile phones at retail agents such as gas stations, farm supply stores and grocery stores than to travel to a bank branch [13]. However, as
banks continue to merge, local businesses tend to view them as non-local entities
[4]; the evidence of which is shown in a demise of small loans to local businesses
[1]. Enter Wonga, a United Kingdom (UK) internet banking loan company that
began as a SameDayCash project, that offers short term, high cost credit, and in
the process has accomplished what was thought to be impossible - loans without
ID or credit agency scoring. By allowing their machine learning algorithm to first
learn from data from their SameDayCash project, Wonga allowed their machine
learning an unheard of 50% default rate. However, they did this knowing that
at some point it would learn to not make those same mistake twice [5]. By the
second year it had issued 100,000 loans amounting to 20 million earning 15 million through interest payments, something traditional banks envied. For 2014,
Wonga is on track to make over 500 million [12]. To discern the ID of the person
applying for the loan: Wonga's machine learning algorithm cross-references 30

T. Andreasen et al. (Eds.): ISMIS 2014, LNAI 8502, pp. 536–541, 2014.
© Springer International Publishing Switzerland 2014

pieces of information against an email address, cookies and IP addresses and in so doing parses another 8,000 online data points that relate to the applicant [15] Wonga then asks the applicant how much and for how long and within ten minutes the applicant receives the loan. Even though they have been criticized for utilizing an extraordinarily high 4,214% APR they have a 95% satisfaction rate and many imitators such as QuickQuid, Lightspeed and others.

2   Rough Sets in Branchless Banking Credit Analysis

In a Wonga recruitment ads they seek applicants experienced in "*non-Microsoft environment*" such as Matlab, LAMP, Java, Python, NServiceBus or other Enterprise Service Bus architectures with core skills in C#, ASP MVC, Unit Testing, Messaging, SOA, TDD and Agile Methodologies" [17]. This shows that Wonga does not require skills in Weka, support vector machine (SVM), domain adaptation, rough set theory or fuzzy sets but rather ask for enterprise services. It is clear that Wonga's machine learning is a statistical-based system that may invoke an element of Bayesian. But given the fact that domain adaptation and rough set practitioners are difficult to find, let alone hire, if Wonga was using them they would have certainly mentioned these skills sets and an analysis of the Wonga's European patent application, EP 2444926A1 also illustrates a statistical architecture. Using rough sets is critically important for Wonga-type models is to move to third world countries in Africa; remember that for 1.1 billion mobile phone users in Africa to access credit, they will only use their mobile or branchless banking records, no ID, no address, no online forms. Also remember that Wonga asks the user to answer 30 questions? The two aforementioned scenarios point away from statistical credit analysis and point directly into a rough set model that will analyze how the user has spent previously earned monies. We will will now explain why such a system will have to incorporate methodologies such as SVM, fuzzy SVM, Variable Precision Rough Set Model (VP-RSM) and other such systems to drill the data down for efficient domain adaptation.

2.1   Experiments Based on Computational Neuroscience

Computational neuroscience is the study of billions of signals occurring in the human brain. Over the last four years the author has focused on rough set theory and domain adaptation to efficiently mine only those signals that procure pathological oscillations[8]. From a computing standpoint, a system analyzes at billions of signals and seeks patterns. Whether these signals stem from, earthquakes, astronomy telescopes, the human brain or financial trends, its still signals. Looking at spending habits as a tree, if a system had to use as its apex node, the instance that the user receives income, then the branches and leaves below it represent the number of transactions and the types of transactions. Similarly to identifying patterns in the brain, a discreet finite automata (DFA) model can represent these spending habits where the state to which the system changes is based upon a transition matrix where columns may represent an expenditure in which current or past parameters have specified values. For example, looking at Figure 1 (b), if we have six mutually exclusive, collectively exhaustive spending categories and we were to apply three limits α, β, and γ of what spending parameters would be characteristic of the previously identified "bad" spending patterns. we would have 6 states in the tree where the last three states would be similar to the first three except that the absolute value of the second derivative $(f''(x) or \frac{d^2 y}{dt^2})$ would be less than γ. A slope $(m = \frac{y_2 - y_1}{t_2 - t_1})$ would also be provided to differentiate between the normal state and the possibility of a bad spending, and likely artifact such as buying alcohol, but its after bills are paid or its a wedding or an event that should not be noted as "bad". See [9] [10] [11]. However, in order to provide a realistic financial model to the proposed DFA tree it needs to accommodate a vast amount of complexity and deal with uncertainty of the billions of types for transactions and patterns. To deal with the uncertainty, in previous work the author has utilized rough sets where these **Three Tenets of Rough Sets** describe the objects and functions that make up an information system: First , U is a nonempty, finite set of objects (object identifiers), second, A is a nonempty, finite set of attributes (functions) i.e. $a : U \to V_a$ for $a \in A$, where V_a is called the domain of a and third, in rough sets one writes (a, v) instead of v, assuming that $v \in V_a$. Information systems can be seen as decision tables and by a decision table we mean an information system $S = (U, A_1 \cup A_2 \cup \{d\})$ where $d \notin A_1 \cup A_2$ is a distinguished attribute called decision.

Domain adaptation is based off of the critical assumption that somewhere out there, there is a perfect classifier that we can adapt on to the present domain. This is called the "Single Good Hypothesis"; $\exists h^*, \epsilon_S(h^*), \epsilon_T(h^*) small$. which is in essence saying, for each person applying for credit on his mobile device that there exists, somewhere out there, a classification rule that is based off of previous persons (S) who were also standing in that person's shoes so to speak, that correctly predicts a particular type of spending pattern in both the training P_n's EEG data and the target P_{n+1}'s EEG data where the error is small. This means that we now have two distributions, a *source* (S) distribution of the previous person's classification rules and a *target* (T) distribution of the present person applying for the bank loan through his mobile device. The *source* distribution, $(x, y) \sim Pr_S[x, y]$, is the *training* distribution where classification rules are derived from many previous *persons* while the *target* (T) distribution, $(x, y) \sim Pr_T[x, y]$, is the *test* distribution which may elicit rules derived from the closest match of the previous *persons* that also incurred very similar spending classification rules. Many methodologies can be used to do this including, Covariate Shift, Representation Learning , Feature Based Supervised Adaptation and Parameter Based Supervised Adaptation, to name a few [3] .

First a *basis of initial statistical based scoring* is rendered for the source data for adaptation processes. Eventually the rough set scoring will also be in the source domain but that is out of the scope of this experiment. using a random number generator a matrix of 2,000 "tuple persons" was distributed with 50 attributes of which a minimum of 20 and a maximum of 40 expenditures were randomly associated with each tuple person. The fifty attributes were weighted

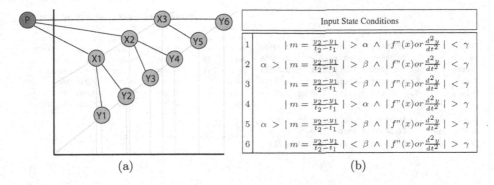

Fig. 1. *Representations* (a) For assessment of the first expenditure (X) after payment (P) the system utilizes a three-pairwise system to procure classifiers for domain adaptation, and (b) Six mutually exclusive, collectively exhaustive input states where α, β, and γ are user selected constants

from good to bad with the first attribute being a best expenditure and the 50th being a worst expenditure. Finally a sequence was randomly generated with 40% guaranteed to pay one of the better expenditures initially after payday which could represent paying rent. This meant that out of the 2,000 people, about 40% first paid rent and thereafter anything could happen. A scoring system counted the number of transactions n and summated the score of each transactions x score by its rank. For example if the first transaction had a score 20 and there were ten transactions then it would count as a twenty, continuing down to the last transaction that may have a score of 5, it is multiplied by 1. Upon processing the scores all patterns along with scores are entered into the source data. This means that we have synthesized the hypothetical where a Telekom entering the credit analysis domain has something to start with and that is statistical, non rough set theory data in a source domain. Now as the person asks for credit we will compare three approaches to rough sets to discern who receives the credit or not. The author randomly selected 100 persons from the source domain and repeated their one month of spending to three months by randomly making slight variations of their original sequence of expenditures. As seen in Figure 1 (a), these experiments only look at the first two expenditures (X and Y respectively) of the ten persons with three options for expenditure 1 that would constitute the system continuing in its analysis and and two leaves for each branch representing the first three "correct" policies. Now the system is set up for the machine learning methodologies to be used.

Four systems would be used for testing. A base brute force statistical scoring method would be compared with three rough systems. The first rough system would focus on identifying outliers and then efficiently decreasing that input's membership. For this the author utilized Wang et al's, new fuzzy SVM [16]. The second rough system would focus on dealing with small degrees of misclassification error in the definition of lower approximations and for this the author utilized the Variable Precision Rough Set Model (VP-RSM) [18][7]. The

	Brute Force	SVM	VP-RSM	VC-DRSA
Person 23	88.20%	86.29%	103.20%	93.29%
Person 27	91.30%	87.07%	104.55%	94.33%
Person 45	76.80%	71.81%	86.50%	77.94%
Person 49	77.20%	76.66%	91.47%	82.77%
Person 51	90.00%	83.70%	100.90%	90.88%
Person 57	94.10%	94.29%	112.35%	101.73%
Person 7U	88.50%	03.72%	100.66%	90.77%
Person 83	88.90%	88.82%	105.88%	95.86%
Outliers FP	11	2	0	1
Outliers FN	19	0	4	1

Person	1	0.232525441	0.41188	0.51667	0.74416
Person	2	0.27616995	0.98224	0.2292	0.35898
Person	3	0.322756225	0.74165	0.95598	0.77501
Person	4	0.443616104	0.37898	0.5652	0.04614
Person	5	0.373296217	0.31171	0.06502	0.23563
.		.	.	.	.
Person	98	0.936492049	0.3453	0.13409	0.95694
Person	99	0.747830233	0.58264	0.79971	0.37467
Person	100	0.24411067	0.16959	0.7787	0.22565

(a)   (b)

Fig. 2. Data Sets: (a) raw data of target domain showing 100 Persons with first four attributes, and (b) the results of the 8 people deemed credit worthy with their corresponding Brute force, SVM, VP-RSM and VC-DRSA scores and the number of false positives and false negatives rendered by each system

third rough system would focus on not using a single class but instead look up and down for the unions between the layers of X and Y and for this the Variable Consistency Dominance-based Rough Set Approach (VC-DRSA) [2] [6] was utilized.

3 Conclusions and Future Work

Out of the 100 persons first drawn down by either brute force or by the one of the rough set approaches and then instantiated into the adapted initial source data, only 8 people were deemed credit worthy, and as seen Figure 2(b) the corresponding scores of Brute force, SVM, VP-RSM and VC-DRSA scores are displayed along with each system's number of false positives and false negatives. Even though it is clear that the brute force, non rough set approach had 11 false positives and 19 false negatives with the rough sets approach barely having any, there are many questions these experimental results raise. First the author is not sure why the VP-RSM has 6 scores above 100% and consistently higher on average than all the other scores. Even though the general outcome is excellent in that out of the 100 persons that applied for credit on the mobile devices without ID, thirty questions and only transaction history, only the best 8 were given loans.

References

1. Alessandrini, P., Presbitero, A.F., Zazzaro, A.: Banks, distances and firms' financing constraints. Review of Finance 13(2), 261–307 (2009)
2. Błaszczyński, J., Greco, S., Słowiński, R., Szelg, M.: Monotonic variable consistency rough set approaches. International Journal of Approximate Reasoning 50(7), 979–999 (2009)

3. Blitzer, J., McDonald, R., Pereira, F.: Domain adaptation with structural correspondence learning. In: Proceedings of the 2006 Conference on Empirical Methods in Natural Language Processing, pp. 120–128. Association for Computational Linguistics (2006)
4. Ely, D.P., Robinson, K.J.: Credit unions and small business lending. Journal of Financial Services Research 35(1), 53–80 (2009)
5. Gelber, R.: The profitability of failure. Datanami, Big Data, Big Analytics, Big Insights, http://www.datanami.com/datanami/2012-02-29/http://the_profitability_of_failure.html (accessed: February 29, 2012)
6. Inuiguchi, M., Yoshioka, Y., Kusunoki, Y.: Variable-precision dominance-based rough set approach and attribute reduction. International Journal of Approximate Reasoning 50(8), 1199–1214 (2009)
7. Kusunoki, Y., Błaszczyński, J., Inuiguchi, M., Słowiński, R.: Empirical risk minimization for variable precision dominance-based rough set approach. In: Lingras, P., Wolski, M., Cornelis, C., Mitra, S., Wasilewski, P. (eds.) RSKT 2013. LNCS, vol. 8171, pp. 133–144. Springer, Heidelberg (2013)
8. Lewis, R., Mello, C.A., White, A.M.: Tracking epileptogenesis progressions with layered fuzzy k-means and k-medoid clustering. Procedia Computer Science 9, 432–438 (2012)
9. Lewis, R., Parks, B., Shmueli, D., White, A.M.: Deterministic finite automata in the detection of epileptogenesis in a noisy domain. In: Proceedings of the Joint Venture of the 18th International Conference Intelligent Information Systems (IIS) and the 25th International Conference on Artificial Intelligence (AI), June 8-10, pp. 207–218 (2010)
10. Lewis, R.A., Parks, B., White, A.M.: Determination of epileptic seizure onset from eeg data using spectral analysis and discrete finite automata. In: 2010 IEEE International Conference on Granular Computing (GrC), pp. 277–282. IEEE (2010)
11. Lewis, R.A., White, A.M.: Multimodal spectral metrics with discrete finite automata for predicting epileptic seizures. In: 2010 IEEE/WIC/ACM International Conference on Web Intelligence and Intelligent Agent Technology (WI-IAT), vol. 2, pp. 445–448. IEEE (2010)
12. Liew, J.: Big data + machine learning = scared banks. Lightspeed Venture Partners, http://pando.com/2012/02/27/big-data-machine-learning-scared-banks/ (accessed: February 24, 2012)
13. Lyman, T., Ivatury, G., Staschen, S.: Use of agents in branchless banking for the poor: Rewards, risks, and regulation. Consultative Group to Assist the Poor, Focus Note 38 (2006)
14. Richard, C.C.: How the us government's market activities can bolster mobile banking abroad. Wash. UL Rev. 88, 765 (2010)
15. Shaw, W.: Wired magazine: Cash machine: Could wonga transform personal finance?, http://www.wired.co.uk/magazine/archive/2011/06/features/wonga/page/4 (accessed: May 5, 2011)
16. Wang, Y., Wang, S., Lai, K.K.: A new fuzzy support vector machine to evaluate credit risk. IEEE Transactions on Fuzzy Systems 13(6), 820–831 (2005)
17. Wonga.com: R&d software developer, wonga labs, https://www.wonga.com/money/r-and-d-software-developer-dublin/ (accessed: February 11, 2014)
18. Ziarko, W.: Variable precision rough set model. Journal of Computer and System Sciences 46(1), 39–59 (1993)

Collective Inference for Handling Autocorrelation in Network Regression

Corrado Loglisci, Annalisa Appice, and Donato Malerba

Dipartimento di Informatica, Università degli Studi di Bari Aldo Moro
via Orabona, 4 - 70126 Bari - Italy
{corrado.loglisci,annalisa.appice,donato.malerba}@uniba.it

Abstract. In predictive data mining tasks, we should account for auto-correlations of both the independent variables and the dependent variable, which we can observe in neighborhood of a target node and that same node. The prediction on a target node should be based on the value of the neighbours which might even be unavailable. To address this problem, the values of the neighbours should be inferred collectively. We present a novel computational solution to perform collective inferences in a network regression task. We define an iterative algorithm, in order to make regression inferences about predictions of multiple nodes simultaneously and feed back the more reliable predictions made by the previous models in the labeled network. Experiments investigate the effectiveness of the proposed algorithm in spatial networks.

1 Introduction

Nowadays, more and more data naturally come in the form of a network (e.g. social and sensor networks) in several application domains. A data network consists of nodes and edges. Nodes are associated to a number of variables that describe characteristics of the entities, edges are associated to a numeric or symbolic information that represents the degree of dependence between (the characteristics of) the linked nodes. This dependence indicates forms of cross-correlation, and, in particular, cross-correlations of a variable with itself (*autocorrelation*) in the data network. In this paper, we focus on the problem of dealing with the property of autocorrelation, between the label and labels of neighboring nodes, in a task of network regression.

Traditional regression algorithms look for predictive models that correlate a dependent variable with the independent variables of the same node only. In this paper, we propose to use collective inferences, in order to accommodate all the three forms of cross-correlation into the regression patterns. We propose a novel algorithm, called CORENA (COllective REgression in-Network Algorithm), that predicts jointly the numeric dependent variables of the inter-linked nodes. Inferences on linked nodes mutually reinforce each other. It exploits an iterative procedure that pools, at each iteration, predictions inferred at neighbor nodes in the description of a node. Inferences made with higher reliability are stably fed back into the network and used to inform subsequent inferences about the inter-linked nodes. The reliability of inferences is measured by looking for the

T. Andreasen et al. (Eds.): ISMIS 2014, LNAI 8502, pp. 542–547, 2014.
© Springer International Publishing Switzerland 2014

principle of the homophily, according to which we expect that similar predictions are more likely to be linked.

The paper is structured as follows. In Section 2 we introduce the problem setting and describe our proposal. An empirical evaluation with real-world networks is reported in Section 3 and conclusions are drawn in Section 4.

2   Iterative Collective Regression

A network N is a set of nodes V connected by edges E. A number (that is usually taken to be positive) called "weight" w is associated to each edge. This weight represents the strength of the connections from one node to another. In this work we consider a network setting in which we have partially labeled nodes L, while the problem is that of providing accurate estimates of unknown labels associated with the unlabeled nodes U ($V = L \cup U$) of a network.

In this paper, we assume that $i)$ nodes of the network are associated with data observations $(\mathbf{x}, y) \in \mathbf{X} \times Y$. $\mathbf{X}$ is a feature space spanned by m descriptor (independent) attributes X_i with $i = 1, \ldots, m$, while Y is the possibly unknown target (dependent) variable (or label) with a range in $\mathbb{R}$; $ii)$ labeled nodes and unlabeled nodes can be part of the same network.

Collective regression is done by building neighboring dependent variables that account for the autocorrelation of a variable with itself in a neighborhood. Neighboring variables can be computed for each independent variable ($X \in \mathbf{X}$), as well as for the dependent variable (Y). We remark here that, while the neighboring information of the independent variables is stable over the learning process, the neighboring information of the dependent variable may change over the learning process as new inferences can be made at the linked nodes. So, neighboring dependent variables that derive from the independent variable are subject to change over the inference process.

The inference process looks for predictive models that correlate the label of a node to the dependent variables associated with the node, as well as to the neighboring variables computed with the neighborhood. Inference is iterative, so that inferences made with high reliability are fed back into the network and used, in order to inform subsequent inferences about linked nodes. In the initialization phase, an initial inference is made by learning a traditional regression model (e.g. a model tree) from L. This model accounts for correlations between the dependent variables and the independent ones, autocorrelations of the independent variables, but still overlook autocorrelations of the dependent variable. This model can be used, in order to initialize unknown labels of U. In the iterative phase, for each node i of V, neighboring variables $yN(i)$, that are associated with the dependent variable Y, can be now computed by aggregating variable values in a neighborhood $N(i)$. A new regression model can be learned from this modified training set. Labels of U, as well as neighboring variables $yN(i)$ are dynamically updated accordingly. A measure of autocorrelation is computed, in order to select the more reliable label among the several ones inferred for the same node by the iterative procedure.

Algorithm 1. CORENA$(L, U, E) \mapsto \hat{U}$

Require: L: labeled nodes set on $\mathbf{X} \times Y$
Require: U: unlabeled nodes set on $\mathbf{X}$
Require: E: edge relation
1: iteration$\leftarrow 0$
2: $\mathbf{X}N \leftarrow$buildingNeighboringVariables$(L \cup U, E, \mathbf{X})$
3: $\hat{U} \leftarrow$ setLabel$(U, \text{learnRegressionModel}(L, \mathbf{X} \times \mathbf{X}N \times Y))$
4: $\mathbf{Y}N \leftarrow$buildingNeighboringVariables$(L \cup \hat{U}, E, Y)$
5: **repeat**
6:    $\hat{U}New \leftarrow$ setLabel$(U, \text{learnRegressionModel}(L, \mathbf{X} \times \mathbf{X}N \times \mathbf{Y}N \times Y))$
7:    $noChange \leftarrow 0$
8:    **for** $i \in U$ **do**
9:       $oldR = $computeReliability$(\text{getLabel}(i, \hat{U}), L)$
10:      $newR = $computeReliability$(\text{getLabel}(i, \hat{U}New), L)$
11:      **if** $newR > oldR$ **then**
12:         setLabel$(\hat{U}, i, \text{getLabel}(i, \hat{U}New))$
13:      **else**
14:         $noChange \leftarrow noChange + 1$
15:      **end if**
16:    **end for**
17:    $iteration \leftarrow iteration + 1$
18:    $\mathbf{Y}N \leftarrow$buildingNeighboringVariables$(L \cup \hat{U}, E, Y)$
19: **until** $(iteration = MAXITER$ or $noChange \geq minNoChange)$

The Algorithm. The top-level description of the iterative collective inference procedure is reported in Algorithm 1. It inputs $L(\mathbf{X} \times Y)$, $U(\mathbf{X})$, E and outputs $\hat{U}(\mathbf{X} \times Y)$. We will call this algorithm CORENA.

The initialization phase (Algorithm 1, lines 2-4) is three stepped: *(a)* for each independent variable $X \in \mathbf{X}$, the associated neighboring dependent variables $\mathbf{X}N$ are built (Algorithm 1, line 2); *(b)* an initial regression model is learned from the training set $L(\mathbf{X} \times \mathbf{X}N \times Y)$, then it is used to initialize the unknown labels of U and store them in $\hat{U}$ (Algorithm 1, line 3); *(c)* for the dependent variable Y, the associated neighboring dependent variables $\mathbf{Y}N$ are built (Algorithm 1, line 4). In steps (a) and (c), neighboring variables are computed by considering the entire node set $(L \cup \hat{U})$ and using one of the schema above defined.

The iterative phase is realized with the main loop (Algorithm 1, lines 5-19). It keeps with the collective theory by looking for new inferences that use "reliable" labels. It stops when the number of labels unchanged in $\hat{U}$ is greater than the user-defined threshold $minNoChange$.

The iterative phase is three stepped:

1. A new regression model is learned from the training set $L(\mathbf{X} \times \mathbf{X}N \times \mathbf{Y}N \times Y)$. It is used to infer new labels for the unlabeled part of the network. These new labels are stored in $\hat{U}New$ (Algorithm 1, line 6).
2. For each node of the unlabeled set U, the reliability of the label estimated at the previous iteration (and stored in $\hat{U}$) and the reliability of the label estimated in the current iteration (and stored in $\hat{U}New$) are computed. Reliability is measured with a measure of local autocorrelation (details are reported below). For each node, the more reliable label is that maintained in $\hat{U}$ for the next iteration (Algorithm 1, lines 8-16).
3. The neighboring variables $\mathbf{Y}N$ are updated according to new labels injected in $\hat{U}$ (Algorithm 1, line 18).

Computing Neighboring Variables. Let us consider a node i of V ($i \in V$), we denote $N(i, d)$ the neighborhood of i over E with size d. It is the set of nodes of V that are connected to i with a path of length less than or equal to d. Formally, $N(i, 1) = \{j \in V | (i, j) \in E\}$, while $N(i, d) = \{j \in V | h \in N(i, d-1) \wedge (h, j) \in E\}$.

Let us consider a variable A ($A \in \mathbf{X}, Y$), we introduce three variable construction schema according to we can build the neighboring variables aN, which are associated with A in the network (V, E).

Variable Schema (Var1). Let us consider the neighborhood $N(i, 1)$. We use this neighborhood, in order to build: (1) the neighboring variable $aN(i, mean)$ as the weighted mean of values of A in $N(i)$ and (2) the neighboring variable $aN(i, stDev)$ as the standard deviation of values of A in $N(i)$. In the weighted computations, weights are that associated with the edges of E.

Variable Schema (Var2). By following the idea reported in [3], we consider here two neighborhoods, namely $N(i, 1)$ and $N(i, 2)$, so that $N(i, 1) \subseteq N(i, 2)$. We compute the weighted means $aN(i, 1, mean)$ and $aN(i, 2, mean)$, as well as the standard deviations $aN(i, 1, stdDev)$ and $aN(i, 2, stdDev)$ of values of A in $N(i, 1)$ and $N(i, 2)$, respectively. In addition, we compute the neighboring variable, defined as $\frac{aN(i, 1, mean)}{aN(i, 2, mean)}$, in order to synthesize the speed at which the variable values change around i over the network.

Variable Schema (Var3). Let us consider the neighborhood $N(i, 1)$. We use this neighborhood, in order to build the neighboring variables $aN(i, [min, max])$ that counts how many values of A fall in the interval $[min, max]$ in $N(i)$. So, the number of neighboring variables associated with A is equal to the number of intervals computed by the equal-frequency discretization of A in the network.

Measuring Reliability of Labels. We measure the reliability of labels predicted at each iteration, in order to select reliable labels that are fed back to the training network for the next iteration. Intuitively, reliable labels should meet the principle of the homophily so that similar labels can be propagated on the linked nodes. The higher the autocorrelation of a label with labels in the neighborhood, the more reliable the homophily of the label in the network. A local measure of autocorrelation can be used to quantify homophily of labels. It returns one numeric value per node; this value measures the degree of homophily of a label in its neighborhood.

The Anselin's local Moran's I [2] is a well known local measure of autocorrelation, largely used in spatial analysis, in order to identify objects that are similar to objects in the neighborhood. The reliable label is chosen by evaluating both the label predicted for each unlabeled node in the present iteration and the label outputted for the same node at the previous iteration. The label having the higher value Anselin's local Moran's I is selected as the more reliable for the next iteration. Operatively, for each unlabeled node, the measure is computed by considering the neighbors that belong to the labeled set. A positive value indicates that the label is surrounded by similar values. The higher the measure, the more reliable with homophily the predicated label.

3   Empirical Evaluation

In the empirical evaluation, we investigate the viability of collective inference solution several network regression problems.

Dataset Description. We consider five data networks computed from real world spatial data sets. The **NWE** dataset concerns 970 censual sections of North West england. We consider the percentage of mortality (dependent variable) and measures of deprivation level in the section according to index scores such as the Jarman Underprivileged Area Score, Townsend score, Carstairs score and the Department of the Environment Index (independent variables). The **FOIXA** dataset contains 420 measurements of the rates of outcrossing (dependent variable) at sample points in the surrounding genetically modified (GM) fields in the Foixa region in Spain. The dependent variables are the number and size of the surrounding GM fields, the ratio of the size of the surrounding GM fields and the size of conventional fields and the average distance between the conventional and the GM fields. The **LAB** dataset contains data measured from 52 sensors (nodes) deployed in the Intel Berkeley Research lab. The measurements concern humidity, light and voltage values (independent variables) and temperature (dependent variable). The **NOAA-clwvi** and **NOAA-pr** datasets concern a NOAA study of climate modeling through 250 measurements of several atmospheric variables. For both datasets, we consider 31 independent variables concerning meteorology, heat flux, pressure, temperature and wind measured at 250 sampling points. In **NOAA-clwvi**, the dependent variable is the content of condensed water in the clouds, while, in **NOAA-pr**, the dependent variable is the precipitation flux.

For all these datasets, the sampling units are associated with spatial coordinates (e.g. latitude and longitude), so they are a spatial distance apart. By considering this spatial information, a network can be obtained. Nodes are the sampling units. Edges are distance-based relations so that two nodes are edged when they are within a distance threshold δ. We use the distance threshold defined as follows: $\delta = \max_{i \in V} (\min_{j \in V, i \neq j} d(i, j))$, where $d()$ is the Euclidean distance between the spatial coordinates. The weights assigned to the edges can be determined in two alternative ways. In the former case, (denoted as **W1**), the weight of an edge (i, j) is equal to the inverse of a power function of $d(i, j)$, that is, $w_{ij} = \frac{1}{d(i,j)^3}$. In the latter case, (denoted as **W2**), we normalize each weight as follows $w_{ij} = \left(\frac{1}{d(i,j)^3} \right) / \left(\sum_{j,(i,h) \in E} \frac{1}{d(i,h)^3} \right)$.

Experimental Setup and Evaluation Measures. We evaluate CORENA by combining the different schemas for the neighboring variables and for weighting the edges. The model tree learner MTSMOTI [1] was used as base learner while the network regression algorithm NCLUS [4] as main competitor. We run CORENA with $minNoChange$ equal to the half of the size of the unlabeled testing set and $maxIt = 15$. The predictive accuracy was evaluated by the root mean squared error ($RMSE$) computed by using a 10-fold cross validation.

Table 1. CORENA with MTSMOTI (as base learner) vs MT-SMOTI and NCLUS. The RMSEs of CORENA are in bold when they are lower than the errors of MTSMOTI. The lower RMSE between all the compared algorithms (CORENA, MTSMOTI and NCLUS) is in italic.

	CORENA(MT-SMOTI)						MT-SMOTI	NCLUS
	W1			W2				
	Var1	Var2	Var3	Var1	Var2	Var3		
FOIXA	**2.535**	**2.533**	**2.528**	*2.522*	*2.522*	**2.526**	2.589	2.969
NWE	**2.36**	**2.36**	*2.35*	**2.36**	**2.36**	*2.35*	2.38	2.55
LAB	**1.573**	**1.684**	*1.349*	**1.419**	**1.459**	**1.407**	5.252	4.627
NOAA-clwvi	**0.222**	**0.222**	**0.223**	**0.203**	*0.197*	**0.201**	0.331	0.322
NOAA-pr	*3.156*	*3.156*	*3.156*	*3.156*	*3.156*	*3.156*	4.074	8.610

Results and Discussion. Table 1 collects errors of CORENA with MTSMOTI as base learner. The baseline is the error computed when labels are estimated by the base learner at the initialization phase. A detailed analysis of these results deserve several considerations. We can observe that in all five data sets CORENA, in all tested configurations, outperforms the baseline MTSMOTI as well as the network competitor NCLUS. This allows us to show that accounting for autocorrelations in a network regression model can really improve accuracy of predictive inferences. Performances of collective inference are not greatly affected by the weighting mechanism, but the choice of the variable construction schema can become a crucial issue in some applications.

4 Conclusions

In this paper, we have investigated the regression task in data networks. The novel contribution of this study is the use of collective inference, in order to accommodate the autocorrelation of the dependent variable in the predictive inference. Experiments, conducted on several real-word data collections, show the algorithm performs, in general, better than both traditional algorithms which do not consider the autocorrelations.

Acknowledgements. This work fulfils the research objectives of the PON 02 00563 3470993 project "VINCENTE - A Virtual collective INtelligenCe ENvironment to develop sustainable Technology Entrepreneurship ecosystems" funded by the Italian Ministry of University and Research.

References

1. Appice, A., Džeroski, S.: Stepwise induction of multi-target model trees. In: Kok, J.N., Koronacki, J., Lopez de Mantaras, R., Matwin, S., Mladenič, D., Skowron, A. (eds.) ECML 2007. LNCS (LNAI), vol. 4701, pp. 502–509. Springer, Heidelberg (2007)
2. Arthur, G.: A history of the concept of spatial autocorrelation: A geographer's perspective. Geographical Analysis 40(3), 297–309 (2008)
3. Ohashi, O., Torgo, L.: Wind speed forecasting using spatio-temporal indicators. In: ECAI 2012, vol. 242, pp. 975–980. IOS Press (2012)
4. Stojanova, D., Ceci, M., Appice, A., Dzeroski, S.: Network regression with predictive clustering trees. Data Min. Knowl. Discov. 25(2), 378–413 (2012)

On Predicting a Call Center's Workload:
A Discretization-Based Approach

Luis Moreira-Matias[1,2,3], Rafael Nunes[3], Michel Ferreira[1,5],
João Mendes-Moreira[2,3], and João Gama[2,4]

[1] Instituto de Telecomunicações, 4200-465 Porto, Portugal
[2] LIAAD-INESC TEC, 4200-465 Porto, Portugal
[3] FEUP, U. Porto, 4200-465 Porto, Portugal
[4] Faculdade de Economia, U. Porto 4200-465 Porto, Portugal
[5] DCC-FCUP, U. Porto, 4169-007 Porto, Portugal
{luis.m.matias,joao.mendes.moreira,jgama}@inescporto.pt,
{rafael.nunes,michel}@dcc.fc.up.pt

Abstract. Agent scheduling in call centers is a major management problem as the optimal ratio between service quality and costs is hardly achieved. In the literature, regression and time series analysis methods have been used to address this problem by predicting the future arrival counts. In this paper, we propose to discretize these target variables into finite intervals. By reducing its domain length, the goal is to accurately mine the demand peaks as these are the main cause for abandoned calls. This was done by employing multi-class classification. This approach was tested on a real-world dataset acquired through a taxi dispatching call center. The results demonstrate that this framework can accurately reduce the number of abandoned calls, while maintaining a reasonable staff-based cost.

Keywords: call centers, arrival forecasting, agent scheduling, discretization, multi-class classification.

1 Introduction

Staffing is one of the major problems in call center management. This paper focuses on predicting workload in call centers in order to improve staff scheduling. By using these predictions, it is possible to formulate the scheduling problem as an optimization problem. The goal is to minimize the number of abandoned calls. This problem can be simply solved by using a heuristic method of interest.

Workload estimation focuses on predicting demand. In the literature, scheduling is formulated according to a **point-wise** prediction of the quantitative target variable. To address this problem, it is possible to divide the predictive models into two types: (a) time series analysis and (b) regression. The first one (a) typically relies on assuming homogeneous or time-varying Poisson processes to *feed* Holt-Winters smoothing models or ARIMA-based models [1,2]. The second type establishes a relationship between the number of arrivals and other explanatory variables such as the day type [3,4].

T. Andreasen et al. (Eds.): ISMIS 2014, LNAI 8502, pp. 548–553, 2014.
© Springer International Publishing Switzerland 2014

Type-a approaches typically assume the future number of arrivals as a linear combination of their historical values. By dealing with large time horizons (days), these models *lose* one of their best assets: the ability to react to *bursty* events [5]. Type-b models are able to establish more complex relationships (e.g. non-linear relationship in Artificial Neural Networks (ANN)). However, many of these models aim at minimizing the root mean squared error (RMSE) between the predicted and the actual arrivals, discarding demand peaks (i.e. outliers). Consequently, these extreme events represent the highest ratio of *abandoned* calls. The compromise between (a) *understaffing* to maintain low costs by losing *some* service demand and (b) *overstaffing* to minimize abandoned calls is hardly done by assuming constant or periodic arrival rates. Consequently, the workload forecasting problem may not be adequately addressed by such methods.

Discretizing continuous variables is a relevant building block of many machine learning algorithms (for instance, C4.5). Hereby, this paper proposes a *local discretization* technique to address this limitation. In this applicational framework, the basic idea consists of determining the number of agents required to meet a demand expressed in **equal-width** intervals. The interval width corresponds to the **expected agent capacity**, which is assumed to be *constant* over time and along the different workers. By dividing the arrival counts into equal-width intervals, the goal is to accomplish two distinct goals: (1) reducing the search space for supervised learning methods and (2) adding the classification methods to the current *pool* of problem solvers. This is a step forward towards solving the problem. The approach that is closest to the one presented here is the one proposed by Shen *et al.* [6]. The authors perform singular value decomposition to reduce the dimensions of the independent variables. However, they still formulate the workload prediction as a numerical forecasting problem.

A small call center with 13 agents running in the city of Porto, Portugal, was chosen as the case study for this paper. Its scheduling is still performed empirically. The results highlight the method's contribution to this problem as it outperforms the numerical prediction methods on the proposed dataset.

This paper is organized as follows: Section 2 describes our methodology to overcome the agent scheduling problem. Section 3 introduces the real-world scenario addressed in this study. The experimental setup and the results obtained are described in Section 4. Finally, conclusions are drawn in the last section.

2   Methodology

Let A_t denote the a stochastic process describing the number of arrivals per period of time. Let $N = \{n_i | n_i \in \mathbb{N} \wedge n_i \leq \Gamma\}$ denote the domain of the arrival counts A_t, where Γ is the maximum admissible value of arrivals per unit of time. This work explores two distinct predictive approaches: (1) one where $A_t \in N$ is used and another (2) where a value *interval* $\pi_i = [b_i, b_{i+1}) \in \Pi$ for A_t is predicted, such that $b_i \leq A_t < b_{i+1}$. Let $\Pi = \{\pi_i | \pi_i = [b_i, b_{i+1}) : b_{i+1} - b_i = b_i - b_{i-1}, \forall b_i \in N\}$ stand for the interval set and $\delta = b_{i+1} - b_i$ define the corresponding **width** interval. By reducing the target variable dimension, the goal is to enhance the detection of future workload peaks.

Let $X = \{X_1, ..., X_\rho\}$ denote the set of ρ *attributes*. Let $x = \{x_1, ..., x_\rho\}$ be a set of attribute values where $x_i \in X_i$. The goal is to infer the function $\hat{f}(x) \sim f(x) \in N : f(x) = A_t, \forall x$. This induction is data driven as it uses a training set T (i.e. a data set where each sample is a pair (x, A_t)) to compute the approximation. The learners usually perform multiple scans over T to iteratively update its models. This cycle stops when it finds a minimal value for a function which establishes the differences between $f(x)$ and $\hat{f}(x)$. This function is known as *objective function*. However, many learners tend to present rough approximations to bursty peak values by *smoothing* their models (e.g. Linear Least Squares). This effect is a major problem in this context. To overcome it, we propose to **discretize** the target variable. Let us redefine the problem as

$$\hat{f}(x) \sim f(x) \in \Pi : f(x) = \pi_i \Rightarrow A_t \in \pi_i, \forall x \qquad (1)$$

Staff scheduling is a *bounded* problem, although the arrival prediction is not. Each call center is constrained to the number of workstations available (denoted as w), and (b) to the number of agents available, i.e. O_m. Let $C_m \in N$ be the maximum workload supported without calls abandoned by each agent per unit of time, and l_t be the number of abandoned calls during the same period. Let $O_w = \min(O_m, w)$ be the maximum number of agents operating per unit of time. The maximum workload supported by the call center without abandoned calls is defined as $\Psi = O_w \times C_m$. Consequently, it is not that relevant to predict whether $A_t > \Psi$ or $A_t \gg \Psi$. Discretizing attributes is a well-known preprocessing technique in machine learning. This paper proposes to discretize the independent variable A to reduce its domain length. In this context, the interval width can be defined as $\delta = C_m$. Therefore, it is possible to redefine the domain of $f(x)$ as $|\Pi| = O_w + 1$. It corresponds to O_w equal-width intervals plus an extra interval where $A_t \in \pi_i = [O_w \times C_m, \Gamma) : l_t = A_t - \Psi > 0$. By predicting an interval instead of an exact value, it is possible to formulate this problem as a *multiclass classification* problem where the label is the interval where A_t will fall into.

Estimating agent capacity C_o in a Call Center is a challenging problem. For the sake of simplicity, it is assumed that this is time-constant for each prediction. Let $Z \subseteq T$ be the dataset describing historical demand peaks defined as

$$Z = \left\{ z_i = \frac{A_t - l_t}{O_t} \middle| \frac{l_t}{A_t} \geq \alpha > 0, \forall t \right\} \qquad (2)$$

where O_t is the number of agents operating in t and α is a user-defined parameter to consider a past arrival count as a bursty event. C_m can be obtained as the **median** number of calls answered by each agent during a bursty event (i.e. $\tilde{Z}$).

The predicted π_t stands for the *desired* workload at time period t. Let H_i denote the number of shifts assigned to an agent i and Ω be its maximum. Let $b_{i+1,t}$ denote the upper-bound value of the predicted arrival count interval π_t. $O(t)$ can be computed as follows

$$\underset{O(t) > 0}{\arg\min} \; b_{i+1,t} - (O(t) \times C_m), \; s.t. : \; H_i \leq \Omega, \; \forall \, i, t \qquad (3)$$

3   Case Study

The case study presented here is based on a taxi dispatching center in Porto, Portugal. The center distributes calls to a fleet of 441 vehicles and employs 13 agents. However, only $O_m = 11$ are available for scheduling due to the existing labor regulations. Their assignments are still performed on a weekly basis. Each agent can only have a maximum number of $\Omega = 5$ shifts assigned per week. However, the number of workstations available is $w = 4$. The call arrivals between June and October 2013 were used as test-bed dataset.

4   Experiments and Results

The first 17 weeks were used as a training set, while the last four were considered a test set. The training set for each week consisted of all the arrival count data available until this point in time. The α value was set as 0.15. An interval resolution of 30 minutes was considered in the experiments (as in [7]).

Two distinct approaches were followed in the experiments: (a) one where the arrival counts were predicted as an exact number and another (b) where they were predicted as an interval. In (a), the Holt-Winters smoothing was employed, as well as the k-Nearest Neighbors (kNN), the Random Forests (RF) and the Projection Pursuit Regression (PPR). In (b), the classification methods employed were the NB, RIPPER, ANN and RF. The classification methods were used with their default parameters. A sensitivity analysis was conducted on the parameter setting of each one of the regression methods employed using a simplified version of the sequential Monte Carlo method [8].

The scheduling problem formulated in the eq. 3 was solved using a Genetic Algorithm. The scheduling performed by the company was compared with the scheduling procedures generated by using the methods in (a,b). Since the output of (a) is an exact count, the π_t was replaced with the exact predicted count A_t.

In (a), the Mean Absolute Error (MAE) and the RMSE were employed as evaluation metrics. In (b), an accuracy-based metric and a user-defined cost-sensitivity matrix (it is more important to predict a peak than a normal arrival count) were used. That matrix expresses the cost of every misclassification case. This is shown in Fig. 1-D. To evaluate the scheduling quality, the number of expected abandoned called was computed in the different schedules proposed.

Table 1 presents an evaluation of the proposed algorithms that were used to predict workload. The results were evaluated based on the estimated capacity of each agent (i.e. C_m) by considering as "lost calls" all the calls besides the scheduled workload. The regression/classification method with the lowest averaged error in each week was used to perform scheduling in the following week. The sensitivity analysis results were used to select the methods for week 1. The scheduling results are provided in Table 2 and in Fig. 1.

Table 1 denotes an excellent performance using both regression/numerical and classification methods. However, the classification approach is almost as flawless as its accuracy $\simeq 1$. The evaluation performed in Table 2 and in Fig. 1 contains

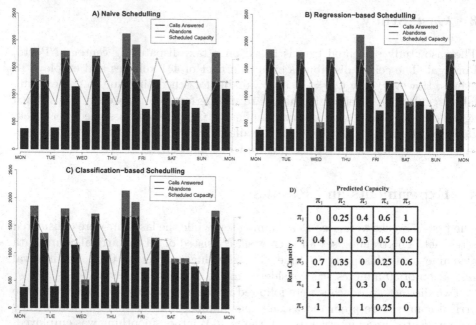

Fig. 1. Detailed Results of week 3 using different methods to estimate the workload (A,B,C) and the error cost matrix employed for the workload misclassification (D). Note the classification's refinement on Thursday evening.

Table 1. Results for the prediction task using both Regression/Numerical and Classification algorithms. The metric values were averaged for all weeks.

Numerical Prediction	Holt-Winters Smoothing	Random Forests	KNN	PPR
RMSE	17.93	19.31	21.31	22.38
MAE	11.52	13.32	13.55	16.14
Error Cost Matrix	0.025	0.029	0.027	0.034

Multi-Class Prediction	NB	Random Forests	RIPPER	ANN
Accuracy	0.938	1.000	1.000	0.805
Error Cost Matrix	0.020	0.000	0.000	0.096

Table 2. Abandoned calls using the different agent schedulings

	Week1	Week2	Week3	Week4	Total
Naive Method	5708	4819	4335	4541	19403
Numerical Prediction	4362	3449	4153	**3951**	15915
Multi-Class Prediction	**4306**	**3421**	**3981**	3961	**15669**

an error since the abandoned calls expressed are merely an expected value based on the scheduled capacity. However, Fig. 1 uncovers the limitations of the naive approach as it clearly overstaffs the night shifts. Not surprisingly, the greatest advantage of employing classification methods in this problem is their capacity to uncover **demand peaks**, as expressed in Fig. 1. These results demonstrate that the **interval-based classification** approach should be regarded as a reliable solution to this problem.

5   Final Remarks

This paper proposes a discretization-based framework to address the workload prediction with the calls made to the call center. The goal with this framework is to accurately predict demand peaks in order to optimize the use of resources. The results obtained show that this problem can be successfully addressed as an interval-based multi-class problem. The authors' goal was to use these findings as proof of concept to open new research lines on this topic.

Acknowledgments. This work was supported by the Project "I-City for Future Mobility" (NORTE-07-0124- FEDER-000064).

References

1. Avramidis, A.N., Deslauriers, A., L'Ecuyer, P.: Modeling daily arrivals to a telephone call center. Management Science 50(7), 896–908 (2004)
2. Taylor, J.W., Snyder, R.D.: Forecasting intraday time series with multiple seasonal cycles using parsimonious seasonal exponential smoothing. Omega 40(6), 748–757 (2012)
3. Weinberg, J., Brown, L.D., Stroud, J.R.: Bayesian forecasting of an inhomogeneous poisson process with applications to call center data. Journal of the American Statistical Association 102(480), 1185–1198 (2007)
4. Millán-Ruiz, D., Hidalgo, J.I.: Forecasting call centre arrivals. Journal of Forecasting 32(7), 628–638 (2013)
5. Moreira-Matias, L., Gama, J., Ferreira, M., Mendes-Moreira, J., Damas, L.: Predicting taxi-passenger demand using streaming data. IEEE Transactions on Intelligent Transportation Systems 14(3), 1393–1402 (2013)
6. Shen, H., Huang, J.Z.: Forecasting time series of inhomogeneous poisson processes with application to call center workforce management. The Annals of Applied Statistics, 601–623 (2008)
7. Aldor-Noiman, S., Feigin, P.D., Mandelbaum, A.: Workload forecasting for a call center: Methodology and a case study. The Annals of Applied Statistics, 1403–1447 (2009)
8. Cappé, O., Godsill, S., Moulines, E.: An overview of existing methods and recent advances in sequential monte carlo. Proceedings of the IEEE 95(5), 899–924 (2007)

Improved Approximation Guarantee for Max Sum Diversification with Parameterised Triangle Inequality

Marcin Sydow[1,2]

[1] Institute of Computer Science, Polish Academy of Sciences, Warsaw, Poland
[2] Web Mining Lab, Polish-Japanese Institute of Information Technology, Warsaw, Poland
msyd@ipipan.waw.pl

Abstract. We present improved $2/\alpha$ approximation guarantee for the problem of selecting *diverse* set of p items when its formulation is based on Max Sum Facility Dispersion problem and the underlying dissimilarity measure satisfies *parameterised triangle inequality* with parameter α.

Diversity-aware approach is gaining interest in many important applications such as web search, recommendation, database querying or summarisation, especially in the context of ambiguous user query or unknown user profile.

In addition, we make some observations on the applicability of these results in practical computations on real data and link to important recent applications in the *result diversification problem* in web search and semantic graph summarisation. The results apply to both relaxed and strengthen variants of the triangle inequality.

Keywords: diversity, max sum facility dispersion, approximation algorithms, parameterised triangle inequality.

1 Introduction

Diversity-aware approach is gaining interest in web search, database querying, recommendation or summarisation (e.g. [3,8]). More precisely, the approach consists in returning to the user the set of items (e.g. search results, or recommended items, etc.) that is not only *relevant* but also *diversified*. The rationale behind diversfying the result set is to reduce potential *result redundance* to cover potentially many different aspects. This is especially important as a tool for maximising the chances that an *unkown actual user intent* behind a potentially *ambiguous query* is at least partially satisfied.

One of possible formulations of such diversification problem is via a variant of an optimisation problem called *Facility Dispersion*. It was originally studied in Operations Research (e.g. [5,7]), in the context of selecting spatial placements for some facilities to make them *mutually distant* to each other (e.g. nuclear plants, amunition dumps, stores of the same brand, etc.).

In information sciences, the spatial distance has been substituted with an abstract concept of pairwise *dissimilarity measure* between the items returned to a user.

Recent examples of such applications include web search [3] or entity summarisation in semantic knowledge graphs [6], among others.

The problem is known to be NP-hard (reduction from the CLIQUE Problem). It is true even when d satisfies triangle inequality, but in such case there exist polynomial-time 2-approximation algorithm for this problem [5].

T. Andreasen et al. (Eds.): ISMIS 2014, LNAI 8502, pp. 554–559, 2014.
© Springer International Publishing Switzerland 2014

In this work, we focus on the most common variant of the problem called *Max Sum Facility Dispersion*. As one of the main results, we present the generalisation of approximation guarantee of 2 for metric case [5] to the value of $2/\alpha$ for the case when the distance satisfies *parameterised triangle inequality* with parameter α.

For example a variant of the *NEM* (Non-linear Elastic Matching) dissimilarity measure used in pattern matching for visual shapes naturally satisfies this property [2].

While parameterising the triangle inequality is not a new topic in the context of approximation algorithms (e.g. [1]) it is usually considered in the context of either *relaxing* or *strengthening* the triangle inequality. Here, we consider the full possible range of the parameter value and, as a consequence, obtain new results that extend existing approximation guarantees in both cases.

As additional contributions, we demonstrate potential applications of this result in practical computations on real datasets (Section 5.1) and explain its impact on already mentioned applications in information sciences (Section 5.2).

2 Max Sum and Max Average Facility Dispersion Problem

In Max Sum Facility Dispersion Problem, the input consists of a full, undirected graph $G(V, E)$, an edge-weight function $d : V^2 \to R^+ \cup \{0\}$ that represents *pairwise distance* between the vertices and a positive natural number $p \leq |V|$. The task is to select a p-element subset $P \subseteq V$ that maximises the following objective function

$$f_{SUM}(P) = \sum_{\{u,v\} \subseteq P} d(u, v) \tag{1}$$

that represents a notion of *dispersion* of the vertices.

This variant of formulation is a special case of the more general *k-dispersion* problem that was studied in [5]. We focus only on the special case (for $k = 1$) since it is useful for the *Result Diversification* problem discussed in Section 5.2. Sometimes, the objective function is formulated as $f_{AVE}(P) = \frac{2}{p(p-1)} \sum_{\{u,v\} \subseteq P} d(u, v)$ [7,4]. Such a variant is called *Max Average* Facility Dispersion. Since the both above formulations are obviously *equivalent* if p is fixed we will refer to them interchangeably in this paper.

3 Parameterised Triangle Inequality

Let V be a non-empty universal set and $d : V^2 \to R^+ \cup \{0\}$ be a distance function. i.e. it is assumed that for all $u, v \in V$ it satisfies *discernibility* $(d(u, v) = 0 \Leftrightarrow u = v)$ and *symmetry* $(d(u, v) = d(v, u))$ conditions. If d additionally satisfies *triangle inequality*: $d(u, v) + d(v, z) \geq d(u, z)$ for all mutually different $u, v, z \in V$, we call d a *metric*.

We introduce the following definition of *parameterised triangle inequality*:

Definition: *Let V be a set, $\alpha \in R, 0 \leq \alpha \leq 2$. A distance function $d : V^2 \to R^+ \cup \{0\}$ satisfies* parameterised triangle inequality *(α-PTI) with parameter α iff for all mutually different $u, v, z \in U : d(u, v) + d(v, z) \geq \alpha d(u, z)$*

α-PTI is a generalisation of the standard triangle inequality (for $\alpha = 1$). 0-PTI is the weakest variant (satisfied by any distance function d) and is equivalent to *semi-metric*.

The higher the value of α the stronger the property. The value of 2 is the highest possible value for α in α-PTI. Notice that since 2-PTI enforces that all non-zero distances are equal (equivalent to *discrete metric*, up to rescaling), the Facility Dispersion Problem becomes trivial for $\alpha = 2$.

4   2/α Approximation Guarantee for Max Sum Dispersion Satisfying α-PTI

Figure 1 shows a heuristic, polynomial-time approximation algorithm, based on computing a maximum-weight matching, that guarantees approximation factor of 2 for Max Average Dispersion when d satisfies triangle inequality [5].

Efficient 2-Approximation Heuristic for Max Average Dispersion Problem:

1. $P = \emptyset$
2. Compute a maximum-weight $\lfloor p/2 \rfloor$-matching M^* in G
3. For each edge in M^*, add its both ends to P
4. In case p is odd, add any node from $V \setminus P$ to P
5. return P

Fig. 1. Approximation Algorithm for Max Sum Dispersion Problem

We now present a theorem that is a generalisation and extension of the 2-factor approximation guarantee [5] of the algorithm presented at Figure 1 for the case when d satisfies α-PTI. Our proof of this theorem, that is a simple adaptation of the one presented in [4] by properly introducing α parameter is presented in the Appendix.

Theorem 1. *Let I be an instance of Max Average Dispersion problem with distance function d satisfying α-PTI for $0 < \alpha < 2$. Let's denote by $OPT(I)$ the value of an optimal solution and by $HRT(I)$ the value of the solution found by the algorithm HRT. It holds that $OPT(I)/HRT(I) < 2/\alpha$.*

As observed in Section 2, the result applies also to the Max Sum Dispersion problem.

It is easy to show that the $2/\alpha$ bound for the algorithm for Max Sum Dispersion with α-PTI presented on Figure 1 is (asymptotically) tight.

5   Practical Applications of the Results

5.1 α-PTI in Practical Computational Problems

Note that in practical applications, the input dataset V is finite, what implies that distance function d *always* satisfies α-PTI for some $0 < \alpha \leq 2$. By checking all triples in $O(|V|^3)$ time it is possible to find the *actual maximum value* of the α parameter so that the α-PTI is satisfied as follows: $\alpha = min_{u,v,z \in V}[d(u,v) + d(v,z)]/d(u,z)$. In this way, it is possible to obtain a constant-factor guarantee for the approximation algorithms considered in this paper even if no prior knowledge is assumed about the distance function. In particular, in the metric case, the guarantee obtained in this way would be *always* better than 2, unless there are no "degenerated" triangles in the data.

In *on-line* problems, when data comes in time, α can be updated regularly.

5.2   Applications to the Result Diversification Problem in Web Search, etc.

Here, we demonstrate in more detail how the result from section 4 impacts some currently important applications in information sciences including web search and others.

For example, in web search, the problem known as *Result Diversification Problem* can be specified in the following way [3]. Given a set V of documents to be potentially relevant to a user query, a number $p \in N^+, p < |V|$, a *document relevance* function $w :$ $V \to R^+$ and pairwise *document dissimilarity* function $d : V^2 \to R^+ \cup \{0\}$, the task is to select a subset $P \subseteq V$ that maximises the value of a properly defined *diversity-aware relevance function*. In [3] the following parameterised, bi-criteria objective function (to be maximised) is proposed as the diversity-aware relevance function:

$$f_{div-sum}(\lambda, P) = (p-1) \sum_{v \in P} w(v) + 2\lambda \sum_{\{u,v\} \subseteq P} d(u,v) \qquad (2)$$

where $\lambda \in R^+ \cup \{0\}$ is a parameter that controls the diversity/relevance-balance.

In the same work it is observed that a proper modification of d to d' (Equation 3) makes the described problem of maximising $f_{div-sum}(\lambda, P)$ equivalent to maximising $\sum_{\{u,v\} \subseteq P} d'_\lambda(u,v)$, where:

$$d'_\lambda(u,v) = w(u) + w(v) + 2\lambda d(u,v) \qquad (3)$$

Thus, it makes the *result diversification* problem described above equivalent to the Max Sum Dispersion problem for d'_λ.

Notice that, due to the observation 5.1, α-PTI, for some $0 \le \alpha \le 2$ is actually satisfied by *all* distance functions in finite sets.

Due to this observation, the following concluding lemma 1 extends the application of the results presented in this paper to the *result diversification* problem.

Lemma 1. *If distance function d satisfies α-PTI, for some $0 < \alpha \le 1$, then the modified distance function d'_λ defined in equation 3 also satisfies α-PTI.*

Proof. $d'_\lambda(u,v) + d'_\lambda(v,z) = 2w(v) + w(u) + w(z) + 2\lambda(d(u,v) + d(v,z)) \ge$

$$\ge 2w(v) + w(u) + w(z) + 2\lambda\alpha d(u,z) \ge$$

$\ge 2w(v) + \alpha[w(u)/\alpha + w(z)/\alpha + 2\lambda d(u,z)] \ge \alpha d'_\lambda(u,z)$ *(Quod erat demonstrandum)* ◊

Besides the mentioned applications in diversification problem in web search, the problem of optimising $f_{div-sum}(\lambda, P)$ objective function has been very recently adapted in [6] in the novel context of *diversified entity summarisation in knowledge graphs* [8].

6   Future Work

In the extended journal version of this paper, that is under preparation, we additionally present recently obtained analogous results for another variant: Max Min Facility Dispersion problem. Another natural direction of research, especially in the context of practical computations, would be to study which common distance metrics (such example is given in [2]) used in important practical applications satisfy α-PTI and with what value of the parameter.

Acknowledgements. Work supported by Polish National Science Centre grant "DIS-QUSS" 2012/07/B/ST6/01239.

References

1. Andreae, T., Bandelt, H.-J.: Performance guarantees for approximation algorithms depending on parametrized triangle inequalities. SIAM Journal of Discrete Mathematics 8, 1–16 (1995)
2. Fagin, R., Stockmeyer, L.: Relaxing the triangle inequality in pattern matching. Int. J. Comput. Vision 30(3), 219–231 (1998)
3. Gollapudi, S., Sharma, A.: An axiomatic approach for result diversification. In: Proceedings of the 18th International Conference on World Wide Web, WWW 2009, pp. 381–390. ACM, New York (2009)
4. Gonzalez, T.F.: Handbook of approx. algorithms and metaheuristics. CRC Press (2007)
5. Hassin, R., Rubinstein, S., Tamir, A.: Approximation algorithms for maximum dispersion. Oper. Res. Lett. 21(3), 133–137 (1997)
6. Kosiński, W., Kuśmierczyk, T., Rembelski, P., Sydow, M.: Application of ant-colony optimisation to compute diversified entity summarisation on semantic knowledge graphs. In: Proc. of International IEEE AAIA 2013/FedCSIS Conference, Annals of Computer Science and Information Systems, vol. 1, pp. 69–76 (2013)
7. Ravi, S.S., Rosenkrantz, D.J., Tayi, G.K.: Heuristic and special case algorithms for dispersion problems. Operations Research 42(2), 299–310 (1994)
8. Sydow, M., Pikula, M., Schenkel, R.: The notion of diversity in graphical entity summarisation on semantic knowledge graphs. Journal of Intelligent Information Systems 41, 109–149 (2013)

APPENDIX: Proof of Theorem 1

Let us introduce some denotations. Let $V' \subseteq V$ be a non-empty subset of vertices. Let $G(V')$ denote the full graph induced on V' and $W(V'), W'(V')$ denote the total weight and average weight of edges in $G(V')$ respectively. By analogy, for a non-empty subset $E' \subseteq E$ of edges, let denote by $W(E')$ and $W'(E') = W(E')/|E'|$ the total and average weight of the edges in E', respectively. We use the following technical Lemma, presented in [5].

Lemma 2. *If $V' \subseteq V$ is a subset of vertices of cardinality at least $p \geq 2$ and M'^{*} is a maximum-weight $\lfloor p/2 \rfloor$-matching in $G(V')$, then $W'(V') \leq W'(M'^{*})$.*

A very short proof, presented in [5] does not assume *anything* on the distance function d, so that we omit it here. The following Lemma, Theorem and their proofs constitute our extensions of the versions presented in [4][pp. 38-8–38-9] (earlier variants were in [5]). The extensions presented here consist in properly introducing the parameter α.

Lemma 3. *Assume that the distance function d satisfies α-PTI for some $0 < \alpha < 2$. If $V' \subseteq V$ is a subset of $p \geq 2$ vertices and M is any $\lfloor p/2 \rfloor$-matching in $G(V')$, then $W'(V') > (\alpha/2)W'(M)$.*

Proof. (of Lemma 3) Let $M = \{\{a_i, b_i\} : 1 \leq i \leq \lfloor p/2 \rfloor\}$ and let denote by V_M the set of all vertices that are ends of the edges in M. For each edge $\{a_i, b_i\} \in M$ let E_i denote the set of edges in $G(V')$ that are incident on a_i or b_i, except the edge $\{a_i, b_i\}$ itself. From α-PTI we get that for any vertex $v \in V_M \setminus \{a_i, b_i\}$ we have $d(v, a_i) + d(v, b_i) \geq \alpha d(a_i, b_i)$. After summing this inequality over all the vertices in $V_M \setminus \{a_i, b_i\}$ we obtain:

$$W(E_i) \geq \alpha(p - 2)d(a_i, b_i) \tag{4}$$

There are two cases:

Case 1: p is even, i.e. $\lfloor p/2 \rfloor = p/2$. After summing up the Inequality 4 above, over all the edge sets E_i, $1 \leq i \leq p/2$, we obtain, on the left-hand side, each edge of $G(V')$ twice, except those in M. Thus, $2[W(V') - W(M)] \geq \alpha(p - 2)W(M)$. If we substitute in the last inequality $W(V') = W'(V')p(p - 1)/2$ and $W(M) = W'(M)p/2$, and divide both sides by p, we can quickly get to $W'(V') \geq (\alpha/2)W'(M)[p - 2 + (2/\alpha)]/(p - 1)$, that is equivalent to $W'(V') > (\alpha/2)W'(M)$ (for $\alpha < 2$). This completes the proof for the Case 1.

Case 2: p is odd, i.e. $\lfloor p/2 \rfloor = (p - 1)/2$. Let x be the only node in $V' \setminus V_M$ and let E_x denote the set of all edges incident on x in $G(V')$. By α-PTI we get:

$$W(E_x) \geq \alpha W(M) \tag{5}$$

Let's again sum up the previous Inequality 4 over all the edges E_i, $1 \leq i \leq \lfloor p/2 \rfloor$. On the left-hand side, each edge in $G(V')$ occurrs twice, except the edges in M (that do not occur at all) and the edges in E_x that occur once, each. Thus, $2[W(V') - W(M)] - W(E_x) \geq \alpha(p - 2)W(M)$. Now, applying the inequality 5, we obtain $2[W(V') - W(M)] \geq \alpha(p - 1)W(M)$. If we now substitute $W(V') = W'(V')p(p - 1)/2$ and $W(M) = W'(M)(p - 1)/2$ and divide both sides by $(p - 1)/2$ we will quickly obtain that $W'(V') \geq (\alpha/2)W'(M)[p - 1 + (2/\alpha)]/p$ that is equivalent to $W'(V') > (\alpha/2)W'(M)$ (for $0 < \alpha < 2$). This completes the Case 2 and the whole proof of the Lemma. *(Quod erat demonstrandum)*$\Diamond$

The following proof of Theorem 1 is an extension of the one proposed in [5] (and later presented in [4]) by properly introducing the parameter α.

Proof. (of Theorem 1 from Section 4) Let P^* and P be the set of nodes in an optimal solution and that in the solution returned by the HRT algorithm for instance I, respectively. By definition, $OPT(I) = W'(P^*)$ and $HRT(I) = W'(P)$. Let M^* and M denote a maximum-weight $\lfloor p/2 \rfloor$-matching in P^* and in P, respectively. By Lemma 2, we get:

$$OPT(I) \leq W'(M^*) \tag{6}$$

In addition, from Lemma 3 we get:

$$HRT(I) > (\alpha/2)W'(M) \tag{7}$$

Now, because the algorithm HRT finds a maximum-weight $\lfloor p/2 \rfloor$-matching in G, we get $W'(M) \geq W'(M^*)$. This, together with the inequalities 6 and 7 implies that $HRT(I) > (\alpha/2)W'(M) \geq (\alpha/2)W'(M^*) \geq OPT(I)/\frac{2}{\alpha}$ that completes the proof of the theorem. *(Quod erat demonstrandum)*$\Diamond$

Learning Diagnostic Diagrams
in Transport-Based Data-Collection Systems

Vu The Tran, Peter Eklund*, and Chris Cook

Faculty of Engineering and Information Science
University of Wollongong
Northfields Avenue, NSW 2522, Australia

Abstract. Insights about service improvement in a transit network can be gained by studying transit service reliability. In this paper, a general procedure for constructing a transit service reliability diagnostic (TSRD) diagram based on a Bayesian network is proposed to automatically build a behavioural model from Automatic Vehicle Location (AVL) and Automatic Passenger Counters (APC) data. Our purpose is to discover the variability of transit service attributes and their effects on traveller behaviour. A TSRD diagram describes and helps to analyse factors affecting public transport by combining domain knowledge with statistical data.

Keywords: AI applications, knowledge discovery, Bayesian networks, transit service reliability

1   Introduction

"The transit industry is in the midst of a revolution from being data poor to data rich. Traditional analysis and decision support tools required little data, not because data has little value, but because traditional management methods had to accommodate a scarcity of data" [4].

Automatic Vehicle Location (AVL) and Automatic Passenger Counters (APC) lead to big data and it is important to investigate how (and if) this can be used to improve transport service reliability. The growth of public transport databases facilitates new approaches to help characterize reliability and – by so doing – improve service planning and operational control. Among knowledge discovery techniques, Bayesian networks – a characterisation of probabilistic knowledge by a graphical diagram – provide a comprehensive method of representing relationships and influences among nodes. Bayesian networks are a fundamental technique in pattern recognition and machine classification [5],

Many studies induce Bayesian networks from data. Oniśko et al. [7] experiment with Bayesian network parameters from small data sets and use Noisy-OR gates to reduce the data requirements in learning conditional probabilities. Nadkarni et al. [6] describe a procedure for constructing Bayesian networks from

---

* On leave in 2014 to the IT University of Copenhagen, Denmark.

T. Andreasen et al. (Eds.): ISMIS 2014, LNAI 8502, pp. 560–566, 2014.
© Springer International Publishing Switzerland 2014

expert domain knowledge using causal mapping. Tungkasthan et al. [8] propose a practical framework for automating the construction of a diagnostic Bayesian network from WWW data sources. In that work, a SMILE (Structural Modeling, Interface, and Learning Engine) Web-based interface allow one to perform Bayesian network diagnosis through the Web.

As can be seen from the literature above, a wide variety of studies have use Bayesian networks for knowledge discovery, however employing Bayesian networks to analyze service reliability using data derived from AVL and APC sources for public transport is novel. To date, the transit industry has lacked a measure of service reliability measured in terms of its impact on customers because traditional measures cannot express how reliability impacts on passengers' perceptions [4]. Our paper focuses on an approach for constructing a transit service reliability diagnostic (TSRD) diagram based on a Bayesian network. A TSRD diagram has the ability to represent cause-effect relationships between transit factors and expresses how each factor will impact on others. A TSRD diagram can used in three ways: (i) as a guide for identifying the causes of service unreliability; (ii) as a learning component for real-time decision making and; (iii) as an offline analysis tool to improve service quality.

The remainder of the paper is organized as follows. The proposed methodology for constructing the TSRD diagram is presented in Section 2. The case study and experimental results are reported and discussed in Section 3.

2   Methodology

Service reliability in a public transport network can be considered as the variability of service attributes and their effect on passenger behaviour. A TSRD diagram based Bayesian network – a prediction-oriented method – is built to provide a better understanding of what causes problems in the transit system, prevent these problems through better service planning and operational management, and develop strategies to correct problems once they appear. A TSRD diagram is represented via a network $\mathcal{N}(\mathcal{G}, \Theta)$, where $\mathcal{G} = < \mathcal{U}, \mathcal{E} >$ is a directed acyclic graph, $\mathcal{U}$ is a set of nodes expressed as $\mathcal{U}\{u_1, u_2, ..., u_n\}$, $\mathcal{E}$ is a set of arcs, and Θ represents a set of conditional probability distributions.

Constructing a TSRD diagram involves of four steps: (1) preparation of transit discovery data set, (2) determining an initial TSRD diagram, (3) learn the TSRD structure and set parameters from training data, (4) assess/test the TSRD diagram. Since data from AVL and APC sources are heterogeneous and uncertain, the initial step combines data from various sources and tables into one dataset which can then be used in the discovery process. The second step is the construction of an initial TSRD diagram, based on cause-effect relationships to draw links between transit variables. The initial TSRD diagram reveals the qualitative relationships between variables in public transit systems. Next, the structure of the initial TSRD diagram and the parameters of variables need to be learned from the dataset. Learning the structure, causal relations, and parameters of variables – which reveal the quantitative relationships between variables – from the dataset is important for an comprehensible and extensible TSRD diagram. The final steps is the assessment and validation of the candidate TSRD diagram.

2.1 Preparation of Transit Discovery Dataset

Step 1: From the original public transit data set variables relevant for the study are considered and selected. The raw AVL and APC data is stored in Tables with the schemes: Stops(StopName, Longitude, Latitude, SegmentID, StopNumber), Buses(BusID, Longitude, Latitude, Timestamp, Speed, SegmentID), Passengers(BusID, Longitude, Latitude, Timestamp, Counts, On/Off).

The raw data is normalized by combining, matching and processing data from the three tables to expose the variables required for analysis; this involves extraction and transformation of the attributes. This process is usually project-specific and the variables may vary, depending on how the TSRD diagram is to be used. In our case, all data is integrated into a single dataset including all of the attributes and their possible states that will be considered for the study of service reliability. Table 1 represents all combined attributes that are used.

Table 1. Description of attributes

No.	Variables	Possible states
1	vehicle Speed $\mathcal{V}$	{Slow, Normal, Fast}
2	vehicle position $\mathcal{X}$	{OnSchedule, OffSchedule}
3	running time $\mathcal{R}$	{OnTime, LessThan5MinLate, MajorLate}
4	passenger alighting $\mathcal{A}$	{Low, Normal, High}
5	passenger boarding $\mathcal{B}$	{Low, Normal, High}
6	dwell time $\mathcal{D}$	{Negligible, Major, Minor}
7	in-vehicle load $\mathcal{L}$	{Normal, Excessive, Unaccepted}
8	passenger wait time $\mathcal{T}_{wait}$	{Negligible, Major, Minor}
9	headway adherence $\mathcal{H}_{adherence}$	{Negligible, Major, Minor}
10	passenger comfort $\xi_{comfort}$	{Good, Accepted, Unaccepted}
11	service reliability $\mathcal{SR}$	{Yes, No}

2.2 Determining an Initial TSRD Diagram

Step 2: In this step we define the goals and understanding of what should be done with the TSRD diagram.

> **Question:** "What causes headway irregularity?"
> **Answer:** "Passengers alighting or boarding and the bus waiting for passengers running to catch the bus"
> **Modeling:** Draw arcs from those nodes to the *Headway adherence* node.

After deciding what variables and states to model, an initial TSRD diagram N^0 is constructed by considering conditional independence by drawing causal links among nodes following question and answer examples such as the one above. To establish the causal relationships, it is helpful to ask direct questions about dependence between variables. Once identified, arcs are added from those causal variables to the affected variable. Probabilities on the edges are obtained initially by subjective estimates. First diagram to the left in Fig. 1 depicts the TSRD diagram N^0 by combining knowledge of bus operations and asking cause-effect questions. The main interest of service reliability diagnosis is to identify causes of unreliability. The context variables in this case are the background information about passengers alighting, passengers boarding, bus position and speed.

2.3 Learning Structure and Parameters from Data Set

Step 3: The initial TSRD diagram is often not good enough because there is often not enough causal knowledge to establish the full topology of the network model. Learning the full structure, causal relations and parameters from a data set are essential for refining and conditioning the TSRD diagram.

The applied structural Expectation Maximization (EM) algorithm requires an initial TSRD diagram N^0 and a dataset D as a starting point for iteration. Learning the probabilities of attributes of the TSRD diagram from data is a form of unsupervised learning. The objective here is to deduce a network that best describes the probability distribution over the training data $\mathcal{D}$.

The structural EM algorithm is an extension to the standard Expectation Maximisation algorithm [1] and is described in [2] and [3]. The algorithm performs a search in the joint space of structure and parameters. At each step, it can either find better parameters for the current structure, or select a new structure. The function Q is the expected score, given by:

Algorithm 1. Learning structure and parameters from dataset

input : $D = \{x^1, ..., x^n\}$: a data set
input : $N^0 = (\mathcal{G}', \Theta^0)$: an initial network
output: $N^* = (\mathcal{G}^*, \Theta^*)$: return the candidate network

begin
   Loop for n= 1,2,... until convergence **begin**
      Find a model G^{n+1} that maximises $Q(\mathcal{G}, \Theta : \mathcal{G}^n, \Theta^n)$
      Let $\Theta^{n+1} = Q(\mathcal{G}^{n+1}, \Theta : \mathcal{G}^n, \Theta^n)$
   return N^*

$$Q(\mathcal{G}, \Theta : \mathcal{G}^*, \Theta^*) = E[\log P(O, h : \mathcal{G}, \Theta) - \text{Penalty}(\mathcal{G}, \Theta)] \qquad (1)$$

where O are the observed variables, h are the values of the hidden variables, and the penalty depends on the dimensionality of $\mathcal{G}$. The procedure converges to a *local* maxima.

2.4 Assess Structure and Parameters

Step 4: Crucial to the methodology is that the structure and parameters of the model are validated. The structure evaluation reveals if important variables have been overlooked, if irrelevant nodes have been included, or if node values are inappropriate. Validation confirms that the model is an accurate representation of the domain. The evaluation and validation consists, in this case, of comparing the behaviour of a network with expert judgements.

3 Results and Discussion

A case study of bus operations on the Gwynneville-Keiraville bus route in the regional city of Wollongong, Australia – population approx. 300,000 – is used to demonstrate and test the proposed method. AVL and APC units are installed for buses on the UniShuttle service to capture the data. As at the end of July 2012 there were a total of 1,844,964 vehicle (bus) location events stored in a MySQL database on our servers. The average monthly number of vehicle events captured is 132,000. There are an average of 4,630 passenger events per month captured and this number will increase 1000% when APC devices are installed on all buses in the fleet as is proposed.

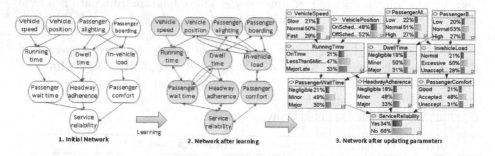

Fig. 1. TSRD diagram: construction process

In diagram centre of Fig. 1, after learning, the topology is modified, with two new arcs added. The new connections are from *VehiclePosition* node to *InvehicleLoad* node and from the *DwellTime* to *PassengerWaitTime* node. After the validation step, the casual connection from *VehiclePosition* to *InvehicleLoad* nodes is eliminated as the expert judgement is that it is inappropriate. In diagram to right of Fig. 1, the conditional probability tables (CPTs) annotate the nodes. These represent how much reliability exists in the current transit network data. Each row in a CPT contains the conditional probability of each node value for a conditioning case.

Based on our results, measures to reduce service unreliability should balance passenger wait time, passenger comfort and headway adherence as the service unreliability is similarly impacted (posterior probability) by each of these three indicators. Transport management would be advised to better control these figures so that service reliability is improved. Of the indirect factors, dwell time has the greatest posterior probability 0.81 (minor and major), as these factors affect passenger wait time and headway adherence. The posterior probability for in-vehicle load and running time is 0.79. These probabilities are high enough to indicate that transportation management should pay more attention to scheduling and planning to improve running time and reduce passenger load. This is a reassuring recommendation that validates the model: namely that service reliability is improved by more buses and fewer passengers.

The use of TSRD diagram represents three aspects: Filtering, Smoothing, and Learning component. Filtering of TSRD diagram is used to compute the belief state of the posterior distribution of transit service reliability over the most recent state, given all the observations (evidence) of the public transit factors made so far. Smoothing of the TSRD diagram is carried out to compute the posterior distribution over a past state, given all the observations (evidence) of the public transit factors made to date. The TSRD diagram is used as a knowledge representation of prior knowledge of real-time control strategies. Learning enables the transit systems to function in initially unfamiliar environments and to become more competent over time than its preliminary knowledge state.

4   Conclusion

Modern scheduling and customer service monitoring is oriented around extreme values (outrider events) rather than traditional mean values. This is mainly because of the large sample sizes produced by automatic data collection and so attention focuses on unusual events.

As these kinds of information are characterized as heterogeneous and uncertain, a TSRD diagram based Bayesian networks is presented in this paper to serve as our knowledge model to analyze automatic data collection. The TSRD diagram has the advantages of an intuitive visual representation with a sound mathematical basis in Bayesian probability and provides an effective approach for analysis of public transit systems to reveal the hidden structure and its relationships, and more importantly, its rules. A case study is used to evaluate and demonstrate the use of TSRD diagram.

References

1. Dempster, A.P., Laird, N.M., Rubin, D.B.: Maximum likelihood from incomplete data via the em algorithm. Journal of the Royal Statistical Society. Series B (Methodological), 1–38 (1977)
2. Friedman, N.: Learning belief networks in the presence of missing values and hidden variables. In: Machine Learning International Worshop then Conference, pp. 125–133. Morgan Kaufmann Publishers, Inc. (1997)
3. Friedman, N.: The bayesian structural em algorithm. In: Proceedings of the Fourteenth Conference on Uncertainty in Artificial Intelligence, pp. 129–138. Morgan Kaufmann Publishers Inc. (1998)
4. Furth, P.G.: Using archived AVL-APC data to improve transit performance and management. Transportation Research Board National Research, vol. 113 (2006)
5. Korb, K.B., Nicholson, A.E.: Bayesian artificial intelligence, vol. 1. CRC Press (2004)

6. Nadkarni, S., Shenoy, P.P.: A causal mapping approach to constructing bayesian networks. Decision Support Systems 38(2), 259–281 (2004)
7. Oniśko, A., Druzdzel, M.J., Wasyluk, H.: Learning bayesian network parameters from small data sets: Application of noisy-or gates. International Journal of Approximate Reasoning 27(2), 165–182 (2001)
8. Tungkasthan, A., Jongsawat, N., Poompuang, P., Intarasema, S., Premchaiswadi, W.: Automatically building diagnostic bayesian networks from on-line data sources and the smile web-based interface. In: Jao, C.S. (ed.) Decision Support Systems, pp. 321–334 (2010)

Author Index

Achemoukh, Farida   486
Ahmed-Ouamer, Rachid   486
Akaichi, Jalel   466
Akdag, Fatih   493
Aldanondo, Michel   144
Andrade, Lúcio Pereira de   1
Andreasen, Troels   264
Andruszkiewicz, Piotr   405
Antônio Alves de Menezes, José   305
Appice, Annalisa   234, 518, 542
Atzmueller, Martin   244

Balke, Wolf-Tilo   274
Barco, Andrés Felipe   144
Barros, Rui   445
Belo, Orlando   445
Bembenik, Robert   425
Benbernou, Salima   315
Bentayeb, Fadila   61
Berchiche-Fellag, Samia   395
Borkowski, Piotr   335
Bouillot, Flavien   345
Bouju, Alain   466
Boussaid, Omar   61
Brocki, Łukasz   355
Brzezinski, Dariusz   10
Bulskov, Henrik   264

Ceci, Michelangelo   50, 365
Chen, Guoning   493
Chen, Jianhua   164
Chung, Cheng-Shiu   174
Ciecierski, Konrad   154
Ciesielski, Krzysztof   335
Cook, Chris   560
Cooper, Rory   174
Correia, Helena   445
Cuzzocrea, Alfredo   456
Czyzewski, Andrzej   224

Dahlbom, Anders   194
Deckert, Magdalena   20
Deguchi, Yutaka   500
Deng, Xiaofei   73
Dimitrova-Grekow, Teodora   506

Ebecken, Nelson Francisco Favilla   1
Eick, Christoph F.   493
Eklund, Peter   560
Espíndola, Rogério Pinto   1

Faber, Pamela   285
Ferreira, Michel   548
Ferreira, Nickerson   456
Firth, Robert   164
Fisher, Robert   174
Fokou, Géraud   512
Fournier-Viger, Philippe   83, 524
Freitas, Fred   305
Fumarola, Fabio   365, 518
Furtado, Pedro   456

Gaborit, Paul   144
Gama, João   548
Grekow, Jacek   184
Grindle, Garrett   174
Gueniche, Ted   524

Hadjali, Allel   512
Homoceanu, Silviu   274

Ienco, Dino   385

Jaeger, Manfred   30
Janning, Ruth   93
Jarczewski, Marcin   506
Jaudoin, Hélène   530
Jean, Stéphane   512
Jensen, Per Anker   264
Jiang, Jiuchuan   30

Kalo, Jan-Christoph   274
Karlsson, Alexander   194
Kashevnik, Alexey   325
Kelleher, Annmarie   174
Kłopotek, Mieczysław A.   335
Koperwas, Jakub   405
Koržinek, Danijel   355
Kostek, Bozena   224
Kozłowski, Marek   405, 415
Krajewski, Robert   415
Kubera, Elżbieta   204

Kucharczyk, Łukasz   435
Kucharski, Waldemar   224
Kursa, Miron B.   214

Lanotte, Pasqua Fabiana   50, 365
Lech, Michal   224
Lettner, Christian   476
Levashova, Tatiana   325
Lewandowski, Jacek   425
Lewis, Rory   536
Liu, Hsinyi   174
Loglisci, Corrado   40, 542
Long, Darrell D.E.   315

Malerba, Donato   40, 50, 234, 365, 518, 542
Malki, Jamal   466
Marasek, Krzysztof   355
Martin-Bautista, Maria J.   285
Mendes-Moreira, João   548
Mezghiche, Mohamed   395
Missaoui, Rokia   61
Moreira-Matias, Luis   548
Moussa, Rim   315

Napierała, Krystyna   123
Nilsson, Jørgen Fischer   264
Nkambou, Roger   524
Nunes, Rafael   548

Pelegrina, Ana Belen   285
Pereira, Luis F. Alves   305
Petersen, Niklas Christoffer   375
Piernik, Maciej   10
Pio, Gianvito   50
Pivert, Olivier   295, 530
Poncelet, Pascal   345
Pravilovic, Sonja   234
Protaziuk, Grzegorz   425
Przybyszewski, Andrzej W.   154

Raghavan, Vijay V.   103
Raś, Zbigniew W.   154, 254
Rauch, Jan   113
Ribeiro de Azevedo, Ryan   305
Rocha, Rodrigo   305
Roche, Mathieu   345
Rodrigues, Paulo   445
Romeo, Salvatore   385
Rybiński, Henryk   405, 415

Sahri, Soror   315
Sakouhi, Thouraya   466
Scarpino, Andrea   365
Schatten, Carlotta   93
Schmidt-Thieme, Lars   93
Scholz, Christoph   244
Sharif, Mohammad Amir   103
Shilov, Nikolay   325
Sid Ali, Selmane   61
Simmons, Reid   174
Šimůnek, Milan   113
Skonieczny, Łukasz   405
Skrzypiec, Magdalena   204
Smirnov, Alexander   325
Smits, Grégory   295, 530
Stefanowski, Jerzy   20, 123
Steinmaurer, Thomas   476
Struk, Wacław   405
Studnicki, James   254
Stumme, Gerd   244
Stumptner, Reinhard   476
Susmaga, Robert   133
Suzuki, Einoshin   500
Sydow, Marcin   554
Szczęch, Izabela   133
Szymański, Julian   435

Tagarelli, Andrea   385
Thion, Virginie   530
Torelli, Michele Damiano   365
Touati, Hakim   254
Tran, Vu The   560
Traxler, Patrick   476
Trzcielińska, Małgorzata   123
Tseng, Vincent S.   83, 524

Vareilles, Elise   144
Villadsen, Jørgen   375

Wannous, Rouaa   466
Wieczorkowska, Alicja A.   204, 214, 254
Wu, Cheng-Wei   83
Wu, Yu Kuang   174

Yao, JingTao   73
Yao, Yiyu   73

Zhong, Hui   194
Zida, Souleymane   83
Zwick, Michael   476